TABLES OF INTEGRALS AND
OTHER MATHEMATICAL DATA

TABLES OF INTEGRALS
AND OTHER
MATHEMATICAL DATA

✦✦   ✦✦   ✦✦

HERBERT BRISTOL DWIGHT

FOURTH EDITION

✦✦

✦✦

THE MACMILLAN COMPANY

Tenth Printing, 1969

Previous editions © copyright 1934, 1947 and 1957 by The Macmillan Company

Library of Congress catalog card number: 61—6419

The Macmillan Company,
Collier-Macmillan Canada, Ltd., Toronto, Ontario

Printed in the United States of America

PREFACE

PREFACE TO THE FIRST EDITION

The first study of any portion of mathematics should not be done from a synopsis of compact results, such as this collection. The references, although they are far from complete, will be helpful, it is hoped, in showing where the derivation of the results is given or where further similar results may be found. A list of numbered references is given at the end of the book. These are referred to in the text as "Ref. 7, p. 32," etc., the page number being that of the publication to which reference is made.

Letters are considered to represent real quantities unless otherwise stated. Where the square root of a quantity is indicated, the positive value is to be taken, unless otherwise indicated. Two vertical lines enclosing a quantity represent the absolute or numerical value of that quantity, that is, the modulus of the quantity. The absolute value is a positive quantity. Thus, $\log |-3| = \log 3$.

The constant of integration is to be understood after each indefinite integral. The indefinite integrals may usually be checked by differentiating.

In algebraic expressions, the symbol log represents natural or Napierian logarithms, that is, logarithms to the base e. When any other base is intended, it will be indicated in the usual manner. When an integral contains the logarithm of a certain quantity, integration should not be carried from a negative to a positive value of that quantity. If the quantity is negative, the logarithm of the absolute value of the quantity may be used, since $\log(-1) = (2k+1)\pi i$ will be part of the constant of integration (see **409.03**). Accordingly, in many cases, the logarithm of an absolute value is shown, in giving

v

an integral, so as to indicate that it applies to real values, both positive and negative.

Inverse trigonometric functions are to be understood as referring to the principal values.

Suggestions and criticisms as to the material of this book and as to errors that may be in it, will be welcomed.

The author desires to acknowledge valuable suggestions from Professors P. Franklin, W. H. Timbie, L. F. Woodruff, and F. S. Woods, of Massachusetts Institute of Technology.

H. B. DWIGHT.

Cambridge, Mass.
December, 1933.

PREFACE TO THE SECOND EDITION

A considerable number of items have been added, including groups of integrals involving

$$(ax^2 + bx + c)^{1/2}, \quad \frac{1}{a + b \sin x} \quad \text{and} \quad \frac{1}{a + b \cos x},$$

also additional material on inverse functions of complex quantities and on Bessel functions. A probability integral table (No. 1045) has been included.

It is desired to express appreciation for valuable suggestions from Professor Wm. R. Smythe of California Institute of Technology and for the continued help and interest of Professor Philip Franklin of the Department of Mathematics, Massachusetts Institute of Technology.

HERBERT B. DWIGHT.

Cambridge, Mass.

PREFACE TO THE THIRD EDITION

In this edition, items **59.1** and **59.2** on determinants have been added. The group (No. 512) of derivatives of inverse trigonometric functions has been made more complete. At the end of Table 1030 material is given, suggested by Dr. Rose M. Ring, which extends the tables of e^x and e^{-x} considerably, and is convenient when a calculating machine is used.

Tables 1015 and 1016 of trigonometric functions of hundredths of degrees are given in this edition. When calculating machines are used, the angles of a problem are usually given in decimals. A great many trigonometric formulas involve addition of angles or multiplication of them by some quantity, and even when the angles are given in degrees, minutes, and seconds, to change the values to decimals of a degree gives the advantages that are always afforded by a decimal system compared with older and more awkward units. In such cases, the tables in hundredths of degrees are advantageous.

HERBERT B. DWIGHT.

Lexington, Mass.

PREFACE TO THE FOURTH EDITION

The group of definite integrals, beginning at No. **850.1**, has been considerably enlarged. A group, **781**, of integrals involving elliptic integrals has been added.

The publishers and printers have taken some pains in this reset edition to avoid the use of nearly illegible letters and numerical characters that are often used in printing mathematical formulas.

Acknowledgment is made to Lt. Richard L. Pratt of the Institute of Technology at Wright-Patterson Air Force Base for some useful suggestions and corrections.

H. B. DWIGHT.

Lincoln Laboratory of M.I.T.,
Lexington, Mass.

CONTENTS

CONTENTS

TABLES OF INTEGRALS AND
OTHER MATHEMATICAL DATA

ALGEBRAIC FUNCTIONS

1. $(1 + x)^n = 1 + nx + \dfrac{n(n-1)}{2!} x^2 + \dfrac{n(n-1)(n-2)}{3!} x^3$

$$+ \cdots + \frac{n!}{(n-r)!\,r!} x^r + \cdots.$$

Note that here and elsewhere we take $0! = 1$. If n is a positive integer, the expression consists of a finite number of terms. If n is not a positive integer, the series is convergent for $x^2 < 1$; and if $n > 0$, the series is convergent also for $x^2 = 1$. [Ref. 21, p. 88.]

2. The coefficient of x^r in No. 1 is denoted by $\dbinom{n}{r}$ or $_nC_r$.

Values are given in the following table.

Table of Binomial Coefficients

$_nC_r$: Values of n in left column; values of r in top row

	0	1	2	3	4	5	6	7	8	9	10
1	1	1									
2	1	2	1				N.B. Sum of any two adja-				
3	1	3	3	1			cent numbers in same row is				
4	1	4	6	4	1		equal to number just below				
5	1	5	10	10	5	1	the right-hand one of them.				
6	1	6	15	20	15	6	1				
7	1	7	21	35	35	21	7	1			
8	1	8	28	56	70	56	28	8	1		
9	1	9	36	84	126	126	84	36	9	1	
10	1	10	45	120	210	252	210	120	45	10	1

For a large table see Ref. 59, v. 1, second section, p. 69.

3. $(1 - x)^n = 1 - nx + \dfrac{n(n-1)}{2!} x^2 - \dfrac{n(n-1)(n-2)}{3!} x^3$

$$+ \cdots + (-1)^r \frac{n!}{(n-r)!\,r!} x^r + \cdots.$$

[See above table and the note under No. **1**.]

1

4. $(a \pm x)^n = a^n \left(1 \pm \dfrac{x}{a} \right)^n.$

4.2. $(1 \pm x)^2 = 1 \pm 2x + x^2.$

4.3. $(1 \pm x)^3 = 1 \pm 3x + 3x^2 \pm x^3.$

4.4. $(1 \pm x)^4 = 1 \pm 4x + 6x^2 \pm 4x^3 + x^4,$

and so forth, using coefficients from the table under No. **2.**

5.1. $(1 \pm x)^{1/4} = 1 \pm \dfrac{1}{4}x - \dfrac{1 \cdot 3}{4 \cdot 8}x^2 \pm \dfrac{1 \cdot 3 \cdot 7}{4 \cdot 8 \cdot 12}x^3$

$$- \dfrac{1 \cdot 3 \cdot 7 \cdot 11}{4 \cdot 8 \cdot 12 \cdot 16}x^4 \pm \cdots, \qquad [x^2 \leqq 1].$$

5.2. $(1 \pm x)^{1/3} = 1 \pm \dfrac{1}{3}x - \dfrac{1 \cdot 2}{3 \cdot 6}x^2 \pm \dfrac{1 \cdot 2 \cdot 5}{3 \cdot 6 \cdot 9}x^3$

$$- \dfrac{1 \cdot 2 \cdot 5 \cdot 8}{3 \cdot 6 \cdot 9 \cdot 12}x^4 \pm \cdots, \qquad [x^2 \leqq 1].$$

5.3. $(1 \pm x)^{1/2} = 1 \pm \dfrac{1}{2}x - \dfrac{1 \cdot 1}{2 \cdot 4}x^2 \pm \dfrac{1 \cdot 1 \cdot 3}{2 \cdot 4 \cdot 6}x^3$

$$- \dfrac{1 \cdot 1 \cdot 3 \cdot 5}{2 \cdot 4 \cdot 6 \cdot 8}x^4 \pm \cdots, \qquad [x^2 \leqq 1].$$

5.4. $(1 \pm x)^{3/2} = 1 \pm \dfrac{3}{2}x + \dfrac{3 \cdot 1}{2 \cdot 4}x^2 \mp \dfrac{3 \cdot 1 \cdot 1}{2 \cdot 4 \cdot 6}x^3$

$$+ \dfrac{3 \cdot 1 \cdot 1 \cdot 3}{2 \cdot 4 \cdot 6 \cdot 8}x^4 \mp \dfrac{3 \cdot 1 \cdot 1 \cdot 3 \cdot 5}{2 \cdot 4 \cdot 6 \cdot 8 \cdot 10}x^5 + \cdots, \quad [x^2 \leqq 1].$$

5.5. $(1 \pm x)^{5/2} = 1 \pm \dfrac{5}{2}x + \dfrac{5 \cdot 3}{2 \cdot 4}x^2 \pm \dfrac{5 \cdot 3 \cdot 1}{2 \cdot 4 \cdot 6}x^3$

$$- \dfrac{5 \cdot 3 \cdot 1 \cdot 1}{2 \cdot 4 \cdot 6 \cdot 8}x^4 \pm \dfrac{5 \cdot 3 \cdot 1 \cdot 1 \cdot 3}{2 \cdot 4 \cdot 6 \cdot 8 \cdot 10}x^5 - \cdots, \quad [x^2 \leqq 1].$$

6. $(1 + x)^{-n} = 1 - nx + \dfrac{n(n + 1)}{2!}x^2 - \dfrac{n(n + 1)(n + 2)}{3!}x^3$

$$+ \cdots + (-1)^r \dfrac{(n + r - 1)!}{(n - 1)!r!}x^r + \cdots, \qquad [x^2 < 1].$$

7. $(1 - x)^{-n} = 1 + nx + \dfrac{n(n+1)}{2!}x^2 + \dfrac{n(n+1)(n+2)}{3!}x^3$

$$+ \cdots + \frac{(n+r-1)!}{(n-1)!r!}x^r + \cdots, \qquad [x^2 < 1].$$

8. $(a \pm x)^{-n} = a^{-n}\left(1 \pm \dfrac{x}{a}\right)^{-n}, \qquad\qquad [x^2 < a^2].$

9.01. $(1 \pm x)^{-1/4} = 1 \mp \dfrac{1}{4}x + \dfrac{1 \cdot 5}{4 \cdot 8}x^2 \mp \dfrac{1 \cdot 5 \cdot 9}{4 \cdot 8 \cdot 12}x^3$

$$+ \frac{1 \cdot 5 \cdot 9 \cdot 13}{4 \cdot 8 \cdot 12 \cdot 16}x^4 \mp \cdots, \qquad [x^2 < 1].$$

9.02. $(1 \pm x)^{-1/3} = 1 \mp \dfrac{1}{3}x + \dfrac{1 \cdot 4}{3 \cdot 6}x^2 \mp \dfrac{1 \cdot 4 \cdot 7}{3 \cdot 6 \cdot 9}x^3$

$$+ \frac{1 \cdot 4 \cdot 7 \cdot 10}{3 \cdot 6 \cdot 9 \cdot 12}x^4 \mp \cdots, \qquad [x^2 < 1].$$

9.03. $(1 \pm x)^{-1/2} = 1 \mp \dfrac{1}{2}x + \dfrac{1 \cdot 3}{2 \cdot 4}x^2 \mp \dfrac{1 \cdot 3 \cdot 5}{2 \cdot 4 \cdot 6}x^3$

$$+ \frac{1 \cdot 3 \cdot 5 \cdot 7}{2 \cdot 4 \cdot 6 \cdot 8}x^4 \mp \cdots, \qquad [x^2 < 1].$$

9.04. $(1 \pm x)^{-1} = 1 \mp x + x^2 \mp x^3 + x^4 \mp \cdots, \qquad [x^2 < 1].$

9.05. $(1 \pm x)^{-3/2} = 1 \mp \dfrac{3}{2}x + \dfrac{3 \cdot 5}{2 \cdot 4}x^2 \mp \dfrac{3 \cdot 5 \cdot 7}{2 \cdot 4 \cdot 6}x^3$

$$+ \frac{3 \cdot 5 \cdot 7 \cdot 9}{2 \cdot 4 \cdot 6 \cdot 8}x^4 \mp \cdots, \qquad [x^2 < 1].$$

9.06. $(1 \pm x)^{-2} = 1 \mp 2x + 3x^2 \mp 4x^3 + 5x^4 \mp \cdots, \qquad [x^2 < 1].$

9.07. $(1 \pm x)^{-5/2} = 1 \mp \dfrac{5}{2}x + \dfrac{5 \cdot 7}{2 \cdot 4}x^2 \mp \dfrac{5 \cdot 7 \cdot 9}{2 \cdot 4 \cdot 6}x^3$

$$+ \frac{5 \cdot 7 \cdot 9 \cdot 11}{2 \cdot 4 \cdot 6 \cdot 8}x^4 \mp \cdots, \qquad [x^2 < 1].$$

9.08. $(1 \pm x)^{-3} = 1 \mp \dfrac{1}{1 \cdot 2}\{2 \cdot 3x \mp 3 \cdot 4x^2 + 4 \cdot 5x^3$

$$\mp 5 \cdot 6x^4 + \cdots \}, \qquad [x^2 < 1].$$

9.09. $(1 \pm x)^{-4} = 1 \mp \dfrac{1}{1 \cdot 2 \cdot 3}\{2 \cdot 3 \cdot 4x \mp 3 \cdot 4 \cdot 5x^2 + 4 \cdot 5 \cdot 6x^3$

$$\mp\ 5 \cdot 6 \cdot 7x^4 + \cdots \},\qquad [x^2 < 1].$$

9.10. $(1 \pm x)^{-5} = 1 \mp \dfrac{1}{1 \cdot 2 \cdot 3 \cdot 4}\{2 \cdot 3 \cdot 4 \cdot 5x \mp 3 \cdot 4 \cdot 5 \cdot 6x^2$

$$+\ 4 \cdot 5 \cdot 6 \cdot 7x^3 \mp 5 \cdot 6 \cdot 7 \cdot 8x^4 + \cdots \},\qquad [x^2 < 1].$$

10.

$2! =$		2	$1/2! = .5$	
$3! =$		6	$1/3! = .16666$	66667
$4! =$		24	$1/4! = .04166$	66667
$5! =$		120	$1/5! = .00833$	33333
$6! =$		720	$1/6! = .00138$	88889
$7! =$		5040	$1/7! = .19841$	26984×10^{-3}
$8! =$		40320	$1/8! = .24801$	58730×10^{-4}
$9! =$	3	62880	$1/9! = .27557$	31922×10^{-5}
$10! =$	36	28800	$1/10! = .27557$	31922×10^{-6}
$11! =$	399	16800	$1/11! = .25052$	10839×10^{-7}
$12! =$	4790	01600	$1/12! = .20876$	75699×10^{-8}

For a large table see Ref. 59 v, 1, second section, pp. 58–68.

11.
$$\lim_{n \to \infty} \frac{n!}{n^n e^{-n} \sqrt{n}} = \sqrt{(2\pi)}.$$

This gives approximate values of $n!$ for large values of n. When $n = 12$ the value given by the formula is $0.007(n!)$ too large and when $n = 20$ it is $0.004(n!)$ too large. [Ref. 21, p. 74. See also **851.4** and **850.4**.]

12.

n	2^n	n	2^n	n	2^{-n}	
2	4	15	32 768	2	.25	
3	8	16	65 536	3	.125	
4	16	17	131 072	4	.0625	
5	32	18	262 144	5	.03125	
6	64	19	524 288	6	.015625	
7	128	20	1 048 576	7	.78125	$\times\ 10^{-2}$
8	256	21	2 097 152	8	.39062 5	$\times\ 10^{-2}$
9	512	22	4 194 304	9	.19531 25	$\times\ 10^{-2}$
10	1 024	23	8 388 608	10	.97656 25	$\times\ 10^{-3}$
11	2 048	24	16 777 216	11	.48828 125	$\times\ 10^{-3}$
12	4 096	25	33 554 432	12	.24414 0625	$\times\ 10^{-3}$
13	8 192	26	67 108 864	13	.12207 03125	$\times\ 10^{-3}$
14	16 384	27	134 217 728	14	.61035 15625	$\times\ 10^{-4}$

15.1. $(a + b + c)^2 \equiv a^2 + b^2 + c^2 + 2ab + 2bc + 2ca.$

[The sign $\equiv$ expresses an identity.]

15.2. $(a + b - c)^2 \equiv a^2 + b^2 + c^2 + 2ab - 2bc - 2ca.$

15.3. $(a - b - c)^2 \equiv a^2 + b^2 + c^2 - 2ab + 2bc - 2ca.$

16. $(a + b + c + d)^2 \equiv a^2 + b^2 + c^2 + d^2 + 2ab + 2ac$
$$+ 2ad + 2bc + 2bd + 2cd.$$

17. $(a + b + c)^3 \equiv a^3 + b^3 + c^3 + 6abc$
$$+ 3(a^2b + ab^2 + b^2c + bc^2 + c^2a + ca^2).$$

20.1 $a + x \equiv (a^2 - x^2)/(a - x).$

20.11. $1 + x \equiv (1 - x^2)/(1 - x).$

20.2. $a^2 + ax + x^2 \equiv (a^3 - x^3)/(a - x).$

20.3. $a^3 + a^2x + ax^2 + x^3 \equiv (a^4 - x^4)/(a - x)$
$$\equiv (a^2 + x^2)(a + x).$$

20.4. $a^4 + a^3x + a^2x^2 + ax^3 + x^4 \equiv (a^5 - x^5)/(a - x).$

20.5. $a^5 + a^4x + a^3x^2 + a^2x^3 + ax^4 + x^5$
$$\equiv (a^6 - x^6)/(a - x) \equiv (a^3 + x^3)(a^2 + ax + x^2).$$

21.1. $a - x \equiv (a^2 - x^2)/(a + x).$

21.2. $a^2 - ax + x^2 \equiv (a^3 + x^3)/(a + x).$

21.3. $a^3 - a^2x + ax^2 - x^3 \equiv (a^4 - x^4)/(a + x)$
$$\equiv (a^2 + x^2)(a - x).$$

21.4. $a^4 - a^3x + a^2x^2 - ax^3 + x^4 \equiv (a^5 + x^5)/(a + x).$

21.5. $a^5 - a^4x + a^3x^2 - a^2x^3 + ax^4 - x^5$
$$\equiv (a^6 - x^6)/(a + x) \equiv (a^3 - x^3)(a^2 - ax + x^2).$$

22. $a^4 + a^2x^2 + x^4 \equiv (a^6 - x^6)/(a^2 - x^2)$
$$\equiv (a^2 + ax + x^2)(a^2 - ax + x^2).$$

22.1. $a^4 - a^2x^2 + x^4 \equiv (a^6 + x^6)/(a^2 + x^2).$

23. $\qquad a^4 + x^4 \equiv (a^2 + x^2)^2 - 2a^2x^2$

$$\equiv (a^2 + ax\sqrt{2} + x^2)(a^2 - ax\sqrt{2} + x^2).$$

25. **Arithmetic Progression** of the first order (first differences constant), to n terms,

$$a + (a + d) + (a + 2d) + (a + 3d) + \cdots + \{a + (n - 1)d\}$$

$$\equiv na + \frac{1}{2}n(n - 1)d$$

$$\equiv \frac{n}{2}(\text{1st term} + n\text{th term}).$$

26. **Geometric Progression,** to n terms,

$$a + ar + ar^2 + ar^3 + \cdots + ar^{n-1} \equiv a(1 - r^n)/(1 - r)$$
$$\equiv a(r^n - 1)/(r - 1).$$

26.1. If $r^2 < 1$, the limit of the sum of an infinite number of terms is $a/(1 - r)$.

27. The reciprocals of the terms of a series in arithmetic progression of the first order are in **Harmonic Progression.** Thus

$$\frac{1}{a}, \qquad \frac{1}{a + d}, \qquad \frac{1}{a + 2d}, \qquad \cdots \qquad \frac{1}{a + (n - 1)d}$$

are in Harmonic Progression.

28.1. **The Arithmetic Mean** of n quantities is

$$\frac{1}{n}(a_1 + a_2 + a_3 + \cdots + a_n).$$

28.2. **The Geometric Mean** of n quantities is

$$(a_1\, a_2\, a_3 \cdots a_n)^{1/n}.$$

28.3. Let the **Harmonic Mean** of n quantities be H. Then

$$\frac{1}{H} = \frac{1}{n}\left(\frac{1}{a_1} + \frac{1}{a_2} + \frac{1}{a_3} + \cdots + \frac{1}{a_n}\right).$$

28.4. The arithmetic mean of a number of positive quantities is $\geqq$ their geometric mean, which in turn is $\geqq$ their harmonic mean.

29. **Arithmetic Progression** of the kth order (kth differences constant).

Series: $u_1, u_2, u_3 \cdots u_n$.

First differences: $d_1', d_2', d_3', \cdots$
  where $d'_1 = u_2 - u_1$, $d'_2 = u_3 - u_2$, etc.

Second differences: $d_1'', d_2'', d_3'', \cdots$
  where $d_1'' = d_2' - d'_1$, etc.

Sum of n terms of the series

$$= \frac{n!}{(n-1)!1!}u_1 + \frac{n!}{(n-2)!2!}d_1' + \frac{n!}{(n-3)!3!}d_1'' + \cdots.$$

29.01. If a numerical table consists of values u_n of a function at equal intervals h of the argument, as follows,

$$f(a) = u_1, \quad f(a+h) = u_2, \quad f(a+2h) = u_3, \quad \text{etc.,}$$

then

$$f(a+ph) = u_1 + pd_1' + \frac{p(p-1)}{2!}d_1''$$

$$+ \frac{p(p-1)(p-2)}{3!}d_1''' + \cdots$$

where $p < 1$ and where d_1', d_1'', etc., are given by 29. The coefficients of d_1', d_1'', d_1''', etc., are called Gregory-Newton Interpolation Coefficients. For numerical values of these coefficients see Ref. 44, vol. 1, pp. 102–109, and Ref. 45.

29.1. $\quad 1 + 2 + 3 + \cdots + n = \dfrac{n}{2}(n+1).$

29.2. $\quad 1^2 + 2^2 + 3^2 + \cdots + n^2 = \dfrac{n}{6}(n+1)(2n+1)$

$$= \frac{n}{6}(2n^2 + 3n + 1).$$

29.3. $\quad 1^3 + 2^3 + 3^3 + \cdots + n^3 = \dfrac{n^2}{4}(n+1)^2$

$$= \frac{n^2}{4}(n^2 + 2n + 1).$$

29.4. $1^4 + 2^4 + 3^4 + \cdots + n^4$

$$= \frac{n}{30}(n + 1)(2n + 1)(3n^2 + 3n - 1)$$

$$= \frac{n}{30}(6n^4 + 15n^3 + 10n^2 - 1).$$

29.9. $\displaystyle\sum_{u=1}^{n} u^p = \frac{n^{p+1}}{p + 1} + \frac{n^p}{2} + \frac{B_1}{2!}pn^{p-1}$

$$- \frac{B_2}{4!}p(p - 1)(p - 2)n^{p-3} + \cdots,$$

omitting terms in n^0 and those that follow.

For values of B_1, B_2, $\cdots$, see **45**.

The above results may be used to find the sum of a series whose nth term is made up of n, n^2, n^3, etc.

30.1. $1 + 3 + 5 + 7 + 9 + \cdots + (2n - 1) = n^2.$

30.2. $1 + 8 + 16 + 24 + 32 + \cdots + 8(n - 1) = (2n - 1)^2.$

33.1. $1 + 3x + 5x^2 + 7x^3 + \cdots = \dfrac{1 + x}{(1 - x)^2}.$

33.2. $1 + ax + (a + b)x^2 + (a + 2b)x^3 + \cdots$

$$= 1 + \frac{ax + (b - a)x^2}{(1 - x)^2}.$$

33.3. $1 + 2^2x + 3^2x^2 + 4^2x^3 + \cdots = \dfrac{1 + x}{(1 - x)^3}.$

33.4. $1 + 3^2x + 5^2x^2 + 7^2x^3 + \cdots = \dfrac{1 + 6x + x^2}{(1 - x)^3}.$

[Contributed by W. V. Lyon.   Ref. 43, p. 448.]

35. $\dfrac{1}{a} - \dfrac{1}{a + b} + \dfrac{1}{a + 2b} - \dfrac{1}{a + 3b} + \cdots = \displaystyle\int_0^1 \frac{x^{a-1}}{1 + x^b}\,dx,$

$$[a, b > 0].$$

35.1. $1 - \dfrac{1}{3} + \dfrac{1}{5} - \dfrac{1}{7} + \dfrac{1}{9} - \cdots = \dfrac{\pi}{4}.$ [See **120** and **48.31**.]

35.2. $1 - \dfrac{1}{4} + \dfrac{1}{7} - \dfrac{1}{10} + \dfrac{1}{13} - \cdots = \dfrac{1}{3}\left(\dfrac{\pi}{\sqrt{3}} + \log_e 2\right).$

[See **165.01**.]

35.3. $\dfrac{1}{2} - \dfrac{1}{5} + \dfrac{1}{8} - \dfrac{1}{11} + \dfrac{1}{14} - \cdots = \dfrac{1}{3}\left(\dfrac{\pi}{\sqrt{3}} - \log_e 2\right).$

[See **165.11**.]

35.4. $1 - \dfrac{1}{5} + \dfrac{1}{9} - \dfrac{1}{13} + \dfrac{1}{17} - \cdots$

$$= \dfrac{1}{4\sqrt{2}}\{\pi + 2 \log_e (\sqrt{2} + 1)\}. \qquad \text{[See 170.]}$$

[Ref. 34, p. 161, Ex. 1.]

38. If there is a **power series** for $f(h)$, it is

$$f(h) = f(0) + hf'(0) + \dfrac{h^2}{2!}f''(0) + \dfrac{h^3}{3!}f'''(0) + \cdots.$$

[MACLAURIN'S SERIES.]

38.1. $f(h) = f(0) + hf'(0) + \dfrac{h^2}{2!}f''(0) + \dfrac{h^3}{3!}f'''(0) + \cdots$

$$+ \dfrac{h^{n-1}}{(n-1)!}f^{(n-1)}(0) + R_n,$$

where, for a suitable value of θ between 0 and 1,

$$R_n = \dfrac{h^n}{n!}f^{(n)}(\theta h), \quad \text{or} \quad \dfrac{h^n}{(n-1)!}(1 - \theta)^{n-1}f^{(n)}(\theta h).$$

39. $f(x + h) = f(x) + hf'(x) + \dfrac{h^2}{2!}f''(x) + \dfrac{h^3}{3!}f'''(x) + \cdots.$

[TAYLOR'S SERIES.]

39.1. $f(x + h) = f(x) + hf'(x) + \dfrac{h^2}{2!}f''(x) + \cdots$

$$+ \dfrac{h^{n-1}}{(n-1)!}f^{(n-1)}(x) + R_n,$$

where, for a suitable value of θ between 0 and 1,

$$R_n = \dfrac{h^n}{n!}f^{(n)}(x + \theta h), \quad \text{or} \quad \dfrac{h^n}{(n-1)!}(1 - \theta)^{n-1}f^{(n)}(x + \theta h).$$

40.
$$f(x + h, y + k) = f(x, y) + \left\{ h\frac{\partial f(x, y)}{\partial x} + k\frac{\partial f(x, y)}{\partial y} \right\}$$

$$+ \frac{1}{2!} \left\{ h^2\frac{\partial^2 f(x, y)}{\partial x^2} + 2hk\frac{\partial^2 f(x, y)}{\partial x\, \partial y} + k^2\frac{\partial^2 f(x, y)}{\partial y^2} \right\}$$

$$+ \frac{1}{3!} \left\{ h^3\frac{\partial^3 f(x, y)}{\partial x^3} + 3h^2k\frac{\partial^3 f(x, y)}{\partial x^2 \partial y} + 3hk^2\frac{\partial^3 f(x, y)}{\partial x\, \partial y^2} \right.$$

$$\left. + k^3\frac{\partial^3 f(x, y)}{\partial y^3} \right\} + \cdots + R_n$$

where, for suitable values of θ_1 and θ_2 between 0 and 1,

$$R_n = \frac{1}{n!} \left\{ h^n\frac{\partial^n}{\partial x^n} + nh^{n-1}k\frac{\partial^n}{\partial x^{n-1}\partial y} \right.$$

$$+ \frac{n(n-1)}{2!}h^{n-2}k^2\frac{\partial^n}{\partial x^{n-2}\partial y^2} + \cdots$$

$$\left. + k^n\frac{\partial^n}{\partial y^n} \right\} f(x + \theta_1 h, y + \theta_2 k). \quad \text{[Ref. 5, No. 886.]}$$

42.1. A number is divisible by 3 if the sum of the figures is divisible by 3.

42.2. A number is divisible by 9 if the sum of the figures is divisible by 9.

42.3. A number is divisible by 2^n if the number consisting of the last n figures is divisible by 2^n.

Bernoulli's Numbers and Euler's Numbers

45.

BERNOULLI'S NUMBERS	$\text{LOG}_{10}\ B_n$	EULER'S NUMBERS	$\text{LOG}_{10}\ E_n$
$B_1 = \dfrac{1}{6}$	$\bar{1}.221\ 8487$	$E_1 = \quad 1$	0
$B_2 = \dfrac{1}{30}$	$\bar{2}.522\ 8787$	$E_2 = \quad 5$	$0.698\ 9700$
$B_3 = \dfrac{1}{42}$	$\bar{2}.376\ 7507$	$E_3 = \quad 61$	$1.785\ 3298$
$B_4 = \dfrac{1}{30}$	$\bar{2}.522\ 8787$	$E_4 = \quad 1{,}385$	$3.141\ 4498$
$B_5 = \dfrac{5}{66}$	$\bar{2}.879\ 4261$	$E_5 = \quad 50{,}521$	$4.703\ 4719$
$B_6 = \dfrac{691}{2730}$	$\bar{1}.403\ 3154$	$E_6 = 2{,}702{,}765$	$6.431\ 8083$
$B_7 = \dfrac{7}{6}$	$0.066\ 9468$	$E_7 = 199{,}360{,}981$	$8.299\ 6402$
$B_8 = \dfrac{3617}{510}$	$0.850\ 7783$		
$B_9 = \dfrac{43{,}867}{798}$	$1.740\ 1350$		
$B_{10} = \dfrac{174{,}611}{330}$	$2.723\ 5577$		
$B_{11} = \dfrac{854{,}513}{138}$	$3.791\ 8396$		

For large tables see Ref. 27, pp. 176, 178; Ref. 34, pp. 234, 260; Ref. 44, vol. 2, pp. 230–242 and 294–302; and Ref. 59, vol. 1, second section, pp. 83–89.

The above notation is used in Refs. 27 and 34 and in "American Standard Mathematical Symbols," *Report of 1928*, Ref. 28. There are several different notations in use and, as stated in the above report, it is desirable when using the letters *B* and *E* for the above series of numbers, to give **47.1** and **47.4** as definitions, or to state explicitly the values of the first few numbers, as $B_1 = 1/6$, $B_2 = 1/30$, $B_3 = 1/42$, etc., $E_1 = 1$, $E_2 = 5$, $E_3 = 61$, etc.

46.1. $E_n = \dfrac{(2n)!}{(2n-2)!2!}\ E_{n-1} - \dfrac{(2n)!}{(2n-4)!4!}\ E_{n-2}$

$$+ \cdots + (-1)^{n-1}$$

taking $0! = 1$ and $E_0 = 1$.

46.2. $\quad B_n = \dfrac{2n}{2^{2n}(2^{2n}-1)}\left[\dfrac{(2n-1)!}{(2n-2)!1!}E_{n-1} - \dfrac{(2n-1)!}{(2n-4)!3!}E_{n-2}\right.$
$$\left. + \cdots + (-1)^{n-1}\right].$$

47.1. $\quad B_n = \dfrac{(2n)!}{\pi^{2n}2^{2n-1}}\left[1 + \dfrac{1}{2^{2n}} + \dfrac{1}{3^{2n}} + \dfrac{1}{4^{2n}} + \cdots\right].$

47.2. $\quad B_n = \dfrac{(2n)!}{\pi^{2n}(2^{2n-1}-1)}\left[1 - \dfrac{1}{2^{2n}} + \dfrac{1}{3^{2n}} - \dfrac{1}{4^{2n}} + \cdots\right].$

47.3. $\quad B_n = \dfrac{2(2n)!}{\pi^{2n}(2^{2n}-1)}\left[1 + \dfrac{1}{3^{2n}} + \dfrac{1}{5^{2n}} + \dfrac{1}{7^{2n}} + \cdots\right].$

47.4. $\quad E_n = \dfrac{2^{2n+2}(2n)!}{\pi^{2n+1}}\left[1 - \dfrac{1}{3^{2n+1}} + \dfrac{1}{5^{2n+1}} - \dfrac{1}{7^{2n+1}} + \cdots\right].$

48.001. $\quad 1 + \dfrac{1}{2} + \dfrac{1}{3} + \dfrac{1}{4} + \cdots = \infty$.

48.002. $\quad 1 + \dfrac{1}{2^2} + \dfrac{1}{3^2} + \dfrac{1}{4^2} + \cdots = \pi^2 B_1 = \dfrac{\pi^2}{6}$ \hfill [See **45.**]
$$= \zeta(2) = 1.64493\ 40668.$$

48.003. $\quad 1 + \dfrac{1}{2^3} + \dfrac{1}{3^3} + \dfrac{1}{4^3} + \cdots = \zeta(3) = 1.20205\ 69032.$

48.004. $\quad 1 + \dfrac{1}{2^4} + \dfrac{1}{3^4} + \dfrac{1}{4^4} + \cdots = \dfrac{\pi^4}{3}B_2 = \dfrac{\pi^4}{90}$ \hfill [See **45.**]
$$= \zeta(4) = 1.08232\ 32337.$$

48.005. $\quad 1 + \dfrac{1}{2^5} + \dfrac{1}{3^5} + \dfrac{1}{4^5} + \cdots = \zeta(5) = 1.03692\ 77551.$

48.006. $\quad 1 + \dfrac{1}{2^6} + \dfrac{1}{3^6} + \dfrac{1}{4^6} + \cdots = \dfrac{2\pi^6}{4^5}B_3 = \dfrac{\pi^6}{945}$ \hfill [See **45.**]
$$= \zeta(6) = 1.01734\ 30620.$$

48.007. $\quad 1 + \dfrac{1}{2^7} + \dfrac{1}{3^7} + \dfrac{1}{4^7} + \cdots = \zeta(7) = 1.00834\ 92774.$

48.008. $\quad 1 + \dfrac{1}{2^8} + \dfrac{1}{3^8} + \dfrac{1}{4^8} + \cdots = \dfrac{\pi^8}{315}B_4 = \dfrac{\pi^8}{9450}$ \hfill [See **45.**]
$$= \zeta(8) = 1.00407\ 73562.$$

48.08. $1 + \dfrac{1}{2^{2n}} + \dfrac{1}{3^{2n}} + \dfrac{1}{4^{2n}} + \cdots = \dfrac{2^{2n-1}\pi^{2n}}{(2n)!} B_n,$

$[n = \text{positive integer}].$ [See **45**, **47.1.**]

48.09. $1 + \dfrac{1}{2^p} + \dfrac{1}{3^p} + \dfrac{1}{4^p} + \cdots = \zeta(p) = \text{Zeta } (p),$ the Riemann

Zeta function. For numerical values of this function, including decimal values of p, see Ref. **45**.

48.11. $1 + \dfrac{1}{3} + \dfrac{1}{5} + \dfrac{1}{7} + \cdots = \infty.$

48.12. $1 + \dfrac{1}{3^2} + \dfrac{1}{5^2} + \dfrac{1}{7^2} + \cdots = \dfrac{3\pi^2}{4} B_1 = \dfrac{\pi^2}{8}.$ [See **45**.]

48.13. $1 + \dfrac{1}{3^3} + \dfrac{1}{5^3} + \dfrac{1}{7^3} + \cdots = \dfrac{7}{8}\zeta(3) = 1.05179\ 9790.$

[See **48.09.**]

48.14. $1 + \dfrac{1}{3^4} + \dfrac{1}{5^4} + \dfrac{1}{7^4} + \cdots = \dfrac{5\pi^4}{16} B_2 = \dfrac{\pi^4}{96}$ [See **45**.]

$= \dfrac{15}{16}\zeta(4) = 1.01467\ 80316.$

48.18. $1 + \dfrac{1}{3^{2n}} + \dfrac{1}{5^{2n}} + \dfrac{1}{7^{2n}} + \cdots = \dfrac{(2^{2n}-1)\pi^{2n}}{2(2n)!} B_n.$ [See **45**, **47.3.**]

48.19. $1 + \dfrac{1}{3^p} + \dfrac{1}{5^p} + \dfrac{1}{7^p} + \cdots = \left(1 - \dfrac{1}{2^p}\right)\zeta(p).$ [See **48.09.**]

48.21. $1 - \dfrac{1}{2} + \dfrac{1}{3} - \dfrac{1}{4} + \cdots = \log_\epsilon 2.$ [See **601.01.**]

48.22. $1 - \dfrac{1}{2^2} + \dfrac{1}{3^2} - \dfrac{1}{4^2} + \cdots = \dfrac{\pi^2}{2} B_1 = \dfrac{\pi^2}{12}.$ [See **45**.]

48.23. $1 - \dfrac{1}{2^3} + \dfrac{1}{3^3} - \dfrac{1}{4^3} + \cdots = \left(1 - \dfrac{2}{2^3}\right)\zeta(3) = 0.90154\ 26774.$

[See **48.09.**]

48.24. $1 - \dfrac{1}{2^4} + \dfrac{1}{3^4} - \dfrac{1}{4^4} + \cdots = \dfrac{(2^3 - 1)\pi^4}{4!} B_2 = \dfrac{7\pi^4}{720}.$ [See **45.**]

$$= \left(1 - \dfrac{2}{2^4}\right)\zeta(4) = 0.94703\ 28295.$$

48.28. $1 - \dfrac{1}{2^{2n}} + \dfrac{1}{3^{2n}} - \dfrac{1}{4^{2n}} + \cdots = \dfrac{(2^{2n-1} - 1)\pi^{2n}}{(2n)!} B_n.$

[See **45, 47.2.**]

48.29. $1 - \dfrac{1}{2^p} + \dfrac{1}{3^p} - \dfrac{1}{4^p} + \cdots = \left(1 - \dfrac{2}{2^p}\right)\zeta(p).$ [See **48.09.**]

[For $p = 1$, see **48.21.**]

48.31. $1 - \dfrac{1}{3} + \dfrac{1}{5} - \dfrac{1}{7} + \cdots = \dfrac{\pi}{4} E_0 = \dfrac{\pi}{4}.$ [$E_0 = 1$ as in **46.1.**]

48.32. $1 - \dfrac{1}{3^2} + \dfrac{1}{5^2} - \dfrac{1}{7^2} + \cdots = G = 0.91596\ 55942.$

[J. W. L. Glaisher, Proc. London Math. Soc., 1876–77.]

[Ref. 69, p. 2.]

48.33. $1 - \dfrac{1}{3^3} + \dfrac{1}{5^3} - \dfrac{1}{7^3} + \cdots = \dfrac{\pi^3}{32} E_1 = \dfrac{\pi^3}{32}.$ [See **45.**]

48.34. $1 - \dfrac{1}{3^4} + \dfrac{1}{5^4} - \dfrac{1}{7^4} + \cdots = 0.98894\ 455.$

48.36. $1 - \dfrac{1}{3^6} + \dfrac{1}{5^6} - \dfrac{1}{7^6} + \cdots = 0.99868\ 522.$

48.38. $1 - \dfrac{1}{3^8} + \dfrac{1}{5^8} - \dfrac{1}{7^8} + \cdots = 0.99985\ 0.$

48.39. $1 - \dfrac{1}{3^{2n+1}} + \dfrac{1}{5^{2n+1}} - \dfrac{1}{7^{2n+1}} + \cdots = \dfrac{\pi^{2n+1}}{2^{2n+2}(2n)!} E_n.$

[See **45, 47.4.**]

Reversion of Series

50. Let a known series be

$$y = ax + bx^2 + cx^3 + dx^4 + ex^5 + fx^6 + gx^7 + \cdots, \quad [a \neq 0],$$

to find the coefficients of the series

$$x = Ay + By^2 + Cy^3 + Dy^4 + Ey^5 + Fy^6 + Gy^7 + \cdots.$$

$$A = \frac{1}{a}. \qquad\qquad B = -\frac{b}{a^3}. \qquad\qquad C = \frac{1}{a^5}(2b^2 - ac).$$

$$D = \frac{1}{a^7}(5abc - a^2d - 5b^3).$$

$$E = \frac{1}{a^9}(6a^2bd + 3a^2c^2 + 14b^4 - a^3e - 21ab^2c).$$

$$F = \frac{1}{a^{11}}(7a^3be + 7a^3cd + 84ab^3c - a^4f - 28a^2b^2d$$
$$- 28a^2bc^2 - 42b^5).$$

$$G = \frac{1}{a^{13}}(8a^4bf + 8a^4ce + 4a^4d^2 + 120a^2b^3d + 180a^2b^2c^2$$
$$+ 132b^6 - a^5g - 36a^3b^2e - 72a^3bcd - 12a^3c^3 - 330ab^4c).$$

[See Ref. 23, p. 11, Ref. 31, p. 116, and *Philosophical Magazine*, vol. 19 (1910), p. 366, for additional coefficients.]

Powers of $S = a + bx + cx^2 + dx^3 + ex^4 + fx^5 \cdots$

51.1. $S^2 = a^2 + 2abx + (b^2 + 2ac)x^2 + 2(ad + bc)x^3$
$$+ (c^2 + 2ae + 2bd)x^4 + 2(af + be + cd)x^5 \cdots.$$

51.2. $S^{1/2} = a^{1/2}\left[1 + \frac{1}{2}\frac{b}{a}x + \left(\frac{1}{2}\frac{c}{a} - \frac{1}{8}\frac{b^2}{a^2}\right)x^2\right.$
$$+ \left(\frac{1}{2}\frac{d}{a} - \frac{1}{4}\frac{bc}{a^2} + \frac{1}{16}\frac{b^3}{a^3}\right)x^3$$
$$\left.+ \left(\frac{1}{2}\frac{e}{a} - \frac{1}{4}\frac{bd}{a^2} - \frac{1}{8}\frac{c^2}{a^2} + \frac{3}{16}\frac{b^2c}{a^3} - \frac{5}{128}\frac{b^4}{a^4}\right)x^4 \cdots\right].$$

51.3. $\quad S^{-1/2} = a^{-1/2}\left[1 - \dfrac{1}{2}\dfrac{b}{a}x + \left(\dfrac{3}{8}\dfrac{b^2}{a^2} - \dfrac{1}{2}\dfrac{c}{a}\right)x^2\right.$

$$+ \left(\dfrac{3}{4}\dfrac{bc}{a^2} - \dfrac{1}{2}\dfrac{d}{a} - \dfrac{5}{16}\dfrac{b^3}{a^3}\right)x^3$$

$$\left. + \left(\dfrac{3}{4}\dfrac{bd}{a^2} + \dfrac{3}{8}\dfrac{c^2}{a^2} - \dfrac{1}{2}\dfrac{e}{a} - \dfrac{15}{16}\dfrac{b^2c}{a^3} + \dfrac{35}{128}\dfrac{b^4}{a^4}\right)x^4 \cdots\right].$$

51.4. $\quad S^{-1} = a^{-1}\left[1 - \dfrac{b}{a}x + \left(\dfrac{b^2}{a^2} - \dfrac{c}{a}\right)x^2 + \left(\dfrac{2bc}{a^2} - \dfrac{d}{a} - \dfrac{b^3}{a^3}\right)x^3\right.$

$$\left. + \left(\dfrac{2bd}{a^2} + \dfrac{c^2}{a^2} - \dfrac{e}{a} - 3\dfrac{b^2c}{a^3} + \dfrac{b^4}{a^4}\right)x^4 \cdots\right].$$

51.5. $\quad S^{-2} = a^{-2}\left[1 - 2\dfrac{b}{a}x + \left(3\dfrac{b^2}{a^2} - 2\dfrac{c}{a}\right)x^2\right.$

$$+ \left(6\dfrac{bc}{a^2} - 2\dfrac{d}{a} - 4\dfrac{b^3}{a^3}\right)x^3$$

$$\left. + \left(6\dfrac{bd}{a^2} - 3\dfrac{c^2}{a^2} - 2\dfrac{e}{a} - 12\dfrac{b^2c}{a^3} + 5\dfrac{b^4}{a^4}\right)x^4 \cdots\right].$$

Roots of Quadratic Equation

55.1. The roots of $ax^2 + bx + c = 0$ are

$$\alpha = \frac{-b + \sqrt{(b^2 - 4ac)}}{2a} = \frac{-2c}{b + \sqrt{(b^2 - 4ac)}},$$

$$\beta = \frac{-b - \sqrt{(b^2 - 4ac)}}{2a} = \frac{-2c}{b - \sqrt{(b^2 - 4ac)}}.$$

The difference of two quantities is inconvenient to compute with precision and in such a case the alternative formula using the numerical sum of two quantities should be used.   [Ref. 41, p. 306.]

55.2. If one root α has been computed precisely, use

$$\beta = -\alpha - \frac{b}{a} \qquad \text{or} \qquad \beta = \frac{c}{a\alpha}.$$

Square Roots of Complex Quantity

58.1. $\sqrt{(x+iy)} = \pm\left[\sqrt{\left(\dfrac{r+x}{2}\right)} + i\sqrt{\left(\dfrac{r-x}{2}\right)}\right].$

58.2. $\sqrt{(x-iy)} = \pm\left[\sqrt{\left(\dfrac{r+x}{2}\right)} - i\sqrt{\left(\dfrac{r-x}{2}\right)}\right],$

where x may be positive or negative,

y is positive
$r = +\sqrt{(x^2 + y^2)}$
$i = \sqrt{(-1)}.$

The positive square roots of $(r+x)/2$ and $(r-x)/2$ are to be used.

[Ref. 61, p. 260.]

58.3. An alternative method is to put $x + iy$ in the form

$$re^{i(\theta + 2\pi k)}$$ [See **604.05.**]

where $r = \sqrt{(x^2 + y^2)}$, $\cos\theta = x/r$, $\sin\theta = y/r$, and k is an integer or 0. Then

$$\sqrt{(x+iy)} = \sqrt{(re^{i\theta})} = \pm\sqrt{r}\,e^{i\theta/2}$$
$$= \pm\sqrt{r}\left(\cos\frac{\theta}{2} + i\sin\frac{\theta}{2}\right).$$

59.1. The determinant

$$\begin{vmatrix} a_{1p} & a_{1q} \\ a_{2p} & a_{2q} \end{vmatrix} \equiv a_{1p}a_{2q} - a_{2p}a_{1q}$$

59.2. The determinant

$$\begin{vmatrix} a_{1p} & a_{1q} & a_{1r} \\ a_{2p} & a_{2q} & a_{2r} \\ a_{3p} & a_{3q} & a_{3r} \end{vmatrix} \equiv a_{1p}\begin{vmatrix} a_{2q} & a_{2r} \\ a_{3q} & a_{3r} \end{vmatrix} - a_{1q}\begin{vmatrix} a_{2p} & a_{2r} \\ a_{3p} & a_{3r} \end{vmatrix} + a_{1r}\begin{vmatrix} a_{2p} & a_{2q} \\ a_{3p} & a_{3q} \end{vmatrix}$$

$$\equiv a_{1p}(a_{2q}a_{3r} - a_{3q}a_{2r}) - a_{1q}(a_{2p}a_{3r} - a_{3p}a_{2r}) + a_{1r}(a_{2p}a_{3q} - a_{3p}a_{2q}).$$

59.3. If there are three simultaneous equations,

$$a_{1p}x + a_{1q}y + a_{1r}z = u$$
$$a_{2p}x + a_{2q}y + a_{2r}z = v$$
$$a_{3p}x + a_{3q}y + a_{3r}z = w$$

then

$$x = \frac{\begin{vmatrix} u & a_{1q} & a_{1r} \\ v & a_{2q} & a_{2r} \\ w & a_{3q} & a_{3r} \end{vmatrix}}{\begin{vmatrix} a_{1p} & a_{1q} & a_{1r} \\ a_{2p} & a_{2q} & a_{2r} \\ a_{3p} & a_{3q} & a_{3r} \end{vmatrix}}, \qquad y = \frac{\begin{vmatrix} a_{1p} & u & a_{1r} \\ a_{2p} & v & a_{2r} \\ a_{3p} & w & a_{3r} \end{vmatrix}}{\begin{vmatrix} a_{1p} & a_{1q} & a_{1r} \\ a_{2p} & a_{2q} & a_{2r} \\ a_{3p} & a_{3q} & a_{3r} \end{vmatrix}},$$

$$z = \frac{\begin{vmatrix} a_{1p} & a_{1q} & u \\ a_{2p} & a_{2q} & v \\ a_{3p} & a_{3q} & w \end{vmatrix}}{\begin{vmatrix} a_{1p} & a_{1q} & a_{1r} \\ a_{2p} & a_{2q} & a_{2r} \\ a_{3p} & a_{3q} & a_{3r} \end{vmatrix}}.$$

For sets of other numbers of independent simultaneous linear equations, the number of equations being equal to the number of unknowns, the roots are given by similar expressions, when the denominator is not 0.

ALGEBRAIC FUNCTIONS—DERIVATIVES

60. $\dfrac{d(au)}{dx} = a\dfrac{du}{dx}$ where a is a constant.

61. $\dfrac{d(u + v)}{dx} = \dfrac{du}{dx} + \dfrac{dv}{dx}.$

62. $\dfrac{d(uv)}{dx} = u\dfrac{dv}{dx} + v\dfrac{du}{dx}.$

63. $\dfrac{d(uvw)}{dx} = uv\dfrac{dw}{dx} + vw\dfrac{du}{dx} + wu\dfrac{dv}{dx}.$

64. $\dfrac{d(x^n)}{dx} = nx^{n-1}.$

64.1. $\dfrac{d\sqrt{x}}{dx} = \dfrac{1}{2\sqrt{x}}.$

64.2. $\dfrac{d(1/x)}{dx} = -\dfrac{1}{x^2}.$

64.3. $\dfrac{d(x^{-n})}{dx} = -nx^{-(n+1)}.$

65. $\dfrac{d(u/v)}{dx} = \dfrac{1}{v}\dfrac{du}{dx} - \dfrac{u}{v^2}\dfrac{dv}{dx} = \dfrac{v\dfrac{du}{dx} - u\dfrac{dv}{dx}}{v^2}.$

66. $\dfrac{df(u)}{dx} = \dfrac{df(u)}{du} \cdot \dfrac{du}{dx}.$

67. $\dfrac{d^2f(u)}{dx^2} = \dfrac{df(u)}{du} \cdot \dfrac{d^2u}{dx^2} + \dfrac{d^2f(u)}{du^2} \cdot \left(\dfrac{du}{dx}\right)^2.$

68. $\dfrac{d^n(uv)}{dx^n} = v\dfrac{d^nu}{dx^n} + n\dfrac{dv}{dx}\dfrac{d^{n-1}u}{dx^{n-1}} + \dfrac{n(n-1)}{2!}\dfrac{d^2v}{dx^2}\dfrac{d^{n-2}u}{dx^{n-2}}$

$$+ \cdots + \dfrac{n!}{(n-k)!k!}\dfrac{d^kv}{dx^k}\dfrac{d^{n-k}u}{dx^{n-k}} + \cdots + \dfrac{ud^nv}{dx^n}.$$

69.1. $\dfrac{d}{dq}\displaystyle\int_p^q f(x)dx = f(q),$ [p constant].

69.2. $\dfrac{d}{dp}\displaystyle\int_{p}^{q} f(x)dx = -f(p),$ [q constant].

69.3. $\dfrac{d}{dc}\displaystyle\int_{p}^{q} f(x, c)dx = \int_{p}^{q} \dfrac{\partial}{\partial c} f(x, c)dx + f(q, c)\dfrac{dq}{dc} - f(p, c)\dfrac{dp}{dc}.$

72.   If $\phi(a) = 0$ and $\psi(a) = 0$, or if $\phi(a) = \infty$ and $\psi(a) = \infty$, then

$$\lim_{x \to a} \frac{\phi(x)}{\psi(x)} = \frac{\phi'(a)}{\psi'(a)}.$$

If, also, $\phi'(a) = 0$ and $\psi'(a) = 0$, or if $\phi'(a) = \infty$ and $\psi'(a) = \infty$, then

$$\lim_{x \to a} \frac{\phi(x)}{\psi(x)} = \frac{\phi''(a)}{\psi''(a)}, \quad \text{and so on.}$$

72.1.   If a function takes the form $0 \times \infty$ or $\infty - \infty$, it may, by an algebraic or other change, be made to take the form $0/0$ or ∞/∞.

72.2.   If a function takes the form 0^0, ∞^0 or 1^∞, it may be made to take the form $0 \times \infty$ and therefore $0/0$ or ∞/∞ by first taking logarithms.   [Ref. 8, Chap. 42.]

79.   **General Formula for Integration by Parts.**

$$\int u\, dv = uv - \int v\, du,$$

or

$$\int u\, dv = uv - \int v\, \frac{du}{dv}\, dv.$$

RATIONAL ALGEBRAIC FUNCTIONS—INTEGRALS

The constant of integration is to be understood with all indefinite integrals.

Integrals Involving x^n

80. $\displaystyle\int dx = x.$ **81.2.** $\displaystyle\int x^2 dx = \frac{x^3}{3}.$

81.1. $\displaystyle\int x\,dx = \frac{x^2}{2}.$ **81.9.** $\displaystyle\int x^n dx = \frac{x^{n+1}}{n+1},$ $[n \neq -1].$

82.1. $\displaystyle\int \frac{dx}{x} = \log_\epsilon |x|.$ [See note preceding **600.**]

Integration in this case should not be carried from a negative to a positive value of x. If x is negative, use $\log |x|$, since $\log(-1) \equiv (2k+1)\pi i$ will be part of the constant of integration.   [See **409.03.**]

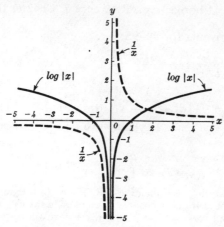

Fig. 82.1. Graphs of $y = 1/x$ and $y = \log |x|$, where x is real.

82.2. $\displaystyle\int \frac{dx}{x^2} = -\frac{1}{x}.$ **82.4.** $\displaystyle\int \frac{dx}{x^4} = -\frac{1}{3x^3}.$

82.3. $\displaystyle\int \frac{dx}{x^3} = -\frac{1}{2x^2}.$ **82.5.** $\displaystyle\int \frac{dx}{x^5} = -\frac{1}{4x^4}.$

82.9. $\displaystyle\int \frac{dx}{x^n} = -\frac{1}{(n-1)x^{n-1}},$ $[n \neq 1].$

Integrals Involving $X = a + bx$

83. $\displaystyle \int (a + bx)^n dx = \frac{1}{b} \int X^n dX = \frac{X^{n+1}}{b(n+1)},$ $\qquad [n \neq -1].$

84.1. $\displaystyle \int x^m (a + bx)^n dx$ may be integrated term by term after expanding $(a + bx)^n$ by the binomial theorem, when n is a positive integer.

84.2. If $m < n$, or if n is fractional, it may be shorter to use

$$\int x^m X^n dx = \frac{1}{b^{m+1}} \int (X - a)^m X^n dX$$

and expand $(X - a)^m$ by the binomial theorem, when m is a positive integer.

85. On integrals of rational algebraic fractions, see the topic partial fractions in text books, *e.g.*, Reference 7, Chapter II.

89. General formula for **90–95**:

$$\int \frac{x^m dx}{X^n} = \frac{1}{b^{m+1}} \int \frac{(X - a)^m dX}{X^n}$$

$$= \frac{1}{b^{m+1}} \left[\sum_{s=0}^{m} \frac{m!(-a)^s X^{m-n-s+1}}{(m-s)!s!(m-n-s+1)} \right],$$

except, where $m - n - s + 1 = 0$, in which case the corresponding term in the square brackets is

$$\frac{m!(-a)^{m-n+1}}{(m-n+1)!(n-1)!} \log |X|,$$

the letters representing real quantities. [Ref. 2, p. 7.] Integration should not be carried from a negative to a positive value of X in the case of $\log |X|$. If X is negative, use $\log |X|$ since $\log(-1) \equiv (2k + 1)\pi i$ will be part of the constant of integration.

90. $\displaystyle \int \frac{dx}{X^n} = \frac{-1}{(n-1)bX^{n-1}},$ $\qquad [n \neq 1].$

90.1. $\displaystyle \int \frac{dx}{X} = \frac{1}{b} \log |X|.$ $\qquad$ [See note on $\log |X|$ under **89.**]

90.2. $\displaystyle \int \frac{dx}{X^2} = -\frac{1}{bX}.$ $\qquad\qquad$ **90.3.** $\displaystyle \int \frac{dx}{X^3} = -\frac{1}{2bX^2}.$

90.4. $\displaystyle\int \frac{dx}{X^4} = -\frac{1}{3bX^3}.$ **90.5.** $\displaystyle\int \frac{dx}{X^5} = -\frac{1}{4bX^4}.$

91. $\displaystyle\int \frac{x\,dx}{X^n} = \frac{1}{b^2}\left[\frac{-1}{(n-2)X^{n-2}} + \frac{a}{(n-1)X^{n-1}}\right],$

[except where any one of the exponents of X is 0, see **89**].

91.1. $\displaystyle\int \frac{x\,dx}{X} = \frac{1}{b^2}[X - a\log|X|].$ [If $X < 0$, use $\log|X|$, see **89**.]

91.2. $\displaystyle\int \frac{x\,dx}{X^2} = \frac{1}{b^2}\left[\log|X| + \frac{a}{X}\right].$

91.3. $\displaystyle\int \frac{x\,dx}{X^3} = \frac{1}{b^2}\left[-\frac{1}{X} + \frac{a}{2X^2}\right].$

91.4. $\displaystyle\int \frac{x\,dx}{X^4} = \frac{1}{b^2}\left[-\frac{1}{2X^2} + \frac{a}{3X^3}\right].$

91.5. $\displaystyle\int \frac{x\,dx}{X^5} = \frac{1}{b^2}\left[-\frac{1}{3X^3} + \frac{a}{4X^4}\right].$

92. $\displaystyle\int \frac{x^2\,dx}{X^n} = \frac{1}{b^3}\left[\frac{-1}{(n-3)X^{n-3}} + \frac{2a}{(n-2)X^{n-2}} - \frac{a^2}{(n-1)X^{n-1}}\right],$

[except where any one of the exponents of X is 0, see **89**.]

92.1. $\displaystyle\int \frac{x^2\,dx}{X} = \frac{1}{b^3}\left[\frac{X^2}{2} - 2aX + a^2\log|X|\right].$

An alternative expression, which differs by a constant, is

$$\frac{x^2}{2b} - \frac{ax}{b^2} + \frac{a^2}{b^3}\log|a+bx|.$$

92.2. $\displaystyle\int \frac{x^2\,dx}{X^2} = \frac{1}{b^3}\left[X - 2a\log|X| - \frac{a^2}{X}\right].$

92.3. $\displaystyle\int \frac{x^2\,dx}{X^3} = \frac{1}{b^3}\left[\log|X| + \frac{2a}{X} - \frac{a^2}{2X^2}\right].$

92.4. $\displaystyle\int \frac{x^2\,dx}{X^4} = \frac{1}{b^3}\left[-\frac{1}{X} + \frac{2a}{2X^2} - \frac{a^2}{3X^3}\right].$

92.5. $\displaystyle\int \frac{x^2\,dx}{X^5} = \frac{1}{b^3}\left[-\frac{1}{2X^2} + \frac{2a}{3X^3} - \frac{a^2}{4X^4}\right].$

92.6. $\displaystyle\int\frac{x^2dx}{X^6} = \frac{1}{b^3}\left[-\frac{1}{3X^3}+\frac{2a}{4X^4}-\frac{a^2}{5X^5}\right].$

92.7. $\displaystyle\int\frac{x^2dx}{X^7} = \frac{1}{b^3}\left[-\frac{1}{4X^4}+\frac{2a}{5X^5}-\frac{a^2}{6X^6}\right].$

93. $\displaystyle\int\frac{x^3dx}{X^n} = \frac{1}{b^4}\left[\frac{-1}{(n-4)X^{n-4}}+\frac{3a}{(n-3)X^{n-3}}\right.$
$$\left.-\frac{3a^2}{(n-2)X^{n-2}}+\frac{a^3}{(n-1)X^{n-1}}\right],$$

[except where any one of the exponents of X is 0, see **89**].

93.1. $\displaystyle\int\frac{x^3dx}{X} = \frac{1}{b^4}\left[\frac{X^3}{3}-\frac{3aX^2}{2}+3a^2X-a^3\log|X|\right]$
$$= \frac{x^3}{3b}-\frac{ax^2}{2b^2}+\frac{a^2x}{b^3}-\frac{a^3}{b^4}\log|a+bx|+\text{constant}.$$

93.2. $\displaystyle\int\frac{x^3dx}{X^2} = \frac{1}{b^4}\left[\frac{X^2}{2}-3aX+3a^2\log|X|+\frac{a^3}{X}\right].$

93.3. $\displaystyle\int\frac{x^3dx}{X^3} = \frac{1}{b^4}\left[X-3a\log|X|-\frac{3a^2}{X}+\frac{a^3}{2X^2}\right].$

93.4. $\displaystyle\int\frac{x^3dx}{X^4} = \frac{1}{b^4}\left[\log|X|+\frac{3a}{X}-\frac{3a^2}{2X^2}+\frac{a^3}{3X^3}\right].$

93.5. $\displaystyle\int\frac{x^3dx}{X^5} = \frac{1}{b^4}\left[-\frac{1}{X}+\frac{3a}{2X^2}-\frac{3a^2}{3X^3}+\frac{a^3}{4X^4}\right].$

93.6. $\displaystyle\int\frac{x^3dx}{X^6} = \frac{1}{b^4}\left[-\frac{1}{2X^2}+\frac{3a}{3X^3}-\frac{3a^2}{4X^4}+\frac{a^3}{5X^5}\right].$

93.7. $\displaystyle\int\frac{x^3dx}{X^7} = \frac{1}{b^4}\left[-\frac{1}{3X^3}+\frac{3a}{4X^4}-\frac{3a^2}{5X^5}+\frac{a^3}{6X^6}\right].$

94. $\displaystyle\int\frac{x^4dx}{X^n} = \frac{1}{b^5}\left[\frac{-1}{(n-5)X^{n-5}}+\frac{4a}{(n-4)X^{n-4}}\right.$
$$\left.-\frac{6a^2}{(n-3)X^{n-3}}+\frac{4a^3}{(n-2)X^{n-2}}-\frac{a^4}{(n-1)X^{n-1}}\right],$$

[except where any one of the exponents of X is 0, see **89**].

94.1. $\displaystyle\int\frac{x^4 dx}{X} = \frac{1}{b^5}\left[\frac{X^4}{4} - \frac{4aX^3}{3} + \frac{6a^2X^2}{2} - 4a^3X + a^4\log|X|\right]$

$\displaystyle = \frac{x^4}{4b} - \frac{ax^3}{3b^2} + \frac{a^2x^2}{2b^3} - \frac{a^3x}{b^4} + \frac{a^4}{b^5}\log|a+bx| + \text{constant.}$

94.2. $\displaystyle\int\frac{x^4 dx}{X^2} = \frac{1}{b^5}\left[\frac{X^3}{3} - \frac{4aX^2}{2} + 6a^2X - 4a^3\log|X| - \frac{a^4}{X}\right].$

94.3. $\displaystyle\int\frac{x^4 dx}{X^3} = \frac{1}{b^5}\left[\frac{X^2}{2} - 4aX + 6a^2\log|X| + \frac{4a^3}{X} - \frac{a^4}{2X^2}\right].$

94.4. $\displaystyle\int\frac{x^4 dx}{X^4} = \frac{1}{b^5}\left[X - 4a\log|X| - \frac{6a^2}{X} + \frac{4a^3}{2X^2} - \frac{a^4}{3X^3}\right].$

94.5. $\displaystyle\int\frac{x^4 dx}{X^5} = \frac{1}{b^5}\left[\log|X| + \frac{4a}{X} - \frac{6a^2}{2X^2} + \frac{4a^3}{3X^3} - \frac{a^4}{4X^4}\right].$

94.6. $\displaystyle\int\frac{x^4 dx}{X^6} = \frac{1}{b^5}\left[-\frac{1}{X} + \frac{4a}{2X^2} - \frac{6a^2}{3X^3} + \frac{4a^3}{4X^4} - \frac{a^4}{5X^5}\right].$

94.7. $\displaystyle\int\frac{x^4 dx}{X^7} = \frac{1}{b^5}\left[-\frac{1}{2X^2} + \frac{4a}{3X^3} - \frac{6a^2}{4X^4} + \frac{4a^3}{5X^5} - \frac{a^4}{6X^6}\right].$

95. $\displaystyle\int\frac{x^5 dx}{X^n} = \frac{1}{b^6}\left[\frac{-1}{(n-6)X^{n-6}} + \frac{5a}{(n-5)X^{n-5}}\right.$

$\displaystyle - \frac{10a^2}{(n-4)X^{n-4}} + \frac{10a^3}{(n-3)X^{n-3}}$

$\displaystyle \left. - \frac{5a^4}{(n-2)X^{n-2}} + \frac{a^5}{(n-1)X^{n-1}}\right],$

[except where any one of the exponents of X is 0, see **89**].

95.1. $\displaystyle\int\frac{x^5 dx}{X} = \frac{1}{b^6}\left[\frac{X^5}{5} - \frac{5aX^4}{4} + \frac{10a^2X^3}{3} - \frac{10a^3X^2}{2}\right.$

$\displaystyle \left. + 5a^4X - a^5\log|X|\right]$

$\displaystyle = \frac{x^5}{5b} - \frac{ax^4}{4b^2} + \frac{a^2x^3}{3b^3} - \frac{a^3x^2}{2b^4} + \frac{a^4x}{b^5}$

$\displaystyle - \frac{a^5}{b^6}\log|a+bx| + \text{constant.}$

[Ref. 1, p. 11.]

95.2. $\displaystyle\int\frac{x^5 dx}{X^2} = \frac{1}{b^6}\left[\frac{X^4}{4} - \frac{5aX^3}{3} + \frac{10a^2X^2}{2} - 10aX^3 \right.$
$$\left. + 5a^4 \log|X| + \frac{a^5}{X}\right].$$

95.3. $\displaystyle\int\frac{x^5 dx}{X^3} = \frac{1}{b^6}\left[\frac{X^3}{3} - \frac{5aX^2}{2} + 10a^2X - 10a^3 \log|X| \right.$
$$\left. - \frac{5a^4}{X} + \frac{a^5}{2X^2}\right].$$

95.4. $\displaystyle\int\frac{x^5 dx}{X^4} = \frac{1}{b^6}\left[\frac{X^2}{2} - 5aX + 10a^2 \log|X| + \frac{10a^3}{X} \right.$
$$\left. - \frac{5a^4}{2X^2} + \frac{a^5}{3X^3}\right].$$

95.5. $\displaystyle\int\frac{x^5 dx}{X^5} = \frac{1}{b^6}\left[X - 5a \log|X| - \frac{10a^2}{X} + \frac{10a^3}{2X^2} - \frac{5a^4}{3X^3} + \frac{a^5}{4X^4}\right].$

95.6. $\displaystyle\int\frac{x^5 dx}{X^6} = \frac{1}{b^6}\left[\log|X| + \frac{5a}{X} - \frac{10a^2}{2X^2} + \frac{10a^3}{3X^3} - \frac{5a^4}{4X^4} + \frac{a^5}{5X^5}\right].$

95.7. $\displaystyle\int\frac{x^5 dx}{X^7} = \frac{1}{b^6}\left[-\frac{1}{X} + \frac{5a}{2X^2} - \frac{10a^2}{3X^3} + \frac{10a^3}{4X^4} - \frac{5a^4}{5X^5} + \frac{a^5}{6X^6}\right].$

95.8. $\displaystyle\int\frac{x^5 dx}{X^8} = \frac{1}{b^6}\left[-\frac{1}{2X^2} + \frac{5a}{3X^3} - \frac{10a^2}{4X^4} + \frac{10a^3}{5X^5} - \frac{5a^4}{6X^6} + \frac{a^5}{7X^7}\right].$

[Ref. 2, pp. 7–11.]

100. General formula for **101–105**:

$$\int\frac{dx}{x^m X^n} = \frac{-1}{a^{m+n-1}}\int\frac{\left(\dfrac{X}{x} - b\right)^{m+n-2}}{\left(\dfrac{X}{x}\right)^n}\,d\left(\frac{X}{x}\right)$$

$$= \frac{-1}{a^{m+n-1}}\left[\sum_{s=0}^{m+n-2}\frac{(m+n-2)!X^{m-s-1}(-b)^s}{(m+n-s-2)!s!(m-s-1)x^{m-s-1}}\right]$$

unless $m - s - 1 = 0$, when the corresponding term in square brackets is

$$\frac{(m+n-2)!}{(m-1)!(n-1)!}(-b)^{m-1}\log\left|\frac{X}{x}\right|.$$

101.1. $\displaystyle\int \frac{dx}{xX} = -\frac{1}{a}\log\left|\frac{X}{x}\right|.$

101.2. $\displaystyle\int \frac{dx}{xX^2} = -\frac{1}{a^2}\left[\log\left|\frac{X}{x}\right| + \frac{bx}{X}\right].$

101.3. $\displaystyle\int \frac{dx}{xX^3} = -\frac{1}{a^3}\left[\log\left|\frac{X}{x}\right| + \frac{2bx}{X} - \frac{b^2x^2}{2X^2}\right].$

101.4. $\displaystyle\int \frac{dx}{xX^4} = -\frac{1}{a^4}\left[\log\left|\frac{X}{x}\right| + \frac{3bx}{X} - \frac{3b^2x^2}{2X^2} + \frac{b^3x^3}{3X^3}\right].$

101.5. $\displaystyle\int \frac{dx}{xX^5} = -\frac{1}{a^5}\left[\log\left|\frac{X}{x}\right| + \frac{4bx}{X} - \frac{6b^2x^2}{2X^2} + \frac{4b^3x^3}{3X^3} - \frac{b^4x^4}{4X^4}\right].$

Alternative solutions, which differ by a constant, are:

101.92. $\displaystyle\int \frac{dx}{xX^2} = \frac{1}{aX} - \frac{1}{a^2}\log\left|\frac{X}{x}\right|.$

101.93. $\displaystyle\int \frac{dx}{xX^3} = \frac{1}{2aX^2} + \frac{1}{a^2X} - \frac{1}{a^3}\log\left|\frac{X}{x}\right|.$

101.94. $\displaystyle\int \frac{dx}{xX^4} = \frac{1}{3aX^3} + \frac{1}{2a^2X^2} + \frac{1}{a^3X} - \frac{1}{a^4}\log\left|\frac{X}{x}\right|.$

101.95. $\displaystyle\int \frac{dx}{xX^5} = \frac{1}{4aX^4} + \frac{1}{3a^2X^3} + \frac{1}{2a^3X^2} + \frac{1}{a^4X} - \frac{1}{a^5}\log\left|\frac{X}{x}\right|.$

[Ref. 2, p. 13.]

102.1. $\displaystyle\int \frac{dx}{x^2X} = -\frac{1}{a^2}\left[\frac{X}{x} - b\log\left|\frac{X}{x}\right|\right].$

102.2. $\displaystyle\int \frac{dx}{x^2X^2} = -\frac{1}{a^3}\left[\frac{X}{x} - 2b\log\left|\frac{X}{x}\right| - \frac{b^2x}{X}\right].$

102.3. $\displaystyle\int \frac{dx}{x^2X^3} = -\frac{1}{a^4}\left[\frac{X}{x} - 3b\log\left|\frac{X}{x}\right| - \frac{3b^2x}{X} + \frac{b^3x^2}{2X^2}\right].$

102.4. $\displaystyle\int \frac{dx}{x^2X^4} = -\frac{1}{a^5}\left[\frac{X}{x} - 4b\log\left|\frac{X}{x}\right| - \frac{6b^2x}{X} + \frac{4b^3x^2}{2X^2} - \frac{b^4x^3}{3X^3}\right].$

Alternative solutions, which differ by a constant, are:

102.91. $\displaystyle\int \frac{dx}{x^2X} = -\frac{1}{ax} + \frac{b}{a^2}\log\left|\frac{X}{x}\right|.$

102.92. $\displaystyle\int\frac{dx}{x^2X^2} = -b\left[\frac{1}{a^2X} + \frac{1}{a^2bx} - \frac{2}{a^3}\log\left|\frac{X}{x}\right|\right].$

102.93. $\displaystyle\int\frac{dx}{x^2X^3} = -b\left[\frac{1}{2a^2X^2} + \frac{2}{a^3X} + \frac{1}{a^3bx} - \frac{3}{a^4}\log\left|\frac{X}{x}\right|\right],$

where $X = a + bx.$

102.94. $\displaystyle\int\frac{dx}{x^2X^4} = -b\left[\frac{1}{3a^2X^3} + \frac{2}{2a^3X^2} + \frac{3}{a^4X} + \frac{1}{a^4bx}\right.$

$$\left. - \frac{4}{a^5}\log\left|\frac{X}{x}\right|\right]. \qquad \text{[Ref. 2, p. 14.]}$$

103.1. $\displaystyle\int\frac{dx}{x^3X} = -\frac{1}{a^3}\left[\frac{X^2}{2x^2} - \frac{2bX}{x} + b^2\log\left|\frac{X}{x}\right|\right]$

$$= -\frac{1}{2ax^2} + \frac{b}{a^2x} - \frac{b^2}{a^3}\log\left|\frac{X}{x}\right| + \text{constant}.$$

103.2. $\displaystyle\int\frac{dx}{x^3X^2} = -\frac{1}{a^4}\left[\frac{X^2}{2x^2} - \frac{3bX}{x} + 3b^2\log\left|\frac{X}{x}\right| + \frac{b^3x}{X}\right].$

103.3. $\displaystyle\int\frac{dx}{x^3X^3} = -\frac{1}{a^5}\left[\frac{X^2}{2x^2} - \frac{4bX}{x} + 6b^2\log\left|\frac{X}{x}\right| + \frac{4b^3x}{X_1} - \frac{b^4x^2}{2X^2}\right].$

104.1. $\displaystyle\int\frac{dx}{x^4X} = -\frac{1}{a^4}\left[\frac{X^3}{3x^3} - \frac{3bX^2}{2x^2} + \frac{3b^2X}{x} - b^3\log\left|\frac{X}{x}\right|\right]$

$$= -\frac{1}{3ax^3} + \frac{b}{2a^2x^2} - \frac{b^2}{a^3x} + \frac{b^3}{a^4}\log\left|\frac{X}{x}\right| + \text{constant}.$$

104.2. $\displaystyle\int\frac{dx}{x^4X^2} = -\frac{1}{a^5}\left[\frac{X^3}{3x^3} - \frac{4bX^2}{2x^2} + \frac{6b^2X}{x} - 4b^3\log\left|\frac{X}{x}\right| - \frac{b^4x}{X}\right].$

105.1. $\displaystyle\int\frac{dx}{x^5X} = -\frac{1}{4ax^4} + \frac{b}{3a^2x^3} - \frac{b^2}{2a^3x^2} + \frac{b^3}{a^4x} - \frac{b^4}{a^5}\log\left|\frac{X}{x}\right|.$

Integrals Involving Linear Factors

110. $\displaystyle\int\frac{(a+x)dx}{(c+x)} = x + (a-c)\log|c+x|.$

110.1. $\displaystyle\int\frac{(a+fx)dx}{(c+gx)} = \frac{fx}{g} + \frac{ag - cf}{g^2}\log|c+gx|.$

111. $\displaystyle\int\frac{dx}{(a+x)(c+x)}=\frac{1}{a-c}\log\left|\frac{c+x}{a+x}\right|,$ $\qquad[a\neq c].$

If $a=c$, see **90.2.**

111.1. $\displaystyle\int\frac{dx}{(a+fx)(c+gx)}=\frac{1}{ag-cf}\log\left|\frac{c+gx}{a+fx}\right|,$ $\qquad[ag\neq cf].$

If $ag=cf$, see **90.2.**

111.2. $\displaystyle\int\frac{x\,dx}{(a+x)(c+x)}=\frac{1}{(a-c)}\left\{a\log|a+x|-c\log|c+x|\right\}.$

112. $\displaystyle\int\frac{dx}{(a+x)(c+x)^2}=\frac{1}{(c-a)(c+x)}+\frac{1}{(c-a)^2}\log\left|\frac{a+x}{c+x}\right|.$

112.1. $\displaystyle\int\frac{x\,dx}{(a+x)(c+x)^2}=\frac{c}{(a-c)(c+x)}-\frac{a}{(a-c)^2}\log\left|\frac{a+x}{c+x}\right|.$

112.2. $\displaystyle\int\frac{x^2dx}{(a+x)(c+x)^2}=\frac{c^2}{(c-a)(c+x)}$

$$+\frac{a^2}{(c-a)^2}\log|a+x|+\frac{c^2-2ac}{(c-a)^2}\log|c+x|.$$

113. $\displaystyle\int\frac{dx}{(a+x)^2(c+x)^2}=\frac{-1}{(a-c)^2}\left(\frac{1}{a+x}+\frac{1}{c+x}\right)$

$$+\frac{2}{(a-c)^3}\log\left|\frac{a+x}{c+x}\right|.$$

113.1. $\displaystyle\int\frac{x\,dx}{(a+x)^2(c+x)^2}=\frac{1}{(a-c)^2}\left(\frac{a}{a+x}+\frac{c}{c+x}\right)$

$$-\frac{a+c}{(a-c)^3}\log\left|\frac{a+x}{c+x}\right|.$$

113.2. $\displaystyle\int\frac{x^2dx}{(a+x)^2(c+x)^2}=\frac{-1}{(a-c)^2}\left(\frac{a^2}{a+x}+\frac{c^2}{c+x}\right)$

$$+\frac{2ac}{(a-c)^3}\log\left|\frac{a+x}{c+x}\right|.$$

[Ref. 1, p. 71.]

Integrals Involving $X = a^2 + x^2$

120. $\displaystyle\int \frac{dx}{1 + x^2} = \tan^{-1} x.$

The principal value of $\tan^{-1} x$ is to be taken, that is,

$$-\frac{\pi}{2} < \tan^{-1} x < \frac{\pi}{2}.$$

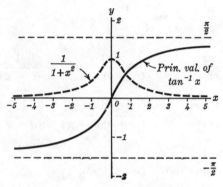

Fig. 120.   Graphs of $1/(1 + x^2)$ and of principal values of $\tan^{-1} x$.

120.01. $\displaystyle\int \frac{dx}{a^2 + b^2 x^2} = \frac{1}{ab} \tan^{-1} \frac{bx}{a}.$

120.1. $\displaystyle\int \frac{dx}{X} = \int \frac{dx}{a^2 + x^2} = \frac{1}{a} \tan^{-1} \frac{x}{a}.$

120.2. $\displaystyle\int \frac{dx}{X^2} = \frac{x}{2a^2 X} + \frac{1}{2a^3} \tan^{-1} \frac{x}{a}.$

120.3. $\displaystyle\int \frac{dx}{X^3} = \frac{x}{4a^2 X^2} + \frac{3x}{8a^4 X} + \frac{3}{8a^5} \tan^{-1} \frac{x}{a}.$

120.4. $\displaystyle\int \frac{dx}{X^4} = \frac{x}{6a^2 X^3} + \frac{5x}{24a^4 X^2} + \frac{5x}{16a^6 X} + \frac{5}{16a^7} \tan^{-1} \frac{x}{a}.$

120.9. $\displaystyle\int \frac{dx}{(a^2 + b^2 x^2)^{n+1}} = \frac{x}{2na^2(a^2 + b^2 x^2)^n} + \frac{2n - 1}{2na^2} \int \frac{dx}{(a^2 + b^2 x^2)^n}.$

[Ref. 2, p. 20.]

121. Integrals of the form

$$\int \frac{x^{2m+1}dx}{(a^2 \pm x^2)^n}$$

by putting $x^2 = z$, become

$$\frac{1}{2}\int \frac{z^m dz}{(a^2 \pm z)^n}$$

for which see **89–105** (m positive, negative, or zero).

121.1. $\displaystyle\int \frac{x\,dx}{X} = \int \frac{x\,dx}{a^2 + x^2} = \frac{1}{2}\log(a^2 + x^2).$

121.2. $\displaystyle\int \frac{x\,dx}{X^2} = -\frac{1}{2X}.$ **121.3.** $\displaystyle\int \frac{x\,dx}{X^3} = -\frac{1}{4X^2}.$

121.4. $\displaystyle\int \frac{x\,dx}{X^4} = -\frac{1}{6X^3}.$

121.9. $\displaystyle\int \frac{x\,dx}{X^{n+1}} = -\frac{1}{2nX^n},$ $[n \neq 0].$

122.1. $\displaystyle\int \frac{x^2 dx}{X} = x - a\tan^{-1}\frac{x}{a}.$

122.2. $\displaystyle\int \frac{x^2 dx}{X^2} = -\frac{x}{2X} + \frac{1}{2a}\tan^{-1}\frac{x}{a}.$

122.3. $\displaystyle\int \frac{x^2 dx}{X^3} = -\frac{x}{4X^2} + \frac{x}{8a^2 X} + \frac{1}{8a^3}\tan^{-1}\frac{x}{a}.$

122.4. $\displaystyle\int \frac{x^2 dx}{X^4} = -\frac{x}{6X^3} + \frac{x}{24a^2 X^2} + \frac{x}{16a^4 X} + \frac{1}{16a^5}\tan^{-1}\frac{x}{a}.$

122.9. $\displaystyle\int \frac{x^2 dx}{X^{n+1}} = \frac{-x}{2nX^n} + \frac{1}{2n}\int \frac{dx}{X^n}.$

123.1. $\displaystyle\int \frac{x^3 dx}{X} = \frac{x^2}{2} - \frac{a^2}{2}\log X.$

123.2. $\displaystyle\int \frac{x^3 dx}{X^2} = \frac{a^2}{2X} + \frac{1}{2}\log X.$

123.3. $\displaystyle\int \frac{x^3 dx}{X^3} = -\frac{1}{2X} + \frac{a^2}{4X^2}.$

123.4. $\displaystyle\int \frac{x^3 dx}{X^4} = -\frac{1}{4X^2} + \frac{a^2}{6X^3}.$

123.9. $\displaystyle\int \frac{x^3 dx}{X^{n+1}} = \frac{-1}{2(n-1)X^{n-1}} + \frac{a^2}{2nX^n},$ $\qquad [n > 1].$

124.1. $\displaystyle\int \frac{x^4 dx}{X} = \frac{x^3}{3} - a^2 x + a^3 \tan^{-1}\frac{x}{a}.$

124.2. $\displaystyle\int \frac{x^4 dx}{X^2} = x + \frac{a^2 x}{2X} - \frac{3a}{2} \tan^{-1}\frac{x}{a}.$

124.3. $\displaystyle\int \frac{x^4 dx}{X^3} = \frac{a^2 x}{4X^2} - \frac{5x}{8X} + \frac{3}{8a} \tan^{-1}\frac{x}{a}.$

124.4. $\displaystyle\int \frac{x^4 dx}{X^4} = \frac{a^2 x}{6X^3} - \frac{7x}{24X^2} + \frac{x}{16a^2 X} + \frac{1}{16a^3} \tan^{-1}\frac{x}{a}.$

125.1. $\displaystyle\int \frac{x^5 dx}{X} = \frac{x^4}{4} - \frac{a^2 x^2}{2} + \frac{a^4}{2} \log X.$

125.2. $\displaystyle\int \frac{x^5 dx}{X^2} = \frac{x^2}{2} - \frac{a^4}{2X} - a^2 \log X.$

125.3. $\displaystyle\int \frac{x^5 dx}{X^3} = \frac{a^2}{X} - \frac{a^4}{4X^2} + \frac{1}{2} \log X.$

125.4. $\displaystyle\int \frac{x^5 dx}{X^4} = -\frac{1}{2X} + \frac{a^2}{2X^2} - \frac{a^4}{6X^3}.$

125.9. $\displaystyle\int \frac{x^5 dx}{X^{n+1}} = \frac{-1}{2(n-2)X^{n-2}} + \frac{a^2}{(n-1)X^{n-1}} - \frac{a^4}{2nX^n},$ $\qquad [n > 2].$

126.1. $\displaystyle\int \frac{x^6 dx}{X} = \frac{x^5}{5} - \frac{a^2 x^3}{3} + a^4 x - a^5 \tan^{-1}\frac{x}{a}.$

127.1. $\displaystyle\int \frac{x^7 dx}{X} = \frac{x^6}{6} - \frac{a^2 x^4}{4} + \frac{a^4 x^2}{2} - \frac{a^6}{2} \log X.$

128.1. $\displaystyle\int \frac{x^8 dx}{X} = \frac{x^7}{7} - \frac{a^2 x^5}{5} + \frac{a^4 x^3}{3} - a^6 x + a^7 \tan^{-1}\frac{x}{a}.$

131.1. $\displaystyle \int \frac{dx}{xX} = \int \frac{dx}{x(a^2 + x^2)} = \frac{1}{2a^2} \log \left(\frac{x^2}{a^2 + x^2} \right).$

131.2. $\displaystyle \int \frac{dx}{xX^2} = \frac{1}{2a^2X} + \frac{1}{2a^4} \log \frac{x^2}{X}.$

131.3. $\displaystyle \int \frac{dx}{xX^3} = \frac{1}{4a^2X^2} + \frac{1}{2a^4X} + \frac{1}{2a^6} \log \frac{x^2}{X}.$

131.4. $\displaystyle \int \frac{dx}{xX^4} = \frac{1}{6a^2X^3} + \frac{1}{4a^4X^2} + \frac{1}{2a^6X} + \frac{1}{2a^8} \log \frac{x^2}{X}.$

132.1. $\displaystyle \int \frac{dx}{x^2X} = -\frac{1}{a^2x} - \frac{1}{a^3} \tan^{-1} \frac{x}{a}.$

132.2. $\displaystyle \int \frac{dx}{x^2X^2} = -\frac{1}{a^4x} - \frac{x}{2a^4X} - \frac{3}{2a^5} \tan^{-1} \frac{x}{a}.$

132.3. $\displaystyle \int \frac{dx}{x^2X^3} = -\frac{1}{a^6x} - \frac{x}{4a^4X^2} - \frac{7x}{8a^6X} - \frac{15}{8a^7} \tan^{-1} \frac{x}{a}.$

133.1. $\displaystyle \int \frac{dx}{x^3X} = -\frac{1}{2a^2x^2} - \frac{1}{2a^4} \log \frac{x^2}{X}.$

133.2. $\displaystyle \int \frac{dx}{x^3X^2} = -\frac{1}{2a^4x^2} - \frac{1}{2a^4X} - \frac{1}{a^6} \log \frac{x^2}{X}.$

133.3. $\displaystyle \int \frac{dx}{x^3X^3} = -\frac{1}{2a^6x^2} - \frac{1}{a^6X} - \frac{1}{4a^4X^2} - \frac{3}{2a^8} \log \frac{x^2}{X}.$

134.1. $\displaystyle \int \frac{dx}{x^4X} = -\frac{1}{3a^2x^3} + \frac{1}{a^4x} + \frac{1}{a^5} \tan^{-1} \frac{x}{a}.$

134.2. $\displaystyle \int \frac{dx}{x^4X^2} = -\frac{1}{3a^4x^3} + \frac{2}{a^6x} + \frac{x}{2a^6X} + \frac{5}{2a^7} \tan^{-1} \frac{x}{a}.$

135.1. $\displaystyle \int \frac{dx}{x^5X} = -\frac{1}{4a^2x^4} + \frac{1}{2a^4x^2} + \frac{1}{2a^6} \log \frac{x^2}{X}.$

135.2. $\displaystyle\int \frac{dx}{x^5 X^2} = -\frac{1}{4a^4 x^4} + \frac{1}{a^6 x^2} + \frac{1}{2a^6 X} + \frac{3}{2a^8} \log \frac{x^2}{X}.$

[See References 1 and 2 for additional integrals of the type of Nos. **120–135**.]

136. $\displaystyle\int \frac{dx}{(f + gx)(a^2 + x^2)} = \frac{1}{(f^2 + a^2 g^2)} \Big[g \log |f + gx|$

$$- \frac{g}{2} \log (a^2 + x^2) + \frac{f}{a} \tan^{-1} \frac{x}{a} \Big].$$

Integrals Involving $X = a^2 - x^2$

140. $\displaystyle\int \frac{dx}{1 - x^2} = \frac{1}{2} \log \left| \frac{1 + x}{1 - x} \right|.$ [See note under **140.1**.]

The function $1/(1 - x^2)$ and its integral can be plotted for negative values of x. See Fig. 140.

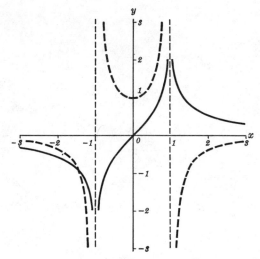

Fig 140.

Dotted graph, $1/(1 - x^2)$.

Full line graph, $\frac{1}{2} \log \left| \dfrac{1 + x}{1 - x} \right|$.

140.01. $\displaystyle\int \frac{dx}{x^2 - 1} = - \int \frac{dx}{1 - x^2}.$ [See **140**.]

140.02. $\displaystyle\int \frac{dx}{a^2 - b^2 x^2} = \frac{1}{2ab} \log \left| \frac{a + bx}{a - bx} \right|.$

Note that

$$\frac{1}{2ab} \log \frac{a + bx}{a - bx} = \frac{1}{ab} \tanh^{-1} \frac{bx}{a}, \qquad [b^2 x^2 < a^2],$$

and

$$\frac{1}{2ab} \log \frac{bx + a}{bx - a} = \frac{1}{ab} \operatorname{ctnh}^{-1} \frac{bx}{a}, \qquad [b^2 x^2 > a^2].$$

140.1. $\displaystyle\int \frac{dx}{X} = \int \frac{dx}{a^2 - x^2} = \frac{1}{2a} \log \left| \frac{a + x}{a - x} \right|.$

Note:
$$\frac{1}{2a} \log \frac{a + x}{a - x} = \frac{1}{a} \tanh^{-1} \frac{x}{a}, \qquad [x^2 < a^2],$$

$$\frac{1}{2a} \log \frac{x + a}{x - a} = \frac{1}{a} \operatorname{ctnh}^{-1} \frac{x}{a}, \qquad [x^2 > a^2].$$

$$[\text{Ref. 8, p. 100, (s) and (s').}]$$

140.2. $\displaystyle\int \frac{dx}{X^2} = \frac{x}{2a^2 X} + \frac{1}{4a^3} \log \left| \frac{a + x}{a - x} \right|.$

140.3. $\displaystyle\int \frac{dx}{X^3} = \frac{x}{4a^2 X^2} + \frac{3x}{8a^4 X} + \frac{3}{16a^5} \log \left| \frac{a + x}{a - x} \right|.$

140.4. $\displaystyle\int \frac{dx}{X^4} = \frac{x}{6a^2 X^3} + \frac{5x}{24a^4 X^2} + \frac{5x}{16a^6 X} + \frac{5}{32a^7} \log \left| \frac{a + x}{a - x} \right|.$

140.9. $\displaystyle\int \frac{dx}{(a^2 - b^2 x^2)^{n+1}} = \frac{x}{2na^2(a^2 - b^2 x^2)^n} + \frac{2n - 1}{2na^2} \int \frac{dx}{(a^2 - b^2 x^2)^n}.$

141.1. $\displaystyle\int \frac{x \, dx}{X} = \int \frac{x \, dx}{a^2 - x^2} = -\frac{1}{2} \log |a^2 - x^2|.$

141.2. $\displaystyle\int \frac{x \, dx}{X^2} = \frac{1}{2X}.$ **141.3.** $\displaystyle\int \frac{x \, dx}{X^3} = \frac{1}{4X^2}.$

141.4. $\displaystyle\int \frac{x \, dx}{X^4} = \frac{1}{6X^3}.$ **141.9.** $\displaystyle\int \frac{x \, dx}{X^{n+1}} = \frac{1}{2nX^n}, \qquad [n \neq 0].$

142.1. $\displaystyle\int \frac{x^2 dx}{X} = -x + \frac{a}{2} \log \left| \frac{a + x}{a - x} \right|.$

142.2. $\displaystyle\int\frac{x^2\,dx}{X^2} = \frac{x}{2X} - \frac{1}{4a}\log\left|\frac{a+x}{a-x}\right|.$

142.3. $\displaystyle\int\frac{x^2\,dx}{X^3} = \frac{x}{4X^2} - \frac{x}{8a^2X} - \frac{1}{16a^3}\log\left|\frac{a+x}{a-x}\right|.$

142.4. $\displaystyle\int\frac{x^2\,dx}{X^4} = \frac{x}{6X^3} - \frac{x}{24a^2X^2} - \frac{x}{16a^4X} - \frac{1}{32a^5}\log\left|\frac{a+x}{a-x}\right|.$

142.9. $\displaystyle\int\frac{x^2\,dx}{X^{n+1}} = \frac{x}{2nX^n} - \frac{1}{2n}\int\frac{dx}{X^n}.$

143.1. $\displaystyle\int\frac{x^3\,dx}{X} = -\frac{x^2}{2} - \frac{a^2}{2}\log|X|.$

143.2. $\displaystyle\int\frac{x^3\,dx}{X^2} = \frac{a^2}{2X} + \frac{1}{2}\log|X|.$

143.3. $\displaystyle\int\frac{x^3\,dx}{X^3} = \frac{-1}{2X} + \frac{a^2}{4X^2}.$ **143.4.** $\displaystyle\int\frac{x^3\,dx}{X^4} = \frac{-1}{4X^2} + \frac{a^2}{6X^3}.$

143.9. $\displaystyle\int\frac{x^3\,dx}{X^{n+1}} = \frac{-1}{2(n-1)X^{n-1}} + \frac{a^2}{2nX^n},$ $[n>1].$

144.1. $\displaystyle\int\frac{x^4\,dx}{X} = -\frac{x^3}{3} - a^2x + \frac{a^3}{2}\log\left|\frac{a+x}{a-x}\right|.$

144.2. $\displaystyle\int\frac{x^4\,dx}{X^2} = x + \frac{a^2x}{2X} - \frac{3a}{4}\log\left|\frac{a+x}{a-x}\right|.$

144.3. $\displaystyle\int\frac{x^4\,dx}{X^3} = \frac{a^2x}{4X^2} - \frac{5x}{8X} + \frac{3}{16a}\log\left|\frac{a+x}{a-x}\right|.$

144.4. $\displaystyle\int\frac{x^4\,dx}{X^4} = \frac{a^2x}{6X^3} - \frac{7x}{24X^2} + \frac{x}{16a^2X} + \frac{1}{32a^3}\log\left|\frac{a+x}{a-x}\right|.$

145.1. $\displaystyle\int\frac{x^5\,dx}{X} = -\frac{x^4}{4} - \frac{a^2x^2}{2} - \frac{a^4}{2}\log|X|.$

145.2. $\displaystyle\int\frac{x^5\,dx}{X^2} = \frac{x^2}{2} + \frac{a^4}{2X} + a^2\log|X|.$

145.3. $\displaystyle\int\frac{x^5\,dx}{X^3} = -\frac{a^2}{X} + \frac{a^4}{4X^2} - \frac{1}{2}\log|X|.$

145.4. $\displaystyle\int \frac{x^5 dx}{X^4} = \frac{1}{2X} - \frac{a^2}{2X^2} + \frac{a^4}{6X^3}.$

145.9. $\displaystyle\int \frac{x^5 dx}{X^{n+1}} = \frac{1}{2(n-2)X^{n-2}} - \frac{a^2}{(n-1)X^{n-1}} + \frac{a^4}{2nX^n}, \qquad [n > 2].$

146.1. $\displaystyle\int \frac{x^6 dx}{X} = -\frac{x^5}{5} - \frac{a^2 x^3}{3} - a^4 x + \frac{a^5}{2} \log\left|\frac{a+x}{a-x}\right|.$

147.1. $\displaystyle\int \frac{x^7 dx}{X} = -\frac{x^6}{6} - \frac{a^2 x^4}{4} - \frac{a^4 x^2}{2} - \frac{a^6}{2} \log|X|.$

148.1. $\displaystyle\int \frac{x^8 dx}{X} = -\frac{x^7}{7} - \frac{a^2 x^5}{5} - \frac{a^4 x^3}{3} - a^6 x + \frac{a}{2} \log\left|\frac{a+x}{a-x}\right|.$

151.1. $\displaystyle\int \frac{dx}{xX} = \int \frac{dx}{x(a^2 - x^2)} = \frac{1}{2a^2} \log\left|\frac{x^2}{a^2 - x^2}\right|.$

151.2. $\displaystyle\int \frac{dx}{xX^2} = \frac{1}{2a^2 X} + \frac{1}{2a^4} \log\left|\frac{x^2}{X}\right|.$

151.3. $\displaystyle\int \frac{dx}{xX^3} = \frac{1}{4a^2 X^2} + \frac{1}{2a^4 X} + \frac{1}{2a^6} \log\left|\frac{x^2}{X}\right|.$

151.4. $\displaystyle\int \frac{dx}{xX^4} = \frac{1}{6a^2 X^3} + \frac{1}{4a^4 X^2} + \frac{1}{2a^6 X} + \frac{1}{2a^8} \log\left|\frac{x^2}{X}\right|.$

152.1. $\displaystyle\int \frac{dx}{x^2 X} = -\frac{1}{a^2 x} + \frac{1}{2a^3} \log\left|\frac{a+x}{a-x}\right|.$

152.2. $\displaystyle\int \frac{dx}{x^2 X^2} = -\frac{1}{a^4 x} + \frac{x}{2a^4 X} + \frac{3}{4a^5} \log\left|\frac{a+x}{a-x}\right|.$

152.3. $\displaystyle\int \frac{dx}{x^2 X^3} = -\frac{1}{a^6 x} + \frac{x}{4a^4 X^2} + \frac{7x}{8a^6 X} + \frac{15}{16a^7} \log\left|\frac{a+x}{a-x}\right|.$

153.1. $\displaystyle\int \frac{dx}{x^3 X} = -\frac{1}{2a^2 x^2} + \frac{1}{2a^4} \log\left|\frac{x^2}{X}\right|.$

153.2. $\displaystyle\int \frac{dx}{x^3 X^2} = -\frac{1}{2a^4 x^2} + \frac{1}{2a^4 X} + \frac{1}{a^6} \log\left|\frac{x^2}{X}\right|.$

153.3. $\displaystyle\int \frac{dx}{x^3 X^3} = -\frac{1}{2a^6 x^2} + \frac{1}{a^6 X} + \frac{1}{4a^4 X^2} + \frac{3}{2a^8} \log\left|\frac{x^2}{X}\right|.$

154.1. $\int \dfrac{dx}{x^4 X} = -\dfrac{1}{3a^2 x^3} - \dfrac{1}{a^4 x} + \dfrac{1}{2a^5} \log \left|\dfrac{a+x}{a-x}\right|.$

154.2. $\int \dfrac{dx}{x^4 X^2} = -\dfrac{1}{3a^4 x^3} - \dfrac{2}{a^6 x} + \dfrac{x}{2a^6 X} + \dfrac{5}{4a^7} \log \left|\dfrac{a+x}{a-x}\right|.$

155.1. $\int \dfrac{dx}{x^5 X} = -\dfrac{1}{4a^2 x^4} - \dfrac{1}{2a^4 x^2} + \dfrac{1}{2a^6} \log \left|\dfrac{x^2}{X}\right|.$

155.2. $\int \dfrac{dx}{x^5 X^2} = -\dfrac{1}{4a^4 x^4} - \dfrac{1}{a^6 x^2} + \dfrac{1}{2a^6 X} + \dfrac{3}{2a^8} \log \left|\dfrac{x^2}{X}\right|.$

[See References 1 and 2 for other integrals of the type of Nos. **140–155.**]

156. $\int \dfrac{dx}{(f+gx)(a^2 - x^2)} = \dfrac{1}{a^2 g^2 - f^2}\bigg[g \log |f + gx|$

$$-\dfrac{g}{2} \log |a^2 - x^2| - \dfrac{f}{2a} \log \left|\dfrac{a+x}{a-x}\right|\bigg].$$

Integrals Involving $X = ax^2 + bx + c$

160.01. $\int \dfrac{dx}{X} = \dfrac{2}{\sqrt{(4ac - b^2)}} \tan^{-1} \dfrac{2ax + b}{\sqrt{(4ac - b^2)}},$ $\qquad [4ac > b^2],$

$$= \dfrac{1}{\sqrt{(b^2 - 4ac)}} \log \left|\dfrac{2ax + b - \sqrt{(b^2 - 4ac)}}{2ax + b + \sqrt{(b^2 - 4ac)}}\right|,$$

$$[b^2 > 4ac],$$

$$= \dfrac{1}{a(p - q)} \log \left|\dfrac{x - p}{x - q}\right|, \qquad\qquad [b^2 > 4ac],$$

where p and q are the roots of $ax^2 + bx + c = 0$,

$$= -\dfrac{2}{\sqrt{(b^2 - 4ac)}} \tanh^{-1} \dfrac{2ax + b}{\sqrt{(b^2 - 4ac)}},$$

$$[b^2 > 4ac, \quad (2ax + b)^2 > b^2 - 4ac],$$

$$= -\dfrac{2}{\sqrt{(b^2 - 4ac)}} \operatorname{ctnh}^{-1} \dfrac{2ax + b}{\sqrt{(b^2 - 4ac)}},$$

$$[b^2 > 4ac, \quad (2ax + b)^2 > b^2 - 4ac],$$

$$= -\dfrac{2}{2ax + b}, \qquad\qquad [b^2 = 4ac].$$

[Put $2ax + b = z$.]

160.02. $\displaystyle\int\frac{dx}{X^2} = \frac{2ax + b}{(4ac - b^2)X} + \frac{2a}{4ac - b^2}\int\frac{dx}{X}.$ [See **160.01.**]

160.03. $\displaystyle\int\frac{dx}{X^3} = \frac{2ax + b}{2(4ac - b^2)X^2} + \frac{3a(2ax + b)}{(4ac - b^2)^2 X} + \frac{6a^2}{(4ac - b^2)^2}\int\frac{dx}{X}.$

160.09. $\displaystyle\int\frac{dx}{X^n} = \frac{2ax + b}{(n - 1)(4ac - b^2)X^{n-1}} + \frac{(2n - 3)2a}{(n - 1)(4ac - b^2)}\int\frac{dx}{X^{n-1}}.$

               [Ref. 1, p. 83.]

160.11. $\displaystyle\int\frac{x\,dx}{X} = \frac{1}{2a}\log|X| - \frac{b}{2a}\int\frac{dx}{X}.$ [See **160.01.**]

160.12. $\displaystyle\int\frac{x\,dx}{X^2} = -\frac{bx + 2c}{(4ac - b^2)X} - \frac{b}{4ac - b^2}\int\frac{dx}{X}.$ [See **160.01.**]

160.19. $\displaystyle\int\frac{x\,dx}{X^n} = -\frac{bx + 2c}{(n - 1)(4ac - b^2)X^{n-1}}$

$$-\frac{b(2n - 3)}{(n - 1)(4ac - b^2)}\int\frac{dx}{X^{n-1}}.$$

160.21. $\displaystyle\int\frac{x^2 dx}{X} = \frac{x}{a} - \frac{b}{2a^2}\log|X| + \frac{b^2 - 2ac}{2a^2}\int\frac{dx}{X}.$ [See **160.01.**]

160.22. $\displaystyle\int\frac{x^2 dx}{X^2} = \frac{(b^2 - 2ac)x + bc}{a(4ac - b^2)X} + \frac{2c}{4ac - b^2}\int\frac{dx}{X}.$ [See **160.01**]

160.27. $\displaystyle\int\frac{x^m dx}{X} = \frac{x^{m-1}}{(m - 1)a} - \frac{c}{a}\int\frac{x^{m-2}dx}{X} - \frac{b}{a}\int\frac{x^{m-1}dx}{X}.$

160.28. $\displaystyle\int\frac{x^m dx}{X^n} = -\frac{x^{m-1}}{(2n - m - 1)aX^{n-1}} + \frac{(m - 1)c}{(2n - m - 1)a}\int\frac{x^{m-2}dx}{X^n}$

$$-\frac{(n - m)b}{(2n - m - 1)a}\int\frac{x^{m-1}dx}{X^n}, \qquad [m \neq 2n - 1].$$

161.11. $\displaystyle\int\frac{dx}{xX} = \frac{1}{2c}\log\frac{x^2}{X} - \frac{b}{2c}\int\frac{dx}{X}.$ [See **160.01**]

161.19. $\displaystyle\int\frac{dx}{xX^n} = \frac{1}{2c(n - 1)X^{n-1}} - \frac{b}{2c}\int\frac{dx}{X^n} + \frac{1}{c}\int\frac{dx}{xX^{n-1}}.$

161.21. $\int \dfrac{dx}{x^2 X} = \dfrac{b}{2c^2} \log \left| \dfrac{X}{x^2} \right| - \dfrac{1}{cx} + \dfrac{b^2 - 2ac}{2c^2} \int \dfrac{dx}{X}.$ [See **160.01.**]

161.29. $\int \dfrac{dx}{x^m X^n} = -\dfrac{1}{(m-1)cx^{m-1}X^{n-1}} - \dfrac{(2n+m-3)a}{(m-1)c} \int \dfrac{dx}{x^{m-2}X^n}$

$\qquad\qquad - \dfrac{(n+m-2)b}{(m-1)c} \int \dfrac{dx}{x^{m-1}X^n},$ $\qquad$ [$m > 1$].

Integrals Involving $a^3 \pm x^3$

165.01. $\int \dfrac{dx}{a^3 + x^3} = \dfrac{1}{6a^2} \log \dfrac{(a+x)^2}{a^2 - ax + x^2} + \dfrac{1}{a^2\sqrt{3}} \tan^{-1} \dfrac{2x - a}{a\sqrt{3}}.$

165.02. $\int \dfrac{dx}{(a^3 + x^3)^2} = \dfrac{x}{3a^3(a^3 + x^3)} + \dfrac{2}{3a^3} \int \dfrac{dx}{a^3 + x^3}.$

165.11. $\int \dfrac{x\,dx}{a^3 + x^3} = \dfrac{1}{6a} \log \dfrac{a^2 - ax + x^2}{(a+x)^2} + \dfrac{1}{a\sqrt{3}} \tan^{-1} \dfrac{2\,x - a}{a\sqrt{3}}.$

165.12. $\int \dfrac{x\,dx}{(a^3 + x^3)^2} = \dfrac{x^2}{3a^3(a^3 + x^3)} + \dfrac{1}{3a^3} \int \dfrac{x\,dx}{a^3 + x^3}.$

165.21. $\int \dfrac{x^2 dx}{a^3 + x^3} = \dfrac{1}{3} \log |a^3 + x^3|.$

165.22. $\int \dfrac{x^2 dx}{(a^3 + x^3)^2} = -\dfrac{1}{3(a^3 + x^3)}.$

165.31. $\int \dfrac{x^3 dx}{a^3 + x^3} = x - a^3 \int \dfrac{dx}{a^3 + x^3}.$ [See **165.01.**]

165.32. $\int \dfrac{x^3 dx}{(a^3 + x^3)^2} = \dfrac{-x}{3(a^3 + x^3)} + \dfrac{1}{3} \int \dfrac{dx}{a^3 + x^3}.$ [See **165.01.**]

165.41. $\int \dfrac{x^4 dx}{a^3 + x^3} = \dfrac{x^2}{2} - a^3 \int \dfrac{x\,dx}{a^3 + x^3}.$ [See **165.11.**]

165.42. $\int \dfrac{x^4 dx}{(a^3 + x^3)^2} = -\dfrac{x^2}{3(a^3 + x^3)} + \dfrac{2}{3} \int \dfrac{x\,dx}{a^3 + x^3}.$ [See **165.11.**]

165.51. $\int \dfrac{x^5 dx}{a^3 + x^3} = \dfrac{x^3}{3} - \dfrac{a^3}{3} \log |a^3 + x^3|.$

165.52. $\displaystyle\int \frac{x^5 dx}{(a^3 + x^3)^2} = \frac{a^3}{3(a^3 + x^3)} + \frac{1}{3} \log |a^3 + x^3|.$

166.11. $\displaystyle\int \frac{dx}{x(a^3 + x^3)} = \frac{1}{3a^3} \log \left| \frac{x^3}{a^3 + x^3} \right|.$

166.12. $\displaystyle\int \frac{dx}{x(a^3 + x^3)^2} = \frac{1}{3a^3(a^3 + x^3)} + \frac{1}{3a^6} \log \left| \frac{x^3}{a^3 + x^3} \right|.$

166.21. $\displaystyle\int \frac{dx}{x^2(a^3 + x^3)} = -\frac{1}{a^3 x} - \frac{1}{a^3} \int \frac{x\, dx}{a^3 + x^3}.$ [See **165.11.**]

166.22. $\displaystyle\int \frac{dx}{x^2(a^3 + x^3)^2} = -\frac{1}{a^6 x} - \frac{x^2}{3a^6(a^3 + x^3)} - \frac{4}{3a^6} \int \frac{x\, dx}{a^3 + x^3}.$

166.31. $\displaystyle\int \frac{dx}{x^3(a^3 + x^3)} = -\frac{1}{2a^3 x^2} - \frac{1}{a^3} \int \frac{dx}{a^3 + x^3}.$ [See **165.01.**]

166.32. $\displaystyle\int \frac{dx}{x^3(a^3 + x^3)^2} = -\frac{1}{2a^6 x^2} - \frac{x}{3a^6(a^3 + x^3)} - \frac{5}{3a^6} \int \frac{dx}{a^3 + x^3}.$

166.41. $\displaystyle\int \frac{dx}{x^4(a^3 + x^3)} = -\frac{1}{3a^3 x^3} + \frac{1}{3a^6} \log \left| \frac{a^3 + x^3}{x^3} \right|.$

166.42. $\displaystyle\int \frac{dx}{x^4(a^3 + x^3)^2} = -\frac{1}{3a^6 x^3} - \frac{1}{3a^6(a^3 + x^3)}$
$$+ \frac{2}{3a^9} \log \left| \frac{a^3 + x^3}{x^3} \right|.$$

168.01. $\displaystyle\int \frac{dx}{a^3 - x^3} = \frac{1}{6a^2} \log \frac{a^2 + ax + x^2}{(a - x)^2} + \frac{1}{a^2 \sqrt{3}} \tan^{-1} \frac{2x + a}{a\sqrt{3}}.$

168.02. $\displaystyle\int \frac{dx}{(a^3 - x^3)^2} = \frac{x}{3a^3(a^3 - x^3)} + \frac{2}{3a^3} \int \frac{dx}{a^3 - x^3}.$

168.11. $\displaystyle\int \frac{x\, dx}{a^3 - x^3} = \frac{1}{6a} \log \frac{a^2 + ax + x^2}{(a - x)^2} - \frac{1}{a\sqrt{3}} \tan^{-1} \frac{2x + a}{a\sqrt{3}}.$

168.12. $\displaystyle\int \frac{x\, dx}{(a^3 - x^3)^2} = \frac{x^2}{3a^3(a^3 - x^3)} + \frac{1}{3a^3} \int \frac{x\, dx}{a^3 - x^3}.$

168.21. $\displaystyle\int \frac{x^2 dx}{a^3 - x^3} = -\frac{1}{3} \log |a^3 - x^3|.$

168.22. $\displaystyle\int \frac{x^2 dx}{(a^3 - x^3)^2} = \frac{1}{3(a^3 - x^3)}.$

168.31. $\displaystyle\int \frac{x^3 dx}{a^3 - x^3} = -x + a^3 \int \frac{dx}{a^3 - x^3}.$　　　　[See **168.01.**]

168.32. $\displaystyle\int \frac{x^3 dx}{(a^3 - x^3)^2} = \frac{x}{3(a^3 - x^3)} - \frac{1}{3}\int \frac{dx}{a^3 - x^3}.$　　　[See **168.01.**]

168.41. $\displaystyle\int \frac{x^4 dx}{a^3 - x^3} = -\frac{x^2}{2} + a^3 \int \frac{x\,dx}{a^3 - x^3}.$　　　[See **168.11.**]

168.42. $\displaystyle\int \frac{x^4 dx}{(a^3 - x^3)^2} = \frac{x^2}{3(a^3 - x^3)} - \frac{2}{3}\int \frac{x\,dx}{a^3 - x^3}.$　　　[See **168.11.**]

168.51. $\displaystyle\int \frac{x^5 dx}{a^3 - x^3} = -\frac{x^3}{3} - \frac{a^3}{3} \log |a^3 - x^3|.$

168.52. $\displaystyle\int \frac{x^5 dx}{(a^3 - x^3)^2} = \frac{a^3}{3(a^3 - x^3)} + \frac{1}{3} \log |a^3 - x^3|.$

169.11. $\displaystyle\int \frac{dx}{x(a^3 - x^3)} = \frac{1}{3a^3} \log \left| \frac{x^3}{a^3 - x^3} \right|.$

169.12. $\displaystyle\int \frac{dx}{x(a^3 - x^3)^2} = \frac{1}{3a^3(a^3 - x^3)} + \frac{1}{3a^6} \log \left| \frac{x^3}{a^3 - x^3} \right|.$

169.21. $\displaystyle\int \frac{dx}{x^2(a^3 - x^3)} = -\frac{1}{a^3 x} + \frac{1}{a^3}\int \frac{x\,dx}{a^3 - x^3}.$　　　[See **168.11.**]

169.22. $\displaystyle\int \frac{dx}{x^2(a^3 - x^3)^2} = -\frac{1}{a^6 x} + \frac{x^2}{3a^6(a^3 - x^3)} + \frac{4}{3a^6}\int \frac{x\,dx}{a^3 - x^3}.$

169.31. $\displaystyle\int \frac{dx}{x^3(a^3 - x^3)} = -\frac{1}{2a^3 x^2} + \frac{1}{a^3}\int \frac{dx}{a^3 - x^3}.$　　　[See **168.01.**]

169.32. $\displaystyle\int \frac{dx}{x^3(a^3 - x^3)^2} = -\frac{1}{2a^6 x^2} + \frac{x}{3a^6(a^3 - x^3)} + \frac{5}{3a^6}\int \frac{dx}{a^3 - x^3}.$

169.41. $\displaystyle\int \frac{dx}{x^4(a^3 - x^3)} = -\frac{1}{3a^3 x^3} + \frac{1}{3a^6} \log \left| \frac{x^3}{a^3 - x^3} \right|.$

169.42. $\displaystyle\int \frac{dx}{x^4(a^3 - x^3)^2} = -\frac{1}{3a^6 x^3} + \frac{1}{3a^6(a^3 - x^3)} + \frac{2}{3a^9} \log \left| \frac{x^3}{a^3 - x^3} \right|.$

Integrals Involving $a^4 \pm x^4$

170.
$$\int \frac{dx}{a^4 + x^4} = \frac{1}{4a^3\sqrt{2}} \log \frac{x^2 + ax\sqrt{2} + a^2}{x^2 - ax\sqrt{2} + a^2}$$
$$+ \frac{1}{2a^3\sqrt{2}} \tan^{-1} \frac{ax\sqrt{2}}{a^2 - x^2}.$$

170.1.
$$\int \frac{x\,dx}{a^4 + x^4} = \frac{1}{2a^2} \tan^{-1} \frac{x^2}{a^2}.$$

170.2.
$$\int \frac{x^2 dx}{a^4 + x^4} = -\frac{1}{4a\sqrt{2}} \log \frac{x^2 + ax\sqrt{2} + a^2}{x^2 - ax\sqrt{2} + a^2}$$
$$+ \frac{1}{2a\sqrt{2}} \tan^{-1} \frac{ax\sqrt{2}}{a^2 - x^2}.$$

170.3.
$$\int \frac{x^3 dx}{a^4 + x^4} = \frac{1}{4} \log (a^4 + x^4).$$

171.
$$\int \frac{dx}{a^4 - x^4} = \frac{1}{4a^3} \log \left| \frac{a + x}{a - x} \right| + \frac{1}{2a^3} \tan^{-1} \frac{x}{a}.$$

171.1.
$$\int \frac{x\,dx}{a^4 - x^4} = \frac{1}{4a^2} \log \left| \frac{a^2 + x^2}{a^2 - x^2} \right|.$$

171.2.
$$\int \frac{x^2 dx}{a^4 - x^4} = \frac{1}{4a} \log \left| \frac{a + x}{a - x} \right| - \frac{1}{2a} \tan^{-1} \frac{x}{a}.$$

171.3.
$$\int \frac{x^3 dx}{a^4 - x^4} = -\frac{1}{4} \log |a^4 - x^4|.$$

173.
$$\int \frac{dx}{x(a + bx^m)} = \frac{1}{am} \log \left| \frac{x^m}{a + bx^m} \right|.$$

IRRATIONAL ALGEBRAIC FUNCTIONS

Integrals Involving $x^{1/2}$

180. $\displaystyle\int x^{p/2}dx = \frac{2}{p+2}\, x^{(p+2)/2}.$

180.1. $\displaystyle\int x^{1/2}dx = \int \sqrt{x}\, dx = \frac{2}{3}\, x^{3/2}.$

180.3. $\displaystyle\int x^{3/2}dx = \frac{2}{5}\, x^{5/2}.$ **180.5.** $\displaystyle\int x^{5/2}dx = \frac{2}{7}\, x^{7/2}.$

181. $\displaystyle\int \frac{dx}{x^{p/2}} = -\frac{2}{(p-2)x^{(p-2)/2}}.$

181.1. $\displaystyle\int \frac{dx}{x^{1/2}} = \int \frac{dx}{\sqrt{x}} = 2x^{1/2}.$ **181.3.** $\displaystyle\int \frac{dx}{x^{3/2}} = -\frac{2}{x^{1/2}}.$

181.5. $\displaystyle\int \frac{dx}{x^{5/2}} = -\frac{2}{3x^{3/2}}.$ **181.7.** $\displaystyle\int \frac{dx}{x^{7/2}} = -\frac{2}{5x^{5/2}}.$

[NOTE—Put $x = u^2$, then $dx = 2u\, du$.]

185.11. $\displaystyle\int \frac{x^{1/2}dx}{a^2 + b^2x} = \frac{2x^{1/2}}{b^2} - \frac{2a}{b^3}\, \tan^{-1} \frac{bx^{1/2}}{a}.$

185.13. $\displaystyle\int \frac{x^{3/2}dx}{a^2 + b^2x} = \frac{2}{3}\frac{x^{3/2}}{b^2} - \frac{2a^2x^{1/2}}{b^4} + \frac{2a^3}{b^5}\, \tan^{-1} \frac{bx^{1/2}}{a}.$

185.21. $\displaystyle\int \frac{x^{1/2}dx}{(a^2 + b^2x)^2} = -\frac{x^{1/2}}{b^2(a^2 + b^2x)} + \frac{1}{ab^3}\, \tan^{-1} \frac{bx^{1/2}}{a}.$

185.23. $\displaystyle\int \frac{x^{3/2}dx}{(a^2 + b^2x)^2} = \frac{2x^{3/2}}{b^2(a^2 + b^2x)} + \frac{3a^2x^{1/2}}{b^4(a^2 + b^2x)} - \frac{3a}{b^5}\, \tan^{-1} \frac{bx^{1/2}}{a}.$

186.11. $\displaystyle\int \frac{dx}{(a^2 + b^2x)x^{1/2}} = \frac{2}{ab}\, \tan^{-1} \frac{bx^{1/2}}{a}.$

186.13. $\displaystyle\int \frac{dx}{(a^2 + b^2x)x^{3/2}} = -\frac{2}{a^2x^{1/2}} - \frac{2b}{a^3}\, \tan^{-1} \frac{bx^{1/2}}{a}.$

186.21. $\displaystyle\int \frac{dx}{(a^2 + b^2x)^2x^{1/2}} = \frac{x^{1/2}}{a^2(a^2 + b^2x)} + \frac{1}{a^3b}\, \tan^{-1} \frac{bx^{1/2}}{a}.$

186.23. $\displaystyle\int \frac{dx}{(a^2 + b^2x)^2 x^{3/2}} = -\frac{2}{a^2(a^2 + b^2x)x^{1/2}} - \frac{3b^2 x^{1/2}}{a^4(a^2 + b^2x)}$
$$-\frac{3b}{a^5} \tan^{-1} \frac{bx^{1/2}}{a}.$$

187.11. $\displaystyle\int \frac{x^{1/2}dx}{a^2 - b^2x} = -\frac{2x^{1/2}}{b^2} + \frac{a}{b^3} \log \left| \frac{a + bx^{1/2}}{a - bx^{1/2}} \right|.$

187.13. $\displaystyle\int \frac{x^{3/2}dx}{a^2 - b^2x} = -\frac{2}{3}\frac{x^{3/2}}{b^2} - \frac{2a^2 x^{1/2}}{b^4} + \frac{a^3}{b^5} \log \left| \frac{a + bx^{1/2}}{a - bx^{1/2}} \right|.$

187.21. $\displaystyle\int \frac{x^{1/2}dx}{(a^2 - b^2x)^2} = \frac{x^{1/2}}{b^2(a^2 - b^2x)} - \frac{1}{2ab^3} \log \left| \frac{a + bx^{1/2}}{a - bx^{1/2}} \right|.$

187.23. $\displaystyle\int \frac{x^{3/2}dx}{(a^2 - b^2x)^2} = \frac{3a^2 x^{1/2} - 2b^2 x^{3/2}}{b^4(a^2 - b^2x)} - \frac{3a}{2b^5} \log \left| \frac{a + bx^{1/2}}{a - bx^{1/2}} \right|.$

188.11. $\displaystyle\int \frac{dx}{(a^2 - b^2x)x^{1/2}} = \frac{1}{ab} \log \left| \frac{a + bx^{1/2}}{a - bx^{1/2}} \right|.$

188.13. $\displaystyle\int \frac{dx}{(a^2 - b^2x)x^{3/2}} = -\frac{2}{a^2 x^{1/2}} + \frac{b}{a^3} \log \left| \frac{a + bx^{1/2}}{a - bx^{1/2}} \right|.$

188.21. $\displaystyle\int \frac{dx}{(a^2 - b^2x)^2 x^{1/2}} = \frac{x^{1/2}}{a^2(a^2 - b^2x)} + \frac{1}{2a^3b} \log \left| \frac{a + bx^{1/2}}{a - bx^{1/2}} \right|.$

188.23. $\displaystyle\int \frac{dx}{(a^2 - b^2x)^2 x^{3/2}} = \frac{-2}{a^2(a^2 - b^2x)x^{1/2}} + \frac{3b^2 x^{1/2}}{a^4(a^2 - b^2x)}$
$$+\frac{3b}{2a^5} \log \left| \frac{a + bx^{1/2}}{a - bx^{1/2}} \right|.$$

189.1. $\displaystyle\int \frac{x^{1/2}dx}{a^4 + x^2} = \frac{-1}{2a\sqrt{2}} \log \frac{x + a\sqrt{(2x)} + a^2}{x - a\sqrt{(2x)} + a^2}$
$$+\frac{1}{a\sqrt{2}} \tan^{-1} \frac{a\sqrt{(2x)}}{a^2 - x}.$$

189.2. $\displaystyle\int \frac{dx}{(a^4 + x^2)x^{1/2}} = \frac{1}{2a^3\sqrt{2}} \log \frac{x + a\sqrt{(2x)} + a^2}{x - a\sqrt{(2x)} + a^2}$
$$+\frac{1}{a^3\sqrt{2}} \tan^{-1} \frac{a\sqrt{(2x)}}{a^2 - x}.$$

189.3. $\displaystyle\int \frac{x^{1/2}dx}{a^4 - x^2} = \frac{1}{2a}\log\left|\frac{a + x^{1/2}}{a - x^{1/2}}\right| - \frac{1}{a}\tan^{-1}\frac{x^{1/2}}{a}.$

189.4. $\displaystyle\int\frac{dx}{(a^4 - x^2)x^{1/2}} = \frac{1}{2a^3}\log\left|\frac{a + x^{1/2}}{a - x^{1/2}}\right| + \frac{1}{a^3}\tan^{-1}\frac{x^{1/2}}{a}.$

[Ref. 4, pp. 149–151.]

Integrals Involving $X^{1/2} = (a + bx)^{1/2}$

190. $\displaystyle\int\frac{x^q dx}{X^{p/2}} = \frac{1}{b^{q+1}}\int\frac{(X - a)^q dX}{X^{p/2}},$　　　　　　　$[q > 0].$

Expand the numerator by the binomial theorem, when q is a positive integer.

191. $\displaystyle\int\frac{dx}{X^{p/2}} = \frac{-2}{(p - 2)bX^{(p-2)/2}}.$ 　　　**191.03.** $\displaystyle\int\frac{dx}{X^{3/2}} = \frac{-2}{bX^{1/2}}.$

191.01. $\displaystyle\int\frac{dx}{X^{1/2}} = \frac{2}{b}X^{1/2}.$ 　　　　　**191.05.** $\displaystyle\int\frac{dx}{X^{5/2}} = \frac{-2}{3bX^{3/2}}.$

191.1. $\displaystyle\int\frac{x\,dx}{X^{p/2}} = \frac{2}{b^2}\left[\frac{-1}{(p - 4)X^{(p-4)/2}} + \frac{a}{(p - 2)X^{(p-2)/2}}\right].$

191.11. $\displaystyle\int\frac{x\,dx}{X^{1/2}} = \frac{2}{b^2}\left(\frac{X^{3/2}}{3} - aX^{1/2}\right).$

191.13. $\displaystyle\int\frac{x\,dx}{X^{3/2}} = \frac{2}{b^2}\left(X^{1/2} + \frac{a}{X^{1/2}}\right).$

191.15. $\displaystyle\int\frac{x\,dx}{X^{5/2}} = \frac{2}{b^2}\left(\frac{-1}{X^{1/2}} + \frac{a}{3X^{3/2}}\right).$

191.17. $\displaystyle\int\frac{x\,dx}{X^{7/2}} = \frac{2}{b^2}\left(\frac{-1}{3X^{3/2}} + \frac{a}{5X^{5/2}}\right).$

191.2. $\displaystyle\int\frac{x^2 dx}{X^{p/2}} = \frac{2}{b^3}\left[\frac{-1}{(p - 6)X^{(p-6)/2}} + \frac{2a}{(p - 4)X^{(p-4)/2}}\right.$
$$\left. - \frac{a^2}{(p - 2)X^{(p-2)/2}}\right].$$

191.21. $\displaystyle\int\frac{x^2 dx}{X^{1/2}} = \frac{2}{b^3}\left(\frac{X^{5/2}}{5} - \frac{2aX^{3/2}}{3} + a^2X^{1/2}\right).$

191.23. $\displaystyle\int \frac{x^2 dx}{X^{3/2}} = \frac{2}{b^3}\left(\frac{X^{3/2}}{3} - 2aX^{1/2} - \frac{a^2}{X^{1/2}}\right).$

191.25. $\displaystyle\int \frac{x^2 dx}{X^{5/2}} = \frac{2}{b^3}\left(X^{1/2} + \frac{2a}{X^{1/2}} - \frac{a^2}{3X^{3/2}}\right).$

191.27. $\displaystyle\int \frac{x^2 dx}{X^{7/2}} = \frac{2}{b^3}\left(\frac{-1}{X^{1/2}} + \frac{2a}{3X^{3/2}} - \frac{a^2}{5X^{5/2}}\right).$

192.1. $\displaystyle\int \frac{dx}{xX^{p/2}} = \frac{2}{(p-2)aX^{(p-2)/2}} + \frac{1}{a}\int \frac{dx}{xX^{(p-2)/2}},$

$$[p > 1]. \qquad \text{[Ref. 2, p. 92.]}$$

192.11. $\displaystyle\int \frac{dx}{xX^{1/2}} = \frac{1}{a^{1/2}}\log\left|\frac{X^{1/2} - a^{1/2}}{X^{1/2} + a^{1/2}}\right|, \qquad [a > 0, X > 0],$

$$= -\frac{2}{a^{1/2}}\tanh^{-1}\frac{X^{1/2}}{a^{1/2}}, \qquad [a > X > 0],$$

$$= -\frac{2}{a^{1/2}}\operatorname{ctnh}^{-1}\frac{X^{1/2}}{a^{1/2}}, \qquad [X > a > 0],$$

$$= \frac{2}{(-a)^{1/2}}\tan^{-1}\frac{X^{1/2}}{(-a)^{1/2}}, \qquad [a < 0, X > 0].$$

$$\text{[Put } X^{1/2} = z. \text{ See Nos. } \mathbf{120.1} \text{ and } \mathbf{140.1}.]$$

192.13. $\displaystyle\int \frac{dx}{xX^{3/2}} = \frac{2}{aX^{1/2}} + \frac{1}{a}\int \frac{dx}{xX^{1/2}}.$ $\qquad$ [See **192.11.**]

192.15. $\displaystyle\int \frac{dx}{xX^{5/2}} = \frac{2}{3aX^{3/2}} + \frac{2}{a^2 X^{1/2}} + \frac{1}{a^2}\int \frac{dx}{xX^{1/2}}.$ $\qquad$ [See **192.11.**]

192.17. $\displaystyle\int \frac{dx}{xX^{7/2}} = \frac{2}{5aX^{5/2}} + \frac{2}{3a^2 X^{3/2}} + \frac{2}{a^3 X^{1/2}} + \frac{1}{a^3}\int \frac{dx}{xX^{1/2}}.$

192.2. $\displaystyle\int \frac{dx}{x^2 X^{p/2}} = \frac{-1}{axX^{(p-2)/2}} - \frac{pb}{2a}\int \frac{dx}{xX^{p/2}}.$ $\qquad$ [Ref. 2, p. 94.]

192.21. $\displaystyle\int \frac{dx}{x^2 X^{1/2}} = \frac{-X^{1/2}}{ax} - \frac{b}{2a}\int \frac{dx}{xX^{1/2}}.$ $\qquad$ [See **192.11.**]

192.23. $\displaystyle\int \frac{dx}{x^2 X^{3/2}} = \frac{-1}{axX^{1/2}} - \frac{3b}{a^2 X^{1/2}} - \frac{3b}{2a^2}\int \frac{dx}{xX^{1/2}}.$ $\qquad$ [See **192.11.**]

192.25. $\displaystyle \int \frac{dx}{x^2 X^{5/2}} = \frac{-1}{axX^{3/2}} - \frac{5b}{3a^2 X^{3/2}} - \frac{5b}{a^3 X^{1/2}} - \frac{5b}{2a^3} \int \frac{dx}{xX^{1/2}}.$

192.9. $\displaystyle \int \frac{dx}{x^p X^{1/2}} = \frac{-X^{1/2}}{(p-1)ax^{p-1}} - \frac{(2p-3)b}{(2p-2)a} \int \frac{dx}{x^{p-1} X^{1/2}}.$

193. $\displaystyle \int X^{p/2} dx = \frac{2X^{(p+2)/2}}{(p+2)b}.$

193.01. $\displaystyle \int X^{1/2} dx = \frac{2X^{3/2}}{3b}.$ **193.03.** $\displaystyle \int X^{3/2} dx = \frac{2X^{5/2}}{5b}.$

193.1. $\displaystyle \int xX^{p/2} dx = \frac{2}{b^2}\left(\frac{X^{(p+4)/2}}{p+4} - \frac{aX^{(p+2)/2}}{p+2}\right).$

193.11. $\displaystyle \int xX^{1/2} dx = \frac{2}{b^2}\left(\frac{X^{5/2}}{5} - \frac{aX^{3/2}}{3}\right).$

193.13. $\displaystyle \int xX^{3/2} dx = \frac{2}{b^2}\left(\frac{X^{7/2}}{7} - \frac{aX^{5/2}}{5}\right).$

193.2. $\displaystyle \int x^2 X^{p/2} dx = \frac{2}{b^3}\left(\frac{X^{(p+6)/2}}{p+6} - \frac{2aX^{(p+4)/2}}{p+4} + \frac{a^2 X^{(p+2)/2}}{p+2}\right).$

193.21. $\displaystyle \int x^2 X^{1/2} dx = \frac{2}{b^3}\left(\frac{X^{7/2}}{7} - \frac{2aX^{5/2}}{5} + \frac{a^2 X^{3/2}}{3}\right).$

194.1. $\displaystyle \int \frac{X^{p/2} dx}{x} = \frac{2X^{p/2}}{p} + a \int \frac{X^{(p-2)/2} dx}{x}.$ [Ref. 2, p. 91.]

194.11. $\displaystyle \int \frac{X^{1/2} dx}{x} = 2X^{1/2} + a \int \frac{dx}{xX^{1/2}}.$ [See **192.11.**]

194.13. $\displaystyle \int \frac{X^{3/2} dx}{x} = \frac{2X^{3/2}}{3} + 2aX^{1/2} + a^2 \int \frac{dx}{xX^{1/2}}.$ [See **192.11.**]

194.15. $\displaystyle \int \frac{X^{5/2} dx}{x} = \frac{2X^{5/2}}{5} + \frac{2aX^{3/2}}{3} + 2a^2 X^{1/2} + a^3 \int \frac{dx}{xX^{1/2}}.$

194.2. $\displaystyle \int \frac{X^{p/2} dx}{x^2} = -\frac{X^{(p+2)/2}}{ax} + \frac{pb}{2a} \int \frac{X^{p/2} dx}{x}.$

194.21. $\displaystyle \int \frac{X^{1/2} dx}{x^2} = -\frac{X^{1/2}}{x} + \frac{b}{2} \int \frac{dx}{xX^{1/2}}.$ [See **192.11.**]

194.31. $\displaystyle\int \frac{X^{1/2}dx}{x^3} = -\frac{(2a + bx)X^{1/2}}{4ax^2} - \frac{b^2}{8a}\int \frac{dx}{xX^{1/2}}.$ [Ref. 1, p. 105.]

Integrals Involving $X^{1/2} = (a + bx)^{1/2}$ and $U^{1/2} = (f + gx)^{1/2}$
Let $k = ag - bf$

195.01. $\displaystyle\int \frac{dx}{X^{1/2}U^{1/2}} = \frac{2}{\sqrt{(-bg)}} \tan^{-1} \sqrt{\left(\frac{-gX}{bU}\right)},$ $\left[\begin{matrix} b > 0 \\ g < 0 \end{matrix}\right],$

$\displaystyle\qquad\qquad = \frac{-1}{\sqrt{(-bg)}} \sin^{-1} \frac{2bgx + ag + bf}{bf - ag},$ $\left[\begin{matrix} b > 0 \\ g < 0 \end{matrix}\right],$

$\displaystyle\qquad\qquad = \frac{2}{\sqrt{(bg)}}\log |\sqrt{(bgX)} + b\sqrt{U}|,$ $[bg > 0].$

195.02. $\displaystyle\int \frac{dx}{X^{1/2}U} = \frac{2}{\sqrt{(-kg)}} \tan^{-1} \frac{gX^{1/2}}{\sqrt{(-kg)}},$ $[kg < 0],$

$\displaystyle\qquad\qquad = \frac{1}{\sqrt{(kg)}} \log \left| \frac{gX^{1/2} - \sqrt{(kg)}}{gX^{1/2} + \sqrt{(kg)}} \right|,$ $[kg > 0].$

195.03. $\displaystyle\int \frac{dx}{X^{1/2}U^{3/2}} = -\frac{2X^{1/2}}{kU^{1/2}}.$

195.04. $\displaystyle\int \frac{U^{1/2}dx}{X^{1/2}} = \frac{X^{1/2}U^{1/2}}{b} - \frac{k}{2b}\int \frac{dx}{X^{1/2}U^{1/2}}.$ [See **195.01.**]

195.09. $\displaystyle\int \frac{U^n dx}{X^{1/2}} = \frac{2}{(2n + 1)b}\left(X^{1/2}U^n - nk\int \frac{U^{n-1}dx}{X^{1/2}}\right).$

196.01. $\displaystyle\int X^{1/2}U^{1/2}dx = \frac{k + 2bU}{4bg} X^{1/2}U^{1/2} - \frac{k^2}{8bg}\int \frac{dx}{X^{1/2}U^{1/2}}.$

196.02. $\displaystyle\int \frac{xdx}{X^{1/2}U^{1/2}} = \frac{X^{1/2}U^{1/2}}{bg} - \frac{ag + bf}{2bg}\int \frac{dx}{X^{1/2}U^{1/2}}.$ [See **195.01.**]

196.03. $\displaystyle\int \frac{dx}{X^{1/2}U^n} = -\frac{1}{(n - 1)k}\left\{\frac{X^{1/2}}{U^{n-1}} + \left(n - \frac{3}{2}\right)b\int \frac{dx}{X^{1/2}U^{n-1}}\right\}.$

196.04. $\displaystyle\int X^{1/2}U^n dx = \frac{1}{(2n + 3)g}\left(2X^{1/2}U^{n+1} + k\int \frac{U^n dx}{X^{1/2}}\right).$

196.05. $\displaystyle\int \frac{X^{1/2}dx}{U^n} = \frac{1}{(n - 1)g}\left(-\frac{X^{1/2}}{U^{n-1}} + \frac{b}{2}\int \frac{dx}{X^{1/2}U^{n-1}}\right).$ [See **196.03.**]

197.
$$\int \frac{f(x^2)dx}{\sqrt{(a + bx^2)}} = \int f\left(\frac{au^2}{1 - bu^2}\right)\frac{du}{(1 - bu^2)}$$

where
$$u = x/\sqrt{(a + bx^2)}.$$

Integrals Involving $r = (x^2 + a^2)^{1/2}$

200.01. $\int \dfrac{dx}{r} = \int \dfrac{dx}{\sqrt{(x^2 + a^2)}} = \log(x + r).$

Note that
$$\log\left(\frac{x + r}{a}\right) = \sinh^{-1}\frac{x}{a} = \frac{1}{2}\log\left(\frac{r + x}{r - x}\right).$$

The positive values of r and a are to be taken.

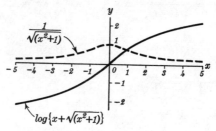

Fig. 200.01. Graphs of $1/\sqrt{(x^2 + 1)}$ and of $\log\{x + \sqrt{(x^2 + 1)}\}$, where x is real.

200.03. $\int \dfrac{dx}{r^3} = \dfrac{1}{a^2}\dfrac{x}{r}.$

200.05. $\int \dfrac{dx}{r^5} = \dfrac{1}{a^4}\left[\dfrac{x}{r} - \dfrac{1}{3}\dfrac{x^3}{r^3}\right].$

200.07. $\int \dfrac{dx}{r^7} = \dfrac{1}{a^6}\left[\dfrac{x}{r} - \dfrac{2}{3}\dfrac{x^3}{r^3} + \dfrac{1}{5}\dfrac{x^5}{r^5}\right].$

200.09. $\int \dfrac{dx}{r^9} = \dfrac{1}{a^8}\left[\dfrac{x}{r} - \dfrac{3}{3}\dfrac{x^3}{r^3} + \dfrac{3}{5}\dfrac{x^5}{r^5} - \dfrac{1}{7}\dfrac{x^7}{r^7}\right].$

200.11. $\int \dfrac{dx}{r^{11}} = \dfrac{1}{a^{10}}\left[\dfrac{x}{r} - \dfrac{4}{3}\dfrac{x^3}{r^3} + \dfrac{6}{5}\dfrac{x^5}{r^5} - \dfrac{4}{7}\dfrac{x^7}{r^7} + \dfrac{1}{9}\dfrac{x^9}{r^9}\right].$

200.13. $\int \dfrac{dx}{r^{13}} = \dfrac{1}{a^{12}}\left[\dfrac{x}{r} - \dfrac{5}{3}\dfrac{x^3}{r^3} + \dfrac{10}{5}\dfrac{x^5}{r^5} - \dfrac{10}{7}\dfrac{x^7}{r^7} + \dfrac{5}{9}\dfrac{x^9}{r^9} - \dfrac{1}{11}\dfrac{x^{11}}{r^{11}}\right].$

200.15. $\displaystyle\int\frac{dx}{r^{15}} = \frac{1}{a^{14}}\left[\frac{x}{r} - \frac{6}{3}\frac{x^3}{r^3} + \frac{15}{5}\frac{x^5}{r^5} - \frac{20}{7}\frac{x^7}{r^7} + \frac{15}{9}\frac{x^9}{r^9}\right.$
$$\left. - \frac{6}{11}\frac{x^{11}}{r^{11}} + \frac{1}{13}\frac{x^{13}}{r^{13}}\right].$$

For **200.03–200.15** let

$$z^2 = \frac{x^2}{x^2+a^2}; \qquad \text{then} \qquad dx = \frac{a\,dz}{(1-z^2)^{3/2}}.$$

201.01. $\displaystyle\int\frac{x\,dx}{r} = r.$ **201.05.** $\displaystyle\int\frac{x\,dx}{r^5} = -\frac{1}{3r^3}.$

201.03. $\displaystyle\int\frac{x\,dx}{r^3} = -\frac{1}{r}.$ **201.07.** $\displaystyle\int\frac{x\,dx}{r^7} = -\frac{1}{5r^5}.$

201.9. $\displaystyle\int\frac{x\,dx}{r^{2p+1}} = -\frac{1}{(2p-1)r^{2p-1}}.$

202.01. $\displaystyle\int\frac{x^2dx}{r} = \frac{xr}{2} - \frac{a^2}{2}\log(x+r).$ [See note under **200.01.**]

202.03. $\displaystyle\int\frac{x^2dx}{r^3} = -\frac{x}{r} + \log(x+r).$

202.05. $\displaystyle\int\frac{x^2dx}{r^5} = \frac{1}{3a^2}\frac{x^3}{r^3}.$

202.07. $\displaystyle\int\frac{x^2dx}{r^7} = \frac{1}{a^4}\left[\frac{1}{3}\frac{x^3}{r^3} - \frac{1}{5}\frac{x^5}{r^5}\right].$

202.09. $\displaystyle\int\frac{x^2dx}{r^9} = \frac{1}{a^6}\left[\frac{1}{3}\frac{x^3}{r^3} - \frac{2}{5}\frac{x^5}{r^5} + \frac{1}{7}\frac{x^7}{r^7}\right].$

202.11. $\displaystyle\int\frac{x^2dx}{r^{11}} = \frac{1}{a^8}\left[\frac{1}{3}\frac{x^3}{r^3} - \frac{3}{5}\frac{x^5}{r^5} + \frac{3}{7}\frac{x^7}{r^7} - \frac{1}{9}\frac{x^9}{r^9}\right].$

202.13. $\displaystyle\int\frac{x^2dx}{r^{13}} = \frac{1}{a^{10}}\left[\frac{1}{3}\frac{x^3}{r^3} - \frac{4}{5}\frac{x^5}{r^5} + \frac{6}{7}\frac{x^7}{r^7} - \frac{4}{9}\frac{x^9}{r^9} + \frac{1}{11}\frac{x^{11}}{r^{11}}\right].$

202.15. $\displaystyle\int\frac{x^2dx}{r^{15}} = \frac{1}{a^{12}}\left[\frac{1}{3}\frac{x^3}{r^3} - \frac{5}{5}\frac{x^5}{r^5} + \frac{10}{7}\frac{x^7}{r^7} - \frac{10}{9}\frac{x^9}{r^9}\right.$
$$\left. + \frac{5}{11}\frac{x^{11}}{r^{11}} - \frac{1}{13}\frac{x^{13}}{r^{13}}\right].$$

203.01. $\displaystyle\int\frac{x^3dx}{r} = \frac{r^3}{3} - a^2r.$

203.03. $\displaystyle\int\frac{x^3dx}{r^3} = r + \frac{a^2}{r}.$

203.05. $\displaystyle\int\frac{x^3dx}{r^5} = -\frac{1}{r} + \frac{a^2}{3r^3}.$

203.07. $\displaystyle\int\frac{x^3dx}{r^7} = -\frac{1}{3r^3} + \frac{a^2}{5r^5}.$

203.9. $\displaystyle\int\frac{x^3dx}{r^{2p+1}} = -\frac{1}{(2p-3)r^{2p-3}} + \frac{a^2}{(2p-1)r^{2p-1}}.$

204.01. $\displaystyle\int\frac{x^4dx}{r} = \frac{x^3r}{4} - \frac{3}{8}a^2xr + \frac{3}{8}a^4\log(x+r).$

[See note under **200.01.**]

204.03. $\displaystyle\int\frac{x^4dx}{r^3} = \frac{xr}{2} + \frac{a^2x}{r} - \frac{3}{2}a^2\log(x+r).$

204.05. $\displaystyle\int\frac{x^4dx}{r^5} = -\frac{x}{r} - \frac{1}{3}\frac{x^3}{r^3} + \log(x+r).$

204.07. $\displaystyle\int\frac{x^4dx}{r^7} = \frac{1}{5a^2}\frac{x^5}{r^5}.$ **204.09.** $\displaystyle\int\frac{x^4dx}{r^9} = \frac{1}{a^4}\left[\frac{1}{5}\frac{x^5}{r^5} - \frac{1}{7}\frac{x^7}{r^7}\right].$

204.11. $\displaystyle\int\frac{x^4dx}{r^{11}} = \frac{1}{a^6}\left[\frac{1}{5}\frac{x^5}{r^5} - \frac{2}{7}\frac{x^7}{r^7} + \frac{1}{9}\frac{x^9}{r^9}\right].$

204.13. $\displaystyle\int\frac{x^4dx}{r^{13}} = \frac{1}{a^8}\left[\frac{1}{5}\frac{x^5}{r^5} - \frac{3}{7}\frac{x^7}{r^7} + \frac{3}{9}\frac{x^9}{r^9} - \frac{1}{11}\frac{x^{11}}{r^{11}}\right].$

204.15. $\displaystyle\int\frac{x^4dx}{r^{15}} = \frac{1}{a^{10}}\left[\frac{1}{5}\frac{x^5}{r^5} - \frac{4}{7}\frac{x^7}{r^7} + \frac{6}{9}\frac{x^9}{r^9} - \frac{4}{11}\frac{x^{11}}{r^{11}} + \frac{1}{13}\frac{x^{13}}{r^{13}}\right].$

205.01. $\displaystyle\int\frac{x^5dx}{r} = \frac{r^5}{5} - \frac{2}{3}a^2r^3 + a^4r.$

205.03. $\displaystyle\int\frac{x^5dx}{r^3} = \frac{r^3}{3} - 2a^2r - \frac{a^4}{r}.$

205.05. $\displaystyle\int\frac{x^5dx}{r^5} = r + \frac{2a^2}{r} - \frac{a^4}{3r^3}.$

205.07. $\displaystyle\int \frac{x^5 dx}{r^7} = -\frac{1}{r} + \frac{2a^2}{3r^3} - \frac{a^4}{5r^5}.$

205.9. $\displaystyle\int \frac{x^5 dx}{r^{2p+1}} = -\frac{1}{(2p-5)r^{2p-5}} + \frac{2a^2}{(2p-3)r^{2p-3}} - \frac{a^4}{(2p-1)r^{2p-1}}.$

206.01. $\displaystyle\int \frac{x^6 dx}{r} = \frac{x^5 r}{6} - \frac{5}{24} a^2 x^3 r + \frac{5}{16} a^4 xr - \frac{5}{16} a^6 \log(x + r).$

[See note under **200.01**.]

206.03. $\displaystyle\int \frac{x^6 dx}{r^3} = \frac{x^5}{4r} - \frac{5}{8}\frac{a^2 x^3}{r} - \frac{15}{8}\frac{a^4 x}{r} + \frac{15}{8} a^4 \log(x + r).$

206.05. $\displaystyle\int \frac{x^6 dx}{r^5} = \frac{x^5}{2r^3} + \frac{10}{3}\frac{a^2 x^3}{r^3} + \frac{5}{2}\frac{a^4 x}{r^3} - \frac{5}{2} a^2 \log(x + r).$

206.07. $\displaystyle\int \frac{x^6 dx}{r^7} = -\frac{23}{15}\frac{x^5}{r^5} - \frac{7}{3}\frac{a^2 x^3}{r^5} - \frac{a^4 x}{r^5} + \log(x + r).$

206.09. $\displaystyle\int \frac{x^6 dx}{r^9} = \frac{1}{7a^2}\frac{x^7}{r^7}.$ **206.11.** $\displaystyle\int \frac{x^6 dx}{r^{11}} = \frac{1}{a^4}\left[\frac{1}{7}\frac{x^7}{r^7} - \frac{1}{9}\frac{x^9}{r^9}\right].$

206.13. $\displaystyle\int \frac{x^6 dx}{r^{13}} = \frac{1}{a^6}\left[\frac{1}{7}\frac{x^7}{r^7} - \frac{2}{9}\frac{x^9}{r^9} + \frac{1}{11}\frac{x^{11}}{r^{11}}\right].$

206.15. $\displaystyle\int \frac{x^6 dx}{r^{15}} = \frac{1}{a^8}\left[\frac{1}{7}\frac{x^7}{r^7} - \frac{3}{9}\frac{x^9}{r^9} + \frac{3}{11}\frac{x^{11}}{r^{11}} - \frac{1}{13}\frac{x^{13}}{r^{13}}\right].$

207.01. $\displaystyle\int \frac{x^7 dx}{r} = \frac{1}{7} r^7 - \frac{3}{5} a^2 r^5 + \frac{3}{3} a^4 r^3 - a^6 r.$

207.03. $\displaystyle\int \frac{x^7 dx}{r^3} = \frac{1}{5} r^5 - \frac{3}{3} a^2 r^3 + 3a^4 r + \frac{a^6}{r}.$

207.05. $\displaystyle\int \frac{x^7 dx}{r^5} = \frac{1}{3} r^3 - 3a^2 r - \frac{3a^4}{r} + \frac{a^6}{3r^3}.$

207.07. $\displaystyle\int \frac{x^7 dx}{r^7} = r + \frac{3a^2}{r} - \frac{3a^4}{3r^3} + \frac{a^6}{5r^5}.$

207.9. $\displaystyle\int \frac{x^7 dx}{r^{2p+1}} = -\frac{1}{(2p-7)r^{2p-7}} + \frac{3a^2}{(2p-5)r^{2p-5}}$
$$-\frac{3a^4}{(2p-3)r^{2p-3}} + \frac{a^6}{(2p-1)r^{2p-1}}.$$

221.01. $\int \dfrac{dx}{xr} = \int \dfrac{dx}{x\sqrt{(x^2 + a^2)}} = -\dfrac{1}{a}\log\left|\dfrac{a + r}{x}\right|.$

Note that

$$-\frac{1}{a}\log\left|\frac{a + r}{x}\right| = -\frac{1}{a}\operatorname{csch}^{-1}\left|\frac{x}{a}\right| = -\frac{1}{a}\sinh^{-1}\left|\frac{a}{x}\right|$$

$$= -\frac{1}{2a}\log\left(\frac{r + a}{r - a}\right).$$

The positive values of a and r are to be taken.

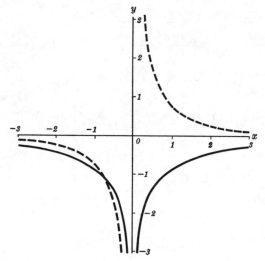

Fig. 221.01.

Dotted graph, $\dfrac{1}{x\sqrt{(x^2 + 1)}}.$

Full line graph, $-\log\left|\dfrac{1 + \sqrt{(x^2 + 1)}}{x}\right|.$

221.03. $\int \dfrac{dx}{xr^3} = \dfrac{1}{a^2 r} - \dfrac{1}{a^3}\log\left|\dfrac{a + r}{x}\right|.$

221.05. $\int \dfrac{dx}{xr^5} = \dfrac{1}{3a^2 r^3} + \dfrac{1}{a^4 r} - \dfrac{1}{a^5}\log\left|\dfrac{a + r}{x}\right|.$

221.07. $\int \dfrac{dx}{xr^7} = \dfrac{1}{5a^2 r^5} + \dfrac{1}{3a^4 r^3} + \dfrac{1}{a^6 r} - \dfrac{1}{a^7}\log\left|\dfrac{a + r}{x}\right|.$

221.09. $\displaystyle\int \frac{dx}{xr^9} = \frac{1}{7a^2r^7} + \frac{1}{5a^4r^5} + \frac{1}{3a^6r^3} + \frac{1}{a^8r} - \frac{1}{a^9}\log\left|\frac{a+r}{x}\right|.$

222.01. $\displaystyle\int \frac{dx}{x^2r} = -\frac{r}{a^2x}.$

222.03. $\displaystyle\int \frac{dx}{x^2r^3} = -\frac{1}{a^4}\left(\frac{r}{x} + \frac{x}{r}\right).$

222.05. $\displaystyle\int \frac{dx}{x^2r^5} = -\frac{1}{a^6}\left(\frac{r}{x} + \frac{2x}{r} - \frac{x^3}{3r^3}\right).$

222.07. $\displaystyle\int \frac{dx}{x^2r^7} = -\frac{1}{a^8}\left(\frac{r}{x} + \frac{3x}{r} - \frac{3x^3}{3r^3} + \frac{x^5}{5r^5}\right).$

222.09. $\displaystyle\int \frac{dx}{x^2r^9} = -\frac{1}{a^{10}}\left(\frac{r}{x} + \frac{4x}{r} - \frac{6x^3}{3r^3} + \frac{4x^5}{5r^5} - \frac{x^7}{7r^7}\right).$

223.01. $\displaystyle\int \frac{dx}{x^3r} = -\frac{r}{2a^2x^2} + \frac{1}{2a^3}\log\left|\frac{a+r}{x}\right|.$

As in **221.01**, we have

$$\log\left|\frac{a+r}{x}\right| = \operatorname{csch}^{-1}\left|\frac{x}{a}\right| = \sinh^{-1}\left|\frac{a}{x}\right|$$
$$= \frac{1}{2}\log\left(\frac{r+a}{r-a}\right).$$

223.03. $\displaystyle\int \frac{dx}{x^3r^3} = -\frac{1}{2a^2x^2r} - \frac{3}{2a^4r} + \frac{3}{2a^5}\log\left|\frac{a+r}{x}\right|.$

223.05. $\displaystyle\int \frac{dx}{x^3r^5} = -\frac{1}{2a^2x^2r^3} - \frac{5}{6a^4r^3} - \frac{5}{2a^6r} + \frac{5}{2a^7}\log\left|\frac{a+r}{x}\right|.$

224.01. $\displaystyle\int \frac{dx}{x^4r} = \frac{1}{a^4}\left(\frac{r}{x} - \frac{r^3}{3x^3}\right).$

224.03. $\displaystyle\int \frac{dx}{x^4r^3} = \frac{1}{a^6}\left(\frac{x}{r} + \frac{2r}{x} - \frac{r^3}{3x^3}\right).$

224.05. $\displaystyle\int \frac{dx}{x^4r^5} = \frac{1}{a^8}\left(-\frac{x^3}{3r^3} + \frac{3x}{r} + \frac{3r}{x} - \frac{r^3}{3x^3}\right).$

For **222** and **224**, use the note following No **200.15**.

225.01. $\displaystyle\int \frac{dx}{x^5 r} = -\frac{r}{4a^2x^4} + \frac{3}{8}\frac{r}{a^4x^2} - \frac{3}{8a^5}\log\left|\frac{a+r}{x}\right|.$

[Ref. 1, p. 121.]

225.03. $\displaystyle\int \frac{dx}{x^5 r^3} = -\frac{1}{4a^2x^4r} + \frac{5}{8a^4x^2r} + \frac{15}{8a^6r} - \frac{15}{8a^7}\log\left|\frac{a+r}{x}\right|.$

226.01. $\displaystyle\int \frac{dx}{x^6 r} = \frac{1}{a^6}\left(-\frac{r}{x} + \frac{2r^3}{3x^3} - \frac{r^5}{5x^5}\right).$

226.03. $\displaystyle\int \frac{dx}{x^6 r^3} = \frac{1}{a^8}\left(-\frac{x}{r} - \frac{3r}{x} + \frac{3r^3}{3x^3} - \frac{r^5}{5x^5}\right).$

. . . .

230.01. $\displaystyle\int r\, dx = \frac{xr}{2} + \frac{a^2}{2}\log(x+r).$ [See note under **200.01.**]

230.03. $\displaystyle\int r^3 dx = \frac{1}{4}xr^3 + \frac{3}{8}a^2xr + \frac{3}{8}a^4\log(x+r).$

230.05. $\displaystyle\int r^5 dx = \frac{1}{6}xr^5 + \frac{5}{24}a^2xr^3 + \frac{5}{16}a^4xr + \frac{5}{16}a^6\log(x+r).$

231.01. $\displaystyle\int xr\, dx = \frac{r^3}{3}.$

231.03. $\displaystyle\int xr^3 dx = \frac{r^5}{5}.$

231.9. $\displaystyle\int xr^{2p+1} dx = \frac{r^{2p+3}}{2p+3}.$

232.01. $\displaystyle\int x^2 r\, dx = \frac{xr^3}{4} - \frac{a^2xr}{8} - \frac{a^4}{8}\log(x+r).$

232.03. $\displaystyle\int x^2 r^3 dx = \frac{xr^5}{6} - \frac{a^2xr^3}{24} - \frac{a^4xr}{16} - \frac{a^6}{16}\log(x+r).$

233.01. $\displaystyle\int x^3 r\, dx = \frac{r^5}{5} - \frac{a^2r^3}{3}.$

233.03. $\displaystyle\int x^3 r^3 dx = \frac{r^7}{7} - \frac{a^2r^5}{5}.$

233.9. $\displaystyle\int x^3 r^{2p+1} dx = \frac{r^{2p+5}}{2p+5} - \frac{a^2 r^{2p+3}}{2p+3}.$

234.01. $\displaystyle\int x^4 r \, dx = \frac{x^3 r^3}{6} - \frac{a^2 x r^3}{8} + \frac{a^4 x r}{16} + \frac{a^6}{16} \log(x+r).$

[See note under 200.01.]

234.03. $\displaystyle\int x^4 r^3 dx = \frac{x^3 r^5}{8} - \frac{a^2 x r^5}{16} + \frac{a^4 x r^3}{64} + \frac{3}{128} a^6 x r$

$$+ \frac{3}{128} a^8 \log(x+r).$$

235.01. $\displaystyle\int x^5 r \, dx = \frac{r^7}{7} - \frac{2a^2 r^5}{5} + \frac{a^4 r^3}{3}.$

235.03. $\displaystyle\int x^5 r^3 dx = \frac{r^9}{9} - \frac{2a^2 r^7}{7} + \frac{a^4 r^5}{5}.$

. . . .

235.9. $\displaystyle\int x^5 r^{2p+1} dx = \frac{r^{2p+7}}{2p+7} - \frac{2a^2 r^{2p+5}}{2p+5} + \frac{a^4 r^{2p+3}}{2p+3}.$

241.01. $\displaystyle\int \frac{r \, dx}{x} = r - a \log \left| \frac{a+r}{x} \right|.$ [See note under **221.01**.]

241.03. $\displaystyle\int \frac{r^3 dx}{x} = \frac{r^3}{3} + a^2 r - a^3 \log \left| \frac{a+r}{x} \right|.$

241.05. $\displaystyle\int \frac{r^5 dx}{x} = \frac{r^5}{5} + \frac{a^2 r^3}{3} + a^4 r - a^5 \log \left| \frac{a+r}{x} \right|.$

241.07. $\displaystyle\int \frac{r^7 dx}{x} = \frac{r^7}{7} + \frac{a^2 r^5}{5} + \frac{a^4 r^3}{3} + a^6 r - a^7 \log \left| \frac{a+r}{x} \right|.$

242.01. $\displaystyle\int \frac{r \, dx}{x^2} = -\frac{r}{x} + \log(x+r).$ [See note under **200.01**.]

242.03. $\displaystyle\int \frac{r^3 dx}{x^2} = -\frac{r^3}{x} + \frac{3}{2} xr + \frac{3}{2} a^2 \log(x+r).$

242.05. $\displaystyle\int \frac{r^5 dx}{x^2} = -\frac{r^5}{x} + \frac{5}{4} xr^3 + \frac{15}{8} a^2 xr + \frac{15}{8} a^4 \log(x+r).$

243.01. $\int \dfrac{r\,dx}{x^3} = -\dfrac{r}{2x^2} - \dfrac{1}{2a} \log \left| \dfrac{a+r}{x} \right|.$ [See note under **221.01.**]

243.03. $\int \dfrac{r^3 dx}{x^3} = -\dfrac{r^3}{2x^2} + \dfrac{3}{2}\,r - \dfrac{3}{2}\,a \log \left| \dfrac{a+r}{x} \right|.$

243.05. $\int \dfrac{r^5 dx}{x^3} = -\dfrac{r^5}{2x^2} + \dfrac{5}{6}\,r^3 + \dfrac{5}{2}\,a^2 r - \dfrac{5}{2}\,a^3 \log \left| \dfrac{a+r}{x} \right|.$

244.01. $\int \dfrac{r\,dx}{x^4} = -\dfrac{r^3}{3a^2 x^3}.$

244.03. $\int \dfrac{r^3 dx}{x^4} = -\dfrac{r^3}{3x^3} - \dfrac{r}{x} + \log (x+r).$ [See note under **200.01.**]

244.05. $\int \dfrac{r^5 dx}{x^4} = -\dfrac{a^2 r^3}{3x^3} - \dfrac{2a^2 r}{x} + \dfrac{xr}{2} + \dfrac{5}{2}\,a^2 \log (x+r).$

245.01. $\int \dfrac{r\,dx}{x^5} = -\dfrac{r}{4x^4} - \dfrac{r}{8a^2 x^2} + \dfrac{1}{8a^3} \log \left| \dfrac{a+r}{x} \right|.$

245.03. $\int \dfrac{r^3 dx}{x^5} = -\dfrac{r^3}{4x^4} - \dfrac{3}{8}\dfrac{r^3}{a^2 x^2} + \dfrac{3}{8}\dfrac{r}{a^2} - \dfrac{3}{8a} \log \left| \dfrac{a+r}{x} \right|.$

246.01. $\int \dfrac{r\,dx}{x^6} = \dfrac{r^3}{5a^2 x^3}\left(\dfrac{2}{3a^2} - \dfrac{1}{x^2} \right).$

246.03. $\int \dfrac{r^3 dx}{x^6} = -\dfrac{r^5}{5a^2 x^5}.$

247.01. $\int \dfrac{r\,dx}{x^7} = -\dfrac{r}{6x^6} - \dfrac{r}{24a^2 x^4} + \dfrac{r}{16a^4 x^2} - \dfrac{1}{16a^5} \log \left| \dfrac{a+r}{x} \right|.$

248.01. $\int \dfrac{r\,dx}{x^8} = \dfrac{r^3}{7a^2 x^3}\left(-\dfrac{1}{x^4} + \dfrac{4}{5a^2 x^2} - \dfrac{8}{15a^4} \right).$

Integrals Involving $s = (x^2 - a^2)^{1/2}$

260.01. $\displaystyle \int \frac{dx}{s} = \int \frac{dx}{\sqrt{(x^2 - a^2)}} = \log|x + s|,$ $[x^2 > a^2].$

Note that

$$\log\left|\frac{x+s}{a}\right| = \frac{1}{2}\log\left(\frac{x+s}{x-s}\right) = \cosh^{-1}\left|\frac{x}{a}\right|.$$

The positive value of $\cosh^{-1}|x/a|$ is to be taken for positive values of x, and the negative value for negative values of x. The positive value of s is to be taken.

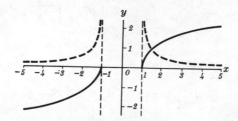

Fig. 260.01.

Dotted graph, $1/\sqrt{(x^2 - 1)}$. Full line graph, $\log|x + \sqrt{(x^2 - 1)}|$.

260.03. $\displaystyle \int \frac{dx}{s^3} = -\frac{1}{a^2}\frac{x}{s}.$

260.05. $\displaystyle \int \frac{dx}{s^5} = \frac{1}{a^4}\left[\frac{x}{s} - \frac{1}{3}\frac{x^3}{s^3}\right].$

260.07. $\displaystyle \int \frac{dx}{s^7} = -\frac{1}{a^6}\left[\frac{x}{s} - \frac{2}{3}\frac{x^3}{s^3} + \frac{1}{5}\frac{x^5}{s^5}\right].$

260.09. $\displaystyle \int \frac{dx}{s^9} = \frac{1}{a^8}\left[\frac{x}{s} - \frac{3}{3}\frac{x^3}{s^3} + \frac{3}{5}\frac{x^5}{s^5} - \frac{1}{7}\frac{x^7}{s^7}\right].$

260.11. $\displaystyle \int \frac{dx}{s^{11}} = -\frac{1}{a^{10}}\left[\frac{x}{s} - \frac{4}{3}\frac{x^3}{s^3} + \frac{6}{5}\frac{x^5}{s^5} - \frac{4}{7}\frac{x^7}{s^7} + \frac{1}{9}\frac{x^9}{s^9}\right].$

260.13. $\displaystyle \int \frac{dx}{s^{13}} = \frac{1}{a^{12}}\left[\frac{x}{s} - \frac{5}{3}\frac{x^3}{s^3} + \frac{10}{5}\frac{x^5}{s^5} - \frac{10}{7}\frac{x^7}{s^7} + \frac{5}{9}\frac{x^9}{s^9} - \frac{1}{11}\frac{x^{11}}{s^{11}}\right].$

260.15.
$$\int \frac{dx}{s^{15}} = -\frac{1}{a^{14}}\left[\frac{x}{s} - \frac{6}{3}\frac{x^3}{s^3} + \frac{15}{5}\frac{x^5}{s^5} - \frac{20}{7}\frac{x^7}{s^7} + \frac{15}{9}\frac{x^9}{s^9}\right.$$
$$\left. -\frac{6}{11}\frac{x^{11}}{s^{11}} + \frac{1}{13}\frac{x^{13}}{s^{13}}\right].$$

For **260.03–260.15**, let

$$z^2 = \frac{x^2}{x^2 - a^2}; \quad \text{then} \quad dx = \frac{-a\,dz}{(z^2 - 1)^{3/2}}.$$

261.01. $\displaystyle\int \frac{x\,dx}{s} = s.$ **261.05.** $\displaystyle\int \frac{x\,dx}{s^5} = -\frac{1}{3s^3}.$

261.03. $\displaystyle\int \frac{x\,dx}{s^3} = -\frac{1}{s}.$ **261.07.** $\displaystyle\int \frac{x\,dx}{s^7} = -\frac{1}{5s^5}.$

261.9. $\displaystyle\int \frac{x\,dx}{s^{2p+1}} = -\frac{1}{(2p-1)s^{2p-1}}.$

262.01. $\displaystyle\int \frac{x^2\,dx}{s} = \frac{xs}{2} + \frac{a^2}{2}\log|x+s|.$ [See note under **260.01.**]

262.03. $\displaystyle\int \frac{x^2\,dx}{s^3} = -\frac{x}{s} + \log|x+s|.$

262.05. $\displaystyle\int \frac{x^2\,dx}{s^5} = -\frac{1}{3a^2}\frac{x^3}{s^3}.$

262.07. $\displaystyle\int \frac{x^2\,dx}{s^7} = \frac{1}{a^4}\left[\frac{1}{3}\frac{x^3}{s^3} - \frac{1}{5}\frac{x^5}{s^5}\right].$

262.09. $\displaystyle\int \frac{x^2\,dx}{s^9} = -\frac{1}{a^6}\left[\frac{1}{3}\frac{x^3}{s^3} - \frac{2}{5}\frac{x^5}{s^5} + \frac{1}{7}\frac{x^7}{s^7}\right].$

262.11. $\displaystyle\int \frac{x^2\,dx}{s^{11}} = \frac{1}{a^8}\left[\frac{1}{3}\frac{x^3}{s^3} - \frac{3}{5}\frac{x^5}{s^5} + \frac{3}{7}\frac{x^7}{s^7} - \frac{1}{9}\frac{x^9}{s^9}\right].$

262.13. $\displaystyle\int \frac{x^2\,dx}{s^{13}} = -\frac{1}{a^{10}}\left[\frac{1}{3}\frac{x^3}{s^3} - \frac{4}{5}\frac{x^5}{s^5} + \frac{6}{7}\frac{x^7}{s^7} - \frac{4}{9}\frac{x^9}{s^9} + \frac{1}{11}\frac{x^{11}}{s^{11}}\right].$

262.15. $\displaystyle\int \frac{x^2\,dx}{s^{15}} = \frac{1}{a^{12}}\left[\frac{1}{3}\frac{x^3}{s^3} - \frac{5}{5}\frac{x^5}{s^5} + \frac{10}{7}\frac{x^7}{s^7} - \frac{10}{9}\frac{x^9}{s^9}\right.$
$$\left. + \frac{5}{11}\frac{x^{11}}{s^{11}} - \frac{1}{13}\frac{x^{13}}{s^{13}}\right].$$

263.01. $\displaystyle\int \frac{x^3 dx}{s} = \frac{s^3}{3} + a^2 s.$ **263.03.** $\displaystyle\int \frac{x^3 dx}{s^3} = s - \frac{a^2}{s}.$

263.05. $\displaystyle\int \frac{x^3 dx}{s^5} = -\frac{1}{s} - \frac{a^2}{3s^3}.$

263.9. $\displaystyle\int \frac{x^3 dx}{s^{2p+1}} = -\frac{1}{(2p-3)s^{2p-3}} - \frac{a^2}{(2p-1)s^{2p-1}}.$

264.01. $\displaystyle\int \frac{x^4 dx}{s} = \frac{x^3 s}{4} + \frac{3}{8} a^2 xs + \frac{3}{8} a^4 \log |x + s|.$

[See note under **260.01.**]

264.03. $\displaystyle\int \frac{x^4 dx}{s^3} = \frac{xs}{2} - \frac{a^2 x}{s} + \frac{3}{2} a^2 \log |x + s|.$

264.05. $\displaystyle\int \frac{x^4 dx}{s^5} = -\frac{x}{s} - \frac{1}{3}\frac{x^3}{s^3} + \log |x + s|.$

264.07. $\displaystyle\int \frac{x^4 dx}{s^7} = -\frac{1}{5a^2}\frac{x^5}{s^5}.$

264.09. $\displaystyle\int \frac{x^4 dx}{s^9} = \frac{1}{a^4}\left[\frac{1}{5}\frac{x^5}{s^5} - \frac{1}{7}\frac{x^7}{s^7} \right].$

264.11. $\displaystyle\int \frac{x^4 dx}{s^{11}} = -\frac{1}{a^6}\left[\frac{1}{5}\frac{x^5}{s^5} - \frac{2}{7}\frac{x^7}{s^7} + \frac{1}{9}\frac{x^9}{s^9} \right].$

264.13. $\displaystyle\int \frac{x^4 dx}{s^{13}} = \frac{1}{a^8}\left[\frac{1}{5}\frac{x^5}{s^5} - \frac{3}{7}\frac{x^7}{s^7} + \frac{3}{9}\frac{x^9}{s^9} - \frac{1}{11}\frac{x^{11}}{s^{11}} \right].$

264.15. $\displaystyle\int \frac{x^4 dx}{s^{15}} = -\frac{1}{a^{10}}\left[\frac{1}{5}\frac{x^5}{s^5} - \frac{4}{7}\frac{x^7}{s^7} + \frac{6}{9}\frac{x^9}{s^9} - \frac{4}{11}\frac{x^{11}}{s^{11}} + \frac{1}{13}\frac{x^{13}}{s^{13}} \right].$

265.01. $\displaystyle\int \frac{x^5 dx}{s} = \frac{s^5}{5} + \frac{2}{3} a^2 s^3 + a^4 s.$

265.03. $\displaystyle\int \frac{x^5 dx}{s^3} = \frac{s^3}{3} + 2a^2 s - \frac{a^4}{s}.$

265.05. $\displaystyle\int \frac{x^5 dx}{s^5} = s - \frac{2a^2}{s} - \frac{a^4}{3s^3}.$

265.07. $\displaystyle\int \frac{x^5 dx}{s^7} = -\frac{1}{s} - \frac{2a^2}{3s^3} - \frac{a^4}{5s^5}.$

265.9. $\int \dfrac{x^5 dx}{s^{2p+1}} = -\dfrac{1}{(2p-5)s^{2p-5}} - \dfrac{2a^2}{(2p-3)s^{2p-3}} - \dfrac{a^4}{(2p-1)s^{2p-1}}.$

266.01. $\int \dfrac{x^6 dx}{s} = \dfrac{x^5 s}{6} + \dfrac{5}{24} a^2 x^3 s + \dfrac{5}{16} a^4 x s + \dfrac{5}{16} a^6 \log |x+s|.$

[See note under **260.01.**]

266.03. $\int \dfrac{x^6 dx}{s^3} = \dfrac{x^5}{4s} + \dfrac{5}{8} \dfrac{a^2 x^3}{s} - \dfrac{15}{8} \dfrac{a^4 x}{s} + \dfrac{15}{8} a^4 \log |x+s|.$

266.05. $\int \dfrac{x^6 dx}{s^5} = \dfrac{x^5}{2s^3} - \dfrac{10}{3} \dfrac{a^2 x^3}{s^3} + \dfrac{5}{2} \dfrac{a^4 x}{s^3} + \dfrac{5}{2} a^2 \log |x+s|.$

266.07. $\int \dfrac{x^6 dx}{s^7} = -\dfrac{23}{15} \dfrac{x^5}{s^5} + \dfrac{7}{3} \dfrac{a^2 x^3}{s^5} - \dfrac{a^4 x}{s^5} + \log |x+s|.$

266.09. $\int \dfrac{x^6 dx}{s^9} = -\dfrac{1}{7a^2} \dfrac{x^7}{s^7}.$

266.11. $\int \dfrac{x^6 dx}{s^{11}} = \dfrac{1}{a^4} \left[\dfrac{1}{7} \dfrac{x^7}{s^7} - \dfrac{1}{9} \dfrac{x^9}{s^9} \right].$

266.13. $\int \dfrac{x^6 dx}{s^{13}} = -\dfrac{1}{a^6} \left[\dfrac{1}{7} \dfrac{x^7}{s^7} - \dfrac{2}{9} \dfrac{x^9}{s^9} + \dfrac{1}{11} \dfrac{x^{11}}{s^{11}} \right].$

266.15. $\int \dfrac{x^6 dx}{s^{15}} = \dfrac{1}{a^8} \left[\dfrac{1}{7} \dfrac{x^7}{s^7} - \dfrac{3}{9} \dfrac{x^9}{s^9} + \dfrac{3}{11} \dfrac{x^{11}}{s^{11}} - \dfrac{1}{13} \dfrac{x^{13}}{s^{13}} \right].$

267.01. $\int \dfrac{x^7 dx}{s} = \dfrac{1}{7} s^7 + \dfrac{3}{5} a^2 s^5 + \dfrac{3}{3} a^4 s^3 + a^6 s.$

267.03. $\int \dfrac{x^7 dx}{s^3} = \dfrac{1}{5} s^5 + \dfrac{3}{3} a^2 s^3 + 3a^4 s - \dfrac{a^6}{s}.$

267.05. $\int \dfrac{x^7 dx}{s^5} = \dfrac{1}{3} s^3 + 3a^2 s - \dfrac{3a^4}{s} - \dfrac{a^6}{3s^3}.$

267.07. $\int \dfrac{x^7 dx}{s^7} = s - \dfrac{3a^2}{s} - \dfrac{3a^4}{3s^3} - \dfrac{a^6}{5s^5}.$

267.9. $\int \dfrac{x^7 dx}{s^{2p+1}} = -\dfrac{1}{(2p-7)s^{2p-7}} - \dfrac{3a^2}{(2p-5)s^{2p-5}}$

$$- \dfrac{3a^4}{(2p-3)s^{2p-3}} - \dfrac{a^6}{(2p-1)s^{2p-1}}.$$

281.01. $\displaystyle\int \frac{dx}{xs} = \int \frac{dx}{x\sqrt{(x^2 - a^2)}} = \frac{1}{a} \cos^{-1}\left|\frac{a}{x}\right| = \frac{1}{a} \sec^{-1}\left|\frac{x}{a}\right|,$

$$[x^2 > a^2].$$

The positive values of s and a are to be taken. The principal values of $\cos^{-1}|a/x|$ are to be taken, that is, they are to be between 0 and $\pi/2$ since $|a/x|$ is a positive quantity.

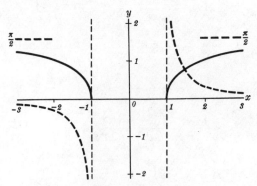

Fig. 281.01.

Dotted graph, $\dfrac{1}{x\sqrt{(x^2 - 1)}}$.

Full line graph, $\cos^{-1}\left|\dfrac{1}{x}\right|$.

281.03. $\displaystyle\int \frac{dx}{xs^3} = -\frac{1}{a^2 s} - \frac{1}{a^3} \cos^{-1}\left|\frac{a}{x}\right|.$

281.05. $\displaystyle\int \frac{dx}{xs^5} = -\frac{1}{3a^2 s^3} + \frac{1}{a^4 s} + \frac{1}{a^5} \cos^{-1}\left|\frac{a}{x}\right|.$

281.07. $\displaystyle\int \frac{dx}{xs^7} = -\frac{1}{5a^2 s^5} + \frac{1}{3a^4 s^3} - \frac{1}{a^6 s} - \frac{1}{a^7} \cos^{-1}\left|\frac{a}{x}\right|.$

281.09. $\displaystyle\int \frac{dx}{xs^9} = -\frac{1}{7a^2 s^7} + \frac{1}{5a^4 s^5} - \frac{1}{3a^6 s^3} + \frac{1}{a^8 s} + \frac{1}{a^9} \cos^{-1}\left|\frac{a}{x}\right|.$

282.01. $\displaystyle\int \frac{dx}{x^2 s} = \frac{s}{a^2 x}.$

282.03. $\displaystyle\int \frac{dx}{x^2 s^3} = -\frac{1}{a^4}\left(\frac{s}{x} + \frac{x}{s}\right).$

282.05. $\int \dfrac{dx}{x^2 s^5} = \dfrac{1}{a^6}\left(\dfrac{s}{x} + \dfrac{2x}{s} - \dfrac{x^3}{3s^3}\right).$

282.07. $\int \dfrac{dx}{x^2 s^7} = -\dfrac{1}{a^8}\left(\dfrac{s}{x} + \dfrac{3x}{s} - \dfrac{3x^3}{3s^3} + \dfrac{x^5}{5s^5}\right).$

282.09. $\int \dfrac{dx}{x^2 s^9} = \dfrac{1}{a^{10}}\left(\dfrac{s}{x} + \dfrac{4x}{s} - \dfrac{6x^3}{3s^3} + \dfrac{4x^5}{5s^5} - \dfrac{x^7}{7s^7}\right).$

283.01. $\int \dfrac{dx}{x^3 s} = \dfrac{s}{2a^2 x^2} + \dfrac{1}{2a^3}\cos^{-1}\left|\dfrac{a}{x}\right|.$ [See note under **281.01.**]

283.03. $\int \dfrac{dx}{x^3 s^3} = \dfrac{1}{2a^2 x^2 s} - \dfrac{3}{2a^4 s} - \dfrac{3}{2a^5}\cos^{-1}\left|\dfrac{a}{x}\right|.$

283.05. $\int \dfrac{dx}{x^3 s^5} = \dfrac{1}{2a^2 x^2 s^3} - \dfrac{5}{6a^4 s^3} + \dfrac{5}{2a^6 s} + \dfrac{5}{2a^7}\cos^{-1}\left|\dfrac{a}{x}\right|.$

284.01. $\int \dfrac{dx}{x^4 s} = \dfrac{1}{a^4}\left(\dfrac{s}{x} - \dfrac{s^3}{3x^3}\right).$

284.03. $\int \dfrac{dx}{x^4 s^3} = -\dfrac{1}{a^6}\left(\dfrac{x}{s} + \dfrac{2s}{x} - \dfrac{s^3}{3x^3}\right).$

284.05. $\int \dfrac{dx}{x^4 s^5} = \dfrac{1}{a^8}\left(-\dfrac{x^3}{3s^3} + \dfrac{3x}{s} + \dfrac{3s}{x} - \dfrac{s^3}{3x^3}\right).$

For **282** and **284**, put

$$z^2 = \frac{x^2}{s^2}; \quad \text{then} \quad dx = \frac{-a\,dz}{(z^2 - 1)^{3/2}}.$$

290.01. $\int s\,dx = \dfrac{xs}{2} - \dfrac{a^2}{2}\log|x + s|.$ [See note under **260.01.**]

290.03. $\int s^3 dx = \dfrac{1}{4}xs^3 - \dfrac{3}{8}a^2 xs + \dfrac{3}{8}a^4 \log|x + s|.$

290.05. $\int s^5 dx = \dfrac{1}{6}xs^5 - \dfrac{5}{24}a^2 xs^3 + \dfrac{5}{16}a^4 xs - \dfrac{5}{16}a^6 \log|x + s|.$

291.01. $\int xs\,dx = \dfrac{s^3}{3}.$ **291.03.** $\int xs^3 dx = \dfrac{s^5}{5}.$

291.9. $\int xs^{2p+1} dx = \dfrac{s^{2p+3}}{2p + 3}.$

292.01. $\displaystyle\int x^2 s\, dx = \frac{xs^3}{4} + \frac{a^2 xs}{8} - \frac{a^4}{8} \log |x + s|.$

[See note under **260.01.**]

292.03. $\displaystyle\int x^2 s^3 dx = \frac{xs^5}{6} + \frac{a^2 xs^3}{24} - \frac{a^4 xs}{16} + \frac{a^6}{16} \log |x + s|.$

293.01. $\displaystyle\int x^3 s\, dx = \frac{s^5}{5} + \frac{a^2 s^3}{3}.$ **293.03.** $\displaystyle\int x^3 s^3 dx = \frac{s^7}{7} + \frac{a^2 s^5}{5}.$

293.9. $\displaystyle\int x^3 s^{2p+1} dx = \frac{s^{2p+5}}{2p + 5} + \frac{a^2 s^{2p+3}}{2p + 3}.$

294.01. $\displaystyle\int x^4 s\, dx = \frac{x^3 s^3}{6} + \frac{a^2 xs^3}{8} + \frac{a^4 xs}{16} - \frac{a^6}{16} \log |x + s|.$

[See note under **260.01.**]

294.03. $\displaystyle\int x^4 s^3 dx = \frac{x^3 s^5}{8} + \frac{a^2 xs^5}{16} + \frac{a^4 xs^3}{64} - \frac{3}{128} a^6 xs$

$$+ \frac{3}{128} a^8 \log |x + s|.$$

295.01. $\displaystyle\int x^5 s\, dx = \frac{s^7}{7} + \frac{2a^2 s^5}{5} + \frac{a^4 s^3}{3}.$

295.03. $\displaystyle\int x^5 s^3 dx = \frac{s^9}{9} + \frac{2a^2 s^7}{7} + \frac{a^4 s^5}{5}.$

295.9. $\displaystyle\int x^5 s^{2p+1} dx = \frac{s^{2p+7}}{2p + 7} + \frac{2a^2 s^{2p+5}}{2p + 5} + \frac{a^4 s^{2p+3}}{2p + 3}.$

301.01. $\displaystyle\int \frac{s\, dx}{x} = s - a \cos^{-1} \left|\frac{a}{x}\right|.$ [See note under **281.01.**]

301.03. $\displaystyle\int \frac{s^3 dx}{x} = \frac{s^3}{3} - a^2 s + a^3 \cos^{-1} \left|\frac{a}{x}\right|.$

301.05. $\displaystyle\int \frac{s^5 dx}{x} = \frac{s^5}{5} - \frac{a^2 s^3}{3} + a^4 s - a^5 \cos^{-1} \left|\frac{a}{x}\right|.$

301.07. $\displaystyle\int \frac{s^7 dx}{x} = \frac{s^7}{7} - \frac{a^2 s^5}{5} + \frac{a^4 s^3}{3} - a^6 s + a^7 \cos^{-1} \left|\frac{a}{x}\right|.$

302.01. $\displaystyle\int\frac{s\,dx}{x^2}=-\frac{s}{x}+\log|x+s|.$ [See note under **260.01.**]

302.03. $\displaystyle\int\frac{s^3dx}{x^2}=-\frac{s^3}{x}+\frac{3}{2}\,xs-\frac{3}{2}\,a^2\log|x+s|.$

302.05. $\displaystyle\int\frac{s^5dx}{x^2}=-\frac{s^5}{x}+\frac{5}{4}\,xs^3-\frac{15}{8}\,a^2xs+\frac{15}{8}\,a^4\log|x+s|.$

303.01. $\displaystyle\int\frac{s\,dx}{x^3}=-\frac{s}{2x^2}+\frac{1}{2a}\,\cos^{-1}\left|\frac{a}{x}\right|.$ [See note under **281.01.**]

303.03. $\displaystyle\int\frac{s^3dx}{x^3}=-\frac{s^3}{2x^2}+\frac{3s}{2}-\frac{3}{2}\,a\cos^{-1}\left|\frac{a}{x}\right|.$

303.05. $\displaystyle\int\frac{s^5dx}{x^3}=-\frac{s^5}{2x^2}+\frac{5}{6}\,s^3-\frac{5}{2}\,a^2s+\frac{5}{2}\,a^3\cos^{-1}\left|\frac{a}{x}\right|.$

304.01. $\displaystyle\int\frac{s\,dx}{x^4}=\frac{s^3}{3a^2x^3}.$

304.03. $\displaystyle\int\frac{s^3dx}{x^4}=-\frac{s^3}{3x^3}-\frac{s}{x}+\log|x+s|.$

[See note under **260.01.**]

304.05. $\displaystyle\int\frac{s^5dx}{x^4}=\frac{a^2s^3}{3x^3}+\frac{2a^2s}{x}+\frac{xs}{2}-\frac{5}{2}\,a^2\log|x+s|.$

305.01. $\displaystyle\int\frac{s\,dx}{x^5}=-\frac{s}{4x^4}+\frac{s}{8a^2x^2}+\frac{1}{8a^3}\,\cos^{-1}\left|\frac{a}{x}\right|.$

305.03. $\displaystyle\int\frac{s^3dx}{x^5}=-\frac{s^3}{4x^4}+\frac{3}{8}\,\frac{s^3}{a^2x^2}-\frac{3}{8}\,\frac{s}{a^2}+\frac{3}{8a}\,\cos^{-1}\left|\frac{a}{x}\right|.$

306.01. $\displaystyle\int\frac{s\,dx}{x^6}=\frac{s^3}{5a^2x^3}\left(\frac{1}{x^2}+\frac{2}{3a^2}\right).$ **306.03.** $\displaystyle\int\frac{s^3dx}{x^6}=\frac{s^5}{5a^2x^5}.$

307.01. $\displaystyle\int\frac{s\,dx}{x^7}=-\frac{s}{6x^6}+\frac{s}{24a^2x^4}+\frac{s}{16a^4x^2}+\frac{1}{16a^5}\,\cos^{-1}\left|\frac{a}{x}\right|.$

308.01. $\displaystyle\int\frac{s\,dx}{x^8}=\frac{s^3}{7a^2x^3}\left(\frac{1}{x^4}+\frac{4}{5a^2x^2}+\frac{8}{15a^4}\right).$

Integrals Involving $t = (a^2 - x^2)^{1/2}$

320.01. $\displaystyle \int \frac{dx}{t} = \int \frac{dx}{\sqrt{(a^2 - x^2)}} = \sin^{-1}\frac{x}{a}$, $[x^2 < a^2]$.

The principal values of $\sin^{-1}(x/a)$ are to be taken, that is, values between $-\pi/2$ and $\pi/2$. The positive values of t and a are to be taken.

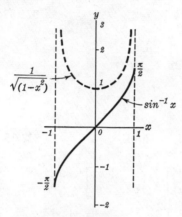

Fig. 320.01.

320.03. $\displaystyle \int \frac{dx}{t^3} = \frac{1}{a^2}\frac{x}{t}.$ **320.05.** $\displaystyle \int \frac{dx}{t^5} = \frac{1}{a^4}\left[\frac{x}{t} + \frac{1}{3}\frac{x^3}{t^3}\right].$

320.07. $\displaystyle \int \frac{dx}{t^7} = \frac{1}{a^6}\left[\frac{x}{t} + \frac{2}{3}\frac{x^3}{t^3} + \frac{1}{5}\frac{x^5}{t^5}\right].$

320.09. $\displaystyle \int \frac{dx}{t^9} = \frac{1}{a^8}\left[\frac{x}{t} + \frac{3}{3}\frac{x^3}{t^3} + \frac{3}{5}\frac{x^5}{t^5} + \frac{1}{7}\frac{x^7}{t^7}\right].$

320.11. $\displaystyle \int \frac{dx}{t^{11}} = \frac{1}{a^{10}}\left[\frac{x}{t} + \frac{4}{3}\frac{x^3}{t^3} + \frac{6}{5}\frac{x^5}{t^5} + \frac{4}{7}\frac{x^7}{t^7} + \frac{1}{9}\frac{x^9}{t^9}\right].$

320.13. $\displaystyle \int \frac{dx}{t^{13}} = \frac{1}{a^{12}}\left[\frac{x}{t} + \frac{5}{3}\frac{x^3}{t^3} + \frac{10}{5}\frac{x^5}{t^5} + \frac{10}{7}\frac{x^7}{t^7} + \frac{5}{9}\frac{x^9}{t^9} + \frac{1}{11}\frac{x^{11}}{t^{11}}\right].$

320.15. $\displaystyle \int \frac{dx}{t^{15}} = \frac{1}{a^{14}}\left[\frac{x}{t} + \frac{6}{3}\frac{x^3}{t^3} + \frac{15}{5}\frac{x^5}{t^5} + \frac{20}{7}\frac{x^7}{t^7} + \frac{15}{9}\frac{x^9}{t^9}\right.$

$$\left. + \frac{6}{11}\frac{x^{11}}{t^{11}} + \frac{1}{13}\frac{x^{13}}{t^{13}}\right].$$

For **320.03–320.15** let

$$z^2 = \frac{x^2}{a^2 - x^2}; \quad \text{then} \quad dx = \frac{a\,dz}{(1 + z^2)^{3/2}}.$$

321.01. $\displaystyle\int \frac{x\,dx}{t} = -t.$

321.05. $\displaystyle\int \frac{x\,dx}{t^5} = \frac{1}{3t^3}.$

321.03. $\displaystyle\int \frac{x\,dx}{t^3} = \frac{1}{t}.$

321.07. $\displaystyle\int \frac{x\,dx}{t^7} = \frac{1}{5t^5}.$

321.9. $\displaystyle\int \frac{x\,dx}{t^{2p+1}} = \frac{1}{(2p-1)t^{2p-1}}.$

322.01. $\displaystyle\int \frac{x^2 dx}{t} = -\frac{xt}{2} + \frac{a^2}{2} \sin^{-1} \frac{x}{a}.$ [See note under **320.01.**]

322.03. $\displaystyle\int \frac{x^2 dx}{t^3} = \frac{x}{t} - \sin^{-1} \frac{x}{a}.$

322.05. $\displaystyle\int \frac{x^2 dx}{t^5} = \frac{1}{3a^2} \frac{x^3}{t^3}.$

322.07. $\displaystyle\int \frac{x^2 dx}{t^7} = \frac{1}{a^4} \left[\frac{1}{3} \frac{x^3}{t^3} + \frac{1}{5} \frac{x^5}{t^5} \right].$

322.09. $\displaystyle\int \frac{x^2 dx}{t^9} = \frac{1}{a^6} \left[\frac{1}{3} \frac{x^3}{t^3} + \frac{2}{5} \frac{x^5}{t^5} + \frac{1}{7} \frac{x^7}{t^7} \right].$

322.11. $\displaystyle\int \frac{x^2 dx}{t^{11}} = \frac{1}{a^8} \left[\frac{1}{3} \frac{x^3}{t^3} + \frac{3}{5} \frac{x^5}{t^5} + \frac{3}{7} \frac{x^7}{t^7} + \frac{1}{9} \frac{x^9}{t^9} \right].$

322.13. $\displaystyle\int \frac{x^2 dx}{t^{13}} = \frac{1}{a^{10}} \left[\frac{1}{3} \frac{x^3}{t^3} + \frac{4}{5} \frac{x^5}{t^5} + \frac{6}{7} \frac{x^7}{t^7} + \frac{4}{9} \frac{x^9}{t^9} + \frac{1}{11} \frac{x^{11}}{t^{11}} \right].$

322.15. $\displaystyle\int \frac{x^2 dx}{t^{15}} = \frac{1}{a^{12}} \left[\frac{1}{3} \frac{x^3}{t^3} + \frac{5}{5} \frac{x^5}{t^5} + \frac{10}{7} \frac{x^7}{t^7} + \frac{10}{9} \frac{x^9}{t^9} \right.$
$$\left. + \frac{5}{11} \frac{x^{11}}{t^{11}} + \frac{1}{13} \frac{x^{13}}{t^{13}} \right].$$

323.01. $\displaystyle\int \frac{x^3 dx}{t} = \frac{t^3}{3} - a^2 t.$

323.03. $\displaystyle\int \frac{x^3 dx}{t^3} = t + \frac{a^2}{t}.$

323.05. $\displaystyle\int \frac{x^3 dx}{t^5} = -\frac{1}{t} + \frac{a^2}{3t^3}.$

323.9. $\displaystyle\int \frac{x^3 dx}{t^{2p+1}} = -\frac{1}{(2p-3)t^{2p-3}} + \frac{a^2}{(2p-1)t^{2p-1}}.$

324.01. $\displaystyle\int \frac{x^4 dx}{t} = -\frac{x^3 t}{4} - \frac{3}{8} a^2 xt + \frac{3}{8} a^4 \sin^{-1} \frac{x}{a}.$

[See note under **320.01**.]

324.03. $\displaystyle\int \frac{x^4 dx}{t^3} = \frac{xt}{2} + \frac{a^2 x}{t} - \frac{3}{2} a^2 \sin^{-1} \frac{x}{a}.$

324.05. $\displaystyle\int \frac{x^4 dx}{t^5} = -\frac{x}{t} + \frac{1}{3} \frac{x^3}{t^3} + \sin^{-1} \frac{x}{a}.$

324.07. $\displaystyle\int \frac{x^4 dx}{t^7} = \frac{1}{5a^2} \frac{x^5}{t^5}.$ **324.09.** $\displaystyle\int \frac{x^4 dx}{t^9} = \frac{1}{a^4} \left[\frac{1}{5} \frac{x^5}{t^5} + \frac{1}{7} \frac{x^7}{t^7} \right].$

324.11. $\displaystyle\int \frac{x^4 dx}{t^{11}} = \frac{1}{a^6} \left[\frac{1}{5} \frac{x^5}{t^5} + \frac{2}{7} \frac{x^7}{t^7} + \frac{1}{9} \frac{x^9}{t^9} \right].$

324.13. $\displaystyle\int \frac{x^4 dx}{t^{13}} = \frac{1}{a^8} \left[\frac{1}{5} \frac{x^5}{t^5} + \frac{3}{7} \frac{x^7}{t^7} + \frac{3}{9} \frac{x^9}{t^9} + \frac{1}{11} \frac{x^{11}}{t^{11}} \right].$

324.15. $\displaystyle\int \frac{x^4 dx}{t^{15}} = \frac{1}{a^{10}} \left[\frac{1}{5} \frac{x^5}{t^5} + \frac{4}{7} \frac{x^7}{t^7} + \frac{6}{9} \frac{x^9}{t^9} + \frac{4}{11} \frac{x^{11}}{t^{11}} + \frac{1}{13} \frac{x^{13}}{t^{13}} \right].$

325.01. $\displaystyle\int \frac{x^5 dx}{t} = -\frac{t^5}{5} + \frac{2a^2 t^3}{3} - a^4 t.$

325.03. $\displaystyle\int \frac{x^5 dx}{t^3} = -\frac{t^3}{3} + 2a^2 t + \frac{a^4}{t}.$

325.05. $\displaystyle\int \frac{x^5 dx}{t^5} = -t - \frac{2a^2}{t} + \frac{a^4}{3t^3}.$

325.07. $\displaystyle\int \frac{x^5 dx}{t^7} = \frac{1}{t} - \frac{2a^2}{3t^3} + \frac{a^4}{5t^5}.$

325.9. $\displaystyle\int \frac{x^5 dx}{t^{2p+1}} = \frac{1}{(2p-5)t^{2p-5}} - \frac{2a^2}{(2p-3)t^{2p-3}} + \frac{a^4}{(2p-1)t^{2p-1}}.$

326.01. $\displaystyle\int \frac{x^6 dx}{t} = -\frac{x^5 t}{6} - \frac{5}{24} a^2 x^3 t - \frac{5}{16} a^4 xt + \frac{5}{16} a^6 \sin^{-1} \frac{x}{a}.$

[See note under **320.01**.]

326.03. $\displaystyle\int \frac{x^6 dx}{t^3} = -\frac{x^5}{4t} - \frac{5}{8} \frac{a^2 x^3}{t} + \frac{15}{8} \frac{a^4 x}{t} - \frac{15}{8} a^4 \sin^{-1} \frac{x}{a}.$

326.05. $\displaystyle\int \frac{x^6 dx}{t^5} = -\frac{x^5}{2t^3} + \frac{10}{3} \frac{a^2 x^3}{t^3} - \frac{5}{2} \frac{a^4 x}{t^3} + \frac{5}{2} a^2 \sin^{-1} \frac{x}{a}.$

326.07. $\int \dfrac{x^6 dx}{t^7} = \dfrac{23}{15}\dfrac{x^5}{t^5} - \dfrac{7}{3}\dfrac{a^2 x^3}{t^5} + \dfrac{a^4 x}{t^5} - \sin^{-1}\dfrac{x}{a}.$

326.09. $\int \dfrac{x^6 dx}{t^9} = \dfrac{1}{7a^2}\dfrac{x^7}{t^7}.$ **326.11.** $\int \dfrac{x^6 dx}{t^{11}} = \dfrac{1}{a^4}\left[\dfrac{1}{7}\dfrac{x^7}{t^7} + \dfrac{1}{9}\dfrac{x^9}{t^9}\right].$

326.13. $\int \dfrac{x^6 dx}{t^{13}} = \dfrac{1}{a^6}\left[\dfrac{1}{7}\dfrac{x^7}{t^7} + \dfrac{2}{9}\dfrac{x^9}{t^9} + \dfrac{1}{11}\dfrac{x^{11}}{t^{11}}\right].$

326.15. $\int \dfrac{x^6 dx}{t^{15}} = \dfrac{1}{a^8}\left[\dfrac{1}{7}\dfrac{x^7}{t^7} + \dfrac{3}{9}\dfrac{x^9}{t^9} + \dfrac{3}{11}\dfrac{x^{11}}{t^{11}} + \dfrac{1}{13}\dfrac{x^{13}}{t^{13}}\right].$

327.01. $\int \dfrac{x^7 dx}{t} = \dfrac{1}{7}t^7 - \dfrac{3}{5}a^2 t^5 + \dfrac{3}{3}a^4 t^3 - a^6 t.$

327.03. $\int \dfrac{x^7 dx}{t^3} = \dfrac{1}{5}t^5 - \dfrac{3}{3}a^2 t^3 + 3a^4 t + \dfrac{a^6}{t}.$

327.05. $\int \dfrac{x^7 dx}{t^5} = \dfrac{1}{3}t^3 - 3a^2 t - \dfrac{3a^4}{t} + \dfrac{a^6}{3t^3}.$

327.07. $\int \dfrac{x^7 dx}{t^7} = t + \dfrac{3a^2}{t} - \dfrac{3a^4}{3t^3} + \dfrac{a^6}{5t^5}.$

327.9. $\int \dfrac{x^7 dx}{t^{2p+1}} = -\dfrac{1}{(2p-7)t^{2p-7}}$

$+ \dfrac{3a^2}{(2p-5)t^{2p-5}} - \dfrac{3a^4}{(2p-3)t^{2p-3}}$

$+ \dfrac{a^6}{(2p-1)t^{2p-1}}.$

341.01. $\int \dfrac{dx}{xt} = \int \dfrac{dx}{x\sqrt{(a^2 - x^2)}}$

$= -\dfrac{1}{a}\log\left|\dfrac{a+t}{x}\right|,\quad [x^2 < a^2].$

Note that

$-\dfrac{1}{a}\log\left|\dfrac{a+t}{x}\right| = -\dfrac{1}{a}\operatorname{sech}^{-1}\left|\dfrac{x}{a}\right|$

$= -\dfrac{1}{a}\cosh^{-1}\left|\dfrac{a}{x}\right|$

$= -\dfrac{1}{2a}\log\left(\dfrac{a+t}{a-t}\right).$

The positive values of $\operatorname{sech}^{-1}|x/a|$, $\cosh^{-1}|a/x|$, a and t are to be taken.

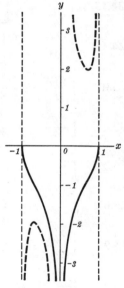

Fig. 341.01

Dotted graph,

$\dfrac{1}{x\sqrt{(1-x^2)}}.$

Full line graph,

$-\log\left|\dfrac{1+\sqrt{(1-x^2)}}{x}\right|.$

341.03. $\int \dfrac{dx}{xt^3} = \dfrac{1}{a^2 t} - \dfrac{1}{a^3} \log \left| \dfrac{a+t}{x} \right|.$

341.05. $\int \dfrac{dx}{xt^5} = \dfrac{1}{3a^2 t^3} + \dfrac{1}{a^4 t} - \dfrac{1}{a^5} \log \left| \dfrac{a+t}{x} \right|.$

341.07. $\int \dfrac{dx}{xt^7} = \dfrac{1}{5a^2 t^5} + \dfrac{1}{3a^4 t^3} + \dfrac{1}{a^6 t} - \dfrac{1}{a^7} \log \left| \dfrac{a+t}{x} \right|.$

341.09. $\int \dfrac{dx}{xt^9} = \dfrac{1}{7a^2 t^7} + \dfrac{1}{5a^4 t^5} + \dfrac{1}{3a^6 t^3} + \dfrac{1}{a^8 t} - \dfrac{1}{a^9} \log \left| \dfrac{a+t}{x} \right|.$

342.01. $\int \dfrac{dx}{x^2 t} = -\dfrac{t}{a^2 x}.$ **342.03.** $\int \dfrac{dx}{x^2 t^3} = \dfrac{1}{a^4} \left(-\dfrac{t}{x} + \dfrac{x}{t} \right).$

342.05. $\int \dfrac{dx}{x^2 t^5} = \dfrac{1}{a^6} \left(-\dfrac{t}{x} + \dfrac{2x}{t} + \dfrac{x^3}{3t^3} \right).$

342.07. $\int \dfrac{dx}{x^2 t^7} = \dfrac{1}{a^8} \left(-\dfrac{t}{x} + \dfrac{3x}{t} + \dfrac{3x^3}{3t^3} + \dfrac{x^5}{5t^5} \right).$

342.09. $\int \dfrac{dx}{x^2 t^9} = \dfrac{1}{a^{10}} \left(-\dfrac{t}{x} + \dfrac{4x}{t} + \dfrac{6x^3}{3t^3} + \dfrac{4x^5}{5t^5} + \dfrac{x^7}{7t^7} \right).$

343.01. $\int \dfrac{dx}{x^3 t} = -\dfrac{t}{2a^2 x^2} - \dfrac{1}{2a^3} \log \left| \dfrac{a+t}{x} \right|.$ [See **341.01.**]

343.03. $\int \dfrac{dx}{x^3 t^3} = -\dfrac{1}{2a^2 x^2 t} + \dfrac{3}{2a^4 t} - \dfrac{3}{2a^5} \log \left| \dfrac{a+t}{x} \right|.$

343.05. $\int \dfrac{dx}{x^3 t^5} = -\dfrac{1}{2a^2 x^2 t^3} + \dfrac{5}{6a^4 t^3} + \dfrac{5}{2a^6 t} - \dfrac{5}{2a^7} \log \left| \dfrac{a+t}{x} \right|.$

344.01. $\int \dfrac{dx}{x^4 t} = -\dfrac{1}{a^4} \left(\dfrac{t}{x} + \dfrac{t^3}{3x^3} \right).$

344.03. $\int \dfrac{dx}{x^4 t^3} = -\dfrac{1}{a^6} \left(-\dfrac{x}{t} + \dfrac{2t}{x} + \dfrac{t^3}{3x^3} \right).$

344.05. $\int \dfrac{dx}{x^4 t^5} = -\dfrac{1}{a^8} \left(-\dfrac{x^3}{3t^3} - \dfrac{3x}{t} + \dfrac{3t}{x} + \dfrac{t^3}{3x^3} \right).$

For **342** and **344**, put $z^2 = \dfrac{x^2}{t^2}$; then $dx = \dfrac{a\, dz}{(1 + z^2)^{3/2}}.$

345.01. $\int \dfrac{dx}{x^5 t} = - \left[\dfrac{t}{4a^2 x^4} + \dfrac{3}{8}\dfrac{t}{a^4 x^2} + \dfrac{3}{8a^5} \log \left| \dfrac{a+t}{x} \right| \right].$

345.03. $\int \dfrac{dx}{x^5 t^3} = - \left[\dfrac{1}{4a^2 x^4 t} + \dfrac{5}{8a^4 x^2 t} - \dfrac{15}{8a^6 t} + \dfrac{15}{8a^7} \log \left| \dfrac{a+t}{x} \right| \right].$

346.01. $\int \dfrac{dx}{x^6 t} = -\dfrac{1}{a^6} \left(\dfrac{t}{x} + \dfrac{2t^3}{3x^3} + \dfrac{t^5}{5x^5} \right).$

346.03. $\int \dfrac{dx}{x^6 t^3} = -\dfrac{1}{a^8} \left(-\dfrac{x}{t} + \dfrac{3t}{x} + \dfrac{3t^3}{3x^3} + \dfrac{t^5}{5x^5} \right).$

350.01. $\int t\, dx = \dfrac{xt}{2} + \dfrac{a^2}{2} \sin^{-1} \dfrac{x}{a}.$ [See note under **320.01.**]

350.03. $\int t^3 dx = \dfrac{xt^3}{4} + \dfrac{3}{8} a^2 xt + \dfrac{3}{8} a^4 \sin^{-1} \dfrac{x}{a}.$

350.05. $\int t^5 dx = \dfrac{xt^5}{6} + \dfrac{5}{24} a^2 xt^3 + \dfrac{5}{16} a^4 xt + \dfrac{5}{16} a^6 \sin^{-1} \dfrac{x}{a}.$

351.01. $\int xt\, dx = -\dfrac{t^3}{3}.$ **351.03.** $\int xt^3 dx = -\dfrac{t^5}{5}.$

351.9. $\int xt^{2p+1} dx = -\dfrac{t^{2p+3}}{2p+3}.$

352.01. $\int x^2 t\, dx = -\dfrac{xt^3}{4} + \dfrac{a^2 xt}{8} + \dfrac{a^4}{8} \sin^{-1} \dfrac{x}{a}.$

[See note under **320.01.**]

352.03. $\int x^2 t^3 dx = -\dfrac{xt^5}{6} + \dfrac{a^2 xt^3}{24} + \dfrac{a^4 xt}{16} + \dfrac{a^6}{16} \sin^{-1} \dfrac{x}{a}.$

353.01. $\int x^3 t\, dx = \dfrac{t^5}{5} - \dfrac{a^2 t^3}{3}.$ **353.03.** $\int x^3 t^3 dx = \dfrac{t^7}{7} - \dfrac{a^2 t^5}{5}.$

353.9. $\int x^3 t^{2p+1} dx = \dfrac{t^{2p+5}}{2p+5} - \dfrac{a^2 t^{2p+3}}{2p+3}.$

354.01. $\int x^4 t\, dx = -\dfrac{x^3 t^3}{6} - \dfrac{a^2 xt^3}{8} + \dfrac{a^4 xt}{16} + \dfrac{a^6}{16} \sin^{-1} \dfrac{x}{a}.$

[See note under **320.01.**]

354.03. $\displaystyle\int x^4 t^3 dx = -\frac{x^3 t^5}{8} - \frac{a^2 x t^5}{16} + \frac{a^4 x t^3}{64} + \frac{3}{128} a^6 x t$

$$+ \frac{3}{128} a^8 \sin^{-1} \frac{x}{a}.$$

355.01. $\displaystyle\int x^5 t\, dx = -\frac{t^7}{7} + \frac{2a^2 t^5}{5} - \frac{a^4 t^3}{3}.$

355.03. $\displaystyle\int x^5 t^3 dx = -\frac{t^9}{9} + \frac{2a^2 t^7}{7} - \frac{a^4 t^5}{5}.$

355.9. $\displaystyle\int x^5 t^{2p+1} dx = -\frac{t^{2p+7}}{2p+7} + \frac{2a^2 t^{2p+5}}{2p+5} - \frac{a^4 t^{2p+3}}{2p+3}.$

361.01. $\displaystyle\int \frac{t\, dx}{x} = t - a \log\left|\frac{a+t}{x}\right|.$ [See note under **341.01.**]

361.03. $\displaystyle\int \frac{t^3 dx}{x} = \frac{t^3}{3} + a^2 t - a^3 \log\left|\frac{a+t}{x}\right|.$

361.05. $\displaystyle\int \frac{t^5 dx}{x} = \frac{t^5}{5} + \frac{a^2 t^3}{3} + a^4 t - a^5 \log\left|\frac{a+t}{x}\right|.$

361.07. $\displaystyle\int \frac{t^7 dx}{x} = \frac{t^7}{7} + \frac{a^2 t^5}{5} + \frac{a^4 t^3}{3} + a^6 t - a^7 \log\left|\frac{a+t}{x}\right|.$

362.01. $\displaystyle\int \frac{t\, dx}{x^2} = -\frac{t}{x} - \sin^{-1} \frac{x}{a}.$ [See note under **320.01.**]

362.03. $\displaystyle\int \frac{t^3 dx}{x^2} = -\frac{t^3}{x} - \frac{3}{2} xt - \frac{3}{2} a^2 \sin^{-1} \frac{x}{a}.$

362.05. $\displaystyle\int \frac{t^5 dx}{x^2} = -\frac{t^5}{x} - \frac{5}{4} xt^3 - \frac{15}{8} a^2 xt - \frac{15}{8} a^4 \sin^{-1} \frac{x}{a}.$

363.01. $\displaystyle\int \frac{t\, dx}{x^3} = -\frac{t}{2x^2} + \frac{1}{2a} \log\left|\frac{a+t}{x}\right|.$ [See note under **341.01.**]

363.03. $\displaystyle\int \frac{t^3 dx}{x^3} = -\frac{t^3}{2x^2} - \frac{3t}{2} + \frac{3a}{2} \log\left|\frac{a+t}{x}\right|.$

363.05. $\displaystyle\int \frac{t^5 dx}{x^3} = -\frac{t^5}{2x^2} - \frac{5}{6} t^3 - \frac{5}{2} a^2 t + \frac{5}{2} a^3 \log\left|\frac{a+t}{x}\right|.$

364.01. $\displaystyle\int \frac{t\, dx}{x^4} = -\frac{t^3}{3a^2 x^3}.$

364.03. $\displaystyle\int\frac{t^3 dx}{x^4} = -\frac{t^3}{3x^3} + \frac{t}{x} + \sin^{-1}\frac{x}{a}.$ [See note under **320.01**.]

364.05. $\displaystyle\int\frac{t^5 dx}{x^4} = -\frac{a^2 t^3}{3x^3} + \frac{2a^2 t}{x} + \frac{xt}{2} + \frac{5}{2}a^2 \sin^{-1}\frac{x}{a}.$

365.01. $\displaystyle\int\frac{t\,dx}{x^5} = -\frac{t}{4x^4} + \frac{t}{8a^2 x^2} + \frac{1}{8a^3}\log\left|\frac{a+t}{x}\right|.$

365.03. $\displaystyle\int\frac{t^3 dx}{x^5} = -\frac{t^3}{4x^4} + \frac{3}{8}\frac{t^3}{a^2 x^2} + \frac{3}{8}\frac{t}{a^2} - \frac{3}{8a}\log\left|\frac{a+t}{x}\right|.$

366.01. $\displaystyle\int\frac{t\,dx}{x^6} = -\frac{t^3}{5a^2 x^3}\left(\frac{1}{x^2} + \frac{2}{3a^2}\right).$ **366.03.** $\displaystyle\int\frac{t^3 dx}{x^6} = -\frac{t^5}{5a^2 x^5}.$

367.01. $\displaystyle\int\frac{t\,dx}{x^7} = -\frac{t}{6x^6} + \frac{t}{24a^2 x^4} + \frac{t}{16a^4 x^2} + \frac{1}{16a^5}\log\left|\frac{a+t}{x}\right|.$

368.01. $\displaystyle\int\frac{t\,dx}{x^8} = -\frac{t^3}{7a^2 x^3}\left(\frac{1}{x^4} + \frac{4}{5a^2 x^2} + \frac{8}{15a^4}\right).$

Integrals of Binomial Differentials
Reduction Formulas

370. $\displaystyle\int x^m(ax^n + b)^p\,dx$

$$= \frac{1}{m+np+1}\left[x^{m+1}u^p + npb\int x^m u^{p-1}\,dx\right].$$

371. $\displaystyle\int x^m(ax^n + b)^p\,dx = \frac{1}{bn(p+1)}\left[-x^{m+1}u^{p+1}\right.$

$$\left. + (m+n+np+1)\int x^m u^{p+1}\,dx\right].$$

372. $\displaystyle\int x^m(ax^n + b)^p\,dx = \frac{1}{(m+1)b}\left[x^{m+1}u^{p+1}\right.$

$$\left. - a(m+n+np+1)\int x^{m+n}u^p\,dx\right].$$

373. $\displaystyle\int x^m(ax^n + b)^p\,dx = \frac{1}{a(m+np+1)}\left[x^{m-n+1}u^{p+1}\right.$

$$\left. - (m-n+1)b\int x^{m-n}u^p\,dx\right].$$

Here $u = ax^n + b$, and a, b, p, m, and n may be any numbers for which no denominator vanishes.

Integrals Involving $X^{1/2} = (ax^2 + bx + c)^{1/2}$

380.001. $\displaystyle\int \frac{dx}{X^{1/2}} = \frac{1}{a^{1/2}} \log |2(aX)^{1/2} + 2ax + b|$, $\qquad\qquad$ $[a > 0]$,

$\displaystyle = \frac{1}{a^{1/2}} \sinh^{-1} \frac{2ax + b}{(4ac - b^2)^{1/2}}$, $\qquad\qquad \begin{bmatrix} a > 0, \\ 4ac > b^2 \end{bmatrix}$,

$\displaystyle = \frac{1}{a^{1/2}} \log |2ax + b|$, $\quad [b^2 = 4ac, a > 0, 2ax + b > 0]$

$\displaystyle = \frac{-1}{a^{1/2}} \log |2ax + b|$, $\quad [b^2 = 4ac, a > 0, 2ax + b < 0]$,

$\displaystyle = \frac{-1}{(-a)^{1/2}} \sin^{-1} \frac{(2ax + b)}{(b^2 - 4ac)^{1/2}}$,

$\qquad\qquad\qquad \begin{bmatrix} a < 0, \quad b^2 > 4ac, \\ |2ax + b| < (b^2 - 4ac)^{1/2} \end{bmatrix}$.

The principal values of $\sin^{-1}$, between $-\pi/2$ and $\pi/2$, are to be taken.

380.003. $\displaystyle\int \frac{dx}{X^{3/2}} = \frac{4ax + 2b}{(4ac - b^2)X^{1/2}}$.

380.005. $\displaystyle\int \frac{dx}{X^{5/2}} = \frac{4ax + 2b}{3(4ac - b^2)X^{1/2}} \left(\frac{1}{X} + \frac{8a}{4ac - b^2} \right)$.

380.009. $\displaystyle\int \frac{dx}{X^{(2n+1)/2}} = \frac{4ax + 2b}{(2n - 1)(4ac - b^2)X^{(2n-1)/2}}$

$\qquad\qquad\qquad + \frac{8a(n - 1)}{(2n - 1)(4ac - b^2)} \int \frac{dx}{X^{(2n-1)/2}}$.

380.011. $\displaystyle\int \frac{xdx}{X^{1/2}} = \frac{X^{1/2}}{a} - \frac{b}{2a} \int \frac{dx}{X^{1/2}}$. $\qquad$ [See **380.001.**]

380.013. $\displaystyle\int \frac{xdx}{X^{3/2}} = -\frac{2bx + 4c}{(4ac - b^2)X^{1/2}}$.

380.019. $\displaystyle\int \frac{xdx}{X^{(2n+1)/2}} = -\frac{1}{(2n - 1)aX^{(2n-1)/2}} - \frac{b}{2a} \int \frac{dx}{X^{(2n+1)/2}}$.

380.021. $\displaystyle\int \frac{x^2dx}{X^{1/2}} = \left(\frac{x}{2a} - \frac{3b}{4a^2} \right) X^{1/2} + \frac{3b^2 - 4ac}{8a^2} \int \frac{dx}{X^{1/2}}$.

[See **380.001.**]

380.111. $\displaystyle\int \frac{dx}{xX^{1/2}} = -\frac{1}{c^{1/2}} \log \left| \frac{2(cX)^{1/2}}{x} + \frac{2c}{x} + b \right|,$ $\qquad [c > 0],$

$\displaystyle\qquad\qquad = -\frac{1}{c^{1/2}} \sinh^{-1} \frac{bx + 2c}{|x|\,(4ac - b^2)^{1/2}},$ $\quad \begin{bmatrix} c > 0, \\ 4ac > b^2 \end{bmatrix},$

$\displaystyle\qquad\qquad = -\frac{1}{c^{1/2}} \log \left| \frac{bx + 2c}{x} \right|,$

$\qquad\qquad\qquad\qquad\qquad [b^2 = 4ac,\ c > 0,\ bx{+}2c > 0],$

$\displaystyle\qquad\qquad = \frac{1}{c^{1/2}} \log \left| \frac{bx + 2c}{x} \right|,$ $\quad [b_2 = 4ac,\ c > 0,\ bx + 2c {<} 0],$

$\displaystyle\qquad\qquad = \frac{1}{(-c)^{1/2}} \sin^{-1} \frac{bx + 2c}{|x|\,(b^2 - 4ac)^{1/2}},$ $\quad \begin{bmatrix} c < 0, \\ b^2 > 4ac \end{bmatrix}.$

380.119. $\displaystyle\int \frac{dx}{xX^{(2n+1)/2}} = \frac{1}{(2n-1)cX^{(2n-1)/2}}$

$\displaystyle\qquad\qquad\qquad\qquad + \frac{1}{c}\int \frac{dx}{xX^{(2n-1)/2}} - \frac{b}{2c}\int \frac{dx}{X^{(2n+1)/2}}.$

380.121. $\displaystyle\int \frac{dx}{x^2X^{1/2}} = -\frac{X^{1/2}}{cx} - \frac{b}{2c}\int \frac{dx}{xX^{1/2}}.$ $\qquad$ [See **380.111.**]

380.201. $\displaystyle\int X^{1/2}dx = \frac{2ax + b}{4a} X^{1/2} + \frac{4ac - b^2}{8a}\int \frac{dx}{X^{1/2}}.$ [See **380.001.**]

380.209. $\displaystyle\int X^{(2n+1)/2}dx = \frac{(2ax + b)X^{(2n+1)/2}}{4a(n+1)}$

$\displaystyle\qquad\qquad\qquad\qquad + \frac{(4ac - b^2)(2n+1)}{8a(n+1)}\int X^{(2n-1)/2}dx.$

380.211. $\displaystyle\int xX^{1/2}dx = \frac{X^{3/2}}{3a} - \frac{b(2ax + b)}{8a^2} X^{1/2}$

$\displaystyle\qquad\qquad\qquad - \frac{b(4ac - b^2)}{16a^2}\int \frac{dx}{X^{1/2}}.$ $\qquad$ [See **380.001.**]

380.219. $\displaystyle\int xX^{(2n+1)/2}dx = \frac{X^{(2n+3)/2}}{(2n+3)a} - \frac{b}{2a}\int X^{(2n+1)/2}dx.$

380.311. $\displaystyle\int \frac{X^{1/2}dx}{x} = X^{1/2} + \frac{b}{2}\int \frac{dx}{X^{1/2}} + c\int \frac{dx}{xX^{1/2}}.$

$\qquad\qquad\qquad\qquad\qquad$ [See **380.001** and **380.111.**]

380.319. $\displaystyle \int \frac{X^{(2n+1)/2}dx}{x} = \frac{X^{(2n+1)/2}}{2n+1} + \frac{b}{2}\int X^{(2n-1)/2}dx + c\int \frac{X^{(2n-1)/2}dx}{x}.$

380.321. $\displaystyle \int \frac{X^{1/2}dx}{x^2} = -\frac{X^{1/2}}{x} + a\int \frac{dx}{X^{1/2}} + \frac{b}{2}\int \frac{dx}{xX^{1/2}},$

where $X = ax^2 + bx + c.$ [See **380.001** and **380.111.**]

383.1. $\displaystyle \int \frac{dx}{x(ax^2 + bx)^{1/2}} = -\frac{2}{bx}(ax^2 + bx)^{1/2}.$

383.2. $\displaystyle \int \frac{dx}{(2ax - x^2)^{1/2}} = \sin^{-1}\frac{x-a}{a}.$

383.3. $\displaystyle \int \frac{xdx}{(2ax - x^2)^{1/2}} = -(2ax - x^2)^{1/2} + a\sin^{-1}\left(\frac{x-a}{a}\right).$

383.4. $\displaystyle \int (2ax - x^2)^{1/2}dx = \frac{x-a}{2}(2ax - x^2)^{1/2} + \frac{a^2}{2}\sin^{-1}\frac{x-a}{a}.$

384.1. $\displaystyle \int \frac{dx}{x(x^n + a^2)^{1/2}} = -\frac{2}{na}\log\left|\frac{a + (x^n + a^2)^{1/2}}{x^{n/2}}\right|.$

384.2. $\displaystyle \int \frac{dx}{x(x^n - a^2)^{1/2}} = \frac{2}{na}\cos^{-1}\left|\frac{a}{x^{n/2}}\right|.$ [See note under **281.01.**]

384.3. $\displaystyle \int \frac{x^{1/2}dx}{(a^3 - x^3)^{1/2}} = \frac{2}{3}\sin^{-1}\left(\frac{x}{a}\right)^{3/2}.$

387. $\displaystyle \int \frac{dx}{(ax^2 + b)\sqrt{(fx^2 + g)}}$

$\displaystyle = \frac{1}{\sqrt{b}\sqrt{(ag - bf)}}\tan^{-1}\frac{x\sqrt{(ag - bf)}}{\sqrt{b}\sqrt{(fx^2 + g)}},$ $[ag > bf],$

$\displaystyle = \frac{1}{2\sqrt{b}\sqrt{(bf - ag)}}\log\left|\frac{\sqrt{b}\sqrt{(fx^2 + g)} + x\sqrt{(bf - ag)}}{\sqrt{b}\sqrt{(fx^2 + g)} - x\sqrt{(bf - ag)}}\right|,$

$[bf > ag].$

⁺⁺ 2 ⁺⁺

TRIGONOMETRIC
FUNCTIONS

400.01. $\sin^2 A + \cos^2 A = 1$.

400.02. $\sin A = \sqrt{(1 - \cos^2 A)}$.

400.03. $\cos A = \sqrt{(1 - \sin^2 A)}$.

400.04. $\tan A = \sin A / \cos A$.

400.05. $\operatorname{ctn} A = \cos A / \sin A = 1/\tan A$.

400.06. $\sec A = 1/\cos A$.

400.07. $\csc A = 1/\sin A$.

400.08. $\sin (- A) = - \sin A$.

400.09. $\cos (- A) = \cos A$.

400.10. $\tan (- A) = - \tan A$.

400.11. $\sec^2 A - \tan^2 A = 1$.

400.12. $\sec A = \sqrt{(1 + \tan^2 A)}$.

400.13. $\tan A = \sqrt{(\sec^2 A - 1)}$.

400.14. $\csc^2 A - \operatorname{ctn}^2 A = 1$.

400.15. $\csc A = \sqrt{(1 + \operatorname{ctn}^2 A)}$.

400.16. $\operatorname{ctn} A = \sqrt{(\csc^2 A - 1)}$.

400.17. $\operatorname{vers} A = 1 - \cos A$.

Note that for real values of A the sign of the above radicals depends on the quadrant in which the angle A lies.

78

401.01. $\sin (A + B) = \sin A \cos B + \cos A \sin B.$

401.02. $\sin (A - B) = \sin A \cos B - \cos A \sin B.$

401.03. $\cos (A + B) = \cos A \cos B - \sin A \sin B.$

401.04. $\cos (A - B) = \cos A \cos B + \sin A \sin B.$

401.05. $2 \sin A \cos B = \sin (A + B) + \sin (A - B).$

401.06. $2 \cos A \cos B = \cos (A + B) + \cos (A - B).$

401.07. $2 \sin A \sin B = \cos (A - B) - \cos (A + B).$

401.08. $\sin A + \sin B = 2 \sin \frac{1}{2}(A + B) \cos \frac{1}{2}(A - B).$

401.09. $\sin A - \sin B = 2 \sin \frac{1}{2}(A - B) \cos \frac{1}{2}(A + B).$

401.10. $\cos A + \cos B = 2 \cos \frac{1}{2}(A + B) \cos \frac{1}{2}(A - B).$

401.11. $\cos A - \cos B = 2 \sin \frac{1}{2}(A + B) \sin \frac{1}{2}(B - A).$

401.12. $\sin^2 A - \sin^2 B = \sin (A + B) \sin (A - B).$

401.13. $\cos^2 A - \cos^2 B = \sin (A + B) \sin (B - A).$

401.14. $\cos^2 A - \sin^2 B = \cos (A + B) \cos (A - B)$
$$= \cos^2 B - \sin^2 A.$$

401.15. $\sec^2 A + \csc^2 A = \sec^2 A \csc^2 \mathrm{A} = \dfrac{1}{\sin^2 A \cos^2 A}.$

401.2. $p \cos A + q \sin A = r \sin (A + \theta),$
where
$$r = \sqrt{(p^2 + q^2)}, \quad \sin \theta = p/r, \quad \cos \theta = q/r$$
or
$$p \cos A + q \sin A = r \cos (A - \phi),$$
where
$$r = \sqrt{(p^2 + q^2)}, \quad \cos \phi = p/r, \quad \sin \phi = q/r.$$

Note that p and q may be positive or negative.

402.01. $\sin (A + B + C)$
$$= \sin A \cos B \cos C + \cos A \sin B \cos C$$
$$+ \cos A \cos B \sin C - \sin A \sin B \sin C.$$

402.02. $\cos (A + B + C)$
$$= \cos A \cos B \cos C - \sin A \sin B \cos C$$
$$- \sin A \cos B \sin C - \cos A \sin B \sin C.$$

402.03. $4 \sin A \sin B \sin C$
$$= \sin (A + B - C) + \sin (B + C - A)$$
$$+ \sin (C + A - B) - \sin (A + B + C).$$

402.04. $4 \sin A \cos B \cos C$
$$= \sin (A + B - C) - \sin (B + C - A)$$
$$+ \sin (C + A - B) + \sin (A + B + C).$$

402.05. $4 \sin A \sin B \cos C$
$$= - \cos (A + B - C) + \cos (B + C - A)$$
$$+ \cos (C + A - B) - \cos (A + B + C).$$

402.06. $4 \cos A \cos B \cos C$
$$= \cos (A + B - C) + \cos (B + C - A)$$
$$+ \cos (C + A - B) + \cos (A + B + C).$$

403.02. $\sin 2A = 2 \sin A \cos A = \dfrac{2 \tan A}{1 + \tan^2 A}.$

403.03. $\sin 3A = 3 \sin A - 4 \sin^3 A.$

403.04. $\sin 4A = \cos A(4 \sin A - 8 \sin^3 A).$

403.05. $\sin 5A = 5 \sin A - 20 \sin^3 A + 16 \sin^5 A.$

403.06. $\sin 6A = \cos A(6 \sin A - 32 \sin^3 A + 32 \sin^5 A).$

403.07. $\sin 7A = 7 \sin A - 56 \sin^3 A + 112 \sin^5 A - 64 \sin^7 A.$

403.10. When n is an even, positive integer,

$$\sin nA = (- 1)^{(n/2)+1} \cos A \left[2^{n-1} \sin^{n-1} A - \frac{(n - 2)}{1!} 2^{n-3} \sin^{n-3} A \right.$$

$$+ \frac{(n - 3)(n - 4)}{2!} 2^{n-5} \sin^{n-5} A$$

$$\left. - \frac{(n - 4)(n - 5)(n - 6)}{3!} 2^{n-7} \sin^{n-7} A + \cdots \right],$$

the series terminating where a coefficient $= 0$.

403.11. An alternative series, giving the same results for numerical values of n, is

$$\sin nA = n \cos A \left[\sin A - \frac{(n^2 - 2^2)}{3!} \sin^3 A \right.$$

$$+ \frac{(n^2 - 2^2)(n^2 - 4^2)}{5!} \sin^5 A$$

$$\left. - \frac{(n^2 - 2^2)(n^2 - 4^2)(n^2 - 6^2)}{7!} \sin^7 A + \cdots \right],$$

$$[n \text{ even and} > 0]. \quad [\text{Ref. 34, p. 181.}]$$

403.12. When n is an odd integer > 1,

$$\sin nA = (-1)^{(n-1)/2} \left[2^{n-1} \sin^n A - \frac{n}{1!} 2^{n-3} \sin^{n-2} A \right.$$

$$+ \frac{n(n-3)}{2!} 2^{n-5} \sin^{n-4} A - \frac{n(n-4)(n-5)}{3!} 2^{n-7} \sin^{n-6} A$$

$$\left. + \frac{n(n-5)(n-6)(n-7)}{4!} 2^{n-9} \sin^{n-8} A - \cdots \right],$$

the series terminating where a coefficient $= 0$.

403.13. An alternative series is

$$\sin nA = n \sin A - \frac{n(n^2 - 1^2)}{3!} \sin^3 A$$

$$+ \frac{n(n^2 - 1^2)(n^2 - 3^2)}{5!} \sin^5 A - \cdots,$$

$$[n \text{ odd and} > 0]. \quad [\text{Ref. 34, p. 180.}]$$

403.22. $\cos 2A = \cos^2 A - \sin^2 A = 2 \cos^2 A - 1 = 1 - 2 \sin^2 A$

$$= \frac{1 - \tan^2 A}{1 + \tan^2 A} = \frac{\operatorname{ctn} A - \tan A}{\operatorname{ctn} A + \tan A}.$$

403.23. $\cos 3A = 4 \cos^3 A - 3 \cos A.$

403.24. $\cos 4A = 8 \cos^4 A - 8 \cos^2 A + 1.$

403.25. $\cos 5A = 16 \cos^5 A - 20 \cos^3 A + 5 \cos A.$

403.26. $\cos 6A = 32 \cos^6 A - 48 \cos^4 A + 18 \cos^2 A - 1.$

403.27. $\cos 7A = 64 \cos^7 A - 112 \cos^5 A + 56 \cos^3 A - 7 \cos A.$

403.3. $\cos nA = 2^{n-1} \cos^n A - \dfrac{n}{1!} 2^{n-3} \cos^{n-2} A$

$+ \dfrac{n(n-3)}{2!} 2^{n-5} \cos^{n-4} A - \dfrac{n(n-4)(n-5)}{3!} 2^{n-7} \cos^{n-6} A$

$+ \dfrac{n(n-5)(n-6)(n-7)}{4!} 2^{n-9} \cos^{n-8} A - \cdots ,$

terminating where a coefficient $= 0$,   [n an integer > 2].

[Ref. 4, pp. 409, 416, and 417, and Ref. 34, p. 177.]

403.4. $\sin \tfrac{1}{2}A = \sqrt{\{\tfrac{1}{2}(1 - \cos A)\}}.$

403.5. $\cos \tfrac{1}{2}A = \sqrt{\{\tfrac{1}{2}(1 + \cos A)\}}.$

404.12. $\sin^2 A = \tfrac{1}{2}(-\cos 2A + 1).$

404.13. $\sin^3 A = \tfrac{1}{4}(-\sin 3A + 3 \sin A).$

404.14. $\sin^4 A = \tfrac{1}{8}(\cos 4A - 4 \cos 2A + \tfrac{6}{2}).$

404.15. $\sin^5 A = \tfrac{1}{16}(\sin 5A - 5 \sin 3A + 10 \sin A).$

404.16. $\sin^6 A = \tfrac{1}{32}(-\cos 6A + 6 \cos 4A - 15 \cos 2A + \tfrac{20}{2}).$

404.17. $\sin^7 A = \tfrac{1}{64}(-\sin 7A + 7 \sin 5A - 21 \sin 3A + 35 \sin A).$

404.22. $\cos^2 A = \tfrac{1}{2}(\cos 2A + 1).$

404.23. $\cos^3 A = \tfrac{1}{4}(\cos 3A + 3 \cos A).$

404.24. $\cos^4 A = \tfrac{1}{8}(\cos 4A + 4 \cos 2A + \tfrac{6}{2}).$

404.25. $\cos^5 A = \tfrac{1}{16}(\cos 5A + 5 \cos 3A + 10 \cos A).$

404.26. $\cos^6 A = \tfrac{1}{32}(\cos 6A + 6 \cos 4A + 15 \cos 2A + \tfrac{20}{2}).$

404.27. $\cos^7 A = \tfrac{1}{64}(\cos 7A + 7 \cos 5A + 21 \cos 3A + 35 \cos A).$

[No. **404** can be extended by inspection by using binomial coefficients.]

405.01. $\tan(A + B) = \dfrac{\tan A + \tan B}{1 - \tan A \tan B} = \dfrac{\operatorname{ctn} A + \operatorname{ctn} B}{\operatorname{ctn} A \operatorname{ctn} B - 1}.$

405.02. $\tan(A - B) = \dfrac{\tan A - \tan B}{1 + \tan A \tan B} = \dfrac{\operatorname{ctn} B - \operatorname{ctn} A}{\operatorname{ctn} A \operatorname{ctn} B + 1}.$

405.03. $\operatorname{ctn}(A + B) = \dfrac{\operatorname{ctn} A \operatorname{ctn} B - 1}{\operatorname{ctn} A + \operatorname{ctn} B} = \dfrac{1 - \tan A \tan B}{\tan A + \tan B}.$

405.04. $\operatorname{ctn}(A - B) = \dfrac{\operatorname{ctn} A \operatorname{ctn} B + 1}{\operatorname{ctn} B - \operatorname{ctn} A} = \dfrac{1 + \tan A \tan B}{\tan A - \tan B}.$

405.05. $\tan A + \tan B = \dfrac{\sin(A + B)}{\cos A \cos B}.$

405.06. $\tan A - \tan B = \dfrac{\sin(A - B)}{\cos A \cos B}.$

405.07. $\operatorname{ctn} A + \operatorname{ctn} B = \dfrac{\sin(A + B)}{\sin A \sin B}.$

405.08. $\operatorname{ctn} A - \operatorname{ctn} B = \dfrac{\sin(B - A)}{\sin A \sin B}.$

405.09. $\tan A + \operatorname{ctn} B = \dfrac{\cos(A - B)}{\cos A \sin B}.$

405.10. $\operatorname{ctn} A - \tan B = \dfrac{\cos(A + B)}{\sin A \cos B}.$

406.02. $\tan 2A = \dfrac{2 \tan A}{1 - \tan^2 A} = \dfrac{2 \operatorname{ctn} A}{\operatorname{ctn}^2 A - 1} = \dfrac{2}{\operatorname{ctn} A - \tan A}.$

406.03. $\tan 3A = \dfrac{3 \tan A - \tan^3 A}{1 - 3 \tan^2 A}.$

406.04. $\tan 4A = \dfrac{4 \tan A - 4 \tan^3 A}{1 - 6 \tan^2 A + \tan^4 A}.$

406.12. $\operatorname{ctn} 2A = \dfrac{\operatorname{ctn}^2 A - 1}{2 \operatorname{ctn} A} = \dfrac{1 - \tan^2 A}{2 \tan A} = \dfrac{\operatorname{ctn} A - \tan A}{2}.$

406.13. $\operatorname{ctn} 3A = \dfrac{\operatorname{ctn}^3 A - 3 \operatorname{ctn} A}{3 \operatorname{ctn}^2 A - 1}.$

406.14. $\operatorname{ctn} 4A = \dfrac{\operatorname{ctn}^4 A - 6 \operatorname{ctn}^2 A + 1}{4 \operatorname{ctn}^3 A - 4 \operatorname{ctn} A}.$

406.2. $\tan \tfrac{1}{2}A = \dfrac{1 - \cos A}{\sin A} = \dfrac{\sin A}{1 + \cos A} = \sqrt{\left(\dfrac{1 - \cos A}{1 + \cos A}\right)}.$

406.3. $\operatorname{ctn} \tfrac{1}{2}A = \dfrac{\sin A}{1 - \cos A} = \dfrac{1 + \cos A}{\sin A} = \sqrt{\left(\dfrac{1 + \cos A}{1 - \cos A}\right)}.$

407. $\sin 0°$ $= 0$ $= \cos 90°.$

$\sin 15° = \sin \dfrac{\pi}{12} = \dfrac{\sqrt{3}-1}{2\sqrt{2}}$ $= \cos 75°.$

$\sin 18° = \sin \dfrac{\pi}{10} = \dfrac{\sqrt{5}-1}{4}$ $= \cos 72°.$

$\sin 30° = \sin \dfrac{\pi}{6} = \dfrac{1}{2}$ $= \cos 60°.$

$\sin 36° = \sin \dfrac{\pi}{5} = \dfrac{\sqrt{(5-\sqrt{5})}}{2\sqrt{2}} = \cos 54°.$

$\sin 45° = \sin \dfrac{\pi}{4} = \dfrac{1}{\sqrt{2}}$ $= \cos 45°.$

$\sin 54° = \sin \dfrac{3\pi}{10} = \dfrac{\sqrt{5}+1}{4}$ $= \cos 36°.$

$\sin 60° = \sin \dfrac{\pi}{3} = \dfrac{\sqrt{3}}{2}$ $= \cos 30°.$

$\sin 72° = \sin \dfrac{2\pi}{5} = \dfrac{\sqrt{(5+\sqrt{5})}}{2\sqrt{2}} = \cos 18°.$

$\sin 75° = \sin \dfrac{5\pi}{12} = \dfrac{\sqrt{3}+1}{2\sqrt{2}}$ $= \cos 15°.$

$\sin 90° = \sin \dfrac{\pi}{2} = 1$ $= \cos 0.$

[Ref. 4, pp. 406–407.]

$\sin 120° = \sin \dfrac{2\pi}{3} = \dfrac{\sqrt{3}}{2}.$ $\qquad$ $\sin 240° = \sin \dfrac{4\pi}{3} = -\dfrac{\sqrt{3}}{2}.$

$\cos 120° = \cos \dfrac{2\pi}{3} = -\dfrac{1}{2}.$ $\qquad$ $\cos 240° = \cos \dfrac{4\pi}{3} = -\dfrac{1}{2}.$

$\sin 180° = \sin \pi = 0.$ $\qquad$ $\sin 270° = \sin \dfrac{3\pi}{2} = -1.$

$\cos 180° = \cos \pi = -1.$ $\qquad$ $\cos 270° = \cos \dfrac{3\pi}{2} = 0.$

408.01. $\sin x = \dfrac{1}{2i}(e^{ix} - e^{-ix})$, where $i = +\sqrt{(-1)}$.

Note that in electrical work the letter j is often used instead of i.

408.02. $\cos x = \dfrac{1}{2}(e^{ix} + e^{-ix}).$

408.03. $\tan x = -i\left(\dfrac{e^{ix} - e^{-ix}}{e^{ix} + e^{-ix}}\right) = -i\left(\dfrac{e^{2ix} - 1}{e^{2ix} + 1}\right).$

408.04. $e^{ix} = \cos x + i \sin x,$ [EULER'S FORMULA].

408.05. $e^{z+ix} = e^z(\cos x + i \sin x).$

408.06. $a^{z+ix} = a^z[\cos(x \log a) + i \sin(x \log a)].$

408.07. $(\cos x + i \sin x)^n = e^{inx} = \cos nx + i \sin nx,$

                 [DE MOIVRE'S FORMULA].

408.08. $(\cos x + i \sin x)^{-n} = \cos nx - i \sin nx.$

408.09. $(\cos x + i \sin x)^{-1} = \cos x - i \sin x.$

408.10. $\sin(ix) = i \sinh x.$ **408.13.** $\operatorname{ctn}(ix) = -i \operatorname{ctnh} x.$

408.11. $\cos(ix) = \cosh x.$ **408.14.** $\sec(ix) = \operatorname{sech} x.$

408.12. $\tan(ix) = i \tanh x.$ **408.15.** $\csc(ix) = -i \operatorname{csch} x.$

408.16. $\sin(x \pm iy) = \sin x \cosh y \pm i \cos x \sinh y.$

408.17. $\cos(x \pm iy) = \cos x \cosh y \mp i \sin x \sinh y.$

408.18. $\tan(x \pm iy) = \dfrac{\sin 2x \pm i \sinh 2y}{\cos 2x + \cosh 2y}.$

408.19. $\operatorname{ctn}(x \pm iy) = \dfrac{\sin 2x \mp i \sinh 2y}{\cosh 2y - \cos 2x}.$

409.01. $ce^{ix} = ce^{i(x+2k\pi)},$ where k is an integer or 0,

         $= c(\cos x + i \sin x) = c\underline{/x}.$ [Ref. 37, p. 51.]

409.02. $1 = e^{0+2k\pi i} = \cos 0 + i \sin 0.$ Note that

         $\cos 2k\pi = \cos 2\pi = \cos 0 = 1.$

409.03. $-1 = e^{0+(2k+1)\pi i} = \cos \pi + i \sin \pi.$ Note that

         $\log(-1) = (2k+1)\pi i.$

409.04. $\sqrt{1} = e^{2k\pi i/2}.$ This has two different values, depending on whether k is even or odd. They are, respectively,

$e^{2r\pi i} = \cos 0 + i \sin 0 = 1,$ $e^{(2r+1)\pi i} = \cos \pi + i \sin \pi = -1,$

where r is an integer or 0.

409.05. $\sqrt{(-1)} = e^{(2r+1)\pi i/2}$. This square root has two different values, depending on whether r is even or odd; they are, respectively,

$$\cos\frac{\pi}{2} + i\sin\frac{\pi}{2} = i, \qquad \cos\frac{3\pi}{2} + i\sin\frac{3\pi}{2} = -i.$$

409.06. $\sqrt[3]{1} = e^{2k\pi i/3}$. This has three different values:

$$e^{2r\pi i} = \cos 0 + i\sin 0 = 1,$$

$$e^{(2r\pi + 2\pi/3)i} = \cos\frac{2\pi}{3} + i\sin\frac{2\pi}{3} = -\frac{1}{2} + i\frac{\sqrt{3}}{2} = \omega,$$

$$e^{(2r\pi + 4\pi/3)i} = \cos\frac{4\pi}{3} + i\sin\frac{4\pi}{3} = -\frac{1}{2} - i\frac{\sqrt{3}}{2} = \omega^2.$$

409.07. $\sqrt[4]{1} = e^{2k\pi i/4}$. This has four different values:

$$e^{2r\pi i} = \cos 0 + i\sin 0 = 1,$$

$$e^{(2r\pi + 2\pi/4)i} = \cos\frac{\pi}{2} + i\sin\frac{\pi}{2} = i,$$

$$e^{(2r\pi + 4\pi/4)i} = \cos\pi + i\sin\pi = -1,$$

$$e^{(2r\pi + 6\pi/4)i} = \cos\frac{3\pi}{2} + i\sin\frac{3\pi}{2} = -i. \quad \text{[See \textbf{409.04} and .05.]}$$

409.08. $\sqrt{i} = e^{(4s+1)\pi i/4}$, from **409.05**, putting $r = 2s$. This has 2 values:

$$e^{\pi i/4} = \cos\frac{\pi}{4} + i\sin\frac{\pi}{4} = \frac{1}{\sqrt{2}} + \frac{i}{\sqrt{2}}, \qquad (s \text{ even}),$$

$$e^{5\pi i/4} = \cos\frac{5\pi}{4} + i\sin\frac{5\pi}{4} = -\left(\frac{1}{\sqrt{2}} + \frac{i}{\sqrt{2}}\right), \qquad (s \text{ odd}).$$

409.09. $\sqrt[n]{1} = e^{2k\pi i/n} = \cos\frac{2k\pi}{n} + i\sin\frac{2k\pi}{n}.$

There are n different values, corresponding to different values of k. The equation $\omega^n = 1$ has n different roots:

$$\omega_0 = \cos 0 + i\sin 0 = 1, \qquad \omega_1 = \cos\frac{2\pi}{n} + i\sin\frac{2\pi}{n},$$

$$\omega_2 = \cos 2\left(\frac{2\pi}{n}\right) + i\sin 2\left(\frac{2\pi}{n}\right), \qquad \cdots \omega_k = \cos k\frac{2\pi}{n} + i\sin k\frac{2\pi}{n},$$

$$\omega_{n-1} = \cos(n-1)\frac{2\pi}{n} + i\sin(n-1)\frac{2\pi}{n}.$$

Note that, by **408.07**,

$$\omega_2 = \omega_1^2, \qquad \omega_3 = \omega_1^3, \qquad \omega_k = \omega_1^k, \qquad \omega_0 = \omega_1^n.$$

409.10. All the nth roots of a quantity may be obtained from any root by multiplying this root by the n roots of unity given in **409.09.**

[Ref. 10, pp. 21–22.]

410. Formulas for Plane Triangles. Let a, b, and c be the sides opposite the angles A, B, and C.

410.01. $a^2 = b^2 + c^2 - 2bc \cos A.$

410.02. $\dfrac{a}{\sin A} = \dfrac{b}{\sin B} = \dfrac{c}{\sin C}.$

410.03. $a = b \cos C + c \cos B.$

410.04. $A + B + C = \pi$ radians $= 180°.$

410.05. $\sin \dfrac{A}{2} = \sqrt{\left(\dfrac{(s - b)(s - c)}{bc} \right)}$, where $s = \frac{1}{2}(a + b + c).$

410.06. $\cos \dfrac{A}{2} = \sqrt{\left(\dfrac{s(s - a)}{bc} \right)}.$

410.07. $\tan \dfrac{A}{2} = \sqrt{\left(\dfrac{(s - b)(s - c)}{s(s - a)} \right)}.$

410.08. $\tan \dfrac{A - B}{2} = \dfrac{a - b}{a + b} \operatorname{ctn} \dfrac{C}{2}.$

410.09. To find c from a, b and C, when using logarithmic trigonometric tables, let

$$\tan \theta = \frac{a + b}{a - b} \tan \frac{C}{2}; \quad \text{then} \quad c = (a - b) \cos \frac{C}{2} \sec \theta.$$

410.10. The area of a triangle is

$$\frac{1}{2} ab \sin C = \sqrt{\{s(s - a)(s - b)(s - c)\}} = \frac{a^2}{2} \frac{\sin B \sin C}{\sin A}.$$

410.11. If $C = 90°$, $c^2 = a^2 + b^2$. To find $c \equiv \sqrt{(a^2 + b^2)}$ when using logarithmic tables, let $\tan \theta = b/a$; then $c = a \sec \theta$.

This is useful also in other types of work. See also Table 1000.

410.12. In a plane triangle,

$$\log a = \log b - \left(\frac{c}{b} \cos A + \frac{c^2}{2b^2} \cos 2A + \cdots \right.$$

$$\left. + \frac{c^n}{nb^n} \cos nA + \cdots \right), \qquad [c < b],$$

$$= \log c - \left(\frac{b}{c} \cos A + \frac{b^2}{2c^2} \cos 2A + \cdots \right.$$

$$\left. + \frac{b^n}{nc^n} \cos nA + \cdots \right), \qquad [b < c.]$$

[See **418.**]

Trigonometric Series

415.01. $\sin x = x - \dfrac{x^3}{3!} + \dfrac{x^5}{5!} - \dfrac{x^7}{7!} + \cdots,$ $[x^2 < \infty].$

415.02. $\cos x = 1 - \dfrac{x^2}{2!} + \dfrac{x^4}{4!} - \dfrac{x^6}{6!} + \cdots,$ $[x^2 < \infty].$

415.03. $\tan x = x + \dfrac{x^3}{3} + \dfrac{2}{15} x^5 + \dfrac{17}{315} x^7 + \dfrac{62}{2835} x^9 + \cdots$

$$\cdots + \frac{2^{2n}(2^{2n} - 1)B_n}{(2n)!} x^{2n-1} + \cdots, \qquad \left[x^2 < \frac{\pi^2}{4}\right].$$

[See **45.**]

415.04. $\operatorname{ctn} x = \dfrac{1}{x} - \dfrac{x}{3} - \dfrac{x^3}{45} - \dfrac{2x^5}{945} - \dfrac{x^7}{4725} - \cdots$

$$\cdots - \frac{2^{2n}B_n}{(2n)!} x^{2n-1} - \cdots, \qquad [x^2 < \pi^2].$$

[See **45.**]

415.05. $\sec x = 1 + \dfrac{x^2}{2} + \dfrac{5}{24} x^4 + \dfrac{61}{720} x^6 + \dfrac{277}{8064} x^8 + \cdots$

$$\cdots + \frac{E_n x^{2n}}{(2n)!} + \cdots, \qquad \left[x^2 < \frac{\pi^2}{4}\right].$$

[See **45.**]

415.06. $\csc x = \dfrac{1}{x} + \dfrac{x}{6} + \dfrac{7}{360} x^3 + \dfrac{31}{15,120} x^5 + \dfrac{127}{604,800} x^7 + \cdots$

$$\cdots + \frac{2(2^{2n-1} - 1)}{(2n)!} B_n x^{2n-1} + \cdots, \qquad [x^2 < \pi^2].$$

[See **45**.]

415.07. $\sin(\theta + x) = \sin\theta + x\cos\theta - \dfrac{x^2 \sin\theta}{2!}$

$$- \frac{x^3 \cos\theta}{3!} + \frac{x^4 \sin\theta}{4!} + \cdots.$$

415.08. $\cos(\theta + x) = \cos\theta - x\sin\theta - \dfrac{x^2 \cos\theta}{2!}$

$$+ \frac{x^3 \sin\theta}{3!} + \frac{x^4 \cos\theta}{4!} - \cdots.$$

416.01. $\dfrac{\pi}{4} = \sin x + \dfrac{\sin 3x}{3} + \dfrac{\sin 5x}{5} + \dfrac{\sin 7x}{7} + \cdots,$

$$[0 < x < \pi, \text{ exclusive}].$$

416.02. c, a constant, $= \dfrac{4c}{\pi}\left(\sin x + \dfrac{\sin 3x}{3} + \dfrac{\sin 5x}{5}\right.$

$$\left. + \frac{\sin 7x}{7} + \cdots \right), \qquad [0 < x < \pi, \text{ exclusive}].$$

416.03. $c = \dfrac{4c}{\pi}\left(\sin \dfrac{\pi x}{a} + \dfrac{1}{3}\sin\dfrac{3\pi x}{a} + \dfrac{1}{5}\sin\dfrac{5\pi x}{a}\right.$

$$\left. + \frac{1}{7}\sin\frac{7\pi x}{a} + \cdots \right), \qquad [0 < x < a, \text{ exclusive}].$$

416.04. $\dfrac{\pi}{4} = \cos x - \dfrac{\cos 3x}{3} + \dfrac{\cos 5x}{5} - \dfrac{\cos 7x}{7} + \cdots,$

$$\left[-\frac{\pi}{2} < x < \frac{\pi}{2}, \text{ exclusive} \right].$$

416.05. c, a constant, $= \dfrac{4c}{\pi}\left(\cos x - \dfrac{\cos 3x}{3} + \dfrac{\cos 5x}{5}\right.$

$$\left. - \frac{\cos 7x}{7} + \cdots \right), \qquad \left[-\frac{\pi}{2} < x < \frac{\pi}{2}, \text{ exclusive} \right].$$

416.06. $c = \dfrac{4c}{\pi}\left(\cos\dfrac{\pi x}{a} - \dfrac{1}{3}\cos\dfrac{3\pi x}{a} + \dfrac{1}{5}\cos\dfrac{5\pi x}{a}\right.$

$\left. - \dfrac{1}{7}\cos\dfrac{7\pi x}{a} + \cdots\right), \qquad \left[-\dfrac{a}{2} < x < \dfrac{a}{2}, \text{exclusive}\right].$

416.07. $x = 2\left(\sin x - \dfrac{\sin 2x}{2} + \dfrac{\sin 3x}{3} - \dfrac{\sin 4x}{4} + \cdots\right),$

$[-\pi < x < \pi, \text{exclusive}].$

416.08. $x = \pi - 2\left(\sin x + \dfrac{\sin 2x}{2} + \dfrac{\sin 3x}{3} + \dfrac{\sin 4x}{4} + \cdots\right),$

$[0 < x < 2\pi, \text{exclusive}].$

416.09. $x = \dfrac{4}{\pi}\left(\sin x - \dfrac{\sin 3x}{3^2} + \dfrac{\sin 5x}{5^2} - \dfrac{\sin 7x}{7^2} + \cdots\right),$

$\left[-\dfrac{\pi}{2} < x < \dfrac{\pi}{2}, \text{inclusive}\right].$

416.10. $x = \dfrac{\pi}{2} - \dfrac{4}{\pi}\left(\cos x + \dfrac{\cos 3x}{3^2} + \dfrac{\cos 5x}{5^2} + \dfrac{\cos 7x}{7^2} + \cdots\right),$

$[0 < x < \pi, \text{inclusive}].$

416.11. $x^2 = \dfrac{\pi^2}{3} - 4\left(\cos x - \dfrac{\cos 2x}{2^2} + \dfrac{\cos 3x}{3^2} - \dfrac{\cos 4x}{4^2} + \cdots\right),$

$[-\pi < x < \pi, \text{inclusive}].$

416.12. $x^2 = \dfrac{\pi^2}{4} - \dfrac{8}{\pi}\left(\cos x - \dfrac{\cos 3x}{3^3} + \dfrac{\cos 5x}{5^3} - \dfrac{\cos 7x}{7^3} + \cdots\right).$

416.13. $x^3 - \pi^2 x = -12\left(\sin x - \dfrac{\sin 2x}{2^3} + \dfrac{\sin 3x}{3^3}\right.$

$\left. - \dfrac{\sin 4x}{4^3} + \cdots\right).$

416.14. $\sin x = \dfrac{4}{\pi}\left(\dfrac{1}{2} - \dfrac{\cos 2x}{1\cdot 3} - \dfrac{\cos 4x}{3\cdot 5} - \dfrac{\cos 6x}{5\cdot 7} - \cdots\right).$

416.15. $\cos x = \dfrac{8}{\pi}\left\{\dfrac{\sin 2x}{1\cdot 3} + \dfrac{2}{3\cdot 5}\sin 4x + \dfrac{3}{5\cdot 7}\sin 6x + \cdots\right.$

$\left. \cdots + \dfrac{n}{(2n-1)(2n+1)}\sin 2nx + \cdots\right\},$

$[0 < x < \pi, \text{exclusive}].$

416.16. $\sin ax = \dfrac{2 \sin a\pi}{\pi} \left\{ \dfrac{\sin x}{1^2 - a^2} - \dfrac{2 \sin 2x}{2^2 - a^2} + \dfrac{3 \sin 3x}{3^2 - a^2} - \cdots \right\}$

where a is not an integer, $[0 < x$, inclusive; $x < \pi$, exclusive$]$.

416.17. $\cos ax = \dfrac{2a \sin a\pi}{\pi} \left\{ \dfrac{1}{2a^2} + \dfrac{\cos x}{1^2 - a^2} - \dfrac{\cos 2x}{2^2 - a^2} \right.$

$$\left. + \dfrac{\cos 3x}{3^2 - a^2} + \cdots \right\}, \qquad [0 < x < \pi, \text{ inclusive}],$$

where a is not an integer.   [Ref. 7, pp. 301–309.]

417.1. $\dfrac{1}{1 - 2a \cos \theta + a^2} = 1 + \dfrac{1}{\sin \theta} (a \sin 2\theta + a^2 \sin 3\theta$

$$+ a^3 \sin 4\theta + \cdots), \qquad [a^2 < 1].$$
$$[\text{Ref. 29, p. 87.}]$$

417.2. $\dfrac{1 - a^2}{1 - 2a \cos \theta + a^2} = 1 + 2(a \cos \theta + a^2 \cos 2\theta$

$$+ a^3 \cos 3\theta + \cdots), \qquad [a^2 < 1].$$

417.3. $\dfrac{1 - a \cos \theta}{1 - 2a \cos \theta + a^2} = 1 + a \cos \theta + a^2 \cos 2\theta$

$$+ a^3 \cos 3\theta + \cdots, \qquad [a^2 < 1].$$

417.4. $\dfrac{\sin \theta}{1 - 2a \cos \theta + a^2} = \sin \theta + a \sin 2\theta + a^2 \sin 3\theta + \cdots,$

$$[a^2 < 1].$$

418. $\log (1 - 2a \cos \theta + a^2)$

$$= -2 \left(a \cos \theta + \dfrac{a^2}{2} \cos 2\theta + \dfrac{a^3}{3} \cos 3\theta + \cdots \right),$$
$$[a^2 < 1],$$
$$= 2 \log |a| - 2 \left(\dfrac{\cos \theta}{a} + \dfrac{\cos 2\theta}{2a^2} + \dfrac{\cos 3\theta}{3a^3} + \cdots \right),$$
$$[a^2 > 1]. \qquad [\text{Ref. 7, Art. 292.}]$$

419.1. $e^{ax} \sin bx = \dfrac{rx \sin \theta}{1!} + \dfrac{r^2 x^2 \sin 2\theta}{2!} + \dfrac{r^3 x^3 \sin 3\theta}{3!} + \cdots,$

where $r = \sqrt{(a^2 + b^2)}, \quad a = r \cos \theta, \quad b = r \sin \theta.$

419.2. $e^{ax} \cos bx = 1 + \dfrac{rx \cos \theta}{1!} + \dfrac{r^2x^2 \cos 2\theta}{2!}$

$$+ \frac{r^3x^3 \cos 3\theta}{3!} + \cdots,$$

where r and θ are as in **419.1.**

420.1. $\sin \alpha + \sin 2\alpha + \sin 3\alpha + \cdots + \sin n\alpha$

$$= \frac{\sin \dfrac{n+1}{2} \alpha \sin \dfrac{n\alpha}{2}}{\sin \dfrac{\alpha}{2}}.$$

420.2. $\cos \alpha + \cos 2\alpha + \cos 3\alpha + \cdots + \cos n\alpha$

$$= \frac{\cos \dfrac{n+1}{2} \alpha \sin \dfrac{n\alpha}{2}}{\sin \dfrac{\alpha}{2}}.$$

420.3. $\sin \alpha + \sin (\alpha + \delta) + \sin (\alpha + 2\delta) + \cdots$

$$+ \sin \{\alpha + (n-1)\delta\} = \frac{\sin \left(\alpha + \dfrac{n-1}{2} \delta\right) \sin \dfrac{n\delta}{2}}{\sin \dfrac{\delta}{2}}.$$

420.4. $\cos \alpha + \cos (\alpha + \delta) + \cos (\alpha + 2\delta) + \cdots$

$$+ \cos \{\alpha + (n-1)\delta\} = \frac{\cos \left(\alpha + \dfrac{n-1}{2} \delta\right) \sin \dfrac{n\delta}{2}}{\sin \dfrac{\delta}{2}}.$$

[Ref. 29, Chap. V.]

421.   If $\sin \theta = x \sin (\theta + \alpha)$,

$\theta + r\pi = x \sin \alpha + \tfrac{1}{2}x^2 \sin 2\alpha + \tfrac{1}{3}x^3 \sin 3\alpha + \cdots,$ $[x^2 < 1]$,

where r is an integer.   [Ref. 29, Art. 78.]

422.1. $\sin \theta = \theta \left(1 - \dfrac{\theta^2}{\pi^2}\right)\left(1 - \dfrac{\theta^2}{2^2\pi^2}\right)\left(1 - \dfrac{\theta^2}{3^2\pi^2}\right) \cdots,$ $[\theta^2 < \infty]$.

422.2. $\cos \theta = \left(1 - \dfrac{4\theta^2}{\pi^2}\right)\left(1 - \dfrac{4\theta^2}{3^2\pi^2}\right)\left(1 - \dfrac{4\theta^2}{5^2\pi^2}\right) \cdots,$ $[\theta^2 < \infty]$.

TRIGONOMETRIC FUNCTIONS—DERIVATIVES

427.1. $\dfrac{d \sin x}{dx} = \cos x.$

427.4. $\dfrac{d \operatorname{ctn} x}{dx} = - \csc^2 x.$

427.2. $\dfrac{d \cos x}{dx} = - \sin x.$

427.5. $\dfrac{d \sec x}{dx} = \sec x \tan x.$

427.3. $\dfrac{d \tan x}{dx} = \sec^2 x.$

427.6. $\dfrac{d \csc x}{dx} = - \csc x \operatorname{ctn} x.$

TRIGONOMETRIC FUNCTIONS—INTEGRALS

In integrating from one point to another, a process of curve plotting is frequently of assistance. Some of the curves, such as the tan curve, have more than one branch. In general, integration should not be carried out from a point on one branch to a point on another branch.

	u	du	$\sin x$	$\cos x$	$\tan x$	x	dx
(1)	$\sin x$	$\cos x\, dx$	u	$\sqrt{1-u^2}$	$\dfrac{u}{\sqrt{1-u^2}}$	$\sin^{-1} u$	$\dfrac{du}{\sqrt{1-u^2}}$
(2)	$\cos x$	$-\sin x\, dx$	$\sqrt{1-u^2}$	u	$\dfrac{\sqrt{1-u^2}}{u}$	$\cos^{-1} u$	$-\dfrac{du}{\sqrt{1-u^2}}$
(3)	$\tan x$	$\sec^2 x\, dx$	$\dfrac{u}{\sqrt{1+u^2}}$	$\dfrac{1}{\sqrt{1+u^2}}$	u	$\tan^{-1} u$	$\dfrac{du}{1+u^2}$
(4)	$\sec x$	$\sec x \tan x\, dx$	$\dfrac{\sqrt{u^2-1}}{u}$	$\dfrac{1}{u}$	$\sqrt{u^2-1}$	$\sec^{-1} u$	$\dfrac{du}{u\sqrt{u^2-1}}$
(5)	$\tan\dfrac{x}{2}$	$\dfrac{1}{2}\sec^2\dfrac{x}{2}\,dx$	$\dfrac{2u}{1+u^2}$	$\dfrac{1-u^2}{1+u^2}$	$\dfrac{2u}{1-u^2}$	$2\tan^{-1} u$	$\dfrac{2\,du}{1+u^2}$

429. Substitutions:*

Replace ctn x, sec x, csc x by $1/\tan x$, $1/\cos x$, $1/\sin x$, respectively.

Notes.   (*a*) $\int F(\sin x) \cos x\, dx$,—use (1).

       (*b*) $\int F(\cos x) \sin x\, dx$,—use (2).

       (*c*) $\int F(\tan x) \sec^2 x\, dx$,—use (3).

       (*d*) Inspection of this table shows desirable substitutions from trigonometric to algebraic, and conversely. Thus, if only tan x, $\sin^2 x$, $\cos^2 x$ appear, use (3).

* From *Macmillan Mathematical Tables.*

Integrals Involving sin x

430.10. $\displaystyle\int \sin x\, dx = -\cos x.$

430.101. $\displaystyle\int \sin (a + bx)dx = -\frac{1}{b} \cos (a + bx).$

430.102. $\displaystyle\int \sin \frac{x}{a}\, dx = -a \cos \frac{x}{a}.$

430.11. $\displaystyle\int x \sin x\, dx = \sin x - x \cos x.$

$\int x \sin a x dx =$

430.12. $\displaystyle\int x^2 \sin x\, dx = 2x \sin x - (x^2 - 2) \cos x.$

430.13. $\displaystyle\int x^3 \sin x\, dx = (3x^2 - 6) \sin x - (x^3 - 6x) \cos x.$

430.14. $\displaystyle\int x^4 \sin x\, dx = (4x^3 - 24x) \sin x - (x^4 - 12x^2 + 24) \cos x.$

430.15. $\displaystyle\int x^5 \sin x\, dx = (5x^4 - 60x^2 + 120) \sin x$
$$- (x^5 - 20x^3 + 120x) \cos x.$$

430.16. $\displaystyle\int x^6 \sin x\, dx = (6x^5 - 120x^3 + 720x) \sin x$
$$- (x^6 - 30x^4 + 360x^2 - 720) \cos x.$$

430.19. $\displaystyle\int x^m \sin x\, dx = -x^m \cos x + m \int x^{m-1} \cos x\, dx.$

[See **440**.]   [Ref. 2, p. 137.]

430.20. $\displaystyle\int \sin^2 x\, dx = \frac{x}{2} - \frac{\sin 2x}{4} = \frac{x}{2} - \frac{\sin x \cos x}{2}.$

note $\int \sin^2 a x dx = \frac{1}{a}[\quad '' \quad] = [\quad '' \quad] \cdot \frac{1}{a}$

430.21. $\displaystyle\int x \sin^2 x\, dx = \frac{x^2}{4} - \frac{x \sin 2x}{4} - \frac{\cos 2x}{8}.$

430.22. $\displaystyle\int x^2 \sin^2 x\, dx = \frac{x^3}{6} - \left(\frac{x^2}{4} - \frac{1}{8}\right) \sin 2x - \frac{x \cos 2x}{4}.$

430.23. $\int x^3 \sin^2 x \, dx = \dfrac{x^4}{8} - \left(\dfrac{x^3}{4} - \dfrac{3x}{8}\right) \sin 2x$

$$- \left(\dfrac{3x^2}{8} - \dfrac{3}{16}\right) \cos 2x.$$

430.30. $\int \sin^3 x \, dx = \dfrac{\cos^3 x}{3} - \cos x.$

430.31. $\int x \sin^3 x \, dx = \dfrac{x \cos 3x}{12} - \dfrac{\sin 3x}{36} - \dfrac{3}{4} x \cos x + \dfrac{3}{4} \sin x.$

[Expand $\sin^3 x$ by **404.13**.]

430.40. $\int \sin^4 x \, dx = \dfrac{3x}{8} - \dfrac{\sin 2x}{4} + \dfrac{\sin 4x}{32}.$

430.50. $\int \sin^5 x \, dx = -\dfrac{5 \cos x}{8} + \dfrac{5 \cos 3x}{48} - \dfrac{\cos 5x}{80}.$

430.60. $\int \sin^6 x \, dx = \dfrac{5x}{16} - \dfrac{15 \sin 2x}{64} + \dfrac{3 \sin 4x}{64} - \dfrac{\sin 6x}{192}.$

430.70. $\int \sin^7 x \, dx = -\dfrac{35 \cos x}{64} + \dfrac{7 \cos 3x}{64} - \dfrac{7 \cos 5x}{320} + \dfrac{\cos 7x}{448}.$

[Ref. 1, p. 239.   Integrate expressions in **404**.]

431.11. $\int \dfrac{\sin x \, dx}{x} = \text{Si}(x) = x - \dfrac{x^3}{3 \cdot 3!} + \dfrac{x^5}{5 \cdot 5!} - \dfrac{x^7}{7 \cdot 7!} + \cdots.$

For table of numerical values, see Ref. 45 and Ref. 55 (f).

431.12. $\int \dfrac{\sin x \, dx}{x^2} = -\dfrac{\sin x}{x} + \int \dfrac{\cos x \, dx}{x}.$ [See **441.11**.]

431.13. $\int \dfrac{\sin x \, dx}{x^3} = -\dfrac{\sin x}{2x^2} - \dfrac{\cos x}{2x} - \dfrac{1}{2} \int \dfrac{\sin x \, dx}{x}.$ [See **431.11**.]

431.14. $\int \dfrac{\sin x \, dx}{x^4} = -\dfrac{\sin x}{3x^3} - \dfrac{\cos x}{6x^2} + \dfrac{\sin x}{6x} - \dfrac{1}{6} \int \dfrac{\cos x \, dx}{x}.$

[See **441.11**.]

431.19. $\int \dfrac{\sin x \, dx}{x^m} = -\dfrac{\sin x}{(m-1)x^{m-1}} + \dfrac{1}{m-1} \int \dfrac{\cos x \, dx}{x^{m-1}}.$

[Ref. 2, p. 138.]

431.21. $\displaystyle\int\frac{\sin^2 x\, dx}{x} = \frac{1}{2}\log|x| - \frac{1}{2}\int\frac{\cos 2x\, d(2x)}{2x}.$ [See **441.11.**]

431.31. $\displaystyle\int\frac{\sin^3 x\, dx}{x} = \frac{3}{4}\int\frac{\sin x\, dx}{x} - \frac{1}{4}\int\frac{\sin 3x\, d(3x)}{3x}.$ [See **431.11.**]

431.9. $\displaystyle\int\frac{\sin^n x\, dx}{x^m}.$ Expand $\sin^n x$ by **404** and integrate each term by **431.1** and **441.1.**

432.10. $\displaystyle\int\frac{dx}{\sin x} = \int\csc x\, dx = \log\left|\tan\frac{x}{2}\right|$

$$= -\frac{1}{2}\log\frac{1+\cos x}{1-\cos x} = \log|\csc x - \operatorname{ctn} x|$$

$$= \lambda\left(x - \frac{\pi}{2}\right),\qquad\qquad \text{(Lambda function)}.$$

[See **603.6.**]

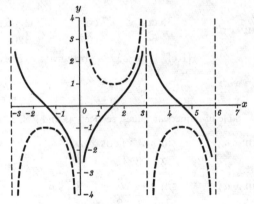

Fig. 432.10. Dotted graph, csc x. Full line graph, $\log\left|\tan\dfrac{x}{2}\right|$.

432.11. $\displaystyle\int\frac{x\, dx}{\sin x} = x + \frac{x^3}{3\cdot 3!} + \frac{7x^5}{3\cdot 5\cdot 5!} + \frac{31x^7}{3\cdot 7\cdot 7!} + \frac{127x^9}{3\cdot 5\cdot 9!}$

$$+ \cdots + \frac{2(2^{2n-1}-1)}{(2n+1)!}B_n x^{2n+1} + \cdots.$$

[See **45.**]

432.12. $\int \dfrac{x^2 dx}{\sin x} = \dfrac{x^2}{2} + \dfrac{x^4}{4 \cdot 3!} + \dfrac{7x^6}{3 \cdot 6 \cdot 5!} + \dfrac{31x^8}{3 \cdot 8 \cdot 7!} + \dfrac{127x^{10}}{5 \cdot 5 \cdot 6 \cdot 8!}$

$$+ \cdots + \dfrac{2(2^{2n-1} - 1)}{(2n + 2)(2n)!} B_n x^{2n+2} + \cdots.$$

[See **45.**]

432.19. $\int \dfrac{x^m dx}{\sin x}.$ Expand $\dfrac{1}{\sin x}$ by **415.06**, multiply by x^m and integrate, $\qquad [m > 0].$

432.20. $\int \dfrac{dx}{\sin^2 x} = \int \csc^2 x\, dx = -\operatorname{ctn} x.$

432.21. $\int \dfrac{x\, dx}{\sin^2 x} = -x \operatorname{ctn} x + \log |\sin x|.$

432.30. $\int \dfrac{dx}{\sin^3 x} = -\dfrac{\cos x}{2 \sin^2 x} + \dfrac{1}{2} \log \left| \tan \dfrac{x}{2} \right|.$

432.31. $\int \dfrac{x\, dx}{\sin^3 x} = -\dfrac{x \cos x}{2 \sin^2 x} - \dfrac{1}{2 \sin x} + \dfrac{1}{2} \int \dfrac{x\, dx}{\sin x}.$ [See **432.11.**]

432.40. $\int \dfrac{dx}{\sin^4 x} = -\dfrac{\cos x}{3 \sin^3 x} - \dfrac{2}{3} \operatorname{ctn} x = -\operatorname{ctn} x - \dfrac{\operatorname{ctn}^3 x}{3}.$

432.41. $\int \dfrac{x\, dx}{\sin^4 x} = -\dfrac{x \cos x}{3 \sin^3 x} - \dfrac{1}{6 \sin^2 x} - \dfrac{2}{3} x \operatorname{ctn} x + \dfrac{2}{3} \log |\sin x|.$

432.50. $\int \dfrac{dx}{\sin^5 x} = -\dfrac{\cos x}{4 \sin^4 x} - \dfrac{3 \cos x}{8 \sin^2 x} + \dfrac{3}{8} \log \left| \tan \dfrac{x}{2} \right|.$

432.60. $\int \dfrac{dx}{\sin^6 x} = -\dfrac{\cos x}{5 \sin^5 x} - \dfrac{4}{15} \dfrac{\cos x}{\sin^3 x} - \dfrac{8}{15} \operatorname{ctn} x.$

432.90. $\int \dfrac{dx}{\sin^n x} = \int \csc^n x\, dx$

$$= -\dfrac{\cos x}{(n - 1) \sin^{n-1} x} + \dfrac{n - 2}{n - 1} \int \dfrac{dx}{\sin^{n-2} x}, \qquad [n > 1].$$

432.91. $\int \dfrac{x\, dx}{\sin^n x} = -\dfrac{x \cos x}{(n - 1) \sin^{n-1} x} - \dfrac{1}{(n - 1)(n - 2) \sin^{n-2} x}$

$$+ \dfrac{n - 2}{n - 1} \int \dfrac{x\, dx}{\sin^{n-2} x}, \qquad [n > 2].$$

433.01. $\displaystyle\int \frac{dx}{1 + \sin x} = -\tan\left(\frac{\pi}{4} - \frac{x}{2}\right).$

433.02. $\displaystyle\int \frac{dx}{1 - \sin x} = \tan\left(\frac{\pi}{4} + \frac{x}{2}\right).$

433.03. $\displaystyle\int \frac{x\,dx}{1 + \sin x} = -x\tan\left(\frac{\pi}{4} - \frac{x}{2}\right) + 2\log\left|\cos\left(\frac{\pi}{4} - \frac{x}{2}\right)\right|.$

433.04. $\displaystyle\int \frac{x\,dx}{1 - \sin x} = x\,\mathrm{ctn}\left(\frac{\pi}{4} - \frac{x}{2}\right) + 2\log\left|\sin\left(\frac{\pi}{4} - \frac{x}{2}\right)\right|.$

433.05. $\displaystyle\int \frac{\sin x\,dx}{1 + \sin x} = x + \tan\left(\frac{\pi}{4} - \frac{x}{2}\right).$

433.06. $\displaystyle\int \frac{\sin x\,dx}{1 - \sin x} = -x + \tan\left(\frac{\pi}{4} + \frac{x}{2}\right).$

433.07. $\displaystyle\int \frac{dx}{\sin x(1 + \sin x)} = \tan\left(\frac{\pi}{4} - \frac{x}{2}\right) + \log\left|\tan\frac{x}{2}\right|.$

433.08. $\displaystyle\int \frac{dx}{\sin x(1 - \sin x)} = \tan\left(\frac{\pi}{4} + \frac{x}{2}\right) + \log\left|\tan\frac{x}{2}\right|.$

434.01. $\displaystyle\int \frac{dx}{(1 + \sin x)^2} = -\frac{1}{2}\tan\left(\frac{\pi}{4} - \frac{x}{2}\right) - \frac{1}{6}\tan^3\left(\frac{\pi}{4} - \frac{x}{2}\right).$

434.02. $\displaystyle\int \frac{dx}{(1 - \sin x)^2} = \frac{1}{2}\,\mathrm{ctn}\left(\frac{\pi}{4} - \frac{x}{2}\right) + \frac{1}{6}\,\mathrm{ctn}^3\left(\frac{\pi}{4} - \frac{x}{2}\right).$

434.03. $\displaystyle\int \frac{\sin x\,dx}{(1 + \sin x)^2} = -\frac{1}{2}\tan\left(\frac{\pi}{4} - \frac{x}{2}\right) + \frac{1}{6}\tan^3\left(\frac{\pi}{4} - \frac{x}{2}\right).$

434.04. $\displaystyle\int \frac{\sin x\,dx}{(1 - \sin x)^2} = -\frac{1}{2}\,\mathrm{ctn}\left(\frac{\pi}{4} - \frac{x}{2}\right) + \frac{1}{6}\,\mathrm{ctn}^3\left(\frac{\pi}{4} - \frac{x}{2}\right).$

434.05. $\displaystyle\int \frac{dx}{1 + \sin^2 x} = \frac{1}{2\sqrt{2}}\sin^{-1}\left(\frac{3\sin^2 x - 1}{\sin^2 x + 1}\right).$ [See **436.6.**]

434.06. $\displaystyle\int \frac{dx}{1 - \sin^2 x} = \int \frac{dx}{\cos^2 x} = \tan x.$ [See **442.20.**]

435. $\displaystyle\int \sin mx \sin nx\,dx = \frac{\sin(m-n)x}{2(m-n)} - \frac{\sin(m+n)x}{2(m+n)},$

$[m^2 \neq n^2.$ If $m^2 = n^2$, see **430.20.**]

436.00. $\displaystyle\int \frac{dx}{a + b \sin x}$

$$= \frac{2}{\sqrt{(a^2 - b^2)}} \tan^{-1} \frac{a \tan (x/2) + b}{\sqrt{(a^2 - b^2)}}, \qquad [a^2 > b^2]$$

$$= \frac{1}{\sqrt{(b^2 - a^2)}} \log \left| \frac{a \tan (x/2) + b - \sqrt{(b^2 - a^2)}}{a \tan (x/2) + b + \sqrt{(b^2 - a^2)}} \right|,$$

$$[b^2 > a^2]$$

$$= \frac{-2}{\sqrt{(b^2 - a^2)}} \tanh^{-1} \frac{a \tan (x/2) + b}{\sqrt{(b^2 - a^2)}},$$

$$[b^2 > a^2, \ |a \tan (x/2) + b| < \sqrt{(b^2 - a^2)}]$$

$$= \frac{-2}{\sqrt{(b^2 - a^2)}} \operatorname{ctnh}^{-1} \frac{a \tan (x/2) + b}{\sqrt{(b^2 - a^2)}},$$

$$[b^2 > a^2, \ |a \tan (x/2) + b| > \sqrt{(b^2 - a^2)}].$$

[See **160.01.** Also Ref. 7, p. 16, and Ref. 5, No. 307.]

The integration should not be carried out from a point on one branch of the curve to a point on another branch. The integrand becomes infinite at $x = \sin^{-1}$ $(- a/b)$, which can occur when $|x| < \pi$.

436.01. $\displaystyle\int \frac{\sin x\, dx}{a + b \sin x} = \frac{x}{b} - \frac{a}{b} \int \frac{dx}{a + b \sin x}.$

436.02. $\displaystyle\int \frac{dx}{\sin x(a + b \sin x)} = \frac{1}{a} \log \left| \tan \frac{x}{2} \right| - \frac{b}{a} \int \frac{dx}{a + b \sin x}.$

436.03. $\displaystyle\int \frac{dx}{(a + b \sin x)^2} = \frac{b \cos x}{(a^2 - b^2)(a + b \sin x)}$

$$+ \frac{a}{a^2 - b^2} \int \frac{dx}{a + b \sin x}.$$

436.04. $\displaystyle\int \frac{\sin x\, dx}{(a + b \sin x)^2} = \frac{a \cos x}{(b^2 - a^2)(a + b \sin x)}$

$$+ \frac{b}{b^2 - a^2} \int \frac{dx}{a + b \sin x}.$$

[For **436.01** to **436.04**, see **436.00.**]

436.5. $\displaystyle\int \frac{dx}{a^2 + b^2 \sin^2 x} = \frac{1}{a\sqrt{(a^2 + b^2)}} \tan^{-1} \frac{\sqrt{(a^2 + b^2)} \tan x}{a},$

436.6.   When $a = b = 1$,

$$\int \frac{dx}{1 + \sin^2 x} = \frac{1}{\sqrt{2}} \tan^{-1} (\sqrt{2} \tan x).$$

See also the alternative solution in **434.05**, which differs by a constant.

436.7. $\displaystyle\int \frac{dx}{a^2 - b^2 \sin^2 x}$

$$= \frac{1}{a\sqrt{(a^2 - b^2)}} \tan^{-1} \frac{\sqrt{(a^2 - b^2)} \tan x}{a},$$

$$[a^2 > b^2],$$

$$= \frac{1}{2a\sqrt{(b^2 - a^2)}} \log \left| \frac{\sqrt{(b^2 - a^2)} \tan x + a}{\sqrt{(b^2 - a^2)} \tan x - a} \right|,$$

$$[b^2 > a^2].$$

If $b^2 = a^2$, see **434.06**.

436.8. $\displaystyle\int \frac{\sin^2 x \, dx}{a^2 - b^2 \sin^2 x} = \frac{a^2}{b^2} \int \frac{dx}{a^2 - b^2 \sin^2 x} - \frac{x}{b^2} \cdot$ [See **436.7.**]

437.1. $\displaystyle\int \frac{\sin x \, dx}{\sqrt{(1 + m^2 \sin^2 x)}} = - \frac{1}{m} \sin^{-1} \frac{m \cos x}{\sqrt{(1 + m^2)}} \cdot$

437.2. $\displaystyle\int \frac{\sin x \, dx}{\sqrt{(1 - m^2 \sin^2 x)}}$

$$= - \frac{1}{m} \log \{ m \cos x + \sqrt{(1 - m^2 \sin^2 x)} \}.$$

437.3. $\displaystyle\int (\sin x)\sqrt{(1 + m^2 \sin^2 x)}dx$

$$= - \frac{\cos x}{2} \sqrt{(1 + m^2 \sin^2 x)} - \frac{1 + m^2}{2m} \sin^{-1} \frac{m \cos x}{\sqrt{(1 + m^2)}} \cdot$$

437.4. $\displaystyle\int (\sin x)\sqrt{(1 - m^2 \sin^2 x)}dx$

$$= - \frac{\cos x}{2} \sqrt{(1 - m^2 \sin^2 x)}$$

$$- \frac{1 - m^2}{2m} \log \{ m \cos x + \sqrt{(1 - m^2 \sin^2 x)} \}.$$

Integrals Involving cos x

440.10. $\displaystyle\int \cos x \, dx = \sin x.$

440.101. $\displaystyle\int \cos (a + bx)dx = \frac{1}{b} \sin (a + bx).$

440.102. $\displaystyle\int \cos \frac{x}{a} \, dx = a \sin \frac{x}{a}.$

440.11. $\displaystyle\int x \cos x \, dx = \cos x + x \sin x.$

440.12. $\displaystyle\int x^2 \cos x \, dx = 2x \cos x + (x^2 - 2) \sin x.$

440.13. $\displaystyle\int x^3 \cos x \, dx = (3x^2 - 6) \cos x + (x^3 - 6x) \sin x.$

440.14. $\displaystyle\int x^4 \cos x \, dx = (4x^3 - 24x) \cos x + (x^4 - 12x^2 + 24) \sin x.$

440.15. $\displaystyle\int x^5 \cos x \, dx = (5x^4 - 60x^2 + 120) \cos x$
$$+ (x^5 - 20x^3 + 120x) \sin x.$$

440.16. $\displaystyle\int x^6 \cos x \, dx = (6x^5 - 120x^3 + 720x) \cos x$
$$+ (x^6 - 30x^4 + 360x^2 - 720) \sin x.$$

440.19. $\displaystyle\int x^m \cos x \, dx = x^m \sin x - m \int x^{m-1} \sin x \, dx.$
$$\text{[See \textbf{430}.]} \qquad \text{[Ref. 2, p. 137.]}$$

440.20. $\displaystyle\int \cos^2 x \, dx = \frac{x}{2} + \frac{\sin 2x}{4} = \frac{x}{2} + \frac{\sin x \cos x}{2}.$

440.21. $\displaystyle\int x \cos^2 x \, dx = \frac{x^2}{4} + \frac{x \sin 2x}{4} + \frac{\cos 2x}{8}.$

440.22. $\displaystyle\int x^2 \cos^2 x \, dx = \frac{x^3}{6} + \left(\frac{x^2}{4} - \frac{1}{8}\right) \sin 2x + \frac{x \cos 2x}{4}.$

440.23. $\displaystyle\int x^3 \cos^2 x \, dx = \frac{x^4}{8} + \left(\frac{x^3}{4} - \frac{3x}{8}\right) \sin 2x$
$$+ \left(\frac{3x^2}{8} - \frac{3}{16}\right) \cos 2x.$$

440.30. $\displaystyle\int \cos^3 x \, dx = \sin x - \frac{\sin^3 x}{3}.$

440.31. $\displaystyle\int x \cos^3 x \, dx = \frac{x \sin 3x}{12} + \frac{\cos 3x}{36} + \frac{3}{4} x \sin x + \frac{3}{4} \cos x.$

[Expand $\cos^3 x$ by **404.23.**]

440.40. $\displaystyle\int \cos^4 x \, dx = \frac{3x}{8} + \frac{\sin 2x}{4} + \frac{\sin 4x}{32}.$

440.50. $\displaystyle\int \cos^5 x \, dx = \frac{5 \sin x}{8} + \frac{5 \sin 3x}{48} + \frac{\sin 5x}{80}.$

440.60. $\displaystyle\int \cos^6 x \, dx = \frac{5x}{16} + \frac{15 \sin 2x}{64} + \frac{3 \sin 4x}{64} + \frac{\sin 6x}{192}.$

440.70. $\displaystyle\int \cos^7 x \, dx = \frac{35 \sin x}{64} + \frac{7 \sin 3x}{64} + \frac{7 \sin 5x}{320} + \frac{\sin 7x}{448}.$

[Ref. 1, p. 240.   Integrate expressions in **404.**]

441.11. $\displaystyle\int \frac{\cos x \, dx}{x} = \log |x| - \frac{x^2}{2 \cdot 2!} + \frac{x^4}{4 \cdot 4!} - \frac{x^6}{6 \cdot 6!} + \cdots.$

For table of numerical values, see Ref. 45 and Ref. 55 f.

441.12. $\displaystyle\int \frac{\cos x \, dx}{x^2} = -\frac{\cos x}{x} - \int \frac{\sin x \, dx}{x}.$ [See **431.11.**]

441.13. $\displaystyle\int \frac{\cos x \, dx}{x^3} = -\frac{\cos x}{2x^2} + \frac{\sin x}{2x} - \frac{1}{2} \int \frac{\cos x \, dx}{x}.$

[See **441.11.**]

441.14. $\displaystyle\int \frac{\cos x \, dx}{x^4} = -\frac{\cos x}{3x^3} + \frac{\sin x}{6x^2} + \frac{\cos x}{6x} + \frac{1}{6} \int \frac{\sin x \, dx}{x}.$

[See **431.11.**]

441.19. $\displaystyle\int \frac{\cos x \, dx}{x^m} = -\frac{\cos x}{(m-1)x^{m-1}} - \frac{1}{m-1} \int \frac{\sin x \, dx}{x^{m-1}}.$

441.21. $\displaystyle\int \frac{\cos^2 x \, dx}{x} = \frac{1}{2} \log |x| + \frac{1}{2} \int \frac{\cos 2x \, d(2x)}{2x}.$ [See **441.11.**]

441.31. $\displaystyle\int \frac{\cos^3 x \, dx}{x} = \frac{3}{4} \int \frac{\cos x \, dx}{x} + \frac{1}{4} \int \frac{\cos 3x \, d(3x)}{3x}.$

[See **441.11.**]

441.9. $\int \dfrac{\cos^n x \, dx}{x^m}.$

Expand $\cos^n x$ by **404** and integrate each term by **441.1.**

442.10. $\int \dfrac{dx}{\cos x} = \int \sec x \, dx = \log \left| \tan \left(\dfrac{\pi}{4} + \dfrac{x}{2} \right) \right|$

$$= \log |\sec x + \tan x| = \frac{1}{2} \log \frac{1 + \sin x}{1 - \sin x}$$

$$= \lambda(x), \quad \text{(Lambda function).} \qquad \text{[See 640.]}$$

442.11. $\int \dfrac{x \, dx}{\cos x} = \dfrac{x^2}{2} + \dfrac{x^4}{4 \cdot 2!} + \dfrac{5x^6}{6 \cdot 4!} + \dfrac{61x^8}{8 \cdot 6!} + \dfrac{1385x^{10}}{10 \cdot 8!} + \cdots$

$$\cdots + \frac{E_n x^{2n+2}}{(2n+2)(2n)!} + \cdots. \qquad \text{[See 45.]}$$

442.12. $\int \dfrac{x^2 dx}{\cos x} = \dfrac{x^3}{3} + \dfrac{x^5}{5 \cdot 2!} + \dfrac{5x^7}{7 \cdot 4!} + \dfrac{61x^9}{9 \cdot 6!} + \dfrac{1385x^{11}}{11 \cdot 8!} + \cdots$

$$\cdots + \frac{E_{n-1} x^{2n+1}}{(2n+1)(2n-2)!} + \cdots. \qquad \text{[See 45.]}$$

442.19. $\int \dfrac{x^m dx}{\cos x}.$ Expand $\dfrac{1}{\cos x}$ by **415.05**, multiply by x^m and

integrate, $[m \neq 0].$

442.20. $\int \dfrac{dx}{\cos^2 x} = \int \sec^2 x \, dx = \tan x.$

442.21. $\int \dfrac{x \, dx}{\cos^2 x} = x \tan x + \log |\cos x|.$

442.30. $\int \dfrac{dx}{\cos^3 x} = \dfrac{\sin x}{2 \cos^2 x} + \dfrac{1}{2} \log \left| \tan \left(\dfrac{\pi}{4} + \dfrac{x}{2} \right) \right|.$

442.31. $\int \dfrac{x \, dx}{\cos^3 x} = \dfrac{x \sin x}{2 \cos^2 x} - \dfrac{1}{2 \cos x} + \dfrac{1}{2} \int \dfrac{x \, dx}{\cos x}.$ [See **442.11.**]

442.40. $\int \dfrac{dx}{\cos^4 x} = \dfrac{\sin x}{3 \cos^3 x} + \dfrac{2}{3} \tan x = \tan x + \dfrac{\tan^3 x}{3}.$

442.41. $\int \dfrac{x \, dx}{\cos^4 x} = \dfrac{x \sin x}{3 \cos^3 x} - \dfrac{1}{6 \cos^2 x} + \dfrac{2}{3} x \tan x + \dfrac{2}{3} \log |\cos x|.$

442.50. $\int \dfrac{dx}{\cos^5 x} = \dfrac{\sin x}{4 \cos^4 x} + \dfrac{3}{8} \dfrac{\sin x}{\cos^2 x} + \dfrac{3}{8} \log \left| \tan \left(\dfrac{\pi}{4} + \dfrac{x}{2} \right) \right|.$

442.60. $\int \dfrac{dx}{\cos^6 x} = \dfrac{\sin x}{5 \cos^5 x} + \dfrac{4}{15} \dfrac{\sin x}{\cos^3 x} + \dfrac{8}{15} \tan x.$

442.90. $\int \dfrac{dx}{\cos^n x} = \int \sec^n x \, dx$

$\qquad = \dfrac{\sin x}{(n-1) \cos^{n-1} x} + \dfrac{n-2}{n-1} \int \dfrac{dx}{\cos^{n-2} x},$ $\qquad\qquad$ $[n > 1]$.

442.91. $\int \dfrac{x \, dx}{\cos^n x} = \dfrac{x \sin x}{(n-1) \cos^{n-1} x} - \dfrac{1}{(n-1)(n-2) \cos^{n-2} x}$

$\qquad\qquad\qquad\qquad + \dfrac{n-2}{n-1} \int \dfrac{x \, dx}{\cos^{n-2} x},$ $\qquad$ $[n > 2]$.

443.01. $\int \dfrac{dx}{1 + \cos x} = \tan \dfrac{x}{2}.$

443.02. $\int \dfrac{dx}{1 - \cos x} = -\operatorname{ctn} \dfrac{x}{2}.$

443.03. $\int \dfrac{x \, dx}{1 + \cos x} = x \tan \dfrac{x}{2} + 2 \log \left| \cos \dfrac{x}{2} \right|.$

443.04. $\int \dfrac{x \, dx}{1 - \cos x} = -x \operatorname{ctn} \dfrac{x}{2} + 2 \log \left| \sin \dfrac{x}{2} \right|.$

443.05. $\int \dfrac{\cos x \, dx}{1 + \cos x} = x - \tan \dfrac{x}{2}.$

443.06. $\int \dfrac{\cos x \, dx}{1 - \cos x} = -x - \operatorname{ctn} \dfrac{x}{2}.$

443.07. $\int \dfrac{dx}{\cos x (1 + \cos x)} = \log \left| \tan \left(\dfrac{\pi}{4} + \dfrac{x}{2} \right) \right| - \tan \dfrac{x}{2}.$

443.08. $\int \dfrac{dx}{\cos x (1 - \cos x)} = \log \left| \tan \left(\dfrac{\pi}{4} + \dfrac{x}{2} \right) \right| - \operatorname{ctn} \dfrac{x}{2}.$

444.01. $\int \dfrac{dx}{(1 + \cos x)^2} = \dfrac{1}{2} \tan \dfrac{x}{2} + \dfrac{1}{6} \tan^3 \dfrac{x}{2}.$

444.02. $\int \dfrac{dx}{(1 - \cos x)^2} = -\dfrac{1}{2} \operatorname{ctn} \dfrac{x}{2} - \dfrac{1}{6} \operatorname{ctn}^3 \dfrac{x}{2}.$

444.03. $\displaystyle\int \frac{\cos x \, dx}{(1 + \cos x)^2} = \frac{1}{2} \tan \frac{x}{2} - \frac{1}{6} \tan^3 \frac{x}{2}.$

444.04. $\displaystyle\int \frac{\cos x \, dx}{(1 - \cos x)^2} = \frac{1}{2} \operatorname{ctn} \frac{x}{2} - \frac{1}{6} \operatorname{ctn}^3 \frac{x}{2}.$

444.05. $\displaystyle\int \frac{dx}{1 + \cos^2 x} = \frac{1}{2\sqrt{2}} \sin^{-1}\left(\frac{1 - 3 \cos^2 x}{1 + \cos^2 x}\right).$ [See **446.6.**]

444.06. $\displaystyle\int \frac{dx}{1 - \cos^2 x} = \int \frac{dx}{\sin^2 x} = -\operatorname{ctn} x.$ [See **432.20.**]

445. $\displaystyle\int \cos mx \cos nx \, dx = \frac{\sin (m - n)x}{2(m - n)} + \frac{\sin (m + n)x}{2(m + n)},$

$$[m^2 \neq n^2. \quad \text{If } m^2 = n^2, \text{ see } \mathbf{440.20.}]$$

446.00. $\displaystyle\int \frac{dx}{a + b \cos x}$

$$= \frac{2}{\sqrt{(a^2 - b^2)}} \tan^{-1} \frac{(a - b) \tan (x/2)}{\sqrt{(a^2 - b^2)}}, \qquad [a^2 > b^2],$$

$$= \frac{1}{\sqrt{(b^2 - a^2)}} \log \left|\frac{(b - a) \tan (x/2) + \sqrt{(b^2 - a^2)}}{(b - a) \tan (x/2) - \sqrt{(b^2 - a^2)}}\right|,$$

$$[b^2 > a^2],$$

$$= \frac{2}{\sqrt{(b^2 - a^2)}} \tanh^{-1} \frac{(b - a) \tan (x/2)}{\sqrt{(b^2 - a^2)}},$$

$$[b^2 > a^2, \quad |(b - a) \tan (x/2)| < \sqrt{(b^2 - a^2)}],$$

$$= \frac{2}{\sqrt{(b^2 - a^2)}} \operatorname{ctnh}^{-1} \frac{(b - a) \tan (x/2)}{\sqrt{(b^2 - a^2)}},$$

$$[b^2 > a^2, \quad |(b - a) \tan (x/2)| > \sqrt{(b^2 - a^2)}].$$

[See Ref. 7, p. 15, and Ref. 5, No. 309.]

The integration should not be carried out from a point on one branch of the curve to a point on another branch. The integrand becomes infinite at $x = \cos^{-1}(-a/b)$ which can occur when $|x| < \pi$.

446.01. $\displaystyle\int \frac{\cos x \, dx}{a + b \cos x} = \frac{x}{b} - \frac{a}{b} \int \frac{dx}{a + b \cos x}.$

446.02. $\displaystyle\int \frac{dx}{\cos x (a + b \cos x)} = \frac{1}{a} \log \left|\tan \left(\frac{x}{2} + \frac{\pi}{4}\right)\right|$

$$- \frac{b}{a} \int \frac{dx}{a + b \cos x}.$$

446.03. $\displaystyle\int \frac{dx}{(a + b \cos x)^2} = \frac{b \sin x}{(b^2 - a^2)(a + b \cos x)}$

$$- \frac{a}{b^2 - a^2} \int \frac{dx}{a + b \cos x}.$$

446.04. $\displaystyle\int \frac{\cos x \, dx}{(a + b \cos x)^2} = \frac{a \sin x}{(a^2 - b^2)(a + b \cos x)}$

$$- \frac{b}{a^2 - b^2} \int \frac{dx}{a + b \cos x}.$$

[For **446.01** to **446.04**, see **446.00**.]

446.2. $\displaystyle\int \frac{dx}{a^2 + b^2 - 2ab \cos x}$

$$= \frac{2}{|a^2 - b^2|} \tan^{-1} \left[\left| \frac{a + b}{a - b} \right| \tan \frac{x}{2} \right], \qquad [a \neq b].$$

[See **446.00**.]   [Ref. 38, p. 52.]

446.5. $\displaystyle\int \frac{dx}{a^2 + b^2 \cos^2 x} = \frac{1}{a\sqrt{(a^2 + b^2)}} \tan^{-1} \frac{a \tan x}{\sqrt{(a^2 + b^2)}}, \quad [a > 0].$

446.6. When $a = b = 1$,

$$\int \frac{dx}{1 + \cos^2 x} = \frac{1}{\sqrt{2}} \tan^{-1} \left(\frac{\tan x}{\sqrt{2}} \right).$$

See also the alternative solution in **444.05**, which differs by a constant.

446.7. $\displaystyle\int \frac{dx}{a^2 - b^2 \cos^2 x}$

$$= \frac{1}{a\sqrt{(a^2 - b^2)}} \tan^{-1} \frac{a \tan x}{\sqrt{(a^2 - b^2)}}, \qquad [a^2 > b^2],$$

$$= \frac{1}{2a\sqrt{(b^2 - a^2)}} \log \left| \frac{a \tan x - \sqrt{(b^2 - a^2)}}{a \tan x + \sqrt{(b^2 - a^2)}} \right|, \qquad [b^2 > a^2].$$

If $b^2 = a^2$, see **444.06**.

Integrals Involving sin x and cos x

450.11. $\displaystyle\int \sin x \cos x \, dx = \frac{\sin^2 x}{2} = - \frac{\cos^2 x}{2} + \text{constant}$

$$= - \frac{\cos 2x}{4} + \text{constant}.$$

450.12. $\int \sin x \cos^2 x \, dx = -\dfrac{\cos^3 x}{3}.$

450.13. $\int \sin x \cos^3 x \, dx = -\dfrac{\cos^4 x}{4}.$

450.19. $\int \sin x \cos^n x \, dx = -\dfrac{\cos^{n+1} x}{n+1}.$

450.21. $\int \sin^2 x \cos x \, dx = \dfrac{\sin^3 x}{3}.$

450.22. $\int \sin^2 x \cos^2 x \, dx = \dfrac{1}{8}\left(x - \dfrac{\sin 4x}{4}\right).$

450.23. $\int \sin^2 x \cos^3 x \, dx = \dfrac{\sin^3 x \cos^2 x}{5} + \dfrac{2}{15} \sin^3 x.$

450.31. $\int \sin^3 x \cos x \, dx = \dfrac{\sin^4 x}{4}.$

450.81. $\int \sin^m x \cos x \, dx = \dfrac{\sin^{m+1} x}{m+1},$ $\qquad\qquad [m \neq -1].$

$\qquad\qquad\qquad\qquad\qquad\qquad\qquad$ [If $m = -1$, see **453.11**.]

450.9. $\int \sin^m x \cos^n x \, dx$

$$= \frac{\sin^{m+1} x \cos^{n-1} x}{m+n} + \frac{n-1}{m+n} \int \sin^m x \cos^{n-2} x \, dx$$

$$= -\frac{\sin^{m-1} x \cos^{n+1} x}{m+n} + \frac{m-1}{m+n} \int \sin^{m-2} x \cos^n x \, dx,$$

$\qquad\quad$ [$m \neq -n$, see **480.9**].$\qquad\qquad$ [See also **461**.]

451.11. $\int \dfrac{dx}{\sin x \cos x} = \log |\tan x|.$

451.12. $\int \dfrac{dx}{\sin x \cos^2 x} = \dfrac{1}{\cos x} + \log \left|\tan \dfrac{x}{2}\right|.$

451.13. $\int \dfrac{dx}{\sin x \cos^3 x} = \dfrac{1}{2 \cos^2 x} + \log |\tan x|.$

451.14. $\displaystyle\int \frac{dx}{\sin x \cos^4 x} = \frac{1}{3 \cos^3 x} + \frac{1}{\cos x} + \log \left|\tan \frac{x}{2}\right|.$

451.15. $\displaystyle\int \frac{dx}{\sin x \cos^5 x} = \frac{1}{4 \cos^4 x} + \frac{1}{2 \cos^2 x} + \log |\tan x|.$

451.19. $\displaystyle\int \frac{dx}{\sin x \cos^n x} = \frac{1}{(n-1)\cos^{n-1} x} + \int \frac{dx}{\sin x \cos^{n-2} x},$

$$[n \neq 1].$$

451.21. $\displaystyle\int \frac{dx}{\sin^2 x \cos x} = -\frac{1}{\sin x} + \log \left|\tan \left(\frac{\pi}{4} + \frac{x}{2}\right)\right|.$

451.22. $\displaystyle\int \frac{dx}{\sin^2 x \cos^2 x} = -2 \operatorname{ctn} 2x = \tan x - \operatorname{ctn} x.$

451.23. $\displaystyle\int \frac{dx}{\sin^2 x \cos^3 x} = \frac{\sin x}{2 \cos^2 x} - \frac{1}{\sin x} + \frac{3}{2} \log \left|\tan \left(\frac{\pi}{4} + \frac{x}{2}\right)\right|.$

451.24. $\displaystyle\int \frac{dx}{\sin^2 x \cos^4 x} = \frac{1}{3 \sin x \cos^3 x} - \frac{8}{3} \operatorname{ctn} 2x.$

451.31. $\displaystyle\int \frac{dx}{\sin^3 x \cos x} = -\frac{1}{2 \sin^2 x} + \log |\tan x|.$

451.32. $\displaystyle\int \frac{dx}{\sin^3 x \cos^2 x} = \frac{1}{\cos x} - \frac{\cos x}{2 \sin^2 x} + \frac{3}{2} \log \left|\tan \frac{x}{2}\right|.$

451.33. $\displaystyle\int \frac{dx}{\sin^3 x \cos^3 x} = -\frac{2 \cos 2x}{\sin^2 2x} + 2 \log |\tan x|.$

451.41. $\displaystyle\int \frac{dx}{\sin^4 x \cos x} = \frac{3 \cos^2 x - 4}{3 \sin^3 x} + \log \left|\tan \left(\frac{\pi}{4} + \frac{x}{2}\right)\right|.$

[Ref. 1, pp. 260–263.]

451.91. $\displaystyle\int \frac{dx}{\sin^m x \cos x} = -\frac{1}{(m-1) \sin^{m-1} x} + \int \frac{dx}{\sin^{m-2} x \cos x},$

$$[m \neq 1].$$

451.92. $\displaystyle\int \frac{dx}{\sin^n x \cos^n x} = 2^{n-1} \int \frac{d(2x)}{\sin^n (2x)}.$ [See **432**.]

451.93. $\displaystyle\int \frac{dx}{\sin^m x \cos^n x}$

$$= \frac{1}{(n-1)\sin^{m-1} x \cos^{n-1} x} + \frac{m+n-2}{n-1}\int \frac{dx}{\sin^m x \cos^{n-2} x},$$
$$[n > 1],$$

$$= -\frac{1}{(m-1)\sin^{m-1} x \cos^{n-1} x} + \frac{m+n-2}{m-1}\int \frac{dx}{\sin^{m-2} x \cos^n x},$$
$$[m > 1].$$

452.11. $\displaystyle\int \frac{\sin x \, dx}{\cos x} = \int \tan x \, dx = -\log |\cos x|$

$$= \log |\sec x|. \qquad\qquad \text{[See } \mathbf{480.1}.]$$

452.12. $\displaystyle\int \frac{\sin x \, dx}{\cos^2 x} = \frac{1}{\cos x} = \sec x.$

452.13. $\displaystyle\int \frac{\sin x \, dx}{\cos^3 x} = \frac{1}{2\cos^2 x} = \frac{1}{2}\tan^2 x + \text{constant}.$

452.14. $\displaystyle\int \frac{\sin x \, dx}{\cos^4 x} = \frac{1}{3\cos^3 x}.$

452.19. $\displaystyle\int \frac{\sin x \, dx}{\cos^n x} = \frac{1}{(n-1)\cos^{n-1} x}, \qquad\qquad [n \neq 1].$

452.21. $\displaystyle\int \frac{\sin^2 x \, dx}{\cos x} = -\sin x + \log \left| \tan\left(\frac{\pi}{4} + \frac{x}{2}\right) \right|.$

452.22. $\displaystyle\int \frac{\sin^2 x \, dx}{\cos^2 x} = \int \tan^2 x \, dx = \tan x - x. \qquad \text{[See } \mathbf{480.2}.]$

452.23. $\displaystyle\int \frac{\sin^2 x \, dx}{\cos^3 x} = \frac{\sin x}{2\cos^2 x} - \frac{1}{2}\log \left| \tan\left(\frac{\pi}{4} + \frac{x}{2}\right) \right|.$

452.24. $\displaystyle\int \frac{\sin^2 x \, dx}{\cos^4 x} = \frac{1}{3}\tan^3 x.$

452.29. $\displaystyle\int \frac{\sin^2 x \, dx}{\cos^n x} = \frac{\sin x}{(n-1)\cos^{n-1} x} - \frac{1}{n-1}\int \frac{dx}{\cos^{n-2} x}, \quad [n \neq 1].$

452.31. $\displaystyle\int \frac{\sin^3 x \, dx}{\cos x} = -\frac{\sin^2 x}{2} - \log |\cos x|.$

452.32. $\displaystyle\int\frac{\sin^3 x\,dx}{\cos^2 x} = \cos x + \sec x.$

452.33. $\displaystyle\int\frac{\sin^3 x\,dx}{\cos^3 x} = \int\tan^3 x\,dx = \frac{1}{2}\tan^2 x + \log|\cos x|.$

[See **480.3**.]

452.34. $\displaystyle\int\frac{\sin^3 x\,dx}{\cos^4 x} = \frac{1}{3\cos^3 x} - \frac{1}{\cos x}.$

452.35. $\displaystyle\int\frac{\sin^3 x\,dx}{\cos^5 x} = \frac{1}{4}\tan^4 x = \frac{1}{4\cos^4 x} - \frac{1}{2\cos^2 x} + \text{constant}.$

452.39. $\displaystyle\int\frac{\sin^3 x\,dx}{\cos^n x} = \frac{1}{(n-1)\cos^{n-1} x} - \frac{1}{(n-3)\cos^{n-3} x},$

$[n \neq 1 \text{ or } 3].$

452.41. $\displaystyle\int\frac{\sin^4 x\,dx}{\cos x} = -\frac{\sin^3 x}{3} - \sin x + \log\left|\tan\left(\frac{\pi}{4}+\frac{x}{2}\right)\right|.$

452.7. $\displaystyle\int\frac{\sin^{n-2} x\,dx}{\cos^n x} = \frac{\tan^{n-1} x}{n-1},$ $[n \neq 1].$

452.8. $\displaystyle\int\frac{\sin^n x\,dx}{\cos^n x} = \int\tan^n x\,dx = \frac{\tan^{n-1} x}{n-1} - \int\tan^{n-2} x\,dx,$

$[n \neq 1. \quad \text{See } 480.9.]$

452.9. $\displaystyle\int\frac{\sin^m x\,dx}{\cos^n x}$

$$= \frac{\sin^{m+1} x}{(n-1)\cos^{n-1} x} - \frac{m-n+2}{n-1}\int\frac{\sin^m x\,dx}{\cos^{n-2} x},$$

$[n \neq 1],$

$$= -\frac{\sin^{m-1} x}{(m-n)\cos^{n-1} x} + \frac{m-1}{m-n}\int\frac{\sin^{m-2} x\,dx}{\cos^n x},$$

$[m \neq n],$

$$= \frac{\sin^{m-1} x}{(n-1)\cos^{n-1} x} - \frac{m-1}{n-1}\int\frac{\sin^{m-2} x\,dx}{\cos^{n-2} x},$$

$[n \neq 1].$

453.11. $\displaystyle\int\frac{\cos x\,dx}{\sin x} = \int\text{ctn } x\,dx = \log|\sin x|.$ [See **490.1**.]

453.12. $\displaystyle\int \frac{\cos x\, dx}{\sin^2 x} = -\frac{1}{\sin x} = -\csc x.$

453.13. $\displaystyle\int \frac{\cos x\, dx}{\sin^3 x} = -\frac{1}{2\sin^2 x} = -\frac{\operatorname{ctn}^2 x}{2} + \text{constant}.$

453.14. $\displaystyle\int \frac{\cos x\, dx}{\sin^4 x} = -\frac{1}{3\sin^3 x}\,.$

453.19. $\displaystyle\int \frac{\cos x\, dx}{\sin^n x} = -\frac{1}{(n-1)\sin^{n-1} x}, \qquad\qquad [n \neq 1].$

453.21. $\displaystyle\int \frac{\cos^2 x\, dx}{\sin x} = \cos x + \log\left|\tan \frac{x}{2}\right|.$

453.22. $\displaystyle\int \frac{\cos^2 x\, dx}{\sin^2 x} = \int \operatorname{ctn}^2 x\, dx = -\operatorname{ctn} x - x. \qquad \text{[See 490.2.]}$

453.23. $\displaystyle\int \frac{\cos^2 x\, dx}{\sin^3 x} = -\frac{\cos x}{2\sin^2 x} - \frac{1}{2}\log\left|\tan \frac{x}{2}\right|.$

453.24. $\displaystyle\int \frac{\cos^2 x\, dx}{\sin^4 x} = -\frac{1}{3}\operatorname{ctn}^3 x.$

453.29. $\displaystyle\int \frac{\cos^2 x\, dx}{\sin^n x} = -\frac{\cos x}{(n-1)\sin^{n-1} x} - \frac{1}{n-1}\int \frac{dx}{\sin^{n-2} x},$
$$[n \neq 1].$$

453.31. $\displaystyle\int \frac{\cos^3 x\, dx}{\sin x} = \frac{\cos^2 x}{2} + \log|\sin x|.$

453.32. $\displaystyle\int \frac{\cos^3 x\, dx}{\sin^2 x} = -\sin x - \csc x.$

453.33. $\displaystyle\int \frac{\cos^3 x\, dx}{\sin^3 x} = \int \operatorname{ctn}^3 x\, dx = -\frac{\operatorname{ctn}^2 x}{2} - \log|\sin x|.$
$$\text{[See 490.3.]}$$

453.34. $\displaystyle\int \frac{\cos^3 x\, dx}{\sin^4 x} = \frac{1}{\sin x} - \frac{1}{3\sin^3 x}\,.$

453.35. $\displaystyle\int \frac{\cos^3 x\, dx}{\sin^5 x} = -\frac{1}{4}\operatorname{ctn}^4 x = \frac{1}{2\sin^2 x} - \frac{1}{4\sin^4 x} + \text{constant}.$

453.39. $\displaystyle\int \frac{\cos^3 x \, dx}{\sin^n x} = \frac{1}{(n-3)\sin^{n-3} x} - \frac{1}{(n-1)\sin^{n-1} x}$,

$$[n \neq 1 \text{ or } 3].$$

453.41. $\displaystyle\int \frac{\cos^4 x \, dx}{\sin x} = \frac{\cos^3 x}{3} + \cos x + \log \left| \tan \frac{x}{2} \right|.$

453.7. $\displaystyle\int \frac{\cos^{n-2} x \, dx}{\sin^n x} = -\frac{\operatorname{ctn}^{n-1} x}{n-1}$, $[n \neq 1]$.

453.8. $\displaystyle\int \frac{\cos^n x \, dx}{\sin^n x} = \int \operatorname{ctn}^n x \, dx$

$$= -\frac{\operatorname{ctn}^{n-1} x}{n-1} - \int \operatorname{ctn}^{n-2} x \, dx,$$

$$[n \neq 1. \quad \text{See } \mathbf{490.9}].$$

453.9. $\displaystyle\int \frac{\cos^n x \, dx}{\sin^m x}$

$$= -\frac{\cos^{n+1} x}{(m-1)\sin^{m-1} x} - \frac{n-m+2}{m-1}\int \frac{\cos^n x \, dx}{\sin^{m-2} x},$$

$$[m \neq 1],$$

$$= \frac{\cos^{n-1} x}{(n-m)\sin^{m-1} x} + \frac{n-1}{n-m}\int \frac{\cos^{n-2} x \, dx}{\sin^m x}, \qquad [m \neq n],$$

$$= -\frac{\cos^{n-1} x}{(m-1)\sin^{m-1} x} - \frac{n-1}{m-1}\int \frac{\cos^{n-2} x \, dx}{\sin^{m-2} x}, \quad [m \neq 1].$$

454.01. $\displaystyle\int \frac{\sin x \, dx}{1 + \cos x} = -\log (1 + \cos x).$

454.02. $\displaystyle\int \frac{\sin x \, dx}{1 - \cos x} = \log (1 - \cos x).$

454.03. $\displaystyle\int \frac{\cos x \, dx}{1 + \sin x} = \log (1 + \sin x).$

454.04. $\displaystyle\int \frac{\cos x \, dx}{1 - \sin x} = -\log (1 - \sin x).$

454.05. $\displaystyle\int \frac{dx}{\sin x(1 + \cos x)} = \frac{1}{2(1 + \cos x)} + \frac{1}{2} \log \left| \tan \frac{x}{2} \right|.$

454.06. $\displaystyle\int \frac{dx}{\sin x(1-\cos x)} = -\frac{1}{2(1-\cos x)} + \frac{1}{2}\log\left|\tan\frac{x}{2}\right|.$

454.07. $\displaystyle\int \frac{dx}{\cos x(1+\sin x)} = -\frac{1}{2(1+\sin x)} + \frac{1}{2}\log\left|\tan\left(\frac{\pi}{4}+\frac{x}{2}\right)\right|.$

454.08. $\displaystyle\int \frac{dx}{\cos x(1-\sin x)} = \frac{1}{2(1-\sin x)} + \frac{1}{2}\log\left|\tan\left(\frac{\pi}{4}+\frac{x}{2}\right)\right|.$

454.09. $\displaystyle\int \frac{\sin x\,dx}{\cos x(1+\cos x)} = \log\left|\frac{1+\cos x}{\cos x}\right|.$

454.10. $\displaystyle\int \frac{\sin x\,dx}{\cos x(1-\cos x)} = \log\left|\frac{1-\cos x}{\cos x}\right|.$

454.11. $\displaystyle\int \frac{\cos x\,dx}{\sin x(1+\sin x)} = -\log\left|\frac{1+\sin x}{\sin x}\right|.$

454.12. $\displaystyle\int \frac{\cos x\,dx}{\sin x(1-\sin x)} = -\log\left|\frac{1-\sin x}{\sin x}\right|.$

454.13. $\displaystyle\int \frac{\sin x\,dx}{\cos x(1+\sin x)} = \frac{1}{2(1+\sin x)} + \frac{1}{2}\log\left|\tan\left(\frac{\pi}{4}+\frac{x}{2}\right)\right|.$

454.14. $\displaystyle\int \frac{\sin x\,dx}{\cos x(1-\sin x)} = \frac{1}{2(1-\sin x)} - \frac{1}{2}\log\left|\tan\left(\frac{\pi}{4}+\frac{x}{2}\right)\right|.$

454.15. $\displaystyle\int \frac{\cos x\,dx}{\sin x(1+\cos x)} = -\frac{1}{2(1+\cos x)} + \frac{1}{2}\log\left|\tan\frac{x}{2}\right|.$

454.16. $\displaystyle\int \frac{\cos x\,dx}{\sin x(1-\cos x)} = -\frac{1}{2(1-\cos x)} - \frac{1}{2}\log\left|\tan\frac{x}{2}\right|.$

455.01. $\displaystyle\int \frac{dx}{\sin x+\cos x} = \frac{1}{\sqrt{2}}\log\left|\tan\left(\frac{x}{2}+\frac{\pi}{8}\right)\right|.$

455.02. $\displaystyle\int \frac{dx}{\sin x-\cos x} = \frac{1}{\sqrt{2}}\log\left|\tan\left(\frac{x}{2}-\frac{\pi}{8}\right)\right|.$

455.03. $\displaystyle\int \frac{\sin x\,dx}{\sin x+\cos x} = \frac{x}{2} - \frac{1}{2}\log|\sin x+\cos x|.$

[See **482.2** and **492.1**.]

455.04. $\displaystyle\int \frac{\sin x \, dx}{\sin x - \cos x} = \frac{x}{2} + \frac{1}{2} \log |\sin x - \cos x|.$

[See **482.2** and **492.1.**]

455.05. $\displaystyle\int \frac{\cos x \, dx}{\sin x + \cos x} = \frac{x}{2} + \frac{1}{2} \log |\sin x + \cos x|.$

[See **482.1** and **492.2.**]

455.06. $\displaystyle\int \frac{\cos x \, dx}{\sin x - \cos x} = -\frac{x}{2} + \frac{1}{2} \log |\sin x - \cos x|.$

[See **482.1** and **492.2.**]

455.07. $\displaystyle\int \frac{dx}{(\sin x + \cos x)^2} = \frac{1}{2} \tan \left(x - \frac{\pi}{4} \right).$

455.08. $\displaystyle\int \frac{dx}{(\sin x - \cos x)^2} = \frac{1}{2} \tan \left(x + \frac{\pi}{4} \right).$

455.09. $\displaystyle\int \frac{dx}{1 + \cos x \pm \sin x} = \pm \log \left| 1 \pm \tan \frac{x}{2} \right|.$

456.1. $\displaystyle\int \frac{dx}{b \cos x + c \sin x} = \frac{1}{r} \log \left| \tan \frac{x + \theta}{2} \right|$

where $r = \sqrt{(b^2 + c^2)}$, $\sin \theta = b/r$, $\cos \theta = c/r$.

[See **401.2** and **432.10.**]

456.2. $\displaystyle\int \frac{dx}{a + b \cos x + c \sin x} = \int \frac{d(x + \theta)}{a + r \sin (x + \theta)}$

where r and θ are given in **456.1.**   [See **436.00.**]

460.1. $\displaystyle\int \frac{dx}{a^2 \cos^2 x + b^2 \sin^2 x} = \frac{1}{ab} \tan^{-1} \left(\frac{b}{a} \tan x \right),$ [See **436.5.**]

460.2. $\displaystyle\int \frac{dx}{a^2 \cos^2 x - b^2 \sin^2 x} = \frac{1}{2ab} \log \left| \frac{b \tan x + a}{b \tan x - a} \right|,$ [See **436.7.**]

461. $\displaystyle\int \sin^m x \cos^n x \, dx.$ If either m or n is a positive odd integer,

the other not necessarily positive nor an integer, put

$$\sin^2 x = 1 - \cos^2 x \quad \text{and} \quad \sin x \, dx = -d \cos x$$

or put

$$\cos^2 x = 1 - \sin^2 x \quad \text{and} \quad \cos x \, dx = d \sin x.$$

If both m and n are positive even integers, put

$$\sin^2 x = \tfrac{1}{2}(1 - \cos 2x), \quad \cos^2 x = \tfrac{1}{2}(1 + \cos 2x)$$

and

$$\sin x \cos x = \tfrac{1}{2} \sin 2x,$$

and similar expressions involving $2x$ instead of x, and so on. See also **450.9**.

465. $\displaystyle \int \sin mx \cos nx \, dx = -\frac{\cos (m-n)x}{2(m-n)} - \frac{\cos (m+n)x}{2(m+n)},$

$$[m^2 \neq n^2]. \qquad [\text{If } m^2 = n^2, \text{ see } \mathbf{450.11}.]$$

470.1. $\displaystyle \int \frac{\cos x \, dx}{\sqrt{(1 + m^2 \sin^2 x)}} = \frac{1}{m} \log\{m \sin x + \sqrt{(1 + m^2 \sin^2 x)}\}.$

470.2. $\displaystyle \int \frac{\cos x \, dx}{\sqrt{(1 - m^2 \sin^2 x)}} = \frac{1}{m} \sin^{-1}(m \sin x).$

470.3. $\displaystyle \int (\cos x)\sqrt{(1 + m^2 \sin^2 x)}\,dx$

$$= \frac{\sin x}{2}\sqrt{(1 + m^2 \sin^2 x)}$$

$$+ \frac{1}{2m}\log\{m \sin x + \sqrt{(1 + m^2 \sin^2 x)}\}.$$

470.4. $\displaystyle \int (\cos x)\sqrt{(1 - m^2 \sin^2 x)}\,dx$

$$= \frac{\sin x}{2}\sqrt{(1 - m^2 \sin^2 x)} + \frac{1}{2m}\sin^{-1}(m \sin x).$$

475.1. $\displaystyle \int f(x, \sin x)dx = -\int f\left(\frac{\pi}{2} - y, \cos y\right) dy,$

where

$$y = \pi/2 - x.$$

475.2. $\displaystyle \int f(x, \cos x)dx = -\int f\left(\frac{\pi}{2} - y, \sin y\right) dy,$

where

$$y = \pi/2 - x.$$

Integrals Involving tan x

480.1. $\displaystyle\int \tan x \, dx = -\log |\cos x| = \log |\sec x|.$

[See **452.11** and **603.4.**]

480.2. $\displaystyle\int \tan^2 x \, dx = \tan x - x.$ [See **452.22.**]

480.3. $\displaystyle\int \tan^3 x \, dx = \tfrac{1}{2} \tan^2 x + \log |\cos x|.$ [See **452.33.**]

480.4. $\displaystyle\int \tan^4 x \, dx = \tfrac{1}{3} \tan^3 x - \tan x + x.$

480.9. $\displaystyle\int \tan^n x \, dx = \frac{\tan^{n-1} x}{n-1} - \int \tan^{n-2} x \, dx,$

$[n \neq 1. \quad \text{See } \mathbf{452.8}].$

481.1. $\displaystyle\int x \tan x \, dx = \frac{x^3}{3} + \frac{x^5}{15} + \frac{2}{105} x^7 + \frac{17}{2835} x^9$

$$+ \frac{62}{11 \times 2835} x^{11} + \cdots + \frac{2^{2n}(2^{2n} - 1)B_n}{(2n+1)!} x^{2n+1} + \cdots,$$

$[x^2 < \pi^2/4. \quad \text{See } \mathbf{415.03} \text{ and } \mathbf{45}].$

481.2. $\displaystyle\int \frac{\tan x \, dx}{x} = x + \frac{x^3}{9} + \frac{2}{75} x^5 + \frac{17}{2205} x^7 + \frac{62}{9 \times 2835} x^9$

$$+ \cdots + \frac{2^{2n}(2^{2n} - 1)B_n}{(2n-1)(2n)!} x^{2n-1} + \cdots,$$

$[x^2 < \pi^2/4. \quad \text{See } \mathbf{415.03} \text{ and } \mathbf{45}].$

482.1. $\displaystyle\int \frac{dx}{\tan x \pm 1} = \pm \frac{x}{2} + \frac{1}{2} \log |\sin x \pm \cos x|.$

[See **455.05** and **.06.**]

482.2. $\displaystyle\int \frac{\tan x \, dx}{\tan x \pm 1} = \int \frac{dx}{1 \pm \text{ctn } x} = \frac{x}{2} \mp \frac{1}{2} \log |\sin x \pm \cos x|.$

[See **455.03, 455.04,** and **492.1.**]

483. $\displaystyle\int \frac{dx}{a + b \tan x} = \int \frac{\cos x \, dx}{a \cos x + b \sin x}$

$$= \frac{1}{(a^2 + b^2)} \{ax + b \log (a \cos x + b \sin x)\}.$$

Integrals Involving ctn x

490.1. $\displaystyle\int \mathrm{ctn}\, x\, dx = \log |\sin x|.$ [See **453.11** and **603.1.**]

490.2. $\displaystyle\int \mathrm{ctn}^2\, x\, dx = -\,\mathrm{ctn}\, x - x.$ [See **453.22.**]

490.3. $\displaystyle\int \mathrm{ctn}^3\, x\, dx = -\tfrac{1}{2}\,\mathrm{ctn}^2\, x - \log |\sin x|.$ [See **453.33.**]

490.4. $\displaystyle\int \mathrm{ctn}^4\, x\, dx = -\tfrac{1}{3}\,\mathrm{ctn}^3\, x + \mathrm{ctn}\, x + x.$

490.9. $\displaystyle\int \mathrm{ctn}^n\, x\, dx = -\frac{\mathrm{ctn}^{n-1} x}{n-1} - \int \mathrm{ctn}^{n-2}\, x\, dx,$

$$[n \neq 1. \quad \text{See } \mathbf{453.8}].$$

491.1. $\displaystyle\int x\, \mathrm{ctn}\, x\, dx = x - \frac{x^3}{9} - \frac{x^5}{225} - \frac{2x^7}{6615} - \frac{x^9}{9 \times 4725}$

$$-\cdots - \frac{2^{2n} B_n}{(2n+1)!}\, x^{2n+1} - \cdots.$$

[See **415.04** and **45.**]

491.2. $\displaystyle\int \frac{\mathrm{ctn}\, x\, dx}{x} = -\frac{1}{x} - \frac{x}{3} - \frac{x^3}{135} - \frac{2x^5}{4725} - \frac{x^7}{7 \times 4725}$

$$-\cdots - \frac{2^{2n} B_n}{(2n-1)(2n)!}\, x^{2n-1} - \cdots.$$

[See **415.04** and **45.**]

492.1. $\displaystyle\int \frac{dx}{1 \pm \mathrm{ctn}\, x} = \int \frac{\tan x\, dx}{\tan x \pm 1}.$ [See **482.2.**]

492.2. $\displaystyle\int \frac{\mathrm{ctn}\, x\, dx}{1 \pm \mathrm{ctn}\, x} = \int \frac{dx}{\tan x \pm 1}.$ [See **482.1.**]

493. $\displaystyle\int \frac{dx}{a + b\,\mathrm{ctn}\, x} = \int \frac{\sin x\, dx}{a \sin x + b \cos x}$

$$= \frac{1}{(a^2 + b^2)} \{ax - b \log (a \sin x + b \cos x)\}.$$

⁛ 3 ⁛

INVERSE TRIGONOMETRIC FUNCTIONS

500.

The following equations do not refer in general to the multiple values of the inverse trigonometric functions, but to the principal values. That is, $\sin^{-1} x$ and $\tan^{-1} x$ lie in the range from $-\pi/2$ to $\pi/2$ and $\cos^{-1} x$ and $\mathrm{ctn}^{-1} x$ in the range from 0 to π. Care should be taken in dealing with inverse functions and in integrating from one point to another. A process of curve plotting is frequently of assistance. Some of the graphs have more than one branch, and in general, integration should not be carried out from a point on one branch to a point on another branch.

501. $\quad \sin^{-1} x = x + \dfrac{x^3}{2 \cdot 3} + \dfrac{1 \cdot 3 x^5}{2 \cdot 4 \cdot 5} + \dfrac{1 \cdot 3 \cdot 5 x^7}{2 \cdot 4 \cdot 6 \cdot 7} + \cdots,$

$$[x^2 < 1. \quad -\pi/2 < \sin^{-1} x < \pi/2].$$

[Expand $1/\sqrt{(1 - x^2)}$ and then integrate it.]

502. $\quad \cos^{-1} x = \dfrac{\pi}{2} - \left(x + \dfrac{x^3}{2 \cdot 3} + \dfrac{1 \cdot 3 x^5}{2 \cdot 4 \cdot 5} + \dfrac{1 \cdot 3 \cdot 5 x^7}{2 \cdot 4 \cdot 6 \cdot 7} + \cdots \right),$

$$[x^2 < 1. \quad 0 < \cos^{-1} x < \pi].$$

503. $\quad \csc^{-1} x = \dfrac{1}{x} + \dfrac{1}{2 \cdot 3 x^3} + \dfrac{1 \cdot 3}{2 \cdot 4 \cdot 5 x^5} + \dfrac{1 \cdot 3 \cdot 5}{2 \cdot 4 \cdot 6 \cdot 7 x^7} + \cdots,$

$$[x^2 > 1. \quad -\pi/2 < \csc^{-1} x < \pi/2].$$

504. $\quad \sec^{-1} x = \dfrac{\pi}{2} - \left(\dfrac{1}{x} + \dfrac{1}{2 \cdot 3 x^3} + \dfrac{1 \cdot 3}{2 \cdot 4 \cdot 5 x^5} + \dfrac{1 \cdot 3 \cdot 5}{2 \cdot 4 \cdot 6 \cdot 7 x^7} + \cdots \right),$

$$[x^2 > 1. \quad 0 < \sec^{-1} x < \pi].$$

505.1. $\quad \tan^{-1} x = x - \dfrac{x^3}{3} + \dfrac{x^5}{5} - \dfrac{x^7}{7} + \cdots, \qquad\qquad [x^2 < 1].$

[Expand $1/(1 + x^2)$ and then integrate it.]

118

505.2. $\tan^{-1} x = \dfrac{\pi}{2} - \dfrac{1}{x} + \dfrac{1}{3x^3} - \dfrac{1}{5x^5} + \dfrac{1}{7x^7} - \cdots,$ $[x > 1].$

505.3. $\tan^{-1} x = -\dfrac{\pi}{2} - \dfrac{1}{x} + \dfrac{1}{3x^3} - \dfrac{1}{5x^5} + \dfrac{1}{7x^7} - \cdots,$

$[x < -1].$

505.4. $\tan^{-1} x = \dfrac{x}{1 + x^2}\left[1 + \dfrac{2}{3}\left(\dfrac{x^2}{1 + x^2}\right) + \dfrac{2\cdot4}{3\cdot5}\left(\dfrac{x^2}{1 + x^2}\right)^2 \right.$

$\left. + \dfrac{2\cdot4\cdot6}{3\cdot5\cdot7}\left(\dfrac{x^2}{1 + x^2}\right)^3 + \cdots\right],$ $[x^2 < \infty].$

[Ref. 31, p. 122.]

For these equations, $\tan^{-1} x$ is between $-\pi/2$ and $\pi/2$.

506.1. $\operatorname{ctn}^{-1} x = \dfrac{\pi}{2} - x + \dfrac{x^3}{3} - \dfrac{x^5}{5} + \dfrac{x^7}{7} - \cdots,$ $[x^2 < 1].$

506.2. $\operatorname{ctn}^{-1} x = \dfrac{1}{x} - \dfrac{1}{3x^3} + \dfrac{1}{5x^5} - \dfrac{1}{7x^7} + \cdots,$ $[x > 1].$

506.3. $\operatorname{ctn}^{-1} x = \pi + \dfrac{1}{x} - \dfrac{1}{3x^3} + \dfrac{1}{5x^5} - \dfrac{1}{7x^7} + \cdots,$ $[x < -1].$

507.10. $\sin^{-1}(x \pm iy) = n\pi + (-1)^n \sin^{-1}\dfrac{2x}{p + q}$

$\pm i(-1)^n \cosh^{-1}\dfrac{p + q}{2}$

taking the principal value of $\sin^{-1}$ (between $-\pi/2$ and $\pi/2$) and the positive values of $\cosh^{-1}$ and of p and q. The quantity $i = \sqrt{-1}$, and n is an integer or 0. The quantity x may be positive or negative but y is positive.

507.11.   The quantity $p = \sqrt{(1 + x)^2 + y^2}$ (positive value),

and

507.12. $q = \sqrt{(1 - x)^2 + y^2}$ (positive value).

Note that if $y = 0$ and $x > 1$, $q = x - 1$ and $p + q = 2x$. If $y = 0$ and $x < 1$, $q = 1 - x$ and $p + q = 2$.

Alternative:

507.13a. $\sin^{-1} A = - i \log_\epsilon (\pm \sqrt{1 - A^2} + iA) + 2k\pi$

or

507.13b. $= i \log_\epsilon (\pm \sqrt{1 - A^2} - iA) + 2k\pi$

where A may be a complex quantity and k is an integer or 0.

For the square root of a complex quantity see **58** and for the logarithm see **604**. The two solutions a and b are identical. The one should be used, in any given case, which involves the numerical sum of two quantities instead of the difference, so as to obtain more convenient precise computation.

507.20. $\cos^{-1} (x + iy)$

$$= \pm \left(\cos^{-1} \frac{2x}{p + q} + 2k\pi - i \cosh^{-1} \frac{p + q}{2} \right).$$

507.21. $\cos^{-1} (x - iy)$

$$= \pm \left(\cos^{-1} \frac{2x}{p + q} + 2k\pi + i \cosh^{-1} \frac{p + q}{2} \right),$$

where y is positive, taking the principal value of $\cos^{-1}$ (between 0 and π) and the positive value of $\cosh^{-1}$. See **507.11** and **507.12**.

Alternative:

507.22a. $\cos^{-1} A = \mp i \log_\epsilon (A + \sqrt{A^2 - 1}) + 2k\pi$

or

507.22b. $= \pm i \log_\epsilon (A - \sqrt{A^2 - 1}) + 2k\pi$

where A may be a complex quantity. See note under **507.13**.

507.30. $\tan^{-1} (x + iy)$

$$= \frac{1}{2} \left\{ (2k + 1)\pi - \tan^{-1} \frac{1 + y}{x} - \tan^{-1} \frac{1 - y}{x} \right\}$$
$$+ \frac{i}{4} \log_\epsilon \frac{(1 + y)^2 + x^2}{(1 - y)^2 + x^2},$$

where the principal values of $\tan^{-1}$ are taken (between $- \pi|2$ and $\pi|2$) and where x and y may be positive or negative.

Alternative:

507.31. $\tan^{-1} (x + iy) = \frac{1}{2} \tan^{-1} \frac{2x}{1 - x^2 - y^2} + \pi k$

$$+ \frac{i}{4} \log_\epsilon \frac{(1 + y)^2 + x^2}{(1 - y)^2 + x^2},$$

where k is 0 or an integer. The proper quadrant for $\tan^{-1}$ is to be taken according to the signs of the values of the numerator and the denominator when they are given in numbers.

507.32. $\tan^{-1}(x + iy) = \dfrac{i}{2} \log_e \dfrac{1 + y - ix}{1 - y + ix} + 2k\pi.$ [See **604.**]

[Ref. 46, Chap. XI.]

508.   For small values of $\cos^{-1} x$,

$$\cos^{-1} x = \left[2(1 - x) + \frac{1}{3}(1 - x)^2 + \frac{4}{45}(1 - x)^3 \right.$$
$$\left. + \frac{1}{35}(1 - x)^4 \cdots \right]^{1/2}$$

The last term used should be practically negligible. The numerical value of the square root may be taken from a large table of square roots, as in Ref. 65.

INVERSE TRIGONOMETRIC FUNCTIONS— DERIVATIVES

512.0. $\dfrac{d}{dx} \sin^{-1} \dfrac{x}{a} = \dfrac{1}{\sqrt{(a^2 - x^2)}},$ [1st and 4th quadrants].

512.1. $\dfrac{d}{dx} \sin^{-1} \dfrac{x}{a} = \dfrac{-1}{\sqrt{(a^2 - x^2)}},$ [2nd and 3rd quadrants].

512.2. $\dfrac{d}{dx} \cos^{-1} \dfrac{x}{a} = \dfrac{-1}{\sqrt{(a^2 - x^2)}},$ [1st and 2nd quadrants].

512.3. $\dfrac{d}{dx} \cos^{-1} \dfrac{x}{a} = \dfrac{1}{\sqrt{(a^2 - x^2)}},$ [3rd and 4th quadrants].

512.4. $\dfrac{d}{dx} \tan^{-1} \dfrac{x}{a} = \dfrac{a}{a^2 + x^2}.$

512.5. $\dfrac{d}{dx} \operatorname{ctn}^{-1} \dfrac{x}{a} = \dfrac{-a}{a^2 + x^2}.$

512.6. $\dfrac{d}{dx} \sec^{-1} \dfrac{x}{a} = \dfrac{a}{x\sqrt{(x^2 - a^2)}},$ [1st and 3rd quadrants].

512.7. $\dfrac{d}{dx} \sec^{-1} \dfrac{x}{a} = \dfrac{-a}{x\sqrt{(x^2 - a^2)}},$ [2nd and 4th quadrants].

512.8. $\dfrac{d}{dx} \csc^{-1} \dfrac{x}{a} = \dfrac{-a}{x\sqrt{(x^2 - a^2)}},$ [1st and 3rd quadrants].

512.9. $\dfrac{d}{dx} \csc^{-1} \dfrac{x}{a} = \dfrac{a}{x\sqrt{(x^2 - a^2)}},$ [2nd and 4th quadrants].

[Except in **512.4** and **512.5**, $a > 0$.]

INVERSE TRIGONOMETRIC FUNCTIONS— INTEGRALS ($a > 0$)

515. $\displaystyle\int \sin^{-1} \dfrac{x}{a}\, dx = x \sin^{-1} \dfrac{x}{a} + \sqrt{(a^2 - x^2)}.$

516. $\displaystyle\int \left(\sin^{-1} \dfrac{x}{a}\right)^2 dx = x \left(\sin^{-1} \dfrac{x}{a}\right)^2 - 2x + 2\sqrt{(a^2 - x^2)} \sin^{-1} \dfrac{x}{a}.$

517.1. $\displaystyle\int x \sin^{-1} \dfrac{x}{a}\, dx = \left(\dfrac{x^2}{2} - \dfrac{a^2}{4}\right) \sin^{-1} \dfrac{x}{a} + \dfrac{x}{4} \sqrt{(a^2 - x^2)}.$

517.2. $\displaystyle\int x^2 \sin^{-1} \dfrac{x}{a}\, dx = \dfrac{x^3}{3} \sin^{-1} \dfrac{x}{a} + \dfrac{1}{9} (x^2 + 2a^2)\sqrt{(a^2 - x^2)}.$

517.3. $\displaystyle\int x^3 \sin^{-1} \dfrac{x}{a}\, dx = \left(\dfrac{x^4}{4} - \dfrac{3a^4}{32}\right) \sin^{-1} \dfrac{x}{a}$

$$+ \dfrac{1}{32} (2x^3 + 3xa^2)\sqrt{(a^2 - x^2)}.$$

517.4. $\displaystyle\int x^4 \sin^{-1} \dfrac{x}{a}\, dx = \dfrac{x^5}{5} \sin^{-1} \dfrac{x}{a}$

$$+ \dfrac{1}{75} (3x^4 + 4x^2a^2 + 8a^4)\sqrt{(a^2 - x^2)}.$$

517.5. $\displaystyle\int x^5 \sin^{-1} \dfrac{x}{a}\, dx = \left(\dfrac{x^6}{6} - \dfrac{5a^6}{96}\right) \sin^{-1} \dfrac{x}{a}$

$$+ \dfrac{1}{288} (8x^5 + 10x^3a^2 + 15xa^4)\sqrt{(a^2 - x^2)}.$$

517.6. $\displaystyle\int x^6 \sin^{-1} \dfrac{x}{a}\, dx = \dfrac{x^7}{7} \sin^{-1} \dfrac{x}{a}$

$$+ \dfrac{1}{245} (5x^6 + 6x^4a^2 + 8x^2a^4 + 16a^6)\sqrt{(a^2 - x^2)}.$$

517.9. $\displaystyle\int x^n \sin^{-1} \frac{x}{a}\, dx = \frac{x^{n+1}}{n+1} \sin^{-1} \frac{x}{a} - \frac{1}{n+1} \int \frac{x^{n+1}dx}{\sqrt{(a^2 - x^2)}},$

$$[n \neq -1]. \qquad [\text{See } 321\text{–}327.]$$

518.1. $\displaystyle\int \frac{1}{x} \sin^{-1} \frac{x}{a}\, dx = \frac{x}{a} + \frac{1}{2\cdot3\cdot3} \frac{x^3}{a^3} + \frac{1\cdot3}{2\cdot4\cdot5\cdot5} \frac{x^5}{a^5}$

$$+ \frac{1\cdot3\cdot5}{2\cdot4\cdot6\cdot7\cdot7} \frac{x^7}{a^7} + \cdots, \qquad [x^2 < a^2].$$

518.2. $\displaystyle\int \frac{1}{x^2} \sin^{-1} \frac{x}{a}\, dx = - \frac{1}{x} \sin^{-1} \frac{x}{a} - \frac{1}{a} \log \left| \frac{a + \sqrt{(a^2 - x^2)}}{x} \right|.$

518.3. $\displaystyle\int \frac{1}{x^3} \sin^{-1} \frac{x}{a}\, dx = - \frac{1}{2x^2} \sin^{-1} \frac{x}{a} - \frac{\sqrt{(a^2 - x^2)}}{2a^2x}.$

518.4. $\displaystyle\int \frac{1}{x^4} \sin^{-1} \frac{x}{a}\, dx = - \frac{1}{3x^3} \sin^{-1} \frac{x}{a} - \frac{\sqrt{(a^2 - x^2)}}{6a^2x^2}$

$$- \frac{1}{6a^3} \log \left| \frac{a + \sqrt{(a^2 - x^2)}}{x} \right|.$$

518.9. $\displaystyle\int \frac{1}{x^n} \sin^{-1} \frac{x}{a}\, dx = - \frac{1}{(n-1)x^{n-1}} \sin^{-1} \frac{x}{a}$

$$+ \frac{1}{n-1} \int \frac{dx}{x^{n-1}\sqrt{(a^2 - x^2)}}, \qquad [n \neq 1].$$

$$[\text{See } 341\text{–}346.]$$

520. $\displaystyle\int \cos^{-1} \frac{x}{a}\, dx = x \cos^{-1} \frac{x}{a} - \sqrt{(a^2 - x^2)}.$

521. $\displaystyle\int \left(\cos^{-1} \frac{x}{a}\right)^2 dx = x \left(\cos^{-1} \frac{x}{a}\right)^2 - 2x - 2\sqrt{(a^2 - x^2)} \cos^{-1} \frac{x}{a}.$

522.1. $\displaystyle\int x \cos^{-1} \frac{x}{a}\, dx = \left(\frac{x^2}{2} - \frac{a^2}{4}\right) \cos^{-1} \frac{x}{a} - \frac{x}{4} \sqrt{(a^2 - x^2)}.$

522.2. $\displaystyle\int x^2 \cos^{-1} \frac{x}{a}\, dx = \frac{x^3}{3} \cos^{-1} \frac{x}{a} - \frac{1}{9} (x^2 + 2a^2)\sqrt{(a^2 - x^2)}.$

522.3. $\displaystyle\int x^3 \cos^{-1} \frac{x}{a}\, dx = \left(\frac{x^4}{4} - \frac{3a^4}{32}\right) \cos^{-1} \frac{x}{a}$

$$- \frac{1}{32} (2x^3 + 3xa^2)\sqrt{(a^2 - x^2)}.$$

522.4. $\displaystyle\int x^4 \cos^{-1}\frac{x}{a}\,dx = \frac{x^5}{5}\cos^{-1}\frac{x}{a}$

$$-\frac{1}{75}\left(3x^4 + 4x^2a^2 + 8a^4\right)\sqrt{(a^2 - x^2)}.$$

522.5. $\displaystyle\int x^5 \cos^{-1}\frac{x}{a}\,dx = \left(\frac{x^6}{6} - \frac{5a^6}{96}\right)\cos^{-1}\frac{x}{a}$

$$-\frac{1}{288}\left(8x^5 + 10x^3a^2 + 15xa^4\right)\sqrt{(a^2 - x^2)}.$$

522.6. $\displaystyle\int x^6 \cos^{-1}\frac{x}{a}\,dx = \frac{x^7}{7}\cos^{-1}\frac{x}{a}$

$$-\frac{1}{245}\left(5x^6 + 6x^4a^2 + 8x^2a^4 + 16a^6\right)\sqrt{(a^2 - x^2)}.$$

522.9. $\displaystyle\int x^n \cos^{-1}\frac{x}{a}\,dx = \frac{x^{n+1}}{n+1}\cos^{-1}\frac{x}{a} + \frac{1}{n+1}\int\frac{x^{n+1}dx}{\sqrt{(a^2 - x^2)}},$

$$[n \neq -1]. \quad \text{[See 321–327.]}$$

523.1. $\displaystyle\int\frac{1}{x}\cos^{-1}\frac{x}{a}\,dx = \frac{\pi}{2}\log|x| - \frac{x}{a} - \frac{1}{2\cdot3\cdot3}\frac{x^3}{a^3}$

$$-\frac{1\cdot3}{2\cdot4\cdot5\cdot5}\frac{x^5}{a^5} - \frac{1\cdot3\cdot5}{2\cdot4\cdot6\cdot7\cdot7}\frac{x^7}{a^7} - \cdots, \quad [x^2 < a^2].$$

523.2. $\displaystyle\int\frac{1}{x^2}\cos^{-1}\frac{x}{a}\,dx = -\frac{1}{x}\cos^{-1}\frac{x}{a} + \frac{1}{a}\log\left|\frac{a + \sqrt{(a^2 - x^2)}}{x}\right|.$

523.3. $\displaystyle\int\frac{1}{x^3}\cos^{-1}\frac{x}{a}\,dx = -\frac{1}{2x^2}\cos^{-1}\frac{x}{a} + \frac{\sqrt{(a^2 - x^2)}}{2a^2x}.$

523.4. $\displaystyle\int\frac{1}{x^4}\cos^{-1}\frac{x}{a}\,dx = -\frac{1}{3x^3}\cos^{-1}\frac{x}{a} + \frac{\sqrt{(a^2 - x^2)}}{6a^2x^2}$

$$+\frac{1}{6a^3}\log\left|\frac{a + \sqrt{(a^2 - x^2)}}{x}\right|.$$

523.9. $\displaystyle\int\frac{1}{x^n}\cos^{-1}\frac{x}{a}\,dx = -\frac{1}{(n-1)x^{n-1}}\cos^{-1}\frac{x}{a}$

$$-\frac{1}{n-1}\int\frac{dx}{x^{n-1}\sqrt{(a^2 - x^2)}}, \quad [n \neq 1].$$

$$\text{[See 341–346.]}$$

525. $\displaystyle\int \tan^{-1}\frac{x}{a}\,dx = x\tan^{-1}\frac{x}{a} - \frac{a}{2}\log(a^2+x^2).$

525.1. $\displaystyle\int x\tan^{-1}\frac{x}{a}\,dx = \frac{1}{2}(x^2+a^2)\tan^{-1}\frac{x}{a} - \frac{ax}{2}.$

525.2. $\displaystyle\int x^2\tan^{-1}\frac{x}{a}\,dx = \frac{x^3}{3}\tan^{-1}\frac{x}{a} - \frac{ax^2}{6} + \frac{a^3}{6}\log(a^2+x^2).$

525.3. $\displaystyle\int x^3\tan^{-1}\frac{x}{a}\,dx = \frac{1}{4}(x^4-a^4)\tan^{-1}\frac{x}{a} - \frac{ax^3}{12} + \frac{a^3x}{4}.$

525.4. $\displaystyle\int x^4\tan^{-1}\frac{x}{a}\,dx = \frac{x^5}{5}\tan^{-1}\frac{x}{a} - \frac{ax^4}{20} + \frac{a^3x^2}{10} - \frac{a^5}{10}\log(a^2+x^2).$

525.5. $\displaystyle\int x^5\tan^{-1}\frac{x}{a}\,dx = \frac{1}{6}(x^6+a^6)\tan^{-1}\frac{x}{a} - \frac{ax^5}{30} + \frac{a^3x^3}{18} - \frac{a^5x}{6}.$

525.6. $\displaystyle\int x^6\tan^{-1}\frac{x}{a}\,dx = \frac{x^7}{7}\tan^{-1}\frac{x}{a} - \frac{ax^6}{42} + \frac{a^3x^4}{28} - \frac{a^5x^2}{14}$
$$+ \frac{a^7}{14}\log(a^2+x^2).$$

525.9. $\displaystyle\int x^n\tan^{-1}\frac{x}{a}\,dx = \frac{x^{n+1}}{n+1}\tan^{-1}\frac{x}{a} - \frac{a}{n+1}\int\frac{x^{n+1}dx}{a^2+x^2},$
$$[n \neq -1). \qquad [\text{See } \mathbf{121\text{-}128}.]$$

526.1. $\displaystyle\int \frac{1}{x}\tan^{-1}\frac{x}{a}\,dx = \frac{x}{a} - \frac{x^3}{3^2a^3} + \frac{x^5}{5^2a^5} - \frac{x^7}{7^2a^7} + \cdots, \quad [x^2 < a^2],$

$$= \frac{\pi}{2}\log|x| + \frac{a}{x} - \frac{a^3}{3^2x^3} + \frac{a^5}{5^2x^5} - \frac{a^7}{7^2x^7} + \cdots,$$
$$[x/a > 1],$$

$$= -\frac{\pi}{2}\log|x| + \frac{a}{x} - \frac{a^3}{3^2x^3} + \frac{a^5}{5^2x^5} - \frac{a^7}{7^2x^7} + \cdots,$$
$$[x/a < -1].$$

For these equations, $\tan^{-1}(x/a)$ is between $-\pi/2$ and $\pi/2$.

526.2. $\displaystyle\int \frac{1}{x^2}\tan^{-1}\frac{x}{a}\,dx = -\frac{1}{x}\tan^{-1}\frac{x}{a} - \frac{1}{2a}\log\frac{a^2+x^2}{x^2}.$

526.3. $\displaystyle\int \frac{1}{x^3}\tan^{-1}\frac{x}{a}\,dx = -\frac{1}{2}\left(\frac{1}{x^2}+\frac{1}{a^2}\right)\tan^{-1}\frac{x}{a} - \frac{1}{2ax}.$

526.4. $\int \dfrac{1}{x^4} \tan^{-1} \dfrac{x}{a}\, dx = -\dfrac{1}{3x^3} \tan^{-1} \dfrac{x}{a} - \dfrac{1}{6ax^2} + \dfrac{1}{6a^3} \log \dfrac{a^2 + x^2}{x^2}.$

526.5. $\int \dfrac{1}{x^5} \tan^{-1} \dfrac{x}{a}\, dx = \dfrac{1}{4}\left(\dfrac{1}{a^4} - \dfrac{1}{x^4}\right) \tan^{-1} \dfrac{x}{a} - \dfrac{1}{12ax^3} + \dfrac{1}{4a^3x}.$

526.9. $\int \dfrac{1}{x^n} \tan^{-1} \dfrac{x}{a}\, dx = -\dfrac{1}{(n-1)x^{n-1}} \tan^{-1} \dfrac{x}{a}$

$$+ \dfrac{a}{n-1} \int \dfrac{dx}{x^{n-1}(a^2 + x^2)}, \qquad [n \neq 1].$$

[See **131–135**.]

528. $\int \operatorname{ctn}^{-1} \dfrac{x}{a}\, dx = x \operatorname{ctn}^{-1} \dfrac{x}{a} + \dfrac{a}{2} \log (a^2 + x^2).$

528.1. $\int x \operatorname{ctn}^{-1} \dfrac{x}{a}\, dx = \dfrac{1}{2} (x^2 + a^2) \operatorname{ctn}^{-1} \dfrac{x}{a} + \dfrac{ax}{2}.$

528.2. $\int x^2 \operatorname{ctn}^{-1} \dfrac{x}{a}\, dx = \dfrac{x^3}{3} \operatorname{ctn}^{-1} \dfrac{x}{a} + \dfrac{ax^2}{6} - \dfrac{a^3}{6} \log (a^2 + x^2).$

528.3. $\int x^3 \operatorname{ctn}^{-1} \dfrac{x}{a}\, dx = \dfrac{1}{4} (x^4 - a^4) \operatorname{ctn}^{-1} \dfrac{x}{a} + \dfrac{ax^3}{12} - \dfrac{a^3 x}{4}.$

528.4. $\int x^4 \operatorname{ctn}^{-1} \dfrac{x}{a}\, dx = \dfrac{x^5}{5} \operatorname{ctn}^{-1} \dfrac{x}{a} + \dfrac{ax^4}{20} - \dfrac{a^3 x^2}{10}$

$$+ \dfrac{a^5}{10} \log (a^2 + x^2).$$

528.5. $\int x^5 \operatorname{ctn}^{-1} \dfrac{x}{a}\, dx = \dfrac{1}{6} (x^6 + a^6) \operatorname{ctn}^{-1} \dfrac{x}{a} + \dfrac{ax^5}{30} - \dfrac{a^3 x^3}{18} + \dfrac{a^5 x}{6}.$

528.6. $\int x^6 \operatorname{ctn}^{-1} \dfrac{x}{a}\, dx = \dfrac{x^7}{7} \operatorname{ctn}^{-1} \dfrac{x}{a} + \dfrac{ax^6}{42} - \dfrac{a^3 x^4}{28} + \dfrac{a^5 x^2}{14}$

$$- \dfrac{a^7}{14} \log (a^2 + x^2).$$

528.9. $\int x^n \operatorname{ctn}^{-1} \dfrac{x}{a}\, dx = \dfrac{x^{n+1}}{n+1} \operatorname{ctn}^{-1} \dfrac{x}{a} + \dfrac{a}{n+1} \int \dfrac{x^{n+1}dx}{a^2 + x^2},$

$[n \neq -1).$ [See **121–128**.]

529.1. $\int \dfrac{1}{x} \operatorname{ctn}^{-1} \dfrac{x}{a} = \dfrac{\pi}{2} \log |x| - \dfrac{x}{a} + \dfrac{x^3}{3^2 a^3} - \dfrac{x^5}{5^2 a^5} + \dfrac{x^7}{7^2 a^7} - \cdots,$

$$[x^2 < a^2],$$

$$= -\dfrac{a}{x} + \dfrac{a^3}{3^2 x^3} - \dfrac{a^5}{5^2 x^5} + \dfrac{a^7}{7^2 x^7} - \cdots,$$

$$[x/a > 1],$$

$$= \pi \log |x| - \dfrac{a}{x} + \dfrac{a^3}{3^2 x^3} - \dfrac{a^5}{5^2 x^5} + \dfrac{a^7}{7^2 x^7} - \cdots,$$

$$[x/a < -1].$$

For these equations, $\operatorname{ctn}^{-1}(x/a)$ is between 0 and π.

529.2. $\int \dfrac{1}{x^2} \operatorname{ctn}^{-1} \dfrac{x}{a} = -\dfrac{1}{x} \operatorname{ctn}^{-1} \dfrac{x}{a} + \dfrac{1}{2a} \log \dfrac{a^2 + x^2}{x^2}.$

529.3. $\int \dfrac{1}{x^3} \operatorname{ctn}^{-1} \dfrac{x}{a} = -\dfrac{1}{2x^2} \operatorname{ctn}^{-1} \dfrac{x}{a} + \dfrac{1}{2ax} + \dfrac{1}{2a^2} \tan^{-1} \dfrac{x}{a}.$

529.4. $\int \dfrac{1}{x^4} \operatorname{ctn}^{-1} \dfrac{x}{a} = -\dfrac{1}{3x^3} \operatorname{ctn}^{-1} \dfrac{x}{a} + \dfrac{1}{6ax^2} - \dfrac{1}{6a^3} \log \dfrac{a^2 + x^2}{x^2}.$

529.5. $\int \dfrac{1}{x^5} \operatorname{ctn}^{-1} \dfrac{x}{a} = -\dfrac{1}{4x^4} \operatorname{ctn}^{-1} \dfrac{x}{a} + \dfrac{1}{12ax^3} - \dfrac{1}{4a^3 x} - \dfrac{1}{4a^4} \tan^{-1} \dfrac{x}{a}.$

529.9. $\int \dfrac{1}{x^n} \operatorname{ctn}^{-1} \dfrac{x}{a} = -\dfrac{1}{(n-1)x^{n-1}} \operatorname{ctn}^{-1} \dfrac{x}{a}$

$$-\dfrac{a}{n-1} \int \dfrac{dx}{x^{n-1}(a^2 + x^2)}, \qquad [n \neq 1).$$

[See **131–135**.]

531. $\int \sec^{-1} \dfrac{x}{a} \, dx = x \sec^{-1} \dfrac{x}{a} - a \log |x + \sqrt{(x^2 - a^2)}|,$

$$[0 < \sec^{-1}(x/a) < \pi/2].$$

$$= x \sec^{-1} \dfrac{x}{a} + a \log |x + \sqrt{(x^2 - a^2)}|,$$

$$[\pi/2 < \sec^{-1}(x/a) < \pi].$$

531.1. $\int x \sec^{-1} \dfrac{x}{a} \, dx = \dfrac{x^2}{2} \sec^{-1} \dfrac{x}{a} - \dfrac{a}{2} \sqrt{(x^2 - a^2)},$

$$[0 < \sec^{-1}(x/a) < \pi/2].$$

$$= \dfrac{x^2}{2} \sec^{-1} \dfrac{x}{a} + \dfrac{a}{2} \sqrt{(x^2 - a^2)},$$

$$[\pi/2 < \sec^{-1}(x/a) < \pi].$$

531.2. $\int x^2 \sec^{-1} \dfrac{x}{a}\, dx$

$$= \frac{x^3}{3} \sec^{-1} \frac{x}{a} - \frac{ax}{6} \sqrt{(x^2 - a^2)} - \frac{a^3}{6} \log |x + \sqrt{(x^2 - a^2)}|,$$

$$[0 < \sec^{-1}(x/a) < \pi/2].$$

$$= \frac{x^3}{3} \sec^{-1} \frac{x}{a} + \frac{ax}{6} \sqrt{(x^2 - a^2)} + \frac{a^3}{6} \log |x + \sqrt{(x^2 - a^2)}|,$$

$$[\pi/2 < \sec^{-1}(x/a) < \pi].$$

531.9. $\int x^n \sec^{-1} \dfrac{x}{a}\, dx = \dfrac{x^{n+1}}{n+1} \sec^{-1} \dfrac{x}{a} - \dfrac{a}{n+1} \int \dfrac{x^n dx}{\sqrt{(x^2 - a^2)}},$

$$[0 < \sec^{-1}(x/a) < \pi/2], \qquad [n \neq -1].$$

$$= \frac{x^{n+1}}{n+1} \sec^{-1} \frac{x}{a} + \frac{a}{n+1} \int \frac{x^n dx}{\sqrt{(x^2 - a^2)}},$$

$$[\pi/2 < \sec^{-1}(x/a) < \pi], \qquad [n \neq -1].$$

532.1. $\int \dfrac{1}{x} \sec^{-1} \dfrac{x}{a}\, dx = \dfrac{\pi}{2} \log |x| + \dfrac{a}{x} + \dfrac{a^3}{2\cdot3\cdot3x^3} + \dfrac{1\cdot3a^5}{2\cdot4\cdot5\cdot5x^5}$

$$+ \frac{1\cdot3\cdot5a^7}{2\cdot4\cdot6\cdot7\cdot7x^7} + \cdots, \qquad [0 < \sec^{-1}(x/a) < \pi].$$

532.2. $\int \dfrac{1}{x^2} \sec^{-1} \dfrac{x}{a}\, dx = -\dfrac{1}{x} \sec^{-1} \dfrac{x}{a} + \dfrac{\sqrt{(x^2 - a^2)}}{ax},$

$$[0 < \sec^{-1}(x/a) < \pi/2].$$

$$= -\frac{1}{x} \sec^{-1} \frac{x}{a} - \frac{\sqrt{(x^2 - a^2)}}{ax},$$

$$[\pi/2 < \sec^{-1}(x/a) < \pi].$$

532.3. $\int \dfrac{1}{x^3} \sec^{-1} \dfrac{x}{a}\, dx$

$$= -\frac{1}{2x^2} \sec^{-1} \frac{x}{a} + \frac{\sqrt{(x^2 - a^2)}}{4ax^2} + \frac{1}{4a^2} \cos^{-1} \left|\frac{a}{x}\right|,$$

$$[0 < \sec^{-1}(x/a) < \pi/2].$$

$$= -\frac{1}{2x^2} \sec^{-1} \frac{x}{a} - \frac{\sqrt{(x^2 - a^2)}}{4ax^2} - \frac{1}{4a^2} \cos^{-1} \left|\frac{a}{x}\right|,$$

$$[\pi/2 < \sec^{-1}(x/a) < \pi].$$

532.4. $\displaystyle\int \frac{1}{x^4} \sec^{-1}\frac{x}{a}\, dx = -\frac{1}{3x^3}\sec^{-1}\frac{x}{a} + \frac{(2x^2 + a^2)}{9a^3x^3}\sqrt{(x^2 - a^2)},$

$$[0 < \sec^{-1}(x/a) < \pi/2].$$

$$= -\frac{1}{3x^3}\sec^{-1}\frac{x}{a} - \frac{(2x^2 + a^2)}{9a^3x^3}\sqrt{(x^2 - a^2)},$$

$$[\pi/2 < \sec^{-1}(x/a) < \pi].$$

532.9. $\displaystyle\int \frac{1}{x^n}\sec^{-1}\frac{x}{a}\, dx$

$$= -\frac{1}{(n-1)x^{n-1}}\sec^{-1}\frac{x}{a} + \frac{a}{n-1}\int \frac{dx}{x^n\sqrt{(x^2 - a^2)}},$$

$$[0 < \sec^{-1}(x/a) < \pi/2], \qquad [n \neq 1].$$

$$= -\frac{1}{(n-1)x^{n-1}}\sec^{-1}\frac{x}{a} - \frac{a}{n-1}\int \frac{dx}{x^n\sqrt{(x^2 - a^2)}},$$

$$[\pi/2 < \sec^{-1}(x/a) < \pi], \qquad [n \neq 1].$$

For **531–532.9**, $x^2 > a^2$.

534. $\displaystyle\int \csc^{-1}\frac{x}{a}\, dx = x\csc^{-1}\frac{x}{a} + a\log|x + \sqrt{(x^2 - a^2)}|,$

$$[0 < \csc^{-1}(x/a) < \pi/2].$$

$$= x\csc^{-1}\frac{x}{a} - a\log|x + \sqrt{(x^2 - a^2)}|,$$

$$[-\pi/2 < \csc^{-1}(x/a) < 0].$$

534.1. $\displaystyle\int x\csc^{-1}\frac{x}{a}\, dx = \frac{x^2}{2}\csc^{-1}\frac{x}{a} + \frac{a}{2}\sqrt{(x^2 - a^2)},$

$$[0 < \csc^{-1}(x/a) < \pi/2].$$

$$= \frac{x^2}{2}\csc^{-1}\frac{x}{a} - \frac{a}{2}\sqrt{(x^2 - a^2)},$$

$$[-\pi/2 < \csc^{-1}(x/a) < 0].$$

534.2. $\displaystyle\int x^2\csc^{-1}\frac{x}{a}\, dx$

$$= \frac{x^3}{3}\csc^{-1}\frac{x}{a} + \frac{ax}{6}\sqrt{(x^2 - a^2)} + \frac{a^3}{6}\log|x + \sqrt{(x^2 - a^2)}|,$$

$$[0 < \csc^{-1}(x/a) < \pi/2].$$

$$= \frac{x^3}{3}\csc^{-1}\frac{x}{a} - \frac{ax}{6}\sqrt{(x^2 - a^2)} - \frac{a^3}{6}\log|x + \sqrt{(x^2 - a^2)}|,$$

$$[-\pi/2 < \csc^{-1}(x/a) < 0].$$

534.9. $\displaystyle\int x^n \csc^{-1} \frac{x}{a} \, dx = \frac{x^{n+1}}{n+1} \csc^{-1} \frac{x}{a} + \frac{a}{n+1} \int \frac{x^n dx}{\sqrt{(x^2 - a^2)}}$,

$$[0 < \csc^{-1}(x/a) < \pi/2], \qquad [n \neq -1].$$

$$= \frac{x^{n+1}}{n+1} \csc^{-1} \frac{x}{a} - \frac{a}{n+1} \int \frac{x^n dx}{\sqrt{(x^2 - a^2)}},$$

$$[-\pi/2 < \csc^{-1}(x/a) < 0], \qquad [n \neq -1].$$

535.1. $\displaystyle\int \frac{1}{x} \csc^{-1} \frac{x}{a} \, dx = -\left(\frac{a}{x} + \frac{1}{2 \cdot 3 \cdot 3} \frac{a^3}{x^3} + \frac{1 \cdot 3}{2 \cdot 4 \cdot 5 \cdot 5} \frac{a^5}{x^5} \right.$

$$\left. + \frac{1 \cdot 3 \cdot 5}{2 \cdot 4 \cdot 6 \cdot 7 \cdot 7} \frac{a^7}{x^7} + \cdots \right),$$

$$[-\pi/2 < \csc^{-1}(x/a) < \pi/2].$$

535.2. $\displaystyle\int \frac{1}{x^2} \csc^{-1} \frac{x}{a} \, dx = -\frac{1}{x} \csc^{-1} \frac{x}{a} - \frac{\sqrt{(x^2 - a^2)}}{ax}$,

$$[0 < \csc^{-1}(x/a) < \pi/2].$$

$$= -\frac{1}{x} \csc^{-1} \frac{x}{a} + \frac{\sqrt{(x^2 - a^2)}}{ax},$$

$$[-\pi/2 < \csc^{-1}(x/a) < 0].$$

535.3. $\displaystyle\int \frac{1}{x^3} \csc^{-1} \frac{x}{a} \, dx$

$$= -\frac{1}{2x^2} \csc^{-1} \frac{x}{a} - \frac{\sqrt{(x^2 - a^2)}}{4ax^2} - \frac{1}{4a^2} \cos^{-1} \left| \frac{a}{x} \right|,$$

$$[0 < \csc^{-1}(x/a) < \pi/2].$$

$$= -\frac{1}{2x^2} \csc^{-1} \frac{x}{a} + \frac{\sqrt{(x^2 - a^2)}}{4ax^2} + \frac{1}{4a^2} \cos^{-1} \left| \frac{a}{x} \right|,$$

$$[-\pi/2 < \csc^{-1}(x/a) < 0].$$

535.4. $\displaystyle\int \frac{1}{x^4} \csc^{-1} \frac{x}{a} \, dx = -\frac{1}{3x^3} \csc^{-1} \frac{x}{a} - \frac{(2x^2 + a^2)}{9a^3x^3} \sqrt{(x^2 - a^2)}$,

$$[0 < \csc^{-1}(x/a) < \pi/2].$$

$$= -\frac{1}{3x^3} \csc^{-1} \frac{x}{a} + \frac{(2x^2 + a^2)}{9a^3x^3} \sqrt{(x^2 - a^2)},$$

$$[-\pi/2 < \csc^{-1}(x/a) < 0].$$

535.9. $\displaystyle\int \frac{1}{x^n} \csc^{-1} \frac{x}{a}\, dx$

$$= -\frac{1}{(n-1)x^{n-1}} \csc^{-1} \frac{x}{a} - \frac{a}{n-1}\int \frac{dx}{x^n \sqrt{(x^2 - a^2)}},$$

$$[0 < \csc^{-1}(x/a) < \pi/2], \quad [n \neq 1].$$

$$= -\frac{1}{(n-1)x^{n-1}} \csc^{-1} \frac{x}{a} + \frac{a}{n-1}\int \frac{dx}{x^n \sqrt{(x^2 - a^2)}},$$

$$[-\pi/2 < \csc^{-1}(x/a) < 0], \quad [n \neq 1].$$

For **534–535.9**, $x^2 > a^2$.

⋅⋅ 4 ⋅⋅

EXPONENTIAL FUNCTIONS

550. $\quad e^x = 1 + \dfrac{x}{1!} + \dfrac{x^2}{2!} + \dfrac{x^3}{3!} + \cdots + \dfrac{x^n}{n!} + \cdots, \qquad [x^2 < \infty].$

550.1. $\quad a^x = e^{x \log a} = 1 + \dfrac{x \log a}{1!} + \dfrac{(x \log a)^2}{2!} + \cdots$

$$+ \dfrac{(x \log a)^n}{n!} + \cdots, \qquad [x^2 < \infty].$$

550.2. $\quad e^{-x} = 1 - \dfrac{x}{1!} + \dfrac{x^2}{2!} - \dfrac{x^3}{3!} + \dfrac{x^4}{4!} - \cdots, \qquad [x^2 < \infty].$

551. $\quad \dfrac{x}{e^x - 1} = 1 - \dfrac{x}{2} + \dfrac{B_1 x^2}{2!} - \dfrac{B_2 x^4}{4!} + \dfrac{B_3 x^6}{6!} - \dfrac{B_4 x^8}{8!} + \cdots,$

$$[x^2 < 4\pi^2. \quad \text{See } \mathbf{45}]. \quad [\text{Ref. 34, p. 234.}]$$

552.1. $\quad e^{\sin u} = 1 + u + \dfrac{u^2}{2!} - \dfrac{3u^4}{4!} - \dfrac{8u^5}{5!} - \dfrac{3u^6}{6!} + \dfrac{56u^7}{7!} + \cdots,$

$$[u^2 < \infty].$$

552.2. $\quad e^{\cos u} = e \left[1 - \dfrac{u^2}{2!} + \dfrac{4u^4}{4!} - \dfrac{31u^6}{6!} + \cdots \right], \qquad [u^2 < \infty].$

552.3. $\quad e^{\tan u} = 1 + u + \dfrac{u^2}{2!} + \dfrac{3u^3}{3!} + \dfrac{9u^4}{4!} + \dfrac{37u^5}{5!} + \cdots,$

$$[u^2 < \pi^2/4].$$

552.4. $\quad e^{\sin^{-1} u} = 1 + u + \dfrac{u^2}{2!} + \dfrac{2u^3}{3!} + \dfrac{5u^4}{4!} + \cdots, \qquad [u^2 < 1].$

[Ref. 5, p. 102.]

552.5. $e^{\tan^{-1}u} = 1 + u + \dfrac{u^2}{2!} - \dfrac{u^3}{3!} - \dfrac{7u^4}{4!} + \dfrac{5u^5}{5!} + \cdots,$ $[u^2 < 1].$

The term in u^n is $a_n u^n/n!$, where $a_{n+1} = a_n - n(n-1)a_{n-1}$.

[Ref. 34, p. 164, No. 19.]

552.6. $e^{-x^2} + e^{-2^2 x^2} + e^{-3^2 x^2} + \cdots$

$$= -\frac{1}{2} + \frac{\sqrt{\pi}}{x}\left[\frac{1}{2} + e^{-\pi^2/x^2} + e^{-2^2\pi^2/x^2} + e^{-3^2\pi^2/x^2} + \cdots\right].$$

The second series may be more rapidly convergent than the first.

[Ref. 31, p. 129.]

553. $\displaystyle\lim_{x \to \infty} x^n e^{-x} = 0$, for all values of n.   [Ref. 8, p. 132.]

EXPONENTIAL FUNCTIONS—DERIVATIVES

563. $\dfrac{de^x}{dx} = e^x.$ **563.1.** $\dfrac{de^{ax}}{dx} = ae^{ax}.$ **563.2.** $\dfrac{da^x}{dx} = a^x \log a.$

563.3. $\dfrac{da^{cx}}{dx} = ca^{cx} \log a.$ **563.4.** $\dfrac{da^y}{dx} = a^y(\log a)\,\dfrac{dy}{dx}.$

$[a = \text{constant}]$

563.5. $\dfrac{du^y}{dx} = yu^{y-1}\dfrac{du}{dx} + u^y(\log u)\,\dfrac{dy}{dx}.$

563.6. $\dfrac{dx^y}{dx} = yx^{y-1} + x^y(\log x)\,\dfrac{dy}{dx}.$

563.7. $\dfrac{dx^x}{dx} = x^x(1 + \log x).$

EXPONENTIAL FUNCTIONS—INTEGRALS

565. $\displaystyle\int e^x dx = e^x.$ **565.1.** $\displaystyle\int e^{ax}dx = \frac{1}{a}\,e^{ax}.$

565.2. $\displaystyle\int e^{-x}dx = -e^{-x}.$ **565.3.** $\displaystyle\int a^x dx = a^x/\log a.$

566. $\displaystyle\int f(e^{ax})dx = \frac{1}{a}\int \frac{f(z)dz}{z}$

where $z = e^{ax}$. Note that

$$a^x = e^{x \log a}, \quad \text{and} \quad a^{cx} = e^{cx \log a}.$$

567.1. $\int xe^{ax}dx = e^{ax}\left[\dfrac{x}{a} - \dfrac{1}{a^2}\right].$

567.2. $\int x^2e^{ax}dx = e^{ax}\left[\dfrac{x^2}{a} - \dfrac{2x}{a^2} + \dfrac{2}{a^3}\right].$

567.3. $\int x^3e^{ax}dx = e^{ax}\left[\dfrac{x^3}{a} - \dfrac{3x^2}{a^2} + \dfrac{6x}{a^3} - \dfrac{6}{a^4}\right].$

567.8. $\int x^ne^{ax}dx = \dfrac{x^ne^{ax}}{a} - \dfrac{n}{a}\int x^{n-1}e^{ax}dx.$

567.9. $\int x^ne^{ax}dx = e^{ax}\left[\dfrac{x^n}{a} - \dfrac{nx^{n-1}}{a^2} + \dfrac{n(n-1)x^{n-2}}{a^3} - \cdots\right.$

$$\left. + (-1)^{n-1}\dfrac{n!x}{a^n} + (-1)^n\dfrac{n!}{a^{n+1}}\right], \qquad [n \geqq 0].$$

568.1. $\int\dfrac{e^{ax}dx}{x} = \log|x| + \dfrac{ax}{1!} + \dfrac{a^2x^2}{2\cdot 2!} + \dfrac{a^3x^3}{3\cdot 3!} + \cdots$

$$\cdots + \dfrac{a^nx^n}{n\cdot n!} + \cdots, \qquad [x^2 < \infty].$$

568.11.   For $\int\dfrac{c^xdx}{x}$, note that $c^x = e^{x\log c}$.

568.2. $\int\dfrac{e^{ax}dx}{x^2} = -\dfrac{e^{ax}}{x} + a\int\dfrac{e^{ax}dx}{x}.$ [See **568.1.**]

568.3. $\int\dfrac{e^{ax}dx}{x^3} = -\dfrac{e^{ax}}{2x^2} - \dfrac{ae^{ax}}{2x} + \dfrac{a^2}{2}\int\dfrac{e^{ax}dx}{x}.$ [See **568.1.**]

568.8. $\int\dfrac{e^{ax}dx}{x^n} = -\dfrac{e^{ax}}{(n-1)x^{n-1}} + \dfrac{a}{n-1}\int\dfrac{e^{ax}dx}{x^{n-1}},$ $[n > 1].$

568.9. $\int\dfrac{e^{ax}dx}{x^n} = -\dfrac{e^{ax}}{(n-1)x^{n-1}} - \dfrac{ae^{ax}}{(n-1)(n-2)x^{n-2}} - \cdots$

$$-\dfrac{a^{n-2}e^{ax}}{(n-1)!x} + \dfrac{a^{n-1}}{(n-1)!}\int\dfrac{e^{ax}dx}{x},$$

$$[n > 1]. \quad [\text{See } \textbf{568.1.}]$$

569. $\int\dfrac{dx}{1+e^x} = x - \log(1+e^x) = \log\dfrac{e^x}{1+e^x}.$

569.1. $\int \dfrac{dx}{a + be^{px}} = \dfrac{x}{a} - \dfrac{1}{ap} \log |a + be^{px}|.$

570. $\int \dfrac{xe^x dx}{(1 + x)^2} = \dfrac{e^x}{1 + x}.$ **570.1.** $\int \dfrac{xe^{ax} dx}{(1 + ax)^2} = \dfrac{e^{ax}}{a^2(1 + ax)}.$

575.1. $\int e^{ax} \sin x \, dx = \dfrac{e^{ax}}{a^2 + 1} (a \sin x - \cos x).$

575.2. $\int e^{ax} \sin^2 x \, dx = \dfrac{e^{ax}}{a^2 + 4} \left(a \sin^2 x - 2 \sin x \cos x + \dfrac{2}{a} \right).$

575.3. $\int e^{ax} \sin^3 x \, dx = \dfrac{e^{ax}}{a^2 + 9} \left[a \sin^3 x - 3 \sin^2 x \cos x \right.$
$$\left. + \dfrac{6(a \sin x - \cos x)}{a^2 + 1} \right].$$

575.9. $\int e^{ax} \sin^n x \, dx = \dfrac{e^{ax} \sin^{n-1} x}{a^2 + n^2} (a \sin x - n \cos x)$
$$+ \dfrac{n(n - 1)}{a^2 + n^2} \int e^{ax} \sin^{n-2} x \, dx.$$

576.1. $\int e^{ax} \cos x \, dx = \dfrac{e^{ax}}{a^2 + 1} (a \cos x + \sin x).$

576.2. $\int e^{ax} \cos^2 x \, dx = \dfrac{e^{ax}}{a^2 + 4} \left(a \cos^2 x + 2 \sin x \cos x + \dfrac{2}{a} \right).$

576.3. $\int e^{ax} \cos^3 x \, dx = \dfrac{e^{ax}}{a^2 + 9} \left[a \cos^3 x + 3 \sin x \cos^2 x \right.$
$$\left. + \dfrac{6(a \cos x + \sin x)}{a^2 + 1} \right].$$

576.9. $\int e^{ax} \cos^n x \, dx = \dfrac{e^{ax} \cos^{n-1} x}{a^2 + n^2} (a \cos x + n \sin x)$
$$+ \dfrac{n(n - 1)}{a^2 + n^2} \int e^{ax} \cos^{n-2} x \, dx. \quad \text{[Ref. 2, p. 141.]}$$

577.1. $\int e^{ax} \sin nx \, dx = \dfrac{e^{ax}}{a^2 + n^2} (a \sin nx - n \cos nx).$

577.2. $\int e^{ax} \cos nx \, dx = \dfrac{e^{ax}}{a^2 + n^2} (a \cos nx + n \sin nx).$
$$\text{[Ref. 7, p. 9.]}$$

⁂ 5 ⁂

PROBABILITY INTEGRALS

585. Normal probability integral $= \dfrac{1}{\sqrt{(2\pi)}} \displaystyle\int_{-x}^{x} e^{-t^2/2} dt$

$$= \operatorname{erf} \frac{x}{\sqrt{2}} \qquad\qquad \text{[See \textbf{590}.]}$$

$$= x \left(\frac{2}{\pi}\right)^{1/2} \left[1 - \frac{x^2}{2\cdot 1\,!3} + \frac{x^4}{2^2\cdot 2\,!5} - \frac{x^6}{2^3\cdot 3\,!7} + \cdots \right]$$

$$[x^2 < \infty]. \quad \text{[See Table \textbf{1045}.]}$$

586. For large values of x, the following asymptotic series may be used:

$$\frac{1}{\sqrt{(2\pi)}}\int_{-x}^{x} e^{-t^2/2} dt$$

$$\approx 1 - \left(\frac{2}{\pi}\right)^{1/2} \frac{e^{-x^2/2}}{x} \left[1 - \frac{1}{x^2} + \frac{1\cdot 3}{x^4} - \frac{1\cdot 3\cdot 5}{x^6} + \frac{1\cdot 3\cdot 5\cdot 7}{x^8} - \cdots \right],$$

where $\approx$ denotes approximate equality. The error is less than the last term used.

590. Error function $= \operatorname{erf} x = \dfrac{2}{\sqrt{\pi}} \displaystyle\int_{0}^{x} e^{-t^2} dt$

$$= \frac{2x}{\sqrt{\pi}} \left[1 - \frac{x^2}{1\,!3} + \frac{x^4}{2\,!5} - \frac{x^6}{3\,!7} + \cdots \right] \qquad [x^2 < \infty].$$

591. $\operatorname{Erf} x \approx 1 - \dfrac{e^{-x^2}}{x\sqrt{\pi}} \left[1 - \dfrac{1}{2x^2} + \dfrac{1\cdot 3}{2^2 x^4} - \dfrac{1\cdot 3\cdot 5}{2^3 x^6} + \cdots \right].$

592. Alternative form of the same series:

$$\operatorname{Erf} x \approx 1 - \frac{e^{-x^2}}{x\sqrt{\pi}} \left[1 - \frac{2!}{1\,!(2x)^2} + \frac{4!}{2\,!(2x)^4} - \frac{6!}{3\,!(2x)^6} + \cdots \right].$$

The error is less than the last term used. [Ref. 9, p. 390.]

For tables of numerical values see Ref. 55e, Vols. I and II; Ref. 5, pp. 128–132; and Ref. 45.

✦ 6 ✦

LOGARITHMIC FUNCTIONS

In these algebraic expressions, log represents natural or Napierian logarithms. Other notations for natural logarithms are logn, ln and $\log_\epsilon$.

600. $\log_\epsilon a = 2.3026 \log_{10} a.$ **600.1.** $\log_{10} a = 0.43429 \log_\epsilon a.$

601. $\log(1 + x) = x - \dfrac{x^2}{2} + \dfrac{x^3}{3} - \dfrac{x^4}{4} + \dfrac{x^5}{5} - \cdots,$

$$[x^2 < 1 \text{ and } x = 1].$$

For $x = 1$, this gives a famous series:

601.01. $\log 2 = 1 - \dfrac{1}{2} + \dfrac{1}{3} - \dfrac{1}{4} + \dfrac{1}{5} - \cdots.$

601.1. $\log(1 - x) = -\left[x + \dfrac{x^2}{2} + \dfrac{x^3}{3} + \dfrac{x^4}{4} + \dfrac{x^5}{5} + \cdots \right],$

$$[x^2 < 1 \text{ and } x = -1].$$

601.2. $\log\left(\dfrac{1+x}{1-x}\right) = 2\left[x + \dfrac{x^3}{3} + \dfrac{x^5}{5} + \dfrac{x^7}{7} + \cdots \right],$

$$= 2 \tanh^{-1} x. \qquad [x^2 < 1]. \quad [\text{See } \mathbf{708}.]$$

601.3. $\log\left(\dfrac{x+1}{x-1}\right) = 2\left[\dfrac{1}{x} + \dfrac{1}{3x^3} + \dfrac{1}{5x^5} + \dfrac{1}{7x^7} + \cdots \right],$

$$= 2 \operatorname{ctnh}^{-1} x. \qquad [x^2 > 1]. \quad [\text{See } \mathbf{709}.]$$

601.4. $\log\left(\dfrac{x+1}{x}\right) = 2\left[\dfrac{1}{2x+1} + \dfrac{1}{3(2x+1)^3} \right.$

$$\left. + \dfrac{1}{5(2x+1)^5} + \cdots \right],$$

$$[(2x+1)^2 > 1]. \qquad [\text{Ref. 29, p. 6.}]$$

137

601.41. $\log (x + a) = \log x + 2\left[\dfrac{a}{2x + a} + \dfrac{a^3}{3(2x + a)^3}\right.$

$$\left. + \dfrac{a^5}{5(2x + a)^5} + \cdots \right], \qquad [a^2 < (2x + a)^2].$$

601.5. $\log x = (x - 1) - \dfrac{(x - 1)^2}{2} + \dfrac{(x - 1)^3}{3}$

$$- \dfrac{(x - 1)^4}{4} + \cdots, \qquad [0 < x \le 2].$$

601.6. $\log x = \dfrac{x - 1}{x} + \dfrac{(x - 1)^2}{2x^2} + \dfrac{(x - 1)^3}{3x^3} + \cdots, \qquad [x > \tfrac{1}{2}].$

601.7. $\log x = 2\left[\dfrac{x - 1}{x + 1} + \dfrac{(x - 1)^3}{3(x + 1)^3} + \dfrac{(x - 1)^5}{5(x + 1)^5} + \cdots\right],$

$$[x > 0].$$

602.1. $\log\left[\dfrac{x}{a} + \sqrt{\left(\dfrac{x^2}{a^2} + 1\right)}\right]$

$$= \dfrac{x}{a} - \dfrac{1}{2\cdot 3}\dfrac{x^3}{a^3} + \dfrac{1\cdot 3}{2\cdot 4\cdot 5}\dfrac{x^5}{a^5} - \dfrac{1\cdot 3\cdot 5}{2\cdot 4\cdot 6\cdot 7}\dfrac{x^7}{a^7} + \cdots,$$

$$[x^2 < a^2].$$

$$= \log \dfrac{2x}{a} + \dfrac{1}{2\cdot 2}\dfrac{a^2}{x^2} - \dfrac{1\cdot 3}{2\cdot 4\cdot 4}\dfrac{a^4}{x^4} + \dfrac{1\cdot 3\cdot 5}{2\cdot 4\cdot 6\cdot 6}\dfrac{a^6}{x^6} - \cdots,$$

$$[x/a > 1].$$

$$= -\log\left|\dfrac{2x}{a}\right| - \dfrac{1}{2\cdot 2}\dfrac{a^2}{x^2} + \dfrac{1\cdot 3}{2\cdot 4\cdot 4}\dfrac{a^4}{x^4} - \dfrac{1\cdot 3\cdot 5}{2\cdot 4\cdot 6\cdot 6}\dfrac{a^6}{x^6} + \cdots,$$

$$[x/a < -1].$$

$$= \sinh^{-1}\dfrac{x}{a} = \operatorname{csch}^{-1}\dfrac{a}{x}. \qquad [\text{See } \mathbf{706}.]$$

602.2. $\log\left[\sqrt{\left(\dfrac{x^2}{a^2} + 1\right)} - \dfrac{x}{a}\right] = -\log\left[\dfrac{x}{a} + \sqrt{\left(\dfrac{x^2}{a^2} + 1\right)}\right].$

Use the series in **602.1** and multiply by -1.

602.3. $\log\left[\dfrac{x}{a} + \sqrt{\left(\dfrac{x^2}{a^2} - 1\right)}\right] = \log\dfrac{2x}{a} - \dfrac{1}{2\cdot 2}\dfrac{a^2}{x^2} - \dfrac{1\cdot 3}{2\cdot 4\cdot 4}\dfrac{a^4}{x^4}$

$$-\dfrac{1\cdot 3\cdot 5}{2\cdot 4\cdot 6\cdot 6}\dfrac{a^6}{x^6} - \cdots, \qquad [x/a > 1].$$

[See **260.01** and **707.**]

602.4. $\log\left[\dfrac{x}{a} - \sqrt{\left(\dfrac{x^2}{a^2} - 1\right)}\right]$

$$= -\log\dfrac{2x}{a} + \dfrac{1}{2\cdot 2}\dfrac{a^2}{x^2} + \dfrac{1\cdot 3}{2\cdot 4\cdot 4}\dfrac{a^4}{x^4} + \dfrac{1\cdot 3\cdot 5}{2\cdot 4\cdot 6\cdot 6}\dfrac{a^6}{x^6} + \cdots,$$

$$[x/a > 1].$$

$$= -\log\left[\dfrac{x}{a} + \sqrt{\left(\dfrac{x^2}{a^2} - 1\right)}\right]. \qquad \text{[See \textbf{602.3} and \textbf{707.}]}$$

602.5. $\log\left[\dfrac{a}{x} + \sqrt{\left(\dfrac{a^2}{x^2} + 1\right)}\right]$

$$= \dfrac{a}{x} - \dfrac{1}{2\cdot 3}\dfrac{a^3}{x^3} + \dfrac{1\cdot 3}{2\cdot 4\cdot 5}\dfrac{a^5}{x^5} - \dfrac{1\cdot 3\cdot 5}{2\cdot 4\cdot 6\cdot 7}\dfrac{a^7}{x^7} + \cdots,$$

$$[x^2 > a^2].$$

$$= \log\dfrac{2a}{x} + \dfrac{1}{2\cdot 2}\dfrac{x^2}{a^2} - \dfrac{1\cdot 3}{2\cdot 4\cdot 4}\dfrac{x^4}{a^4} + \dfrac{1\cdot 3\cdot 5}{2\cdot 4\cdot 6\cdot 6}\dfrac{x^6}{a^6} - \cdots,$$

$$[a/x > 1].$$

$$= -\log\left|\dfrac{2a}{x}\right| - \dfrac{1}{2\cdot 2}\dfrac{x^2}{a^2} + \dfrac{1\cdot 3}{2\cdot 4\cdot 4}\dfrac{x^4}{a^4} - \dfrac{1\cdot 3\cdot 5}{2\cdot 4\cdot 6\cdot 6}\dfrac{x^6}{a^6} + \cdots,$$

$$[a/x < -1].$$

$$= \operatorname{csch}^{-1}\dfrac{x}{a} = \sinh^{-1}\dfrac{a}{x}. \qquad \text{[See \textbf{602.1} and \textbf{711.}]}$$

602.6. $\log\left[\sqrt{\left(\dfrac{a^2}{x^2} + 1\right)} - \dfrac{a}{x}\right] = -\log\left[\dfrac{a}{x} + \sqrt{\left(\dfrac{a^2}{x^2} + 1\right)}\right].$

Use the series in **602.5** and multiply by -1.

602.7. $\log\left[\dfrac{a}{x} + \sqrt{\left(\dfrac{a^2}{x^2} - 1\right)}\right] = \log\dfrac{2a}{x} - \dfrac{1}{2\cdot 2}\dfrac{x^2}{a^2} - \dfrac{1\cdot 3}{2\cdot 4\cdot 4}\dfrac{x^4}{a^4}$

$$-\dfrac{1\cdot 3\cdot 5}{2\cdot 4\cdot 6\cdot 6}\dfrac{x^6}{a^6} - \cdots, \qquad [a/x > 1].$$

602.8. $\log\left[\dfrac{a}{x} - \sqrt{\left(\dfrac{a^2}{x^2} - 1\right)}\right]$

$$= -\log\frac{2a}{x} + \frac{1}{2\cdot 2}\frac{x^2}{a^2} + \frac{1\cdot 3}{2\cdot 4\cdot 4}\frac{x^4}{a^4} + \frac{1\cdot 3\cdot 5}{2\cdot 4\cdot 6\cdot 6}\frac{x^6}{a^6} + \cdots,$$

$$[a/x > 1].$$

$$= -\log\left[\frac{a}{x} + \sqrt{\left(\frac{a^2}{x^2} - 1\right)}\right]. \qquad \text{[See 710.]}$$

603.1. $\log|\sin x| = \log|x| - \dfrac{x^2}{6} - \dfrac{x^4}{180} - \dfrac{x^6}{2835} - \cdots$

$$\cdots - \frac{2^{2n-1}B_n x^{2n}}{n(2n)!} - \cdots, \qquad [x^2 < \pi^2].$$

[Integrate **415.04**.   See **490.1** and **45**.]

603.2. $\log|\sin x| = -\log 2 - \cos 2x - \dfrac{\cos 4x}{2} - \dfrac{\cos 6x}{3} - \cdots,$

[Ref. 38, p. 275.] $\qquad$ [$\sin x \neq 0$].

603.3. $\log\cos x = -\dfrac{x^2}{2} - \dfrac{x^4}{12} - \dfrac{x^6}{45} - \dfrac{17x^8}{2520} - \cdots$

$$\cdots - \frac{2^{2n-1}(2^{2n} - 1)B_n x^{2n}}{n(2n)!} - \cdots,$$

$[x^2 < \pi^2/4]$.   [Integrate **415.03**.   See **480.1** and **45**.]

603.4. $\log|\cos x| = -\log 2 + \cos 2x - \dfrac{\cos 4x}{2} + \dfrac{\cos 6x}{3} - \cdots,$

[Ref. 38, p. 275.] $\qquad$ [$\cos x \neq 0$].

603.5. $\log\cos x = -\dfrac{1}{2}\left[\sin^2 x + \dfrac{\sin^4 x}{2\cdot} + \dfrac{\sin^6 x}{3}\right.$

$$\left. + \frac{\sin^8 x}{4} + \cdots \right], \quad [x^2 < \pi^2/4].$$

603.6. $\log|\tan x| = \log|x| + \dfrac{x^2}{3} + \dfrac{7}{90}x^4 + \dfrac{62}{2835}x^6 + \cdots$

$$\cdots + \frac{2^{2n}(2^{2n-1} - 1)B_n x^{2n}}{n(2n)!} + \cdots,$$

$[x^2 < \pi^2/4]$.   [See **415.06**, **432.10**, and **45**.]

604. $\log (x + iy) = \log r + i(\theta + 2\pi k)$,
where $r = \sqrt{(x^2 + y^2)}$, $\cos \theta = x/r$, $\sin \theta = y/r$, k is an integer
or 0, r is positive, $i = \sqrt{(-1)}$.   [Ref. 5, p. 3.]

604.05. $x + iy = re^{i(\theta + 2\pi k)}$.   [$\theta$ in radians.]   [See **604.**]

604.1. $\log (-1) = \log 1 + (2k + 1)\pi i$
$= (2k + 1)\pi i$.   [See **409.03.**]

605. $\lim_{x \to 0} x \log x = 0$.   [See **72.**]

LOGARITHMIC FUNCTIONS—INTEGRALS
Except where noted, x and $a > 0$.

610. $\int \log x \, dx = x \log x - x$.

610.01. $\int \log (ax) dx = x \log (ax) - x$.

610.1. $\int x \log x \, dx = \dfrac{x^2}{2} \log x - \dfrac{x^2}{4}$.

610.2. $\int x^2 \log x \, dx = \dfrac{x^3}{3} \log x - \dfrac{x^3}{9}$.

610.3. $\int x^3 \log x \, dx = \dfrac{x^4}{4} \log x - \dfrac{x^4}{16}$.

610.9. $\int x^p \log (ax) dx = \dfrac{x^{p+1}}{p+1} \log (ax) - \dfrac{x^{p+1}}{(p+1)^2}$, $[p \neq -1]$.

611.1. $\int \dfrac{\log x}{x} \, dx = \dfrac{(\log x)^2}{2}$.

611.11. $\int \dfrac{\log (ax)}{x} \, dx = \dfrac{1}{2} \{\log (ax)\}^2$.

611.2. $\int \dfrac{\log x}{x^2} \, dx = -\dfrac{\log x}{x} - \dfrac{1}{x}$.

611.3. $\int \dfrac{\log x}{x^3} \, dx = -\dfrac{\log x}{2x^2} - \dfrac{1}{4x^2}$.

611.9. $\displaystyle\int\frac{\log{(ax)}}{x^p}\,dx = -\,\frac{\log{(ax)}}{(p-1)x^{p-1}} - \frac{1}{(p-1)^2x^{p-1}},\qquad [p\neq 1].$

612. $\displaystyle\int(\log x)^2dx = x(\log x)^2 - 2x\log x + 2x.$

612.1. $\displaystyle\int x(\log x)^2dx = \frac{x^2}{2}(\log x)^2 - \frac{x^2}{2}\log x + \frac{x^2}{4}.$

612.2. $\displaystyle\int x^2(\log x)^2dx = \frac{x^3}{3}(\log x)^2 - \frac{2x^3}{9}\log x + \frac{2x^3}{27}.$

612.9. $\displaystyle\int x^p(\log x)^2dx = \frac{x^{p+1}}{p+1}(\log x)^2 - \frac{2x^{p+1}}{(p+1)^2}\log x$
$$+ \frac{2x^{p+1}}{(p+1)^3},\quad [p\neq -1].$$

613.1. $\displaystyle\int\frac{(\log x)^2dx}{x} = \frac{(\log x)^3}{3}.$

613.2. $\displaystyle\int\frac{(\log x)^2dx}{x^2} = -\,\frac{(\log x)^2}{x} - \frac{2\log x}{x} - \frac{2}{x}.$

613.3. $\displaystyle\int\frac{(\log x)^2dx}{x^3} = -\,\frac{(\log x)^2}{2x^2} - \frac{\log x}{2x^2} - \frac{1}{4x^2}.$

613.9. $\displaystyle\int\frac{(\log x)^2dx}{x^p} = -\,\frac{(\log x)^2}{(p-1)x^{p-1}} - \frac{2\log x}{(p-1)^2x^{p-1}}$
$$-\,\frac{2}{(p-1)^3x^{p-1}},\quad [p\neq 1].$$

614. $\displaystyle\int(\log x)^3dx = x(\log x)^3 - 3x(\log x)^2 + 6x\log x - 6x.$

615. $\displaystyle\int(\log x)^qdx = x(\log x)^q - q\int(\log x)^{q-1}dx,\qquad [q\neq -1].$

616.1. $\displaystyle\int\frac{(\log x)^qdx}{x} = \frac{(\log x)^{q+1}}{q+1},\qquad [q\neq -1].$

616.2. $\displaystyle\int x^p(\log x)^qdx = \frac{x^{p+1}(\log x)^q}{p+1} - \frac{q}{p+1}\int x^p(\log x)^{q-1}dx,$
$$[p,q\neq -1].$$

616.3. $\displaystyle\int\frac{(\log x)^q dx}{x^p} = \frac{-(\log x)^q}{(p-1)x^{p-1}} + \frac{q}{p-1}\int\frac{(\log x)^{q-1}dx}{x^p},$

$$[p, -q \neq 1].$$

617. $\displaystyle\int\frac{dx}{\log x} = \log|\log x| + \log x + \frac{(\log x)^2}{2\cdot 2!} + \frac{(\log x)^3}{3\cdot 3!} + \cdots.$

617.1. $\displaystyle\int\frac{x\,dx}{\log x} = \log|\log x| + 2\log x + \frac{(2\log x)^2}{2\cdot 2!}$

$$+ \frac{(2\log x)^3}{3\cdot 3!} + \cdots.$$

617.2. $\displaystyle\int\frac{x^2 dx}{\log x} = \log|\log x| + 3\log x + \frac{(3\log x)^2}{2\cdot 2!}$

$$+ \frac{(3\log x)^3}{3\cdot 3!} + \cdots.$$

617.9. $\displaystyle\int\frac{x^p dx}{\log x} = \log|\log x| + (p+1)\log x$

$$+ \frac{(p+1)^2(\log x)^2}{2\cdot 2!} + \frac{(p+1)^3(\log x)^3}{3\cdot 3!} + \cdots,$$

$$\left[= \int\frac{e^y dy}{y} \text{ where } y = (p+1)\log x. \quad \text{See } \mathbf{568.1}\right].$$

618.1. $\displaystyle\int\frac{dx}{x\log x} = \log|\log x|.$ $\qquad\qquad$ [Put $\log x = y$, $x = e^y$.]

618.2. $\displaystyle\int\frac{dx}{x^2\log x} = \log|\log x| - \log x + \frac{(\log x)^2}{2\cdot 2!}$

$$- \frac{(\log x)^3}{3\cdot 3!} + \cdots.$$

618.3. $\displaystyle\int\frac{dx}{x^3\log x} = \log|\log x| - 2\log x + \frac{(2\log x)^2}{2\cdot 2!}$

$$- \frac{(2\log x)^3}{3\cdot 3!} + \cdots.$$

618.9. $\displaystyle\int\frac{dx}{x^p\log x} = \log|\log x| - (p-1)\log x$

$$+ \frac{(p-1)^2(\log x)^2}{2\cdot 2!} - \frac{(p-1)^3(\log x)^3}{3\cdot 3!} + \cdots.$$

619.1. $\displaystyle\int \frac{dx}{x(\log x)^q} = \frac{-1}{(q-1)(\log x)^{q-1}},$ $\qquad [q \neq 1].$

619.2. $\displaystyle\int \frac{x^p dx}{(\log x)^q} = \frac{-x^{p+1}}{(q-1)(\log x)^{q-1}} + \frac{p+1}{q-1}\int \frac{x^p dx}{(\log x)^{q-1}},$ $[q \neq 1].$

619.3. $\displaystyle\int \frac{dx}{x^p(\log x)^q} = \frac{-1}{x^{p-1}(q-1)(\log x)^{q-1}}$

$$-\frac{p-1}{q-1}\int \frac{dx}{x^p(\log x)^{q-1}}, \qquad [q \neq 1].$$

620. $\displaystyle\int \log(a+bx)dx = \frac{a+bx}{b}\log(a+bx) - x.$

620.1. $\displaystyle\int x\log(a+bx)dx = \frac{b^2x^2 - a^2}{2b^2}\log(a+bx) + \frac{ax}{2b} - \frac{x^2}{4}.$

621.1. $\displaystyle\int \frac{\log(a+bx)dx}{x}$

$$= (\log a)\log x + \frac{bx}{a} - \frac{b^2x^2}{2^2a^2} + \frac{b^3x^3}{3^2a^3} - \frac{b^4x^4}{4^2a^4} + \cdots,$$
$$[b^2x^2 < a^2].$$

$$= \frac{(\log bx)^2}{2} - \frac{a}{bx} + \frac{a^2}{2^2b^2x^2} - \frac{a^3}{3^2b^3x^3} + \frac{a^4}{4^2b^4x^4} - \cdots,$$
$$[b^2x^2 > a^2].$$

621.2. $\displaystyle\int \frac{\log(a+bx)dx}{x^2} = \frac{b}{a}\log x - \left(\frac{1}{x} + \frac{b}{a}\right)\log(a+bx).$

621.9. $\displaystyle\int \frac{\log(a+bx)dx}{x^p} = -\frac{\log(a+bx)}{(p-1)x^{p-1}}$

$$+ \int \frac{b\,dx}{(p-1)(a+bx)x^{p-1}},$$
$$[p \neq 1]. \qquad [\text{See } 101\text{–}105.]$$

622. $\displaystyle\int \frac{\log x\,dx}{a+bx} = \frac{(\log x)\log(a+bx)}{b} - \int \frac{\log(a+bx)dx}{bx}.$
$$[\text{See } 621.1.]$$

623. $\displaystyle\int \log(x^2+a^2)dx = x\log(x^2+a^2) - 2x + 2a\tan^{-1}\frac{x}{a}.$

623.1. $\int x \log (x^2 + a^2)dx = \frac{1}{2} [(x^2 + a^2) \log (x^2 + a^2) - x^2].$

623.2. $\int x^2 \log (x^2 + a^2)dx = \frac{1}{3} \left[x^3 \log (x^2 + a^2) - \frac{2}{3} x^3 \right.$
$$\left. + 2xa^2 - 2a^3 \tan^{-1} \frac{x}{a} \right].$$

623.3. $\int x^3 \log (x^2 + a^2)dx = \frac{1}{4} \left[(x^4 - a^4) \log (x^2 + a^2) \right.$
$$\left. - \frac{x^4}{2} + x^2a^2 \right].$$

623.4. $\int x^4 \log (x^2 + a^2)dx = \frac{1}{5} \left[x^5 \log (x^2 + a^2) - \frac{2}{5} x^5 \right.$
$$\left. + \frac{2}{3} x^3a^2 - 2xa^4 + 2a^5 \tan^{-1} \frac{x}{a} \right].$$

623.5. $\int x^5 \log (x^2 + a^2)dx = \frac{1}{6} \left[(x^6 + a^6) \log (x^2 + a^2) \right.$
$$\left. - \frac{x^6}{3} + \frac{x^4a^2}{2} - x^2a^4 \right].$$

623.6. $\int x^6 \log (x^2 + a^2)dx = \frac{1}{7} \left[x^7 \log (x^2 + a^2) - \frac{2}{7} x^7 \right.$
$$\left. + \frac{2}{5} x^5a^2 - \frac{2}{3} x^3a^4 + 2xa^6 - 2a^7 \tan^{-1} \frac{x}{a} \right].$$

623.7. $\int x^7 \log (x^2 + a^2)dx = \frac{1}{8} \left[(x^8 - a^8) \log (x^2 + a^2) \right.$
$$\left. - \frac{x^8}{4} + \frac{x^6a^2}{3} - \frac{x^4a^4}{2} + x^2a^6 \right].$$

624. $\int \log |x^2 - a^2|dx = x \log |x^2 - a^2| - 2x + a \log \left| \frac{x + a}{x - a} \right|.$

624.1. $\int x \log |x^2 - a^2|dx = \frac{1}{2} [(x^2 - a^2) \log |x^2 - a^2| - x^2].$

624.2. $\int x^2 \log |x^2 - a^2|dx = \frac{1}{3} \left[x^3 \log |x^2 - a^2| - \frac{2}{3} x^3 \right.$
$$\left. - 2xa^2 + a^3 \log \left| \frac{x + a}{x - a} \right| \right].$$

624.3. $\int x^3 \log |x^2 - a^2| dx = \dfrac{1}{4} \Bigg[(x^4 - a^4) \log |x^2 - a^2|$

$$- \dfrac{x^4}{2} - x^2 a^2 \Bigg].$$

624.4. $\int x^4 \log |x^2 - a^2| dx = \dfrac{1}{5} \Bigg[x^5 \log |x^2 - a^2| - \dfrac{2}{5} x^5$

$$- \dfrac{2}{3} x^3 a^2 - 2xa^4 + a^5 \log \left| \dfrac{x+a}{x-a} \right| \Bigg].$$

624.5. $\int x^5 \log |x^2 - a^2| dx = \dfrac{1}{6} \Bigg[(x^6 - a^6) \log |x^2 - a^2|$

$$- \dfrac{x^6}{3} - \dfrac{x^4 a^2}{2} - x^2 a^4 \Bigg].$$

624.6. $\int x^6 \log |x^2 - a^2| dx = \dfrac{1}{7} \Bigg[x^7 \log |x^2 - a^2| - \dfrac{2}{7} x^7$

$$- \dfrac{2}{5} x^5 a^2 - \dfrac{2}{3} x^3 a^4 - 2xa^6 + a^7 \log \left| \dfrac{x+a}{x-a} \right| \Bigg].$$

624.7. $\int x^7 \log |x^2 - a^2| dx = \dfrac{1}{8} \Bigg[(x^8 - a^8) \log |x^2 - a^2|$

$$- \dfrac{x^8}{4} - \dfrac{x^6 a^2}{3} - \dfrac{x^4 a^4}{2} - x^2 a^6 \Bigg].$$

When integrals of the type $\int x^p \log (a^2 - x^2) dx$ are required, these expressions can be used.

Integrals Involving $r = (x^2 + a^2)^{1/2}$

625. $\int \log (x + r) dx = x \log (x + r) - r.$ [See **730**.]

The positive value of r is to be taken.

625.1. $\int x \log (x + r) dx = \left(\dfrac{x^2}{2} + \dfrac{a^2}{4} \right) \log (x + r) - \dfrac{xr}{4}.$

625.2. $\int x^2 \log (x + r) dx = \dfrac{x^3}{3} \log (x + r) - \dfrac{r^3}{9} + \dfrac{a^2 r}{3}.$

[See **730.2**.]

625.3. $\int x^3 \log(x + r)dx = \left(\dfrac{x^4}{4} - \dfrac{3a^4}{32}\right) \log(x + r)$

$$- \dfrac{x^3 r}{16} + \dfrac{3}{32}\, a^2 xr. \qquad \text{[See 730.3.]}$$

625.4. $\int x^4 \log(x + r)dx = \dfrac{x^5}{5} \log(x + r) - \dfrac{r^5}{25} + \dfrac{2}{15}\, a^2 r^3 - \dfrac{a^4 r}{5}.$

625.9. $\int x^p \log(x + r)dx = \dfrac{x^{p+1}}{p + 1} \log(x + r) - \dfrac{1}{p + 1} \int \dfrac{x^{p+1}dx}{r},$

$$[p \neq -1]. \quad \text{[See 201.01–207.01 and 730.9.]}$$

626.1. $\int \dfrac{1}{x} \log\left[\dfrac{x}{a} + \sqrt{\left(\dfrac{x^2}{a^2} + 1\right)}\right] dx$

$$= \dfrac{x}{a} - \dfrac{1}{2 \cdot 3 \cdot 3} \dfrac{x^3}{a^3} + \dfrac{1 \cdot 3}{2 \cdot 4 \cdot 5 \cdot 5} \dfrac{x^5}{a^5} - \dfrac{1 \cdot 3 \cdot 5}{2 \cdot 4 \cdot 6 \cdot 7 \cdot 7} \dfrac{x^7}{a^7} + \cdots,$$

$$[x^2 < a^2].$$

$$= \dfrac{1}{2}\left(\log \dfrac{2x}{a}\right)^2 - \dfrac{1}{2^3} \dfrac{a^2}{x^2} + \dfrac{1 \cdot 3}{2 \cdot 4^3} \dfrac{a^4}{x^4} - \dfrac{1 \cdot 3 \cdot 5}{2 \cdot 4 \cdot 6^3} \dfrac{a^6}{x^6} + \cdots,$$

$$[x/a > 1].$$

$$= -\dfrac{1}{2}\left(\log \left|\dfrac{2x}{a}\right|\right)^2 + \dfrac{1}{2^3} \dfrac{a^2}{x^2} - \dfrac{1 \cdot 3}{2 \cdot 4^3} \dfrac{a^4}{x^4} + \dfrac{1 \cdot 3 \cdot 5}{2 \cdot 4 \cdot 6^3} \dfrac{a^6}{x^6} - \cdots,$$

$$[x/a < -1]. \quad \text{[See 731.1.]}$$

626.2. $\int \dfrac{\log(x + r)}{x^2} = -\dfrac{\log(x + r)}{x} - \dfrac{1}{a} \log\left|\dfrac{a + r}{x}\right|,$

where $r = (x^2 + a^2)^{1/2}$. $\qquad$ [See 731.2.]

626.3. $\int \dfrac{\log(x + r)}{x^3} = -\dfrac{\log(x + r)}{2x^2} - \dfrac{r}{2a^2 x}.$ $\qquad$ [See 731.3.]

626.9. $\int \dfrac{\log(x + r)}{x^p} = -\dfrac{\log(x + r)}{(p - 1)x^{p-1}} + \dfrac{1}{p - 1} \int \dfrac{dx}{x^{p-1}r},$

$$[p \neq 1]. \quad \text{[See 221.01–226.01 and 731.9.]}$$

Integrals Involving $s = (x^2 - a^2)^{1/2}$

627. $\int \log (x + s)dx = x \log (x + s) - s.$ [See **732.**]

The positive value of s is to be taken.

627.1. $\int x \log (x + s)dx = \left(\dfrac{x^2}{2} - \dfrac{a^2}{4}\right) \log (x + s) - \dfrac{xs}{4}.$

627.2. $\int x^2 \log (x + s)dx = \dfrac{x^3}{3} \log (x + s) - \dfrac{s^3}{9} - \dfrac{a^2s}{3}.$ [See **732.2.**]

627.3. $\int x^3 \log (x + s)dx = \left(\dfrac{x^4}{4} - \dfrac{3a^4}{32}\right) \log (x + s)$

$$- \dfrac{x^3s}{16} - \dfrac{3}{32} a^2xs.$$ [See **732.3.**]

627.4. $\int x^4 \log (x + s)dx = \dfrac{x^5}{5} \log (x + s) - \dfrac{s^5}{25} - \dfrac{2}{15} a^2s^3 - \dfrac{a^4s}{5}.$

627.9. $\int x^p \log (x + s)dx = \dfrac{x^{p+1}}{p + 1} \log (x + s)$

$$- \dfrac{1}{p + 1} \int \dfrac{x^{p+1}dx}{s}, \qquad [p \neq -1].$$
[See **261.01–267.01** and **732.9.**]

628.1. $\int \dfrac{1}{x} \log \left[\dfrac{x}{a} + \sqrt{\left(\dfrac{x^2}{a^2} - 1\right)}\right] dx$

$$= \dfrac{1}{2}\left(\log \dfrac{2x}{a}\right)^2 + \dfrac{1}{2^3} \dfrac{a^2}{x^2} + \dfrac{1 \cdot 3}{2 \cdot 4^3} \dfrac{a^4}{x^4} + \dfrac{1 \cdot 3 \cdot 5}{2 \cdot 4 \cdot 6^3} \dfrac{a^6}{x^6} + \cdots,$$
$$[x/a > 1]. \qquad [\text{See } \textbf{733.1.}]$$

628.2. $\int \dfrac{\log (x + s)}{x^2} dx = - \dfrac{\log (x + s)}{x} + \dfrac{1}{a} \sec^{-1} \left|\dfrac{x}{a}\right|,$

$$[0 < \sec^{-1} |x/a| < \pi/2]. \qquad [\text{See } \textbf{733.2.}]$$

628.3. $\int \dfrac{\log (x + s)}{x^3} dx = - \dfrac{\log (x + s)}{2x^2} + \dfrac{s}{2a^2x}.$ [See **733.3.**]

628.9. $\displaystyle\int \frac{\log (x + s)}{x^p} \, dx = - \frac{\log (x + s)}{(p-1)x^{p-1}} + \frac{1}{p-1} \int \frac{dx}{x^{p-1}s},$

$$[p \neq 1]. \qquad \text{[See \textbf{281.01}–\textbf{284.01} and \textbf{733.9}.]}$$

630.1. $\displaystyle\int \log \sin x \, dx = x \log x - x - \frac{x^3}{18} - \frac{x^5}{900}$

$$- \frac{x^7}{19845} - \cdots - \frac{2^{2n-1}B_n x^{2n+1}}{n(2n+1)!} - \cdots, \qquad [0 < x < \pi].$$

$$\text{[See \textbf{45}.]} \qquad \text{[Integrate \textbf{603.1}.]}$$

$$= - x \log 2 - \frac{\sin 2x}{2} - \frac{\sin 4x}{2 \cdot 2^2} - \frac{\sin 6x}{2 \cdot 3^2} - \cdots,$$

$$[0 < x < \pi]. \qquad \text{[Integrate \textbf{603.2}.]}$$

630.2. $\displaystyle\int \log \cos x \, dx = - \frac{x^3}{6} - \frac{x^5}{60} - \frac{x^7}{315}$

$$- \frac{17x^9}{22680} - \cdots - \frac{2^{2n-1}(2^{2n} - 1)B_n}{n(2n+1)!} x^{2n+1} - \cdots,$$

$$[x^2 < \pi^2/4]. \qquad \text{[See \textbf{45}.]} \qquad \text{[Integrate \textbf{603.3}.]}$$

$$= - x \log 2 + \frac{\sin 2x}{2} - \frac{\sin 4x}{2 \cdot 2^2} + \frac{\sin 6x}{2 \cdot 3^2} - \cdots,$$

$$[x^2 < \pi^2/4]. \qquad \text{[Integrate \textbf{603.4}.]}$$

630.3. $\displaystyle\int \log \tan x \, dx = x \log x - x + \frac{x^3}{9} + \frac{7x^5}{450}$

$$+ \frac{62x^7}{19845} + \cdots + \frac{2^{2n}(2^{2n-1} - 1)B_n}{n(2n+1)!} x^{2n+1} + \cdots,$$

$$[0 < x < \pi/2]. \qquad \text{[See \textbf{45}.]} \qquad \text{[Integrate \textbf{603.6}.]}$$

631.1. $\displaystyle\int \sin \log x \, dx = \frac{1}{2} x \sin \log x - \frac{1}{2} x \cos \log x.$

631.2. $\displaystyle\int \cos \log x \, dx = \frac{1}{2} x \sin \log x + \frac{1}{2} x \cos \log x.$

632. $\displaystyle\int e^{ax} \log x \, dx = \frac{1}{a} e^{ax} \log x - \frac{1}{a} \int \frac{e^{ax}}{x} \, dx.$ \qquad [See \textbf{568.1}.]

$$\text{[Ref. 20, p. 46, No. 106.]}$$

Lambda Function and Gudermannian

635.1. $\int \log (a + r) \, dx = x \log (a + r) + a \log (x + r) - x$,

where $r = (x^2 + a^2)^{1/2}$. Contributed by Philip J. Hart.

635.2. $\int x \log (a + r) \, dx = \frac{x^2}{2} \log (a + r) + \frac{1}{2} ar - \frac{x^2}{4}$.

635.3. $\int \log (a + t) \, dx = x \log (a + t) + a \sin^{-1} \frac{x}{a} - x$,

where $t = (a^2 - x^2)^{1/2}$. The principal values of $\sin^{-1} (x/a)$ are to be taken, that is, values between $-\pi/2$ and $\pi/2$. The positive values of r, t, and a are to be taken. [See **320.01.**]

635.4. $\int x \log (a + t) \, dx = \frac{x^2}{2} \log (a + t) - \frac{1}{2} at - \frac{x^2}{4}$.

640.   If $x = \log \tan \left(\frac{\pi}{4} + \frac{\theta}{2} \right) = \log (\sec \theta + \tan \theta)$

$\theta = \text{gd } x = $ the **gudermannian** of $x = 2 \tan^{-1} e^x - \frac{\pi}{2}$.

641. $x = gd^{-1} \theta = \lambda(\theta)$, the **lambda** function.

642.1. $\sinh x = \tan \theta$.   **642.2.** $\cosh x = \sec \theta$.

642.3. $\tanh x = \sin \theta$.   **642.4.** $\tanh (x/2) = \tan (\theta/2)$.

642.5. $\dfrac{d \text{ gd } x}{dx} = \text{sech } x$.   **642.6.** $\dfrac{d \text{ gd}^{-1} x}{dx} = \sec x$,

$$[- \pi/2 < \theta < \pi/2].$$

If θ is tabulated for values of x, the hyperbolic functions may be obtained from a table of circular functions.

HYPERBOLIC FUNCTIONS

650.01. $\cosh^2 x - \sinh^2 x = 1$.

650.02. $\sinh x = \sqrt{(\cosh^2 x - 1)}$, $[x > 0]$.

         $= - \sqrt{(\cosh^2 x - 1)}$, $[x < 0]$.

650.03. $\cosh x = \sqrt{(1 + \sinh^2 x)}$.   **650.05.** $\operatorname{sech} x = 1/\cosh x$.

650.04. $\tanh x = \sinh x/\cosh x$.   **650.06.** $\operatorname{csch} x = 1/\sinh x$.

650.07. $\tanh^2 x + \operatorname{sech}^2 x = 1$.

650.08. $\operatorname{ctnh}^2 x - \operatorname{csch}^2 x = 1$.

650.09. $\sinh(-x) = -\sinh x$.

650.10. $\cosh(-x) = \cosh x$.

650.11. $\tanh(-x) = -\tanh x$.

651.01. $\sinh(x \pm y) = \sinh x \cosh y \pm \cosh x \sinh y$.

651.02. $\cosh(x \pm y) = \cosh x \cosh y \pm \sinh x \sinh y$.

651.03. $2 \sinh x \cosh y = \sinh(x + y) + \sinh(x - y)$.

651.04. $2 \cosh x \cosh y = \cosh(x + y) + \cosh(x - y)$.

651.05. $2 \sinh x \sinh y = \cosh(x + y) - \cosh(x - y)$.

651.06. $\sinh x + \sinh y = 2 \sinh \dfrac{x + y}{2} \cosh \dfrac{x - y}{2}$.

651.07. $\sinh x - \sinh y = 2 \sinh \dfrac{x - y}{2} \cosh \dfrac{x + y}{2}$.

651.08. $\cosh x + \cosh y = 2 \cosh \dfrac{x + y}{2} \cosh \dfrac{x - y}{2}$.

651.09. $\cosh x - \cosh y = 2 \sinh \dfrac{x + y}{2} \sinh \dfrac{x - y}{2}$.

151

651.10. $\sinh^2 x - \sinh^2 y = \sinh(x + y)\sinh(x - y)$
$$= \cosh^2 x - \cosh^2 y.$$

651.11. $\sinh^2 x + \cosh^2 y = \cosh(x + y)\cosh(x - y)$
$$= \cosh^2 x + \sinh^2 y.$$

651.12. $\operatorname{csch}^2 x - \operatorname{sech}^2 x = \operatorname{csch}^2 x \operatorname{sech}^2 x - \dfrac{1}{\sinh^2 x \cosh^2 x}.$

651.13. $(\sinh x + \cosh x)^n = \sinh nx + \cosh nx.$

651.14. $\dfrac{1}{\sinh x + \cosh x} = \cosh x - \sinh x.$

652.12. $\sinh 2x = 2\sinh x \cosh x.$

652.13. $\sinh 3x = 3\sinh x + 4\sinh^3 x.$

652.22. $\cosh 2x = \cosh^2 x + \sinh^2 x$
$$= 2\sinh^2 x + 1 = 2\cosh^2 x - 1.$$

652.23. $\cosh 3x = 4\cosh^3 x - 3\cosh x.$

652.3. $\sinh^2 x = \frac{1}{2}(\cosh 2x - 1).$

652.4. $\cosh^2 x = \frac{1}{2}(\cosh 2x + 1).$

652.5. $\sinh \dfrac{x}{2} = \sqrt{\{\frac{1}{2}(\cosh x - 1)\}},$　　　　　　　　$[x > 0].$

$$= -\sqrt{\{\frac{1}{2}(\cosh x - 1)\}},$$　　　　　　　　$[x < 0].$

652.6. $\cosh \dfrac{x}{2} = \sqrt{\{\frac{1}{2}(\cosh x + 1)\}}.$

653.1. $\tanh(x \pm y) = \dfrac{\tanh x \pm \tanh y}{1 \pm \tanh x \tanh y}.$

653.2. $\tanh\left(\dfrac{x \pm y}{2}\right) = \dfrac{\sinh x \pm \sinh y}{\cosh x + \cosh y}.$

653.3. $\tanh 2x = \dfrac{2\tanh x}{1 + \tanh^2 x}.$

653.4. $\tanh x \pm \tanh y = \dfrac{\sinh(x \pm y)}{\cosh x \cosh y}.$

653.5. $\tanh \dfrac{x}{2} = \dfrac{\cosh x - 1}{\sinh x} = \dfrac{\sinh x}{\cosh x + 1}.$

653.6. $\quad \text{ctnh}\,(x \pm y) = \dfrac{\text{ctnh}\,x\,\text{ctnh}\,y \pm 1}{\text{ctnh}\,y \pm \text{ctnh}\,x}.$

653.7. $\quad \text{ctnh}\,2x = \dfrac{\text{ctnh}^2\,x + 1}{2\,\text{ctnh}\,x}.$

653.8. $\quad \text{ctnh}\,\dfrac{x}{2} = \dfrac{\sinh x}{\cosh x - 1} = \dfrac{\cosh x + 1}{\sinh x}.$

654.1. $\quad \sinh x = \dfrac{1}{2}\,(e^x - e^{-x})$

$$= \frac{1}{2}\left(\log_\epsilon^{-1} x - \frac{1}{\log_\epsilon^{-1} x}\right),$$

where $\log_\epsilon^{-1}$ denotes the natural anti-logarithm. This may be taken from a table of natural logarithms if series 550 is slowly convergent as with large values of x. By noting that $\log_\epsilon^{-1} x = \log_{10}^{-1}\,(.4343x)$, a table of common logarithms can be used.

654.2. $\quad \cosh x = \dfrac{1}{2}\,(e^x + e^{-x})$

$$= \frac{1}{2}\left(\log_\epsilon^{-1} x + \frac{1}{\log_\epsilon^{-1} x}\right).$$

[See note under **654.1.**]

654.3. $\quad \tanh x = \dfrac{e^x - e^{-x}}{e^x + e^{-x}} = \dfrac{e^{2x} - 1}{e^{2x} + 1} = \dfrac{1 - e^{-2x}}{1 + e^{-2x}}$

The last form was suggested by H. A. Wheeler.

654.4. $\quad \cosh x + \sinh x = e^x.$ **654.5.** $\quad \cosh x - \sinh x = e^{-x}.$

654.6. $\quad \sinh\,(ix) = i\sin x.$ **654.7.** $\quad \cosh\,(ix) = \cos x.$

654.8. $\quad \tanh\,(ix) = i\tan x.$

655.1. $\quad \sinh\,(x \pm iy) = \sinh x \cos y \pm i\cosh x \sin y.$

655.2. $\quad \cosh\,(x \pm iy) = \cosh x \cos y \pm i\sinh x \sin y.$

655.3. $\quad \tanh\,(x \pm iy) = \dfrac{\sinh 2x \pm i\sin 2y}{\cosh 2x + \cos 2y}.$

655.4. $\quad \text{ctnh}\,(x \pm iy) = \dfrac{\sinh 2x \mp i\sin 2y}{\cosh 2x - \cos 2y}.$

656.1. $\quad \sinh 0 = 0.$ **656.2.** $\quad \cosh 0 = 1.$ **656.3.** $\quad \tanh 0 = 0.$

657.1. $\quad \sinh x = x + \dfrac{x^3}{3!} + \dfrac{x^5}{5!} + \dfrac{x^7}{7!} + \cdots,$ $\qquad [x^2 < \infty].$

657.2. $\cosh x = 1 + \dfrac{x^2}{2!} + \dfrac{x^4}{4!} + \dfrac{x^6}{6!} + \cdots,$ $[x^2 < \infty].$

657.3. $\tanh x = x - \dfrac{x^3}{3} + \dfrac{2}{15} x^5 - \dfrac{17}{315} x^7 + \dfrac{62}{2835} x^9$

$$- \cdots + \frac{(-1)^{n-1} 2^{2n} (2^{2n} - 1)}{(2n)!} B_n x^{2n-1} + \cdots,$$

$$[x^2 < \pi^2/4. \quad \text{See } \mathbf{45}].$$

657.4. $\operatorname{ctnh} x = \dfrac{1}{x} + \dfrac{x}{3} - \dfrac{x^3}{45} + \dfrac{2x^5}{945} - \dfrac{x^7}{4725} + \cdots$

$$+ \frac{(-1)^{n-1} 2^{2n}}{(2n)!} B_n x^{2n-1} + \cdots,$$

$$[x^2 < \pi^2. \quad \text{See } \mathbf{45}].$$

657.5. $\operatorname{sech} x = 1 - \dfrac{x^2}{2!} + \dfrac{5}{4!} x^4 - \dfrac{61}{6!} x^6 + \dfrac{1385}{8!} x^8 - \cdots$

$$+ \frac{(-1)^n}{(2n)!} E_n x^{2n} + \cdots,$$

$$[x^2 < \pi^2/4. \quad \text{See } \mathbf{45}].$$

657.6. $\operatorname{csch} x = \dfrac{1}{x} - \dfrac{x}{6} + \dfrac{7x^3}{360} - \dfrac{31x^5}{15120} + \cdots$

$$+ \frac{2(-1)^n (2^{2n-1} - 1)}{(2n)!} B_n x^{2n-1} + \cdots,$$

$$[x^2 < \pi^2. \quad \text{See } \mathbf{45}].$$

658.1. For large, positive values of x,
$$\tanh x = 1 - 2 \left(e^{-2x} - e^{-4x} + e^{-6x} \cdots \right).$$
For negative arguments, use $\tanh(-x) = -\tanh x$, where x is positive.

658.2. For large, positive values of x,
$$\operatorname{ctnh} x = 1 + 2 \left(e^{-2x} + e^{-4x} + e^{-6x} + \cdots \right).$$
For negative arguments, use $\operatorname{ctnh}(-x) = -\operatorname{ctnh} x$, where x is positive.

658.3 For large, positive values of x,
$$\operatorname{sech} x = 2 \left(e^{-x} - e^{-3x} + e^{-5x} - \cdots \right).$$
For negative arguments, use $\operatorname{sech}(-x) = \operatorname{sech} x$, where x is positive.

658.4 For large, positive values of x,

$$\text{csch } x = 2\, (e^{-x} + e^{-3x} + e^{-5x} + \cdots).$$

For negative arguments, use csch $(-x) = -$ csch x, where x is positive.

HYPERBOLIC FUNCTIONS—DERIVATIVES

667.1. $\dfrac{d \sinh x}{dx} = \cosh x.$

667.3. $\dfrac{d \tanh x}{dx} = \text{sech}^2 x.$

667.2. $\dfrac{d \cosh x}{dx} = \sinh x.$

667.4. $\dfrac{d \text{ ctnh } x}{dx} = - \text{ csch}^2 x.$

667.5. $\dfrac{d \text{ sech } x}{dx} = - \text{ sech } x \tanh x.$

667.6. $\dfrac{d \text{ csch } x}{dx} = - \text{ csch } x \text{ ctnh } x.$

HYPERBOLIC FUNCTIONS—INTEGRALS

670. An integral of a trigonometric function often can be changed into the corresponding integral of a hyperbolic function by changing x to ix and substituting

$$\sin (ix) = i \sinh x, \quad \cos (ix) = \cosh x, \quad \tan (ix) = i \tanh x, \text{ etc.}$$

[See **408.10–.15.**]

This substitution is useful also with other classes of formulas.

Integrals Involving sinh x

671.10. $\displaystyle\int \sinh x \, dx = \cosh x.$

671.101. $\displaystyle\int \sinh \frac{x}{a} \, dx = a \cosh \frac{x}{a}.$

671.11. $\displaystyle\int x \sinh x \, dx = x \cosh x - \sinh x.$

671.12. $\displaystyle\int x^2 \sinh x \, dx = (x^2 + 2) \cosh x - 2x \sinh x.$

671.13. $\int x^3 \sinh x \, dx = (x^3 + 6x) \cosh x - (3x^2 + 6) \sinh x.$

671.19. $\int x^p \sinh x \, dx = x^p \cosh x - p \int x^{p-1} \cosh x \, dx.$ [See **677.1.**]

671.20. $\int \sinh^2 x \, dx = \dfrac{\sinh 2x}{4} - \dfrac{x}{2}.$

671.21. $\int x \sinh^2 x \, dx = \dfrac{x \sinh 2x}{4} - \dfrac{\cosh 2x}{8} - \dfrac{x^2}{4}.$

671.30. $\int \sinh^3 x \, dx = \dfrac{\cosh^3 x}{3} - \cosh x.$

671.40. $\int \sinh^4 x \, dx = \dfrac{\sinh 4x}{32} - \dfrac{\sinh 2x}{4} + \dfrac{3x}{8}.$

671.90. $\int \sinh^p x \, dx = \dfrac{1}{p} \sinh^{p-1} x \cosh x - \dfrac{p-1}{p} \int \sinh^{p-2} x \, dx.$

672.11. $\int \dfrac{\sinh x}{x} \, dx = x + \dfrac{x^3}{3 \cdot 3!} + \dfrac{x^5}{5 \cdot 5!} + \dfrac{x^7}{7 \cdot 7!} + \cdots .$

672.12. $\int \dfrac{\sinh x}{x^2} \, dx = -\dfrac{\sinh x}{x} + \int \dfrac{\cosh x}{x} \, dx.$ [See **678.11.**]

672.21. $\int \dfrac{\sinh^2 x}{x} \, dx = -\dfrac{1}{2} \log |x| + \dfrac{1}{2} \int \dfrac{\cosh 2x}{2x} \, d(2x).$ [See **678.11.**]

673.10. $\int \dfrac{dx}{\sinh x} = \int \operatorname{csch} x \, dx = \log \left| \tanh \dfrac{x}{2} \right| = \log \dfrac{e^x - 1}{e^x + 1}$

$$= -\dfrac{1}{2} \log \dfrac{\cosh x + 1}{\cosh x - 1}.$$

673.11. $\int \dfrac{x \, dx}{\sinh x} = x - \dfrac{x^3}{3 \cdot 3!} + \dfrac{7x^5}{3 \cdot 5 \cdot 5!} - \dfrac{31x^7}{3 \cdot 7 \cdot 7!} + \dfrac{127x^9}{3 \cdot 5 \cdot 9!}$

$$- \cdots + (-1)^n \dfrac{2(2^{2n-1} - 1)}{(2n + 1)!} B_n x^{2n+1} + \cdots ,$$

$$[x^2 < \pi^2. \quad \text{See } \mathbf{45}.]$$

673.19. $\int \dfrac{x^p dx}{\sinh x}$. Expand $\dfrac{1}{\sinh x}$ by **657.6**, multiply by x^p and

integrate, $[p \neq 0]$.

673.20. $\int \dfrac{dx}{\sinh^2 x} = \int \operatorname{csch}^2 x \, dx = -\operatorname{ctnh} x.$

673.21. $\int \dfrac{x \, dx}{\sinh^2 x} = -x \operatorname{ctnh} x + \log |\sinh x|.$

673.30. $\int \dfrac{dx}{\sinh^3 x} = \int \operatorname{csch}^3 x \, dx = -\dfrac{\cosh x}{2 \sinh^2 x} - \dfrac{1}{2} \log \left| \tanh \dfrac{x}{2} \right|.$

673.40. $\int \dfrac{dx}{\sinh^4 x} = \operatorname{ctnh} x - \dfrac{\operatorname{ctnh}^3 x}{3}.$

673.90. $\int \dfrac{dx}{\sinh^p x} = -\dfrac{\cosh x}{(p-1) \sinh^{p-1} x} - \dfrac{p-2}{p-1} \int \dfrac{dx}{\sinh^{p-2} x},$

$[p > 1].$

675. $\int \sinh mx \sinh nx \, dx = \dfrac{\sinh (m+n)x}{2(m+n)} - \dfrac{\sinh (m-n)x}{2(m-n)},$

$[m^2 \neq n^2.$ If $m^2 = n^2$, see **671.20**].

Integrals Involving cosh x

677.10. $\int \cosh x \, dx = \sinh x.$

677.101. $\int \cosh \dfrac{x}{a} \, dx = a \sinh \dfrac{x}{a}.$

677.11. $\int x \cosh x \, dx = x \sinh x - \cosh x.$

677.12. $\int x^2 \cosh x \, dx = (x^2 + 2) \sinh x - 2x \cosh x.$

677.13. $\int x^3 \cosh x \, dx = (x^3 + 6x) \sinh x - (3x^2 + 6) \cosh x.$

677.19. $\int x^p \cosh x \, dx = x^p \sinh x - p \int x^{p-1} \sinh x \, dx.$

[See **671.1**.]

677.20. $\displaystyle\int \cosh^2 x\, dx = \frac{\sinh 2x}{4} + \frac{x}{2}.$

677.21. $\displaystyle\int x \cosh^2 x\, dx = \frac{x \sinh 2x}{4} - \frac{\cosh 2x}{8} + \frac{x^2}{4}.$

677.30. $\displaystyle\int \cosh^3 x\, dx = \frac{\sinh^3 x}{3} + \sinh x.$

677.40. $\displaystyle\int \cosh^4 x\, dx = \frac{\sinh 4x}{32} + \frac{\sinh 2x}{4} + \frac{3x}{8}.$

677.90. $\displaystyle\int \cosh^p x\, dx = \frac{1}{p} \sinh x \cosh^{p-1} x + \frac{p-1}{p} \int \cosh^{p-2} x\, dx.$

678.11. $\displaystyle\int \frac{\cosh x}{x}\, dx = \log |x| + \frac{x^2}{2\cdot 2!} + \frac{x^4}{4\cdot 4!} + \frac{x^6}{6\cdot 6!} + \cdots.$

678.12. $\displaystyle\int \frac{\cosh x}{x^2}\, dx = -\frac{\cosh x}{x} + \int \frac{\sinh x}{x}\, dx.$ [See **672.11.**]

678.21. $\displaystyle\int \frac{\cosh^2 x\, dx}{x} = \frac{1}{2} \log |x| + \frac{1}{2} \int \frac{\cosh 2x}{2x}\, d(2x).$ [See **678.11.**]

679.10. $\displaystyle\int \frac{dx}{\cosh x} = \int \operatorname{sech} x\, dx = \tan^{-1} (\sinh x)$

$$= 2 \tan^{-1} e^x + \text{constant}.$$

679.11. $\displaystyle\int \frac{x\, dx}{\cosh x} = \frac{x^2}{2} - \frac{x^4}{4\cdot 2!} + \frac{5x^6}{6\cdot 4!} - \frac{61x^8}{8\cdot 6!} + \frac{1385x^{10}}{10\cdot 8!}$

$$- \cdots + \frac{(-1)^n E_n}{(2n+2)(2n)!} x^{2n+2} + \cdots,$$

$$[x^2 < \pi^2/4. \quad \text{See } \mathbf{45}].$$

679.19. $\displaystyle\int \frac{x^p dx}{\cosh x}.$ Expand $\dfrac{1}{\cosh x}$ by **657.5,** multiply by x^p and integrate, $[p \neq 0].$

679.20. $\displaystyle\int \frac{dx}{\cosh^2 x} = \int \operatorname{sech}^2 x\, dx = \tanh x.$

679.21. $\displaystyle\int \frac{x\, dx}{\cosh^2 x} = x \tanh x - \log \cosh x.$

679.30. $\displaystyle\int \frac{dx}{\cosh^3 x} = \frac{\sinh x}{2 \cosh^2 x} + \frac{1}{2} \tan^{-1}(\sinh x).$

679.40. $\displaystyle\int \frac{dx}{\cosh^4 x} = \tanh x - \frac{\tanh^3 x}{3}.$

679.90. $\displaystyle\int \frac{dx}{\cosh^p x} = \frac{\sinh x}{(p-1)\cosh^{p-1} x} + \frac{p-2}{p-1}\int \frac{dx}{\cosh^{p-2} x},\ [p > 1].$

681. $\displaystyle\int \cosh mx \cosh nx\, dx = \frac{\sinh(m+n)x}{2(m+n)} + \frac{\sinh(m-n)x}{2(m-n)},$
$$[m^2 \neq n^2]. \quad [\text{If } m^2 = n^2, \text{ see } \mathbf{677.20}.]$$

682.01. $\displaystyle\int \frac{dx}{\cosh x + 1} = \tanh \frac{x}{2}.$

682.02. $\displaystyle\int \frac{dx}{\cosh x - 1} = -\operatorname{ctnh} \frac{x}{2}.$

682.03. $\displaystyle\int \frac{x\, dx}{\cosh x + 1} = x \tanh \frac{x}{2} - 2 \log \cosh \frac{x}{2}.$

682.04. $\displaystyle\int \frac{x\, dx}{\cosh x - 1} = -x \operatorname{ctnh} \frac{x}{2} + 2 \log \left|\sinh \frac{x}{2}\right|.$

682.05. $\displaystyle\int \frac{\cosh x\, dx}{\cosh x + 1} = x - \tanh \frac{x}{2}.$

682.06. $\displaystyle\int \frac{\cosh x\, dx}{\cosh x - 1} = x - \operatorname{ctnh} \frac{x}{2}.$

682.07. $\displaystyle\int \frac{dx}{\cosh x(\cosh x + 1)} = \tan^{-1}(\sinh x) - \tanh \frac{x}{2}.$

682.08. $\displaystyle\int \frac{dx}{\cosh x(\cosh x - 1)} = -\tan^{-1}(\sinh x) - \operatorname{ctnh} \frac{x}{2}.$

682.09. $\displaystyle\int \frac{dx}{(\cosh x + 1)^2} = \frac{1}{2} \tanh \frac{x}{2} - \frac{1}{6} \tanh^3 \frac{x}{2}.$

682.10. $\displaystyle\int \frac{dx}{(\cosh x - 1)^2} = \frac{1}{2} \operatorname{ctnh} \frac{x}{2} - \frac{1}{6} \operatorname{ctnh}^3 \frac{x}{2}.$

682.11. $\int \dfrac{dx}{\cosh^2 x + 1} = \dfrac{1}{2\sqrt{2}} \cosh^{-1}\left(\dfrac{3\cosh^2 x - 1}{\cosh^2 x + 1}\right).$

Use the positive value of the inverse cosh for positive x and the negative value for negative x.

682.12. $\int \dfrac{dx}{\cosh^2 x - 1} = \int \dfrac{dx}{\sinh^2 x} = -\operatorname{ctnh} x.$ [See **673.20.**]

Integrals Involving sinh x and cosh x

685.11. $\int \sinh x \cosh x \, dx = \dfrac{\sinh^2 x}{2} = \dfrac{\cosh^2 x}{2} + \text{constant}$

$$= \dfrac{\cosh 2x}{4} + \text{constant}.$$

685.12. $\int \sinh x \cosh^2 x \, dx = \dfrac{\cosh^3 x}{3}.$

685.13. $\int \sinh x \cosh^3 x \, dx = \dfrac{\cosh^4 x}{4}.$

685.19. $\int \sinh x \cosh^p x \, dx = \dfrac{\cosh^{p+1} x}{p+1},$ $[p \neq -1].$

685.21. $\int \sinh^2 x \cosh x \, dx = \dfrac{\sinh^3 x}{3}.$

685.22. $\int \sinh^2 x \cosh^2 x \, dx = \dfrac{\sinh 4x}{32} - \dfrac{x}{8}.$

685.31. $\int \sinh^3 x \cosh x \, dx = \dfrac{\sinh^4 x}{4}.$

685.91. $\int \sinh^p x \cosh x \, dx = \dfrac{\sinh^{p+1} x}{p+1},$ $[p \neq -1].$

686.11. $\int \dfrac{dx}{\sinh x \cosh x} = \log |\tanh x|.$

686.12. $\int \dfrac{dx}{\sinh x \cosh^2 x} = \dfrac{1}{\cosh x} + \log \left|\tanh \dfrac{x}{2}\right|.$

686.13. $\displaystyle\int\frac{dx}{\sinh x \cosh^3 x} = \frac{1}{2\cosh^2 x} + \log|\tanh x|.$

686.19. $\displaystyle\int\frac{dx}{\sinh x \cosh^p x} = \frac{1}{(p-1)\cosh^{p-1} x}$

$$+ \int\frac{dx}{\sinh x \cosh^{p-2} x}, \qquad [p \neq 1].$$

686.21. $\displaystyle\int\frac{dx}{\sinh^2 x \cosh x} = -\frac{1}{\sinh x} - \tan^{-1}(\sinh x).$

686.22. $\displaystyle\int\frac{dx}{\sinh^2 x \cosh^2 x} = -2\,\text{ctnh}\,2x.$

686.31. $\displaystyle\int\frac{dx}{\sinh^3 x \cosh x} = -\frac{1}{2\sinh^2 x} - \log|\tanh x|.$

686.91. $\displaystyle\int\frac{dx}{\sinh^p x \cosh x} = -\frac{1}{(p-1)\sinh^{p-1} x}$

$$-\int\frac{dx}{\sinh^{p-2} x \cosh x}, \qquad [p \neq 1].$$

687.11. $\displaystyle\int\frac{\sinh x\,dx}{\cosh x} = \int\tanh x\,dx = \log\cosh x.$ [See **691.01.**]

687.12. $\displaystyle\int\frac{\sinh x\,dx}{\cosh^2 x} = -\frac{1}{\cosh x} = -\operatorname{sech} x.$

687.13. $\displaystyle\int\frac{\sinh x\,dx}{\cosh^3 x} = -\frac{1}{2\cosh^2 x} = \frac{\tanh^2 x}{2} + \text{constant}.$

687.19. $\displaystyle\int\frac{\sinh x\,dx}{\cosh^p x} = -\frac{1}{(p-1)\cosh^{p-1} x}, \qquad [p \neq 1].$

687.21. $\displaystyle\int\frac{\sinh^2 x}{\cosh x}\,dx = \sinh x - \tan^{-1}(\sinh x).$

687.22. $\displaystyle\int\frac{\sinh^2 x}{\cosh^2 x}\,dx = \int\tanh^2 x = x - \tanh x.$ [See **691.02.**]

687.29. $\displaystyle\int\frac{\sinh^2 x}{\cosh^p x}\,dx = -\frac{\sinh x}{(p-1)\cosh^{p-1} x}$

$$+ \frac{1}{p-1}\int\frac{dx}{\cosh^{p-2} x}, \qquad [p \neq 1].$$

687.31. $\displaystyle\int \frac{\sinh^3 x}{\cosh x}\, dx = \frac{\sinh^2 x}{2} - \log \cosh x.$

687.32. $\displaystyle\int \frac{\sinh^3 x}{\cosh^2 x}\, dx = \cosh x + \operatorname{sech} x.$

687.33. $\displaystyle\int \frac{\sinh^3 x}{\cosh^3 x}\, dx = \int \tanh^3 x\, dx = -\frac{\tanh^2 x}{2} + \log \cosh x.$

[See **691.03**.]

687.34. $\displaystyle\int \frac{\sinh^3 x}{\cosh^4 x}\, dx = \frac{1}{3 \cosh^3 x} - \frac{1}{\cosh x}.$

687.39. $\displaystyle\int \frac{\sinh^3 x}{\cosh^p x}\, dx = \frac{1}{(p-1)\cosh^{p-1} x} - \frac{1}{(p-3)\cosh^{p-3} x},$

$[p \neq 1 \text{ or } 3].$

687.7. $\displaystyle\int \frac{\sinh^{p-2} x}{\cosh^p x}\, dx = \frac{\tanh^{p-1} x}{p-1},$ $[p \neq 1].$

688.11. $\displaystyle\int \frac{\cosh x}{\sinh x}\, dx = \int \operatorname{ctnh} x\, dx = \log |\sinh x|.$ [See **692.01**.]

688.12. $\displaystyle\int \frac{\cosh x}{\sinh^2 x}\, dx = -\frac{1}{\sinh x} = -\operatorname{csch} x.$

688.13. $\displaystyle\int \frac{\cosh x}{\sinh^3 x}\, dx = -\frac{1}{2 \sinh^2 x} = -\frac{\operatorname{ctnh}^2 x}{2} + \text{constant}.$

688.19. $\displaystyle\int \frac{\cosh x}{\sinh^p x}\, dx = -\frac{1}{(p-1)\sinh^{p-1} x},$ $[p \neq 1].$

688.21. $\displaystyle\int \frac{\cosh^2 x}{\sinh x}\, dx = \cosh x + \log \left| \tanh \frac{x}{2} \right|.$

688.22. $\displaystyle\int \frac{\cosh^2 x}{\sinh^2 x}\, dx = \int \operatorname{ctnh}^2 x\, dx = x - \operatorname{ctnh} x.$ [See **692.02**.]

688.29. $\displaystyle\int \frac{\cosh^2 x}{\sinh^p x}\, dx = -\frac{\cosh x}{(p-1)\sinh^{p-1} x}$

$$+ \frac{1}{p-1} \int \frac{dx}{\sinh^{p-2} x}, \qquad [p \neq 1].$$

688.31. $\int \dfrac{\cosh^3 x}{\sinh x} \, dx = \dfrac{\cosh^2 x}{2} + \log |\sinh x|.$

688.32. $\int \dfrac{\cosh^3 x}{\sinh^2 x} \, dx = \sinh x - \operatorname{csch} x.$

688.33. $\int \dfrac{\cosh^3 x}{\sinh^3 x} \, dx = \int \operatorname{ctnh}^3 x \, dx \quad = -\dfrac{\operatorname{ctnh}^2 x}{2} + \log |\sinh x|.$

[See **692.03.**]

688.34. $\int \dfrac{\cosh^3 x}{\sinh^4 x} \, dx = -\dfrac{1}{3 \sinh^3 x} - \dfrac{1}{\sinh x}.$

688.39. $\int \dfrac{\cosh^3 x}{\sinh^p x} \, dx = -\dfrac{1}{(p-1) \sinh^{p-1} x} - \dfrac{1}{(p-3) \sinh^{p-3} x},$

$[p \neq 1 \text{ or } 3].$

688.7. $\int \dfrac{\cosh^{p-2} x}{\sinh^p x} \, dx = -\dfrac{\operatorname{ctnh}^{p-1} x}{p-1},$ $\qquad [p \neq 1].$

689.01. $\int \dfrac{\sinh x \, dx}{\cosh x + 1} = \log (\cosh x + 1).$

689.02. $\int \dfrac{\sinh x \, dx}{\cosh x - 1} = \log (\cosh x - 1).$

689.03. $\int \dfrac{dx}{\sinh x(\cosh x + 1)} = -\dfrac{1}{2(\cosh x + 1)} + \dfrac{1}{2} \log \left| \tanh \dfrac{x}{2} \right|.$

689.04. $\int \dfrac{dx}{\sinh x(\cosh x - 1)} = \dfrac{1}{2(\cosh x - 1)} - \dfrac{1}{2} \log \left| \tanh \dfrac{x}{2} \right|.$

689.05. $\int \dfrac{\sinh x \, dx}{\cosh x(\cosh x + 1)} = \log \left(\dfrac{\cosh x}{\cosh x + 1} \right).$

689.06. $\int \dfrac{\sinh x \, dx}{\cosh x(\cosh x - 1)} = \log \left(\dfrac{\cosh x - 1}{\cosh x} \right).$

689.07. $\int \sinh mx \cosh nx \, dx = \dfrac{\cosh (m+n)x}{2(m+n)} + \dfrac{\cosh (m-n)x}{2(m-n)},$

$[m^2 \neq n^2. \qquad \text{If } m^2 = n^2, \text{ see } \mathbf{685.11.}]$

Integrals Involving tanh x and ctnh x

691.01. $\displaystyle\int \tanh x \, dx = \log \cosh x.$ [See **687.11.**]

691.02. $\displaystyle\int \tanh^2 x \, dx = x - \tanh x.$ [See **687.22.**]

691.03. $\displaystyle\int \tanh^3 x \, dx = -\frac{\tanh^2 x}{2} + \log \cosh x.$ [See **687.33.**]

691.09. $\displaystyle\int \tanh^p x \, dx = -\frac{\tanh^{p-1} x}{p - 1} + \int \tanh^{p-2} x \, dx,$ $[p \neq 1].$

693.01. $\displaystyle\int \operatorname{ctnh} x \, dx = \log |\sinh x|.$ [See **688.11.**]

693.02. $\displaystyle\int \operatorname{ctnh}^2 x \, dx = x - \operatorname{ctnh} x.$ [See **688.22.**]

693.03. $\displaystyle\int \operatorname{ctnh}^3 x \, dx = -\frac{\operatorname{ctnh}^2 x}{2} + \log |\sinh x|.$ [See **688.33.**]

693.09. $\displaystyle\int \operatorname{ctnh}^p x \, dx = -\frac{\operatorname{ctnh}^{p-1} x}{p - 1} + \int \operatorname{ctnh}^{p-2} x \, dx,$ $[p \neq 1].$

⋆⋆ 8 ⋆⋆

INVERSE
HYPERBOLIC FUNCTIONS

700. $\sinh^{-1} x = \cosh^{-1} \sqrt{(x^2 + 1)}$.

Use the positive value of $\cosh^{-1}$ when x is positive and the negative value when x is negative.

700.1. $\sinh^{-1} x = \tanh^{-1} \dfrac{x}{\sqrt{(x^2 + 1)}} = \operatorname{csch}^{-1} \dfrac{1}{x}$

$$= - \sinh^{-1} (- x) = \log \{x + \sqrt{(x^2 + 1)}\}.$$

[See **602.1** and **706.**]

701. $\cosh^{-1} x = \pm \sinh^{-1} \sqrt{(x^2 - 1)} = \pm \tanh^{-1} \dfrac{\sqrt{(x^2 - 1)}}{x}$

$$= \operatorname{sech}^{-1} \dfrac{1}{x} = \pm \log \{x + \sqrt{(x^2 - 1)}\},$$

$[x > 1]$.   [See **602.3** and **707.**]

702. $\tanh^{-1} x = \operatorname{ctnh}^{-1} \dfrac{1}{x} = \dfrac{1}{2} \log \dfrac{1 + x}{1 - x}$, $[x^2 < 1]$.   [See **708.**]

703. $\operatorname{ctnh}^{-1} x = \tanh^{-1} \dfrac{1}{x} = \dfrac{1}{2} \log \dfrac{x + 1}{x - 1}$, $[x^2 > 1]$.   [See **709.**]

704. $\operatorname{sech}^{-1} x = \pm \log \left\{ \dfrac{1}{x} + \sqrt{\left(\dfrac{1}{x^2} - 1 \right)} \right\}$,

$[0 < x < 1]$.   [See **710.**]

705. $\operatorname{csch}^{-1} x = \log \left\{ \dfrac{1}{x} + \sqrt{\left(\dfrac{1}{x^2} + 1 \right)} \right\}$.   [See **711.**]

165

706. $\sinh^{-1} x$

$$= x - \frac{1}{2\cdot3}\, x^3 + \frac{1\cdot3}{2\cdot4\cdot5}\, x^5 - \frac{1\cdot3\cdot5}{2\cdot4\cdot6\cdot7}\, x^7 + \cdots, \quad [x^2 < 1].$$

$$= \log (2x) + \frac{1}{2\cdot2x^2} - \frac{1\cdot3}{2\cdot4\cdot4x^4} + \frac{1\cdot3\cdot5}{2\cdot4\cdot6\cdot6x^6} - \cdots,$$
$$[x > 1].$$

$$= -\log |2x| - \frac{1}{2\cdot2x^2} + \frac{1\cdot3}{2\cdot4\cdot4x^4} - \frac{1\cdot3\cdot5}{2\cdot4\cdot6\cdot6x^6} + \cdots,$$
$$[x < -1]. \quad \text{[See 602.1.]}$$

707. $\cosh^{-1} x = \pm \left[\log (2x) - \frac{1}{2\cdot2x^2} - \frac{1\cdot3}{2\cdot4\cdot4x^4} \right.$

$$\left. - \frac{1\cdot3\cdot5}{2\cdot4\cdot6\cdot6x^6} - \cdots \right], \quad [x > 1].$$
$$\text{[See 602.3 and 602.4.]}$$

708. $\tanh^{-1} x = x + \dfrac{x^3}{3} + \dfrac{x^5}{5} + \dfrac{x^7}{7} + \cdots, \qquad [x^2 < 1].$
$$\text{[See 601.2.]}$$

709. $\operatorname{ctnh}^{-1} x = \dfrac{1}{x} + \dfrac{1}{3x^3} + \dfrac{1}{5x^5} + \dfrac{1}{7x^7} + \cdots, \qquad [x^2 > 1].$
$$\text{[See 601.3.]}$$

710. $\operatorname{sech}^{-1} x = \pm \left[\log \dfrac{2}{x} - \dfrac{1}{2\cdot2}\, x^2 - \dfrac{1\cdot3}{2\cdot4\cdot4}\, x^4 \right.$

$$\left. - \frac{1\cdot3\cdot5}{2\cdot4\cdot6\cdot6}\, x^6 - \cdots \right], \qquad [0 < x < 1].$$
$$\text{[See 602.7 and 602.8.]}$$

711. $\operatorname{csch}^{-1} x$

$$= \frac{1}{x} - \frac{1}{2\cdot3x^3} + \frac{1\cdot3}{2\cdot4\cdot5x^5} - \frac{1\cdot3\cdot5}{2\cdot4\cdot6\cdot7x^7} + \cdots, \qquad [x^2 > 1].$$

$$= \log \frac{2}{x} + \frac{1}{2\cdot2}\, x^2 - \frac{1\cdot3}{2\cdot4\cdot4}\, x^4 + \frac{1\cdot3\cdot5}{2\cdot4\cdot6\cdot6}\, x^6 - \cdots,$$
$$[0 < x < 1].$$

$$= -\log \left| \frac{2}{x} \right| - \frac{1}{2\cdot2}\, x^2 + \frac{1\cdot3}{2\cdot4\cdot4}\, x^4 - \frac{1\cdot3\cdot5}{2\cdot4\cdot6\cdot6}\, x^6 + \cdots,$$
$$[-1 < x < 0]. \quad \text{[See 602.5.]}$$

720. $\sinh^{-1}(\pm x + iy) = \pm(-1)^n \cosh^{-1}\dfrac{s + t}{2}$

$$+ i(-1)^n \sin^{-1}\dfrac{2y}{s + t} + in\pi,$$

where the principal value of $\sin^{-1}$ (between $-\pi/2$ and $\pi/2$) and the positive value of $\cosh^{-1}$ are taken,

n is an integer or 0,

x is positive,

y is positive or negative,

and where

720.1. $s = \sqrt{(1 + y)^2 + x^2}$ (positive value),

720.2. $t = \sqrt{(1 - y)^2 + x^2}$ (positive value).

Note that if $x = 0$ and $y > 1$, $t = y - 1$ and $s + t = 2y$.
If $x = 0$ and $y < 1$, $t = 1 - y$ and $s + t = 2$.

Alternative:

720.3a. $\sinh^{-1} A = \log_e(\pm\sqrt{1 + A^2} + A) + i2k\pi$

or

720.3b. $= -\log_e(\pm\sqrt{1 + A^2} - A) + i2k\pi$

where A may be a complex quantity and k is an integer or 0.

For the square root of a complex quantity see **58**, and for the logarithm see **604**. The two solutions a and b are identical. In any given case, the one should be used which involves the numerical sum of two quantities instead of the difference, so as to obtain more convenient precise computation.

721.1. $\cosh^{-1}(x + iy)$

$$= \pm\left(\cosh^{-1}\dfrac{p + q}{2} + i\cos^{-1}\dfrac{2x}{p + q} + i2k\pi\right).$$

721.2. $\cosh^{-1}(x - iy)$

$$= \pm\left(\cosh^{-1}\dfrac{p + q}{2} - i\cos^{-1}\dfrac{2x}{p + q} + i2k\pi\right),$$

where the positive value of $\cosh^{-1}$ and the principal value of $\cos^{-1}$ (between 0 and π) are taken,

x is positive or negative,

y is positive,

721.3. $p = \sqrt{(1 + x)^2 + y^2}$ (positive value).

721.4. $q = \sqrt{(1 - x)^2 + y^2}$ (positive value).

Alternative:

721.5a. $\cosh^{-1} A = \pm \log_e (A + \sqrt{A^2 - 1}) + i2k\pi$

or

721.5b. $= \mp \log_e (A - \sqrt{A^2 - 1}) + i2k\pi.$

[See note following **720.3.**]

722.1. $\tanh^{-1} (x + iy) = \dfrac{1}{4} \log_e \dfrac{(1 + x)^2 + y^2}{(1 - x)^2 + y^2}$

$$+ \frac{i}{2} \left\{ (2k + 1)\pi - \tan^{-1} \frac{1 + x}{y} - \tan^{-1} \frac{1 - x}{y} \right\},$$

where the principal values of $\tan^{-1}$ (between $-\pi/2$ and $\pi/2$) are
taken and where x and y may be positive or negative.

[See formula for $\tanh^{-1} (x + iy)$ in Ref. 24, p. 115.]

Alternative:

722.2. $\tanh^{-1} (x + iy) = \dfrac{1}{4} \log_e \dfrac{(1 + x)^2 + y^2}{(1 - x)^2 + y^2}$

$$+ \frac{i}{2} \tan^{-1} \frac{2y}{1 - x^2 - y^2} + i\pi k$$

where k is 0 or an integer. The proper quadrant for $\tan^{-1}$ is to be
taken according to the signs of the values of the numerator and the
denominator when they are given in numbers.

722.3. $\tanh^{-1} (x + iy) = \dfrac{1}{2} \log_e \dfrac{1 + x + iy}{1 - x - iy}.$ [See **604.**]

[Ref. 46, Chap. XI.]

INVERSE HYPERBOLIC FUNCTIONS—DERIVATIVES,

728.1. $\dfrac{d}{dx} \sinh^{-1} \dfrac{x}{a} = \dfrac{1}{\sqrt{(x^2 + a^2)}}.$

728.2. $\dfrac{d}{dx} \cosh^{-1} \dfrac{x}{a} = \dfrac{1}{\sqrt{(x^2 - a^2)}},$ $\left[\cosh^{-1} \dfrac{x}{a} > 0, \quad \dfrac{x}{a} > 1 \right].$

728.3. $\dfrac{d}{dx} \cosh^{-1} \dfrac{x}{a} = \dfrac{-1}{\sqrt{(x^2 - a^2)}},$ $\left[\cosh^{-1} \dfrac{x}{a} < 0, \quad \dfrac{x}{a} > 1\right].$

728.4. $\dfrac{d}{dx} \tanh^{-1} \dfrac{x}{a} = \dfrac{a}{a^2 - x^2},$ $[x^2 < a^2].$

728.5. $\dfrac{d}{dx} \operatorname{ctnh}^{-1} \dfrac{x}{a} = \dfrac{a}{a^2 - x^2},$ $[x^2 > a^2].$

728.6. $\dfrac{d}{dx} \operatorname{sech}^{-1} \dfrac{x}{a} = \dfrac{-a}{x\sqrt{(a^2 - x^2)}},$

$[\operatorname{sech}^{-1}(x/a) > 0, \quad 0 < x/a < 1].$

728.7. $\dfrac{d}{dx} \operatorname{sech}^{-1} \dfrac{x}{a} = \dfrac{a}{x\sqrt{(a^2 - x^2)}},$

$[\operatorname{sech}^{-1}(x/a) < 0, \quad 0 < x/a < 1].$

728.8. $\dfrac{d}{dx} \operatorname{csch}^{-1} \dfrac{x}{a} = \dfrac{-a}{|x|\sqrt{(x^2 + a^2)}}.$

[Except in **728.4** and **728.5**, $a > 0$.]

INVERSE HYPERBOLIC FUNCTIONS: INTEGRALS—($a > 0$)

730. $\displaystyle\int \sinh^{-1} \dfrac{x}{a}\, dx = x \sinh^{-1} \dfrac{x}{a} - \sqrt{(x^2 + a^2)}.$

730.1. $\displaystyle\int x \sinh^{-1} \dfrac{x}{a}\, dx = \left(\dfrac{x^2}{2} + \dfrac{a^2}{4}\right) \sinh^{-1} \dfrac{x}{a} - \dfrac{x}{4} \sqrt{(x^2 + a^2)}.$

730.2. $\displaystyle\int x^2 \sinh^{-1} \dfrac{x}{a}\, dx = \dfrac{x^3}{3} \sinh^{-1} \dfrac{x}{a} + \dfrac{2a^2 - x^2}{9} \sqrt{(x^2 + a^2)}.$

730.3. $\displaystyle\int x^3 \sinh^{-1} \dfrac{x}{a}\, dx = \left(\dfrac{x^4}{4} - \dfrac{3a^4}{32}\right) \sinh^{-1} \dfrac{x}{a}$

$$+ \dfrac{3a^2 x - 2x^3}{32} \sqrt{(x^2 + a^2)}.$$

730.4. $\displaystyle\int x^4 \sinh^{-1} \dfrac{x}{a}\, dx = \dfrac{x^5}{5} \sinh^{-1} \dfrac{x}{a}$

$$- \dfrac{8a^4 - 4a^2 x^2 + 3x^4}{75} \sqrt{(x^2 + a^2)}.$$

[See **625–625.4**.]

730.9. $\int x^p \sinh^{-1} \dfrac{x}{a}\, dx = \dfrac{x^{p+1}}{p+1} \sinh^{-1} \dfrac{x}{a} - \dfrac{1}{p+1} \int \dfrac{x^{p+1}dx}{\sqrt{(x^2 + a^2)}},$

$$[p \neq -1]. \quad \text{[See } \mathbf{201.01\text{--}207.01} \text{ and } \mathbf{625.9}.]$$

731.1. $\int \dfrac{1}{x} \sinh^{-1} \dfrac{x}{a}\, dx$

$$= \dfrac{x}{a} - \dfrac{1}{2\cdot3\cdot3} \dfrac{x^3}{a^3} + \dfrac{1\cdot3}{2\cdot4\cdot5\cdot5} \dfrac{x^5}{a^5} - \dfrac{1\cdot3\cdot5}{2\cdot4\cdot6\cdot7\cdot7} \dfrac{x^7}{a^7} + \cdots,$$

$$[x^2 < a^2].$$

$$= \dfrac{1}{2}\left(\log \dfrac{2x}{a}\right)^2 - \dfrac{1}{2^3} \dfrac{a^2}{x^2} + \dfrac{1\cdot3}{2\cdot4^3} \dfrac{a^4}{x^4} - \dfrac{1\cdot3\cdot5}{2\cdot4\cdot6^3} \dfrac{a^6}{x^6} + \cdots,$$

$$[x/a > 1].$$

$$= -\dfrac{1}{2}\left(\log \left|\dfrac{2x}{a}\right|\right)^2 + \dfrac{1}{2^3} \dfrac{a^2}{x^2} - \dfrac{1\cdot3}{2\cdot4^3} \dfrac{a^4}{x^4} + \dfrac{1\cdot3\cdot5}{2\cdot4\cdot6^3} \dfrac{a^6}{x^6} - \cdots,$$

$$[x/a < -1].$$

731.2. $\int \dfrac{1}{x^2} \sinh^{-1} \dfrac{x}{a}\, dx = -\dfrac{1}{x} \sinh^{-1} \dfrac{x}{a} - \dfrac{1}{a} \log \left|\dfrac{a + \sqrt{(x^2 + a^2)}}{x}\right|.$

731.3. $\int \dfrac{1}{x^3} \sinh^{-1} \dfrac{x}{a}\, dx = -\dfrac{1}{2x^2} \sinh^{-1} \dfrac{x}{a} - \dfrac{\sqrt{(x^2 + a^2)}}{2a^2x}.$

$$\text{[See } \mathbf{626.1} \text{ to } \mathbf{.3}.]$$

731.9. $\int \dfrac{1}{x^p} \sinh^{-1} \dfrac{x}{a}\, dx = -\dfrac{1}{(p-1)x^{p-1}} \sinh^{-1} \dfrac{x}{a}$

$$+ \dfrac{1}{p-1} \int \dfrac{dx}{x^{p-1}\sqrt{(x^2 + a^2)}},$$

$$[p \neq 1]. \quad \text{[See } \mathbf{221.01\text{--}226.01} \text{ and } \mathbf{626.9}.]$$

732. $\int \cosh^{-1} \dfrac{x}{a}\, dx = x \cosh^{-1} \dfrac{x}{a} - \sqrt{(x^2 - a^2)},$

$$[\cosh^{-1}(x/a) > 0],$$

$$= x \cosh^{-1} \dfrac{x}{a} + \sqrt{(x^2 - a^2)},$$

$$[\cosh^{-1}(x/a) < 0].$$

732.1. $\int x \cosh^{-1} \dfrac{x}{a}\, dx = \left(\dfrac{x^2}{2} - \dfrac{a^2}{4}\right) \cosh^{-1} \dfrac{x}{a} - \dfrac{x}{4} \sqrt{(x^2 - a^2)},$

$$[\cosh^{-1}(x/a) > 0],$$

$$= \left(\dfrac{x^2}{2} - \dfrac{a^2}{4}\right) \cosh^{-1} \dfrac{x}{a} + \dfrac{x}{4} \sqrt{(x^2 - a^2)},$$

$$[\cosh^{-1}(x/a) < 0].$$

732.2. $\displaystyle\int x^2 \cosh^{-1}\frac{x}{a}\, dx = \frac{x^3}{3}\cosh^{-1}\frac{x}{a} - \frac{2a^2 + x^2}{9}\sqrt{(x^2 - a^2)},$

$$[\cosh^{-1}(x/a) > 0].$$

$$= \frac{x^3}{3}\cosh^{-1}\frac{x}{a} + \frac{2a^2 + x^2}{9}\sqrt{(x^2 - a^2)},$$

$$[\cosh^{-1}(x/a) < 0].$$

732.3. $\displaystyle\int x^3 \cosh^{-1}\frac{x}{a}\, dx$

$$= \left(\frac{x^4}{4} - \frac{3a^4}{32}\right)\cosh^{-1}\frac{x}{a} - \frac{3a^2 x + 2x^3}{32}\sqrt{(x^2 - a^2)},$$

$$[\cosh^{-1}(x/a) > 0],$$

$$= \left(\frac{x^4}{4} - \frac{3a^4}{32}\right)\cosh^{-1}\frac{x}{a} + \frac{3a^2 x + 2x^3}{32}\sqrt{(x^2 - a^2)},$$

$$[\cosh^{-1}(x/a) < 0].$$

732.4. $\displaystyle\int x^4 \cosh^{-1}\frac{x}{a}\, dx$

$$= \frac{x^5}{5}\cosh^{-1}\frac{x}{a} - \frac{8a^4 + 4a^2 x^2 + 3x^4}{75}\sqrt{(x^2 - a^2)},$$

$$[\cosh^{-1}(x/a) > 0],$$

$$= \frac{x^5}{5}\cosh^{-1}\frac{x}{a} + \frac{8a^4 + 4a^2 x^2 + 3x^4}{75}\sqrt{(x^2 - a^2)},$$

$$[\cosh^{-1}(x/a) < 0]. \quad [\text{See } 627\text{–}627.4.]$$

732.9. $\displaystyle\int x^p \cosh^{-1}\frac{x}{a}\, dx = \frac{x^{p+1}}{p+1}\cosh^{-1}\frac{x}{a} - \frac{1}{p+1}\int \frac{x^{p+1}dx}{\sqrt{(x^2 - a^2)}},$

$$[\cosh^{-1}(x/a) > 0,\quad p \neq -1].$$

$$= \frac{x^{p+1}}{p+1}\cosh^{-1}\frac{x}{a} + \frac{1}{p+1}\int \frac{x^{p+1}dx}{\sqrt{(x^2 - a^2)}},$$

$$[\cosh^{-1}(x/a) < 0,\quad p \neq -1]. \qquad [\text{See } 261.01\text{–}267.01 \text{ and } 627.9.]$$

733.1. $\displaystyle\int \frac{1}{x}\cosh^{-1}\frac{x}{a}\, dx$

$$= \frac{1}{2}\left(\log\frac{2x}{a}\right)^2 + \frac{1}{2^3}\frac{a^2}{x^2} + \frac{1\cdot 3}{2\cdot 4^3}\frac{a^4}{x^4} + \frac{1\cdot 3\cdot 5}{2\cdot 4\cdot 6^3}\frac{a^6}{x^6} + \cdots,$$

$$[\cosh^{-1}(x/a) > 0],$$

$$= -\left[\frac{1}{2}\left(\log\frac{2x}{a}\right)^2 + \frac{1}{2^3}\frac{a^2}{x^2} + \frac{1\cdot 3}{2\cdot 4^3}\frac{a^4}{x^4} + \frac{1\cdot 3\cdot 5}{2\cdot 4\cdot 6^3}\frac{a^6}{x^6} + \cdots\right],$$

$$[\cosh^{-1}(x/a) < 0].$$

733.2. $\displaystyle\int \frac{1}{x^2} \cosh^{-1} \frac{x}{a}\, dx = -\frac{1}{x} \cosh^{-1} \frac{x}{a} + \frac{1}{a} \sec^{-1} \left|\frac{x}{a}\right|,$

$$[\cosh^{-1} (x/a) > 0, \quad 0 < \sec^{-1} |x/a| < \pi/2],$$

$$= -\frac{1}{x} \cosh^{-1} \frac{x}{a} - \frac{1}{a} \sec^{-1} \left|\frac{x}{a}\right|,$$

$$[\cosh^{-1} (x/a) < 0, \quad 0 < \sec^{-1} |x/a| < \pi/2].$$

733.3. $\displaystyle\int \frac{1}{x^3} \cosh^{-1} \frac{x}{a}\, dx = -\frac{1}{2x^2} \cosh^{-1} \frac{x}{a} + \frac{\sqrt{(x^2 - a^2)}}{2a^2 x},$

$$[\cosh^{-1} (x/a) > 0],$$

$$= -\frac{1}{2x^2} \cosh^{-1} \frac{x}{a} - \frac{\sqrt{(x^2 - a^2)}}{2a^2 x},$$

$$[\cosh^{-1} (x/a) < 0]. \quad [\text{See } \mathbf{628.1}\text{-}\mathbf{.3}.]$$

733.9. $\displaystyle\int \frac{1}{x^p} \cosh^{-1} \frac{x}{a}\, dx$

$$= -\frac{1}{(p-1)x^{p-1}} \cosh^{-1} \frac{x}{a} + \frac{1}{p-1} \int \frac{dx}{x^{p-1}\sqrt{(x^2 - a^2)}},$$

$$[\cosh^{-1} (x/a) > 0, \quad p \neq 1],$$

$$= -\frac{1}{(p-1)x^{p-1}} \cosh^{-1} \frac{x}{a} - \frac{1}{p-1} \int \frac{dx}{x^{p-1}\sqrt{(x^2 - a^2)}},$$

$$[\cosh^{-1} (x/a) < 0, \quad p \neq 1]. \qquad [\text{See } \mathbf{281.01}\text{-}\mathbf{284.01} \text{ and } \mathbf{628.9}.]$$

For **732–733.9**, $\dfrac{x}{a} > 1$.

734. $\displaystyle\int \tanh^{-1} \frac{x}{a}\, dx = x \tanh^{-1} \frac{x}{a} + \frac{a}{2} \log (a^2 - x^2).$

734.1. $\displaystyle\int x \tanh^{-1} \frac{x}{a}\, dx = \frac{x^2 - a^2}{2} \tanh^{-1} \frac{x}{a} + \frac{ax}{2}.$

734.2. $\displaystyle\int x^2 \tanh^{-1} \frac{x}{a}\, dx = \frac{x^3}{3} \tanh^{-1} \frac{x}{a} + \frac{ax^2}{6} + \frac{a^3}{6} \log (a^2 - x^2).$

734.3. $\displaystyle\int x^3 \tanh^{-1} \frac{x}{a}\, dx = \frac{x^4 - a^4}{4} \tanh^{-1} \frac{x}{a} + \frac{ax^3}{12} + \frac{a^3 x}{4}.$

734.9. $\displaystyle\int x^p \tanh^{-1} \frac{x}{a}\, dx = \frac{x^{p+1}}{p+1} \tanh^{-1} \frac{x}{a} - \frac{a}{p+1} \int \frac{x^{p+1}dx}{a^2 - x^2},$

$$[p \neq -1]. \qquad [\text{See } \mathbf{141.1}\text{-}\mathbf{148.1}.]$$

735.1. $\int \frac{1}{x} \tanh^{-1} \frac{x}{a} \, dx = \frac{x}{a} + \frac{x^3}{3^2 a^3} + \frac{x^5}{5^2 a^5} + \frac{x^7}{7^2 a^7} + \cdots$.

735.2. $\int \frac{1}{x^2} \tanh^{-1} \frac{x}{a} \, dx = -\frac{1}{x} \tanh^{-1} \frac{x}{a} - \frac{1}{2a} \log \left(\frac{a^2 - x^2}{x^2} \right)$.

735.3. $\int \frac{1}{x^3} \tanh^{-1} \frac{x}{a} \, dx = \frac{1}{2} \left(\frac{1}{a^2} - \frac{1}{x^2} \right) \tanh^{-1} \frac{x}{a} - \frac{1}{2ax}$.

735.4. $\int \frac{1}{x^4} \tanh^{-1} \frac{x}{a} \, dx = -\frac{1}{3x^3} \tanh^{-1} \frac{x}{a} - \frac{1}{6ax^2}$
$$- \frac{1}{6a^3} \log \left(\frac{a^2 - x^2}{x^2} \right).$$

735.5. $\int \frac{1}{x^5} \tanh^{-1} \frac{x}{a} \, dx = \frac{1}{4} \left(\frac{1}{a^4} - \frac{1}{x^4} \right) \tanh^{-1} \frac{x}{a} - \frac{1}{12ax^3} - \frac{1}{4a^3 x}$.

735.9. $\int \frac{1}{x^p} \tanh^{-1} \frac{x}{a} \, dx = -\frac{1}{(p-1)x^{p-1}} \tanh^{-1} \frac{x}{a}$
$$+ \frac{a}{p-1} \int \frac{dx}{x^{p-1}(a^2 - x^2)}, \qquad [p \neq 1].$$
$$\text{[See } \mathbf{151.1 - 155.1.}]$$

For **734–735.9**, $x^2 < a^2$.

736. $\int \operatorname{ctnh}^{-1} \frac{x}{a} \, dx = x \operatorname{ctnh}^{-1} \frac{x}{a} + \frac{a}{2} \log (x^2 - a^2)$.

736.1. $\int x \operatorname{ctnh}^{-1} \frac{x}{a} \, dx = \frac{x^2 - a^2}{2} \operatorname{ctnh}^{-1} \frac{x}{a} + \frac{ax}{2}$.

736.2. $\int x^2 \operatorname{ctnh}^{-1} \frac{x}{a} \, dx = \frac{x^3}{3} \operatorname{ctnh}^{-1} \frac{x}{a} + \frac{ax^2}{6} + \frac{a^3}{6} \log (x^2 - a^2)$.

736.3. $\int x^3 \operatorname{ctnh}^{-1} \frac{x}{a} \, dx = \frac{x^4 - a^4}{4} \operatorname{ctnh}^{-1} \frac{x}{a} + \frac{ax^3}{12} + \frac{a^3 x}{4}$.

736.9. $\int x^p \operatorname{ctnh}^{-1} \frac{x}{a} \, dx = \frac{x^{p+1}}{p+1} \operatorname{ctnh}^{-1} \frac{x}{a} - \frac{a}{p+1} \int \frac{x^{p+1} dx}{a^2 - x^2}$,
$$[p \neq -1]. \qquad \text{[See } \mathbf{141.1 - 148.1.}]$$

737.1. $\int \frac{1}{x} \operatorname{ctnh}^{-1} \frac{x}{a} \, dx = -\frac{a}{x} - \frac{a^3}{3^2 x^3} - \frac{a^5}{5^2 x^5} - \frac{a^7}{7^2 x^7} - \cdots$.

737.2. $\int \dfrac{1}{x^2} \operatorname{ctnh}^{-1} \dfrac{x}{a}\, dx = -\dfrac{1}{x} \operatorname{ctnh}^{-1} \dfrac{x}{a} - \dfrac{1}{2a} \log\left(\dfrac{x^2 - a^2}{x^2}\right).$

737.3. $\int \dfrac{1}{x^3} \operatorname{ctnh}^{-1} \dfrac{x}{a}\, dx = \dfrac{1}{2}\left(\dfrac{1}{a^2} - \dfrac{1}{x^2}\right) \operatorname{ctnh}^{-1} \dfrac{x}{a} - \dfrac{1}{2ax}.$

737.4. $\int \dfrac{1}{x^4} \operatorname{ctnh}^{-1} \dfrac{x}{a}\, dx = -\dfrac{1}{3x^3} \operatorname{ctnh}^{-1} \dfrac{x}{a}$

$$- \dfrac{1}{6ax^2} - \dfrac{1}{6a^3} \log\left(\dfrac{x^2 - a^2}{x^2}\right).$$

737.5. $\int \dfrac{1}{x^5} \operatorname{ctnh}^{-1} \dfrac{x}{a}\, dx = \dfrac{1}{4}\left(\dfrac{1}{a^4} - \dfrac{1}{x^4}\right) \operatorname{ctnh}^{-1} \dfrac{x}{a} - \dfrac{1}{12ax^3} - \dfrac{1}{4a^3x}.$

737.9. $\int \dfrac{1}{x^p} \operatorname{ctnh}^{-1} \dfrac{x}{a}\, dx = -\dfrac{1}{(p-1)x^{p-1}} \operatorname{ctnh}^{-1} \dfrac{x}{a}$

$$+ \dfrac{a}{p-1} \int \dfrac{dx}{x^{p-1}(a^2 - x^2)}, \qquad [p \neq 1].$$

For 736–737.9, $x^2 > a^2$.   [See **151.1–155.1.**]

738. $\int \operatorname{sech}^{-1} \dfrac{x}{a}\, dx = x \operatorname{sech}^{-1} \dfrac{x}{a} + a \sin^{-1} \dfrac{x}{a},$ $\quad [\operatorname{sech}^{-1}(x/a) > 0],$

$\qquad\qquad = x \operatorname{sech}^{-1} \dfrac{x}{a} - a \sin^{-1} \dfrac{x}{a},$ $\quad [\operatorname{sech}^{-1}(x/a) < 0].$

738.1. $\int x \operatorname{sech}^{-1} \dfrac{x}{a}\, dx = \dfrac{x^2}{2} \operatorname{sech}^{-1} \dfrac{x}{a} - \dfrac{a}{2} \sqrt{(a^2 - x^2)},$

$$[\operatorname{sech}^{-1}(x/a) > 0],$$

$$= \dfrac{x^2}{2} \operatorname{sech}^{-1} \dfrac{x}{a} + \dfrac{a}{2} \sqrt{(a^2 - x^2)},$$

$$[\operatorname{sech}^{-1}(x/a) < 0].$$

738.2. $\int x^2 \operatorname{sech}^{-1} \dfrac{x}{a}\, dx = \dfrac{x^3}{3} \operatorname{sech}^{-1} \dfrac{x}{a} - \dfrac{ax}{6} \sqrt{(a^2 - x^2)}$

$$+ \dfrac{a^3}{6} \sin^{-1} \dfrac{x}{a}, \quad [\operatorname{sech}^{-1}(x/a) > 0],$$

$$= \dfrac{x^3}{3} \operatorname{sech}^{-1} \dfrac{x}{a} + \dfrac{ax}{6} \sqrt{(a^2 - x^2)}$$

$$- \dfrac{a^3}{6} \sin^{-1} \dfrac{x}{a}, \quad [\operatorname{sech}^{-1}(x/a) < 0].$$

738.9. $\int x^p \operatorname{sech}^{-1} \dfrac{x}{a}\, dx = \dfrac{x^{p+1}}{p+1} \operatorname{sech}^{-1} \dfrac{x}{a} + \dfrac{a}{p+1} \int \dfrac{x^p dx}{\sqrt{(a^2 - x^2)}},$

$$[\operatorname{sech}^{-1}(x/a) > 0, \quad p \neq -1],$$

$$= \dfrac{x^{p+1}}{p+1} \operatorname{sech}^{-1} \dfrac{x}{a} - \dfrac{a}{p+1} \int \dfrac{x^p dx}{\sqrt{(a^2 - x^2)}},$$

$$[\operatorname{sech}^{-1}(x/a) < 0, \quad p \neq -1].$$

$$[\text{See } \mathbf{320.01\text{--}327.01}.]$$

739.1. $\int \dfrac{1}{x} \operatorname{sech}^{-1} \dfrac{x}{a}\, dx = -\dfrac{1}{2}\left(\log \dfrac{a}{x}\right) \log \dfrac{4a}{x} - \dfrac{1}{2^3} \dfrac{x^2}{a^2} - \dfrac{1\cdot 3}{2\cdot 4^3} \dfrac{x^4}{a^4}$

$$- \dfrac{1\cdot 3\cdot 5}{2\cdot 4\cdot 6^3} \dfrac{x^6}{a^6} - \cdots, \qquad [\operatorname{sech}^{-1}(x/a) > 0],$$

$$= \dfrac{1}{2}\left(\log \dfrac{a}{x}\right) \log \dfrac{4a}{x} + \dfrac{1}{2^3} \dfrac{x^2}{a^2} + \dfrac{1\cdot 3}{2\cdot 4^3} \dfrac{x^4}{a^4}$$

$$+ \dfrac{1\cdot 3\cdot 5}{2\cdot 4\cdot 6^3} \dfrac{x^6}{a^6} + \cdots, \qquad [\operatorname{sech}^{-1}(x/a) < 0].$$

739.2. $\int \dfrac{1}{x^2} \operatorname{sech}^{-1} \dfrac{x}{a}\, dx = -\dfrac{1}{x} \operatorname{sech}^{-1} \dfrac{x}{a} + \dfrac{\sqrt{(a^2 - x^2)}}{ax},$

$$[\operatorname{sech}^{-1}(x/a) > 0],$$

$$= -\dfrac{1}{x} \operatorname{sech}^{-1} \dfrac{x}{a} - \dfrac{\sqrt{(a^2 - x^2)}}{ax},$$

$$[\operatorname{sech}^{-1}(x/a) < 0].$$

739.9. $\int \dfrac{1}{x^p} \operatorname{sech}^{-1} \dfrac{x}{a}\, dx$

$$= -\dfrac{1}{(p-1)x^{p-1}} \operatorname{sech}^{-1} \dfrac{x}{a} - \dfrac{a}{p-1} \int \dfrac{dx}{x^p \sqrt{(a^2 - x^2)}},$$

$$[\operatorname{sech}^{-1}(x/a) > 0, \quad p \neq 1],$$

$$= -\dfrac{1}{(p-1)x^{p-1}} \operatorname{sech}^{-1} \dfrac{x}{a} + \dfrac{a}{p-1} \int \dfrac{dx}{x^p \sqrt{(a^2 - x^2)}},$$

$$[\operatorname{sech}^{-1}(x/a) < 0, \quad p \neq 1]. \quad [\text{See } \mathbf{342.01\text{--}346.01}.]$$

For **738–739.9**, $0 < x/a < 1$.

In the following items $a > 0$ and, except in **741.1**, $x > 0$.

740. $\int \operatorname{csch}^{-1} \dfrac{x}{a}\, dx = x \operatorname{csch}^{-1} \dfrac{x}{a} + a \sinh^{-1} \dfrac{x}{a}.$

740.1. $\displaystyle\int x\,\operatorname{csch}^{-1}\frac{x}{a}\,dx = \frac{x^2}{2}\operatorname{csch}^{-1}\frac{x}{a} + \frac{a}{2}\sqrt{(x^2 + a^2)}.$

740.9. $\displaystyle\int x^p\,\operatorname{csch}^{-1}\frac{x}{a}\,dx = \frac{x^{p+1}}{p+1}\operatorname{csch}^{-1}\frac{x}{a} + \frac{a}{p+1}\int\frac{x^p\,dx}{\sqrt{(x^2 + a^2)}},$

$$[p \neq -1].\quad [\text{See } \mathbf{200.01\text{--}207.01.}]$$

741.1. $\displaystyle\int\frac{1}{x}\operatorname{csch}^{-1}\frac{x}{a}\,dx = -\frac{a}{x} + \frac{1}{2\cdot3\cdot3}\frac{a^3}{x^3} - \frac{1\cdot3}{2\cdot4\cdot5\cdot5}\frac{a^5}{x^5}$

$$+ \frac{1\cdot3\cdot5}{2\cdot4\cdot6\cdot7\cdot7}\frac{a^7}{x^7} - \cdots,\quad [x^2 > a^2],$$

$$= -\frac{1}{2}\left(\log\frac{a}{x}\right)\log\frac{4a}{x} + \frac{1}{2^3}\frac{x^2}{a^2} - \frac{1\cdot3}{2\cdot4^3}\frac{x^4}{a^4}$$

$$+ \frac{1\cdot3\cdot5}{2\cdot4\cdot6^3}\frac{x^6}{a^6} - \cdots,\quad [0 < x/a < 1],$$

$$= \frac{1}{2}\log\left|\frac{a}{x}\right|\log\left|\frac{4a}{x}\right| - \frac{1}{2^3}\frac{x^2}{a^2} + \frac{1\cdot3}{2\cdot4^3}\frac{x^4}{a^4}$$

$$- \frac{1\cdot3\cdot5}{2\cdot4\cdot6^3}\frac{x^6}{a^6} + \cdots,\quad [-1 < x/a < 0].$$

741.9. $\displaystyle\int\frac{1}{x^p}\operatorname{csch}^{-1}\frac{x}{a}\,dx$

$$= -\frac{1}{(p-1)x^{p-1}}\operatorname{csch}^{-1}\frac{x}{a} - \frac{a}{p-1}\int\frac{dx}{x^p\sqrt{(x^2 + a^2)}},$$

$$[p \neq 1].\quad [\text{See } \mathbf{222.01\text{--}226.01.}]$$

⁘ 9 ⁘

ELLIPTIC FUNCTIONS

750.   Let $u = \int_0^{\phi} \dfrac{d\phi}{\sqrt{(1 - k^2 \sin^2 \phi)}}$, $[k^2 < 1]$,

$ = \int_0^{x} \dfrac{dx}{\sqrt{(1 - x^2)}\sqrt{(1 - k^2 x^2)}}$, $[x = \sin \phi]$,

$ = F(\phi, k) =$ **elliptic integral of the first kind.**

[See **770.**]

751.1. ϕ is the *amplitude*, and k the *modulus*.

751.2. $\phi = \operatorname{am} u$.

751.3. $\sin \phi = \operatorname{sn} u = x$.

751.4. $\cos \phi = \operatorname{cn} u = \sqrt{(1 - x^2)}$.

751.5. $\Delta\phi$ or $\Delta(\phi, k) = \sqrt{(1 - k^2 \sin^2 \phi)} = \operatorname{dn} u = \sqrt{(1 - k^2 x^2)}$.

751.6. $\tan \phi = \operatorname{tn} u = \dfrac{x}{\sqrt{(1 - x^2)}}$.

751.7.   The *complementary modulus* $= k' = \sqrt{(1 - k^2)}$.

752. $u = \operatorname{am}^{-1}(\phi, k) = \operatorname{sn}^{-1}(x, k) = \operatorname{cn}^{-1}\{\sqrt{(1 - x^2)}, k)\}$

$ = \operatorname{dn}^{-1}\{\sqrt{(1 - k^2 x^2)}, k\} = \operatorname{tn}^{-1}\left[\dfrac{x}{\sqrt{(1 - x^2)}}, k\right]$.

753.1.	$\operatorname{am}(-u) = -\operatorname{am} u$.	**754.2.**	$\operatorname{sn} 0 = 0$.
753.2.	$\operatorname{sn}(-u) = -\operatorname{sn} u$.	**754.3.**	$\operatorname{cn} 0 = 1$.
753.3.	$\operatorname{cn}(-u) = \operatorname{cn} u$.	**754.4.**	$\operatorname{dn} 0 = 1$.
753.4.	$\operatorname{dn}(-u) = \operatorname{dn} u$.	**755.1.**	$\operatorname{sn}^2 u + \operatorname{cn}^2 u = 1$.
753.5.	$\operatorname{tn}(-u) = -\operatorname{tn} u$.	**755.2.**	$\operatorname{dn}^2 u + k^2 \operatorname{sn}^2 u = 1$.
754.1.	$\operatorname{am} 0 = 0$.	**755.3.**	$\operatorname{dn}^2 u - k^2 \operatorname{cn}^2 u = k'^2$.

756.1. $\operatorname{sn}(u \pm v) = \dfrac{\operatorname{sn} u \operatorname{cn} v \operatorname{dn} v \pm \operatorname{cn} u \operatorname{sn} v \operatorname{dn} u}{1 - k^2 \operatorname{sn}^2 u \operatorname{sn}^2 v}.$

756.2. $\operatorname{cn}(u \pm v) = \dfrac{\operatorname{cn} u \operatorname{cn} v \mp \operatorname{sn} u \operatorname{sn} v \operatorname{dn} u \operatorname{dn} v}{1 - k^2 \operatorname{sn}^2 u \operatorname{sn}^2 v}.$

756.3. $\operatorname{dn}(u \pm v) = \dfrac{\operatorname{dn} u \operatorname{dn} v \mp k^2 \operatorname{sn} u \operatorname{sn} v \operatorname{cn} u \operatorname{cn} v}{1 - k^2 \operatorname{sn}^2 u \operatorname{sn}^2 v}.$

756.4. $\operatorname{tn}(u \pm v) = \dfrac{\operatorname{tn} u \operatorname{dn} v \pm \operatorname{tn} v \operatorname{dn} u}{1 \mp \operatorname{tn} u \operatorname{tn} v \operatorname{dn} u \operatorname{dn} v}.$

757.1. $\operatorname{sn} 2u = \dfrac{2 \operatorname{sn} u \operatorname{cn} u \operatorname{dn} u}{1 - k^2 \operatorname{sn}^4 u}.$

757.2. $\operatorname{cn} 2u = \dfrac{\operatorname{cn}^2 u - \operatorname{sn}^2 u \operatorname{dn}^2 u}{1 - k^2 \operatorname{sn}^4 u} = \dfrac{2 \operatorname{cn}^2 u}{1 - k^2 \operatorname{sn}^4 u} - 1.$

757.3. $\operatorname{dn} 2u = \dfrac{\operatorname{dn}^2 u - k^2 \operatorname{sn}^2 u \operatorname{cn}^2 u}{1 - k^2 \operatorname{sn}^4 u} = \dfrac{2 \operatorname{dn}^2 u}{1 - k^2 \operatorname{sn}^4 u} - 1.$

757.4. $\operatorname{tn} 2u = \dfrac{2 \operatorname{tn} u \operatorname{dn} u}{1 - \operatorname{tn}^2 u \operatorname{dn}^2 u}.$

758.1. $\operatorname{sn} \dfrac{u}{2} = \sqrt{\left(\dfrac{1 - \operatorname{cn} u}{1 + \operatorname{dn} u}\right)}.$

758.2. $\operatorname{cn} \dfrac{u}{2} = \sqrt{\left(\dfrac{\operatorname{cn} u + \operatorname{dn} u}{1 + \operatorname{dn} u}\right)}.$

758.3. $\operatorname{dn} \dfrac{u}{2} = \sqrt{\left(\dfrac{\operatorname{cn} u + \operatorname{dn} u}{1 + \operatorname{cn} u}\right)}.$

759.1. $\operatorname{sn}(iu, k) = i \operatorname{tn}(u, k').$

759.2. $\operatorname{cn}(iu, k) = \dfrac{1}{\operatorname{cn}(u, k')}.$

759.3. $\operatorname{dn}(iu, k) = \dfrac{\operatorname{dn}(u, k')}{\operatorname{cn}(u, k')}.$

760.1. $\operatorname{sn} u = u - (1 + k^2) \dfrac{u^3}{3!} + (1 + 14k^2 + k^4) \dfrac{u^5}{5!}$

$$- (1 + 135k^2 + 135k^4 + k^6) \dfrac{u^7}{7!} + \cdots.$$

760.2. $\operatorname{cn} u = 1 - \dfrac{u^2}{2!} + (1 + 4k^2) \dfrac{u^4}{4!} - (1 + 44k^2 + 16k^4) \dfrac{u^6}{6!}$

$$+ (1 + 408k^2 + 912k^4 + 64k^6) \dfrac{u^8}{8!} - \cdots .$$

760.3. $\operatorname{dn} u = 1 - k^2 \dfrac{u^2}{2!} + (4 + k^2)k^2 \dfrac{u^4}{4!} - (16 + 44k^2 + k^4)k^2 \dfrac{u^6}{6!}$

$$+ (64 + 912k^2 + 408k^4 + k^6)k^2 \dfrac{u^8}{8!} - \cdots .$$

760.4. $\operatorname{am} u = u - k^2 \dfrac{u^3}{3!} + (4 + k^2)k^2 \dfrac{u^5}{5!} - (16 + 44k^2 + k^4)k^2 \dfrac{u^7}{7!}$

$$+ (64 + 912k^2 + 408k^4 + k^6)k^2 \dfrac{u^9}{9!} - \cdots .$$

[Ref. 21, p. 156.]

ELLIPTIC FUNCTIONS—DERIVATIVES

768.1. $\dfrac{d}{du} \operatorname{sn} u = \operatorname{cn} u \operatorname{dn} u.$

768.2. $\dfrac{d}{du} \operatorname{cn} u = - \operatorname{sn} u \operatorname{dn} u.$

768.3. $\dfrac{d}{du} \operatorname{dn} u = - k^2 \operatorname{sn} u \operatorname{cn} u.$ [Ref. 36, p. 25.]

ELLIPTIC FUNCTIONS—INTEGRALS

770.   Elliptic Integral of the First Kind.

$$F(\phi, k) = \int_0^\phi \frac{d\phi}{\sqrt{(1 - k^2 \sin^2 \phi)}}, \qquad [k^2 < 1],$$

$$= \int_0^x \frac{dx}{\sqrt{(1 - x^2)}\sqrt{(1 - k^2 x^2)}}, \qquad [x = \sin \phi].$$

[See **750.**]

771.   Elliptic Integral of the Second Kind.

$$E(\phi, k) = \int_0^\phi \sqrt{(1 - k^2 \sin^2 \phi)}\, d\phi$$

$$= \int_0^x \frac{\sqrt{(1 - k^2 x^2)}}{\sqrt{(1 - x^2)}}\, dx, \qquad [x = \sin \phi].$$

772.   Elliptic Integral of the Third Kind.

$$\Pi(\phi, n, k) = \int_0^\phi \frac{d\phi}{(1 + n \sin^2 \phi)\sqrt{(1 - k^2 \sin^2 \phi)}}$$

$$= \int_0^x \frac{dx}{(1 + nx^2)\sqrt{(1 - x^2)}\sqrt{(1 - k^2 x^2)}},$$

$$[x = \sin \phi].$$

The letter n is called the parameter.

Complete Elliptic Integrals (See Tables 1040–1041)

773.1. $K = \int_0^{\pi/2} \frac{d\phi}{\sqrt{(1 - k^2 \sin^2 \phi)}}$

$$= \frac{\pi}{2} \left(1 + \frac{1^2}{2^2} k^2 + \frac{1^2 \cdot 3^2}{2^2 \cdot 4^2} k^4 + \frac{1^2 \cdot 3^2 \cdot 5^2}{2^2 \cdot 4^2 \cdot 6^2} k^6 + \cdots \right),$$

$$[k^2 < 1].$$

773.2. $K = \frac{\pi}{2}(1 + m)\left[1 + \frac{1^2}{2^2} m^2 + \frac{1^2 \cdot 3^2}{2^2 \cdot 4^2} m^4 + \frac{1^2 \cdot 3^2 \cdot 5^2}{2^2 \cdot 4^2 \cdot 6^2} m^6 + \cdots \right],$

where $m = (1 - k')/(1 + k')$.   [Ref. 31, p. 135.]

This series is more rapidly convergent than **773.1** since $m^2 < k^2$.

777.3. $K = \log \frac{4}{k'} + \frac{1^2}{2^2} \left(\log \frac{4}{k'} - \frac{2}{1 \cdot 2} \right) k'^2$

$$+ \frac{1^2 \cdot 3^2}{2^2 \cdot 4^2} \left(\log \frac{4}{k'} - \frac{2}{1 \cdot 2} - \frac{2}{3 \cdot 4} \right) k'^4$$

$$+ \frac{1^2 \cdot 3^2 \cdot 5^2}{2^2 \cdot 4^2 \cdot 6^2} \left(\log \frac{4}{k'} - \frac{2}{1 \cdot 2} - \frac{2}{3 \cdot 4} - \frac{2}{5 \cdot 6} \right) k'^6 + \cdots,$$

where $k' = \sqrt{(1 - k^2)}$, and log denotes natural logarithm.

[Ref. 33, pp. 46 and 54.]

774.1. $E = \int_0^{\pi/2} \sqrt{(1 - k^2 \sin^2 \phi)} d\phi$

$$= \frac{\pi}{2} \left(1 - \frac{1}{2^2} k^2 - \frac{1^2 \cdot 3}{2^2 \cdot 4^2} k^4 - \frac{1^2 \cdot 3^2 \cdot 5}{2^2 \cdot 4^2 \cdot 6^2} k^6 - \cdots \right),$$

$$[k^2 < 1].$$

774.2. $E = \dfrac{\pi}{2(1 + m)} \left[1 + \dfrac{m^2}{2^2} + \dfrac{1^2}{2^2 \cdot 4^2} m^4 + \dfrac{1^2 \cdot 3^2}{2^2 \cdot 4^2 \cdot 6^2} m^6 + \cdots \right]$

where $m = (1 - k')/(1 + k')$.   [Ref. 31, p. 136.]

This series is more rapidly convergent than **774.1** since $m^2 < k^2$.

774.3. $E = 1 + \dfrac{1}{2} \left(\log \dfrac{4}{k'} - \dfrac{1}{1 \cdot 2} \right) k'^2$

$\qquad + \dfrac{1^2 \cdot 3}{2^2 \cdot 4} \left(\log \dfrac{4}{k'} - \dfrac{2}{1 \cdot 2} - \dfrac{1}{3 \cdot 4} \right) k'^4$

$\qquad + \dfrac{1^2 \cdot 3^2 \cdot 5}{2^2 \cdot 4^2 \cdot 6} \left(\log \dfrac{4}{k'} - \dfrac{2}{1 \cdot 2} - \dfrac{2}{3 \cdot 4} - \dfrac{1}{5 \cdot 6} \right) k'^6 + \cdots .$

[Ref. 33, pp. 46 and 54.]

775. $F(\phi, k) = \displaystyle\int_0^\phi \dfrac{d\phi}{\sqrt{(1 - k^2 \sin^2 \phi)}}$

$\qquad = \dfrac{2\phi}{\pi} K - \sin \phi \cos \phi \left(\dfrac{1}{2} A_2 k^2 + \dfrac{1 \cdot 3}{2 \cdot 4} A_4 k^4 \right.$

$\qquad\qquad\qquad\qquad\qquad \left. + \dfrac{1 \cdot 3 \cdot 5}{2 \cdot 4 \cdot 6} A_6 k^6 + \cdots \right),$

where

$\qquad A_2 = \dfrac{1}{2}, \qquad\qquad A_4 = \dfrac{3}{2 \cdot 4} + \dfrac{1}{4} \sin^2 \phi,$

$\qquad A_6 = \dfrac{3 \cdot 5}{2 \cdot 4 \cdot 6} + \dfrac{5}{4 \cdot 6} \sin^2 \phi + \dfrac{1}{6} \sin^4 \phi,$

$\qquad A_8 = \dfrac{3 \cdot 5 \cdot 7}{2 \cdot 4 \cdot 6 \cdot 8} + \dfrac{5 \cdot 7}{4 \cdot 6 \cdot 8} \sin^2 \phi + \dfrac{7}{6 \cdot 8} \sin^4 \phi + \dfrac{1}{8} \sin^6 \phi,$

and K is found by **773** or from tables.   [Ref. 5, No. 545.]

776. $F(\phi, k) = \phi + \dfrac{1}{2} v_2 k^2 + \dfrac{1 \cdot 3}{2 \cdot 4} v_4 k^4 + \dfrac{1 \cdot 3 \cdot 5}{2 \cdot 4 \cdot 6} v_6 k^6 + \cdots ,$

where

$\qquad v_{2n} = \displaystyle\int \sin^{2n} \phi \, d\phi.$ [See **430**.]   [Ref. 36, p. 26.]

777. $E(\phi, k)\displaystyle\int_0^\phi = \sqrt{(1 - k^2 \sin^2 \phi)}d\phi$

$$= \frac{2\phi}{\pi} E + \sin \phi \cos \phi \left(\frac{1}{2} A_2 k^2 + \frac{1}{2\cdot4} A_4 k^4\right.$$

$$\left. + \frac{1\cdot3}{2\cdot4\cdot6} A_6 k^6 + \cdots\right)$$

where A_2, A_4, $\cdots$ are given in **775**, and where E may be obtained by **774** or from tables.   [Ref. 5, No. 546.]

780.1. $\displaystyle\int_0^x \frac{dx}{\sqrt{(1 + x^2)}\sqrt{(1 + k'^2 x^2)}} = \text{tn}^{-1}(x, k) = F(\tan^{-1} x, k),$

$$[0 < x]. \quad [\text{Ref. 36, p. 42, eq. (4).}]$$

781.01. $\displaystyle\int_0^x \frac{dx}{\sqrt{(a^2 - x^2)}\sqrt{(b^2 - x^2)}} = \frac{1}{a} \text{sn}^{-1}\left(\frac{x}{b}, \frac{b}{a}\right) = \frac{1}{a} F(\phi, k),$

$$\left[\phi = \sin^{-1}\left(\frac{x}{b}\right), k = \frac{b}{a}\right], \qquad [0 < x < b < a].$$

781.02. $\displaystyle\int_b^x \frac{dx}{\sqrt{(a^2 - x^2)}\sqrt{(x^2 - b^2)}} = \frac{1}{a}\{K(k) - F(\phi, k)\},$

$$\left[\phi = \sin^{-1}\left\{\frac{\sqrt{(a^2 - x^2)}}{\sqrt{(a^2 - b^2)}}\right\}, k = \frac{\sqrt{(a^2 - b^2)}}{a}\right],$$

$$[0 < b < x < a].$$

$K(k) = F(\pi/2, k)$, the complete elliptic integral (Table 1040).

As is usual, integration from x_1 to x_2 can be carried out by taking the difference of the integrals from b to x_2 and b to x_1. Numerical tables of $F(\phi, k)$ and $E(\phi, k)$ may be found in References 47, 17, 21, and 5, where $\sin \alpha$ or $\sin \theta = k$.

781.03. $\displaystyle\int_a^x \frac{dx}{\sqrt{(x^2 - a^2)}\sqrt{(x^2 - b^2)}} = \frac{1}{a}\{K(k) - F(\phi, k)\},$

$$\left[\phi = \sin^{-1}\left(\frac{a}{x}\right), k = \frac{b}{a}\right], \qquad [0 < b < a < x].$$

781.04. $\displaystyle\int_0^x \frac{dx}{\sqrt{(a^2 + x^2)}\sqrt{(b^2 + x^2)}} = \frac{1}{a} \text{tn}^{-1}\left\{\frac{x}{b}, \frac{\sqrt{(a^2 - b^2)}}{a}\right\}$

$$= \frac{1}{a} F(\phi, k),$$

$$\left[\phi = \tan^{-1}\left(\frac{x}{b}\right), k = \frac{\sqrt{(a^2 - b^2)}}{a}\right], \quad [0 < b < a; 0 < x].$$

781.05. $\int_0^x \dfrac{dx}{\sqrt{(a^2 + x^2)}\sqrt{(b^2 - x^2)}} = \dfrac{1}{\sqrt{(a^2 + b^2)}} \{K(k) - F(\phi, k)\},$

$$\left[\phi = \cos^{-1}\left(\frac{x}{b}\right), k = \frac{b}{\sqrt{(a^2 + b^2)}}\right], \qquad [0 < x < b].$$

781.06. $\int_b^x \dfrac{dx}{\sqrt{(a^2 + x^2)}\sqrt{(x^2 - b^2)}} = \dfrac{1}{\sqrt{(a^2 + b^2)}} F(\phi, k),$

$$\left[\phi = \cos^{-1}\left(\frac{b}{x}\right), k = \frac{a}{\sqrt{(a^2 + b^2)}}\right], \qquad [0 < b < x; 0 < a].$$

[For **781.01–781.06**, see Ref. 66, Chap. 2.]

781.11. $\int_0^x \dfrac{x^2 dx}{\sqrt{(a^2 - x^2)}\sqrt{(b^2 - x^2)}} = a F(\phi, k) - a E(\phi, k),$

$$\left[\phi = \sin^{-1}\left(\frac{x}{b}\right), k = \frac{b}{a}\right], \qquad [0 < x < b < a].$$

[See Ref. 39, p. 293.]

781.12. $\int_b^x \dfrac{x^2 dx}{\sqrt{(a^2 - x^2)}\sqrt{(x^2 - b^2)}} = a E\left(\frac{\pi}{2}, k\right) - a E(\phi, k),$

$$\left[\phi = \sin^{-1}\left\{\frac{\sqrt{(a^2 - x^2)}}{\sqrt{(a^2 - b^2)}}\right\}, k = \frac{\sqrt{(a^2 - b^2)}}{a}\right], \qquad [0 < b < x < a].$$

[See '*Integraltafeln* · · · ,' by W. Meyer zur Capellen, Ref. 68, p. 122. This book is a large collection of indefinite integrals.]

781.13. $\int_a^x \dfrac{x^2 dx}{\sqrt{(x^2 - a^2)}\sqrt{(x^2 - b^2)}}$

$$= \frac{\sqrt{(x^2 - a^2)}\sqrt{(x^2 - b^2)}}{x} + a K(k) - a F(\phi, k)$$

$$- a E\left(\frac{\pi}{2}, k\right) + a E(\phi, k),$$

$$\left[\phi = \sin^{-1}\left(\frac{a}{x}\right), k = \frac{b}{a}\right], \qquad [0 < b < a < x].$$

781.14. $\int_0^x \dfrac{x^2 dx}{\sqrt{(a^2 + x^2)}\sqrt{(b^2 + x^2)}} = \dfrac{x\sqrt{(a^2 + x^2)}}{\sqrt{(b^2 + x^2)}} - a E(\phi, k),$

$$\left[\phi = \tan^{-1}\left(\frac{x}{b}\right), k = \frac{\sqrt{(a^2 - b^2)}}{a}\right], \qquad [0 < x; 0 < b < a].$$

781.15. $\displaystyle\int_0^x \frac{x^2 dx}{\sqrt{(a^2 + x^2)}\sqrt{(b^2 - x^2)}}$

$$= \sqrt{(a^2 + b^2)}\left[E\left(\frac{\pi}{2}, k\right) - E(\phi, k) \right] - \frac{a^2}{\sqrt{(a^2 + b^2)}}$$

$$\left[K(k) - F(\phi, k) \right], \left[\phi = \cos^{-1}\left(\frac{x}{b}\right), k = \frac{b}{\sqrt{(a^2 + b^2)}} \right],$$

$$[0 < x < b; 0 < a]. \qquad \text{[Ref. 68, p. 121.]}$$

781.16. $\displaystyle\int_b^x \frac{x^2 dx}{\sqrt{(a^2 + x^2)}\sqrt{(x^2 - b^2)}}$

$$= \frac{\sqrt{(a^2 + x^2)}\sqrt{(x^2 - b^2)}}{x} + \frac{b^2}{\sqrt{(a^2 + b^2)}} F(\phi, k)$$

$$- \sqrt{(a^2 + b^2)}\, E(\phi, k), \left[\phi = \cos^{-1}\left(\frac{b}{x}\right), k = \frac{a}{\sqrt{(a^2 + b^2)}} \right],$$

$$[0 < b < x; 0 < a].$$

781.21. $\displaystyle\int_0^x \frac{\sqrt{(a^2 - x^2)}}{\sqrt{(b^2 - x^2)}}\, dx = a\, E(\phi, k), \qquad [0 < x < b < a],$

$$\left[\phi = \sin^{-1}\left(\frac{x}{b}\right), k = \frac{b}{a} \right]. \qquad \text{[Ref. 68, p. 113.]}$$

781.22. $\displaystyle\int_0^x \frac{\sqrt{(b^2 - x^2)}}{\sqrt{(a^2 - x^2)}}\, dx = a\, E(\phi, k) - \left(\frac{a^2 - b^2}{a}\right) F(\phi, k),$

$$[0 < x < b < a],$$

$$\left[\phi = \sin^{-1}\left(\frac{x}{b}\right), k = \frac{b}{a} \right]. \qquad \text{[Ref. 68, p. 113.]}$$

781.23. $\displaystyle\int_0^x \frac{\sqrt{(a^2 + x^2)}}{\sqrt{(b^2 + x^2)}}\, dx = \frac{x\sqrt{(a^2 + x^2)}}{\sqrt{(b^2 + x^2)}} + a\, F(\phi, k) - a\, E(\phi, k),$

$$\left[\phi = \tan^{-1}\left(\frac{x}{b}\right), k = \frac{\sqrt{(a^2 - b^2)}}{a} \right], \quad [0 < x; 0 < b < a].$$

781.24. $\displaystyle\int_b^x \frac{\sqrt{(a^2 + x^2)}}{\sqrt{(x^2 - b^2)}}\, dx$

$$= \frac{\sqrt{(a^2 + x^2)}\sqrt{(x^2 - b^2)}}{x} + \sqrt{(a^2 + b^2)}\, [F(\phi, k) - E(\phi, k)],$$

$$\left[\phi = \cos^{-1}\left(\frac{b}{x}\right), k = \frac{a}{\sqrt{(a^2 + b^2)}} \right], \quad [0 < b < x; 0 < a].$$

When this integral is multiplied above and below by its numerator, it can be expressed in terms of **781.06** and **781.16**, and similarly for many other cases.

781.51. $\int_0^x \dfrac{x^4 dx}{\sqrt{(a^2 - x^2)}\sqrt{(b^2 - x^2)}}$

$$= \frac{x\sqrt{(a^2 - x^2)}\sqrt{(b^2 - x^2)}}{3} + \frac{(2 + k^2)b^3}{3k^3} F(\phi, k)$$

$$- \frac{2(1 + k^2)b^3}{3k^3} E(\phi, k), \qquad \left[\phi = \sin^{-1}\left(\frac{x}{b}\right), k = \frac{b}{a}\right].$$

781.61. $\int_0^x \sqrt{(a^2 - x^2)}\sqrt{(b^2 - x^2)}dx = \dfrac{x\sqrt{(a^2 - x^2)}\sqrt{(b^2 - x^2)}}{3}$

$$+ \left(\frac{b^3}{3k} + \frac{2b^3}{3k^3} - a^3\right) F(\phi, k)$$

$$+ \left(a^3 + ab^2 - \frac{2b^3}{3k} - \frac{2b^3}{3k^3}\right) E(\phi, k),$$

$$\left[\phi = \sin^{-1}\left(\frac{x}{b}\right), k = \frac{b}{a}\right].$$

Multiply the integral above and below by its integrand and use **781.01, .11,** and **.51.**

782.01. $\int_0^\phi \dfrac{\sin^2 \phi \, d\phi}{\sqrt{(1 - k^2 \sin^2 \phi)}} = \int_0^u \text{sn}^2 u \, du$

$$= \int_0^x \frac{x^2 dx}{\sqrt{(1 - x^2)}\sqrt{(1 - k^2 x^2)}}$$

$$= \frac{1}{k^2}\left\{F(\phi, k) - E(\phi, k)\right\}, \qquad [x = \sin \phi].$$

782.02. $\int_0^\phi \dfrac{\cos^2 \phi \, d\phi}{\sqrt{(1 - k^2 \sin^2 \phi)}} = \int_0^u \text{cn}^2 u \, du$

$$= \int_0^x \frac{\sqrt{(1 - x^2)}}{\sqrt{(1 - k^2 x^2)}} \, dx$$

$$= \frac{E(\phi, k)}{k^2} - \frac{1 - k^2}{k^2} F(\phi, k), \quad [x = \sin \phi].$$

782.03. $\int_0^\phi \dfrac{\tan^2 \phi \, d\phi}{\sqrt{(1 - k^2 \sin^2 \phi)}} = \int_0^u \text{tn}^2 u \, du$

$$= \frac{\tan \phi}{(1 - k^2)} \sqrt{(1 - k^2 \sin^2 \phi)} - \frac{E(\phi, k)}{1 - k^2}.$$

782.04. $\displaystyle\int_0^\phi \frac{d\phi}{(\cos^2 \phi)\sqrt{(1 - k^2 \sin^2 \phi)}} = \frac{\sqrt{(1 - k^2 \sin^2 \phi)}}{1 - k^2} \tan \phi$
$$+ F(\phi,k) - \frac{E(\phi,k)}{1 - k^2}.$$

782.05. $\displaystyle\int_0^\phi \frac{\sqrt{(1 - k^2 \sin^2 \phi)}}{\cos^2 \phi} \, d\phi = (\tan \phi)\sqrt{(1 - k^2 \sin^2 \phi)}$
$$+ F(\phi,k) - E(\phi, k).$$

782.06. $\displaystyle\int_0^\phi (\tan^2 \phi)\sqrt{(1 - k^2 \sin^2 \phi)}d\phi = (\tan \phi)\sqrt{(1 - k^2 \sin^2 \phi)}$
$$+ F(\phi, k) - 2E(\phi, k.)$$

785.1. $\displaystyle\int \text{sn } u \, du = - \frac{1}{k} \cosh^{-1}\left(\frac{\text{dn } u}{k'}\right).$ [Ref. 36, p. 58.]

785.2. $\displaystyle\int \text{cn } u \, du = \frac{1}{k} \cos^{-1} (\text{dn } u).$

785.3. $\displaystyle\int \text{dn } u \, du = \sin^{-1} (\text{sn } u) = \text{am } u.$

786.1. $\displaystyle\int \frac{du}{\text{sn } u} = \log \left(\frac{\text{sn } u}{\text{cn } u + \text{dn } u}\right).$

786.2. $\displaystyle\int \frac{du}{\text{cn } u} = \frac{1}{k'} \log \left(\frac{k' \text{ sn } u + \text{dn } u}{\text{cn } u}\right).$

786.3. $\displaystyle\int \frac{du}{\text{dn } u} = \frac{1}{k'} \tan^{-1} \left(\frac{k' \text{ sn } u - \text{cn } u}{k' \text{ sn } u + \text{cn } u}\right).$ [Ref. 5, No. 583.]

787.1. $\displaystyle\int \text{sn}^{-1} x \, dx = x \, \text{sn}^{-1} x + \frac{1}{k} \cosh \left[\frac{\sqrt{(1 - k^2 x^2)}}{k'}\right].$

787.2. $\displaystyle\int \text{cn}^{-1} x \, dx = x \, \text{cn}^{-1} x - \frac{1}{k} \cos^{-1} \sqrt{(k'^2 + k^2 x^2)}.$

787.3. $\displaystyle\int \text{dn}^{-1} x \, dx = x \, \text{dn}^{-1} x - \sin^{-1} \left[\frac{\sqrt{(1 - x^2)}}{k}\right].$

[Ref. 36, Chap. III.]

788.1. $\displaystyle\frac{\partial E}{\partial k} = \frac{1}{k} (E - K).$

788.2. $\displaystyle\frac{\partial K}{\partial k} = \frac{1}{k} \left(\frac{E}{k'^2} - K\right).$

✦ 10 ✦

BESSEL FUNCTIONS

800. Bessel's differential equation is

$$\frac{d^2u}{dx^2} + \frac{1}{x}\frac{du}{dx} + \left(1 - \frac{n^2}{x^2}\right)u = 0.$$

[Ref. 12, p. 7, eq. (7).]

Bessel Function of the First Kind, $J_n(x)$

Denote $\dfrac{d}{dx}J_n(x)$ by $J_n{}'$, etc.

801.1. $xJ_n{}' = nJ_n - xJ_{n+1}.$ **801.3.** $2nJ_n = xJ_{n-1} + xJ_{n+1}.$

801.2. $xJ_n{}' = -nJ_n + xJ_{n-1}.$ **801.4.** $2J_n{}' = J_{n-1} - J_{n+1}.$

801.5. $4J_n{}'' = J_{n-2} - 2J_n + J_{n+2}.$

801.6. $\dfrac{d}{dx}(x^nJ_n) = x^nJ_{n-1}.$ **801.7.** $\dfrac{d}{dx}(x^{-n}J_n) = -x^{-n}J_{n+1}.$

801.82. $J_2 = \dfrac{2J_1}{x} - J_0.$

801.83. $J_3 = \left(\dfrac{8}{x^2} - 1\right)J_1 - \dfrac{4J_0}{x}.$

801.84. $J_4 = \left(1 - \dfrac{24}{x^2}\right)J_0 + \dfrac{8}{x}\left(\dfrac{6}{x^2} - 1\right)J_1.$

801.85. $J_5 = \dfrac{12}{x}\left(1 - \dfrac{16}{x^2}\right)J_0 + \left(\dfrac{384}{x^4} - \dfrac{72}{x^2} + 1\right)J_1.$

801.90. $J_0{}' = -J_1.$

801.91. $J_1' = J_0 - \dfrac{J_1}{x}.$

801.92. $J_2' = \dfrac{2J_0}{x} + \left(1 - \dfrac{4}{x^2}\right)J_1.$

801.93. $J_3' = \left(\dfrac{12}{x^2} - 1\right)J_0 + \left(5 - \dfrac{24}{x^2}\right)\dfrac{J_1}{x}.$

801.94. $J_4' = \dfrac{8}{x}\left(\dfrac{12}{x^2} - 1\right)J_0 - \left(\dfrac{192}{x^4} - \dfrac{40}{x^2} + 1\right)J_1.$

801.95. $J_5' = \left(\dfrac{960}{x^4} - \dfrac{84}{x^2} + 1\right)J_0 - \left(\dfrac{1920}{x^4} - \dfrac{408}{x^2} + 13\right)\dfrac{J_1}{x}.$

For tables of $J_0(x)$ and $J_1(x)$ see Ref. 50; Ref. 12, p.267; Ref. 13, p. 666; and Ref. 63.

Bessel Function of the Second Kind, $N_n(x)$

$N_n(x)$ as in Ref. 17 and Ref. 62, pp. 357–358, and same as $Y_n(x)$ in Ref. 13 (not boldface Y_n) and Ref. 50.

802.1. $xN_n' = nN_n - xN_{n+1}.$

802.2. $xN_n' = -nN_n + xN_{n-1}.$

802.3. $2nN_n = xN_{n-1} + xN_{n+1}.$

802.4. $2N_n' = N_{n-1} - N_{n+1}.$

802.5. $4N_n'' = N_{n-2} - 2N_n + N_{n+2}.$

802.6. $\dfrac{d}{dx}(x^n N_n) = x^n N_{n-1}.$

802.7. $\dfrac{d}{dx}(x^{-n} N_n) = -x^{-n} N_{n+1}.$

802.82. $N_2 = \dfrac{2N_1}{x} - N_0.$

802.83. $N_3 = \left(\dfrac{8}{x^2} - 1\right)N_1 - \dfrac{4N_0}{x}.$

802.84. $N_4 = \left(1 - \dfrac{24}{x^2}\right)N_0 + \dfrac{8}{x}\left(\dfrac{6}{x^2} - 1\right)N_1.$

802.85. $N_5 = \dfrac{12}{x}\left(1 - \dfrac{16}{x^2}\right)N_0 + \left(\dfrac{384}{x^4} - \dfrac{72}{x^2} + 1\right)N_1.$

802.90. $N_0' = -N_1.$

802.91. $N_1' = N_0 - \dfrac{N_1}{x}.$

802.92. $N_2' = \dfrac{2N_0}{x} + \left(1 - \dfrac{4}{x^2}\right)N_1.$

802.93. $N_3' = \left(\dfrac{12}{x^2} - 1\right)N_0 + \left(5 - \dfrac{24}{x^2}\right)\dfrac{N_1}{x}.$

802.94. $N_4' = \dfrac{8}{x}\left(\dfrac{12}{x^2} - 1\right)N_0 - \left(\dfrac{192}{x^4} - \dfrac{40}{x^2} + 1\right)N_1.$

802.95. $N_5' = \left(\dfrac{960}{x^4} - \dfrac{84}{x^2} + 1\right)N_0 - \left(\dfrac{1920}{x^4} - \dfrac{408}{x^2} + 13\right)\dfrac{N_1}{x}.$

For tables of $N_0(x)$ and $N_1(x)$ see Ref. 50, Ref. 13, p. 666, and Ref. 17.

Modified Bessel Function of the First Kind, $I_n(x)$

803.1. $xI_n' = nI_n + xI_{n+1}.$ **803.3.** $2nI_n = xI_{n-1} - xI_{n+1}.$

803.2. $xI_n' = -nI_n + xI_{n-1}.$ **803.4.** $2I_n' = I_{n-1} + I_{n+1}.$

803.5. $4I_n'' = I_{n-2} + 2I_n + I_{n+2}.$

803.6. $\dfrac{d}{dx}(x^n I_n) = x^n I_{n-1}.$ **803.7.** $\dfrac{d}{dx}(x^{-n} I_n) = x^{-n} I_{n+1}.$

803.82. $I_2 = I_0 - \dfrac{2I_1}{x}.$

803.83. $I_3 = \left(\dfrac{8}{x^2} + 1\right)I_1 - \dfrac{4I_0}{x}.$

803.84. $I_4 = \left(\dfrac{24}{x^2} + 1\right)I_0 - \dfrac{8}{x}\left(\dfrac{6}{x^2} + 1\right)I_1.$

803.85. $I_5 = \left(\dfrac{384}{x^4} + \dfrac{72}{x^2} + 1\right)I_1 - \dfrac{12}{x}\left(\dfrac{16}{x^2} + 1\right)I_0.$

803.90. $I_0' = I_1.$

803.91. $I_1' = I_0 - \dfrac{I_1}{x}.$

803.92. $I_2' = I_1 \left(\dfrac{4}{x^2} + 1 \right) - \dfrac{2I_0}{x}.$

803.93. $I_3' = \left(\dfrac{12}{x^2} + 1 \right) I_0 - \left(\dfrac{24}{x^2} + 5 \right) \dfrac{I_1}{x}.$

803.94. $I_4' = \left(\dfrac{192}{x^4} + \dfrac{40}{x^2} + 1 \right) I_1 - \dfrac{8}{x} \left(\dfrac{12}{x^2} + 1 \right) I_0.$

803.95. $I_5' = \left(\dfrac{960}{x^4} + \dfrac{84}{x^2} + 1 \right) I_0 - \left(\dfrac{1920}{x^4} + \dfrac{408}{x^2} + 13 \right) \dfrac{I_1}{x}.$

For tables of $I_0(x)$ and $I_1(x)$ see Ref. 50, p. 214, Ref. 12, p. 303, and Ref. 17. Tables of $e^{-x}I_0(x)$ and $e^{-x}I_1(x)$, Ref. 13.

Modified Bessel Function of the Second Kind, $K_n(x)$

804.1. $xK_n' = nK_n - xK_{n+1}.$

804.2. $xK_n' = -nK_n - xK_{n-1}.$

804.3. $2nK_n = xK_{n+1} - xK_{n-1}.$

804.4. $2K_n' = -K_{n-1} - K_{n+1}.$

804.5. $4K_n'' = K_{n-2} + 2K_n + K_{n+2}.$

804.6. $\dfrac{d}{dx} (x^n K_n) = -x^n K_{n-1}.$

804.7. $\dfrac{d}{dx} (x^{-n} K_n) = -x^{-n} K_{n+1}.$

804.82. $K_2 = K_0 + \dfrac{2K_1}{x}.$

804.83. $K_3 = \dfrac{4K_0}{x} + \left(\dfrac{8}{x^2} + 1 \right) K_1.$

804.84. $K_4 = \left(\dfrac{24}{x^2} + 1 \right) K_0 + \dfrac{8}{x} \left(\dfrac{6}{x^2} + 1 \right) K_1.$

804.85. $K_5 = \dfrac{12}{x} \left(\dfrac{16}{x^2} + 1 \right) K_0 + \left(\dfrac{384}{x^4} + \dfrac{72}{x^2} + 1 \right) K_1.$

804.90. $K_0' = - K_1.$

804.91. $K_1' = - K_0 - \dfrac{K_1}{x}.$

804.92. $K_2' = - \dfrac{2K_0}{x} - \left(\dfrac{4}{x^2} + 1\right) K_1.$

804.93. $K_3' = - \left(\dfrac{12}{x^2} + 1\right) K_0 - \left(\dfrac{24}{x^2} + 5\right) \dfrac{K_1}{x}.$

804.94. $K_4' = - \dfrac{8}{x}\left(\dfrac{12}{x^2} + 1\right) K_0 - \left(\dfrac{192}{x^4} + \dfrac{40}{x^2} + 1\right) K_1.$

804.95. $K_5' = - \left(\dfrac{960}{x^4} + \dfrac{84}{x^2} + 1\right) K_0 - \left(\dfrac{1920}{x^4} + \dfrac{408}{x^2} + 13\right) \dfrac{K_1}{x}.$

For tables of $K_0(x)$ and $K_1(x)$ see Ref. 50, p. 266, and Ref. 12, p. 313. Tables of $e^x K_0(x)$ and $e^x K_1(x)$, Ref. 13.

807.1. $J_0(x) = 1 - (\tfrac{1}{2}x)^2 + \dfrac{(\tfrac{1}{2}x)^4}{1^2 \cdot 2^2} - \dfrac{(\tfrac{1}{2}x)^6}{1^2 \cdot 2^2 \cdot 3^2} + \cdots .$

807.21. $J_1(x) = - J_0'(x) = \tfrac{1}{2}x - \dfrac{(\tfrac{1}{2}x)^3}{1^2 \cdot 2} + \dfrac{(\tfrac{1}{2}x)^5}{1^2 \cdot 2^2 \cdot 3} - \cdots .$

807.22. $J_2(x) = \dfrac{x^2}{2^2 2!} - \dfrac{x^4}{2^4 1! 3!} + \dfrac{x^6}{2^6 2! 4!} - \dfrac{x^8}{2^8 3! 5!} + \cdots .$

807.3. When n is a positive integer,

$$J_n(x) = \dfrac{(\tfrac{1}{2}x)^n}{n!}\left[1 - \dfrac{(\tfrac{1}{2}x)^2}{1(n + 1)} + \dfrac{(\tfrac{1}{2}x)^4}{1 \cdot 2(n + 1)(n + 2)} - \cdots \right].$$

807.4. When n is an integer,

$$J_{-n}(x) = (- 1)^n J_n(x).$$

807.5. When n is not a positive integer, replace $n!$ in **807.3** by $\Pi(n)$. [See **853.1**.] [Ref. 12, p. 14, cq. (16).]

807.61. $J_1'(x) = \dfrac{1}{2} - \dfrac{3x^2}{2^3 1! 2!} + \dfrac{5x^4}{2^5 2! 3!} - \dfrac{7x^6}{2^7 3! 4!} + \cdots .$

807.62. $J_2'(x) = \dfrac{x}{4} - \dfrac{4x^3}{2^4 1!3!} + \dfrac{6x^5}{2^6 2!4!} - \dfrac{8x^7}{2^8 3!5!} + \cdots$.

807.69. $J_n'(x) = \dfrac{x^{n-1}}{2^n(n-1)!} - \dfrac{(n+2)x^{n+1}}{2^{n+2}1!(n+1)!}$

$$+ \dfrac{(n+4)x^{n+3}}{2^{n+4}2!(n+2)!} - \dfrac{(n+6)x^{n+5}}{2^{n+6}3!(n+3)!} + \cdots,$$

$$[n \text{ an integer} > 0].$$

Asymptotic Series for Large Values of x

808.1. $J_0(x) = \left(\dfrac{2}{\pi x}\right)^{1/2}\left[P_0(x) \cos\left(x - \dfrac{\pi}{4}\right) - Q_0(x) \sin\left(x - \dfrac{\pi}{4}\right)\right]$,

where

808.11. $P_0(x) \approx 1 - \dfrac{1^2 \cdot 3^2}{2!(8x)^2} + \dfrac{1^2 \cdot 3^2 \cdot 5^2 \cdot 7^2}{4!(8x)^4}$

$$- \dfrac{1^2 \cdot 3^2 \cdot 5^2 \cdot 7^2 \cdot 9^2 \cdot 11^2}{6!(8x)^6} + \cdots.$$

808.12. $Q_0(x) \approx -\dfrac{1^2}{1!8x} + \dfrac{1^2 \cdot 3^2 \cdot 5^2}{3!(8x)^3} - \dfrac{1^2 \cdot 3^2 \cdot 5^2 \cdot 7^2 \cdot 9^2}{5!(8x)^5} + \cdots$.

808.2. $J_1(x) = \left(\dfrac{2}{\pi x}\right)^{1/2}\left[P_1(x) \cos\left(x - \dfrac{3\pi}{4}\right)\right.$

$$\left. - Q_1(x) \sin\left(x - \dfrac{3\pi}{4}\right)\right],$$

where

808.21. $P_1(x) \approx 1 + \dfrac{1^2 \cdot 3 \cdot 5}{2!(8x)^2} - \dfrac{1^2 \cdot 3^2 \cdot 5^2 \cdot 7 \cdot 9}{4!(8x)^4}$

$$+ \dfrac{1^2 \cdot 3^2 \cdot 5^2 \cdot 7^2 \cdot 9^2 \cdot 11 \cdot 13}{6!(8x)^6} - \cdots.$$

The signs are alternately $+$ and $-$ after the first term.

808.22. $Q_1(x) \approx \dfrac{1 \cdot 3}{1!8x} - \dfrac{1^2 \cdot 3^2 \cdot 5 \cdot 7}{3!(8x)^3} + \dfrac{1^2 \cdot 3^2 \cdot 5^2 \cdot 7^2 \cdot 9 \cdot 11}{5!(8x)^5} - \cdots$.

808.3. $\quad J_n(x) = \left(\dfrac{2}{\pi x}\right)^{1/2}\left[P_n(x)\cos\left(x - \dfrac{n\pi}{2} - \dfrac{\pi}{4}\right)\right.$

$$\left. - Q_n(x)\sin\left(x - \dfrac{n\pi}{2} - \dfrac{\pi}{4}\right)\right],$$

where

808.31. $\quad P_n(x) \approx 1 - \dfrac{(4n^2 - 1^2)(4n^2 - 3^2)}{2!(8x)^2}$

$$+ \dfrac{(4n^2 - 1^2)(4n^2 - 3^2)(4n^2 - 5^2)(4n^2 - 7^2)}{4!(8x)^4} - \cdots.$$

808.32. $\quad Q_n(x) \approx \dfrac{4n^2 - 1^2}{1!8x}$

$$- \dfrac{(4n^2 - 1^2)(4n^2 - 3^2)(4n^2 - 5^2)}{3!(8x)^3} + \cdots.$$

808.4. $\quad J_n'(x) = -\left(\dfrac{2}{\pi x}\right)^{1/2}\left[P_n{}^{(1)}(x)\sin\left(x - \dfrac{n\pi}{2} - \dfrac{\pi}{4}\right)\right.$

$$\left. + Q_n{}^{(1)}(x)\cos\left(x - \dfrac{n\pi}{2} - \dfrac{\pi}{4}\right)\right],$$

where, from **801.4,**

808.41. $\quad P_n{}^{(1)}(x) \approx 1 - \dfrac{(4n^2 - 1^2)(4n^2 + 3 \times 5)}{2!(8x)^2}$

$$+ \dfrac{(4n^2 - 1^2)\,(4n^2 - 3^2)\,(4n^2 - 5^2)\,(4n^2 + 7 \times 9)}{4!(8x)^4} - \cdots.$$

808.42. $\quad Q_n{}^{(1)}(x) \approx \dfrac{4n^2 + 1 \times 3}{1!8x}$

$$- \dfrac{(4n^2 - 1^2)(4n^2 - 3^2)(4n^2 + 5 \times 7)}{3!(8x)^3} + \cdots.$$

Extension of these series can be made by inspection. The sign $\approx$ denotes approximate equality. Note that the various series for large values of x are asymptotic expansions and there is a limit to the amount of precision which they will give.

809.01. $\quad J_{\frac{1}{2}}(x) = \left(\dfrac{2}{\pi x}\right)^{1/2}\sin x.$

809.03. $\quad J_{\frac{3}{2}}(x) = \left(\dfrac{2}{\pi x}\right)^{1/2}\left(\dfrac{\sin x}{x} - \cos x\right).$

809.05. $J_{\frac{5}{2}}(x) = \left(\dfrac{2}{\pi x}\right)^{1/2} \left\{\left(\dfrac{3}{x^2} - 1\right) \sin x - \dfrac{3}{x} \cos x\right\}.$

809.21. $J_{-\frac{1}{2}}(x) = \left(\dfrac{2}{\pi x}\right)^{1/2} \cos x.$

809.23. $J_{-\frac{3}{2}}(x) = \left(\dfrac{2}{\pi x}\right)^{1/2}\left(- \sin x - \dfrac{\cos x}{x}\right).$

809.25. $J_{-\frac{5}{2}}(x) = \left(\dfrac{2}{\pi x}\right)^{1/2} \left\{\dfrac{3}{x} \sin x + \left(\dfrac{3}{x^2} - 1\right) \cos x\right\}.$

[For higher orders see Ref. 12, p. 17.]

811.1. $N_0(x) = \dfrac{2}{\pi}\left(\gamma + \log_\epsilon \dfrac{x}{2}\right) J_0(x) + \dfrac{2\,(\frac{1}{2}x)^2}{\pi\,(1!)^2}$

$\qquad\qquad - \dfrac{2\,(\frac{1}{2}x)^4}{\pi\,(2!)^2}\,(1 + \tfrac{1}{2}) + \dfrac{2\,(\frac{1}{2}x)^6}{\pi\,(3!)^2}\,(1 + \tfrac{1}{2} + \tfrac{1}{3}) - \cdots,$

where γ is Euler's constant 0.577 2157.   [See **851.1.**]

[See note preceding **802.1.**]

811.2. $N_1(x) = \dfrac{2}{\pi}\left(\gamma + \log_\epsilon \dfrac{x}{2}\right) J_1(x) - \dfrac{2}{\pi x}$

$\qquad - \dfrac{1}{\pi}\sum_{p=0}^{\infty}\dfrac{(-1)^p}{p!(p+1)!}\left(\dfrac{x}{2}\right)^{2p+1}\left\{2\left(1 + \tfrac{1}{2} + \cdots + \dfrac{1}{p}\right) + \dfrac{1}{p+1}\right\}.$

811.3. $N_n(x) = \dfrac{2}{\pi}\left(\gamma + \log_\epsilon \dfrac{x}{2}\right) J_n(x)$

$\qquad - \dfrac{1}{\pi}\sum_{p=0}^{n-1}\dfrac{(n-p-1)!}{p!}\left(\dfrac{x}{2}\right)^{2p-n} - \dfrac{1}{\pi}\sum_{p=0}^{\infty}\dfrac{(-1)^p}{p!(n+p)!}\left(\dfrac{x}{2}\right)^{2p+n}$

$\qquad\qquad \times \left(1 + \dfrac{1}{2} + \dfrac{1}{3} + \cdots + \dfrac{1}{p} + 1 + \dfrac{1}{2} + \cdots + \dfrac{1}{n+p}\right),$

where n is a positive integer. The last quantity in parentheses is $\left(1 + \dfrac{1}{2} + \cdots + \dfrac{1}{n}\right)$ when $p = 0$.

[Ref. 49, p. 161, eq. (61),
and Ref. 50, p. 174.]

Asymptotic Series for Large Values of x

812.1. $N_0(x) = \left(\dfrac{2}{\pi x}\right)^{1/2} \left[P_0(x) \sin\left(x - \dfrac{\pi}{4}\right) + Q_0(x) \cos\left(x - \dfrac{\pi}{4}\right) \right].$

812.2. $N_1(x) = \left(\dfrac{2}{\pi x}\right)^{1/2} \left[P_1(x) \sin\left(x - \dfrac{3\pi}{4}\right) \right.$

$$\left. + Q_1(x) \cos\left(x - \dfrac{3\pi}{4}\right) \right].$$

812.3. $N_n(x) = \left(\dfrac{2}{\pi x}\right)^{1/2} \left[P_n(x) \sin\left(x - \dfrac{n\pi}{2} - \dfrac{\pi}{4}\right) \right.$

$$\left. + Q_n(x) \cos\left(x - \dfrac{n\pi}{2} - \dfrac{\pi}{4}\right) \right].$$

[For the P and Q series see **808**.]

812.4. $N_n'(x) = \left(\dfrac{2}{\pi x}\right)^{1/2} \left[P_n^{(1)}(x) \cos\left(x - \dfrac{n\pi}{2} - \dfrac{\pi}{4}\right) \right.$

$$\left. - Q_n^{(1)}(x) \sin\left(x - \dfrac{n\pi}{2} - \dfrac{\pi}{4}\right) \right].$$

[For $P_n^{(1)}(x)$ and $Q_n^{(1)}(x)$ see **808.41** and **808.42**.]

813.1. $I_0(x) = J_0(ix) = 1 + (\tfrac{1}{2}x)^2 + \dfrac{(\tfrac{1}{2}x)^4}{1^2 \cdot 2^2} + \dfrac{(\tfrac{1}{2}x)^6}{1^2 \cdot 2^2 \cdot 3^2} + \cdots$

where $i = \sqrt{(-1)}$.

813.2. $I_1(x) = i^{-1}J_1(ix) = I_0'(x) = \tfrac{1}{2}x + \dfrac{(\tfrac{1}{2}x)^3}{1^2 \cdot 2} + \dfrac{(\tfrac{1}{2}x)^5}{1^2 \cdot 2^2 \cdot 3} + \cdots.$

813.3. When n is a positive integer,

$I_n(x) = i^{-n}J_n(ix)$

$$= \dfrac{(\tfrac{1}{2}x)^n}{n!} \left[1 + \dfrac{(\tfrac{1}{2}x)^2}{1(n+1)} + \dfrac{(\tfrac{1}{2}x)^4}{1 \cdot 2(n+1)(n+2)} + \cdots \right]$$

$$= \sum_{p=0}^{\infty} \dfrac{(\tfrac{1}{2}x)^{n+2p}}{p!(n+p)!}.$$

813.4. When n is an integer,

$$I_{-n}(x) = I_n(x).$$

813.5. When n is not a positive integer, replace $n!$ in **813.3** by $\Pi(n)$.

[See **853.1**.] [Ref. 12, p. 20.]

Asymptotic Series for Large Values of x

814.1. $I_0(x) \approx \dfrac{e^x}{\sqrt{(2\pi x)}} \left[1 + \dfrac{1^2}{1!\,8x} + \dfrac{1^2 \cdot 3^2}{2!\,(8x)^2} + \cdots \right].$

814.2. $I_n(x) \approx \dfrac{e^x}{\sqrt{(2\pi x)}} \left[1 - \dfrac{4n^2 - 1^2}{1!\,8x} \right.$

$$\left. + \dfrac{(4n^2 - 1^2)(4n^2 - 3^2)}{2!\,(8x)^2} - \cdots \right].$$

814.3. $I_n{}'(x) \approx \dfrac{e^x}{\sqrt{(2\pi x)}}$

$$\times \left[1 - \dfrac{4n^2 + 1 \times 3}{1!\,8x} + \dfrac{(4n^2 - 1^2)(4n^2 + 3 \times 5)}{2!\,(8x)^2} \right.$$

$$\left. - \dfrac{(4n^2 - 1^2)(4n^2 - 3^2)(4n^2 + 5 \times 7)}{3!\,(8x)^3} + \cdots \right].$$

The terms of the series in **814.3** are similar to those in **808.41** and **808.42**.

815.1. $K_0(x) = - \left(\gamma + \log_e \dfrac{x}{2} \right) I_0(x)$

$$+ \dfrac{(\frac{1}{2}x)^2}{(1!)^2} + \dfrac{(\frac{1}{2}x)^4}{(2!)^2} (1 + \tfrac{1}{2}) + \dfrac{(\frac{1}{2}x)^6}{(3!)^2} (1 + \tfrac{1}{2} + \tfrac{1}{3}) + \cdots,$$

where γ is Euler's constant, 0.577 2157.

815.2. $K_n(x) = (-1)^{n+1} \left(\gamma + \log_e \dfrac{x}{2} \right) I_n(x)$

$$+ \dfrac{1}{2} \sum_{p=0}^{n-1} \dfrac{(-1)^p (n - p - 1)!}{p!} \left(\dfrac{x}{2} \right)^{2p-n}$$

$$+ \dfrac{(-1)^n}{2} \sum_{p=0}^{\infty} \dfrac{1}{p!\,(n + p)!} \left(\dfrac{x}{2} \right)^{2p+n}$$

$$\times \left(1 + \dfrac{1}{2} + \dfrac{1}{3} + \cdots + \dfrac{1}{p} + 1 + \dfrac{1}{2} + \cdots + \dfrac{1}{n + p} \right),$$

where n is a positive integer. The last quantity in parentheses is $\left(1 + \dfrac{1}{2} + \cdots + \dfrac{1}{n} \right)$ when $p = 0$.

[Ref. 13, p. 80, and Ref. 50, p. 264.]

Note that the letter K is sometimes, particularly in earlier writings, used to denote other expressions in connection with Bessel functions.

815.3. When n is an integer,

$$K_{-n}(x) = K_n(x).$$

815.4. When n is not an integer,

$$K_n(x) = \frac{\pi}{2 \sin n\pi} \{I_{-n}(x) - I_n(x)\}.$$

Asymptotic Series for Large Values of x

816.1.

$$K_0(x) \approx \left(\frac{\pi}{2x}\right)^{1/2} e^{-x} \left[1 - \frac{1^2}{1!8x} + \frac{1^2 \cdot 3^2}{2!(8x)^2} - \frac{1^2 \cdot 3^2 \cdot 5^2}{3!(8x)^3} + \cdots \right]$$

where $\approx$ denotes approximate equality.

816.2. $K_n(x) \approx \left(\frac{\pi}{2x}\right)^{1/2} e^{-x} \left[1 + \frac{4n^2 - 1^2}{1!8x} \right.$

$$\left. + \frac{(4n^2 - 1^2)(4n^2 - 3^2)}{2!(8x)^2} + \cdots \right].$$

[Ref. 12, p. 55, eq. (50).]

816.3. $K_n{}'(x) \approx - \left(\frac{\pi}{2x}\right)^{1/2} e^{-x}$

$$\times \left[1 + \frac{4n^2 + 1 \times 3}{1!8x} + \frac{(4n^2 - 1^2)(4n^2 + 3 \times 5)}{2!(8x)^2} \right.$$

$$\left. + \frac{(4n^2 - 1^2)(4n^2 - 3^2)(4n^2 + 5 \times 7)}{3!(8x)^3} + \cdots \right]$$

[from **804.4.**]

The series can be extended by inspection.

817.1. $H_0{}^{(1)}(z) = J_0(z) + iN_0(z).$

817.2. $K_0(z) = \frac{\pi i}{2} H_0{}^{(1)}(iz).$

817.3. $H_n{}^{(1)}(z) = J_n(z) + iN_n(z).$

817.4. $H_n{}^{(2)}(z) = J_n(z) - iN_n(z).$ [Ref. 13, p. 73.]

817.5. $K_n(z) = \frac{\pi i}{2} e^{in\pi/2} H_n{}^{(1)}(iz).$ [Ref. 13, p. 78.]

For all values of x and ϕ,

818.1. $\cos(x \sin \phi) = J_0(x) + 2J_2(x) \cos 2\phi + 2J_4(x) \cos 4\phi + \cdots$.

818.2. $\sin(x \sin \phi) = 2J_1(x) \sin \phi + 2J_3(x) \sin 3\phi$
$$+ 2J_5(x) \sin 5\phi + \cdots.$$

818.3. $\cos(x \cos \phi) = J_0(x) - 2J_2(x) \cos 2\phi$
$$+ 2J_4(x) \cos 4\phi - \cdots.$$

818.4. $\sin(x \cos \phi) = 2J_1(x) \cos \phi - 2J_3(x) \cos 3\phi$
$$+ 2J_5(x) \cos 5\phi - \cdots.$$
$$[\text{Ref. 12, p. 32.}]$$

Bessel Functions of Argument $xi\sqrt{i}$, of the First Kind

(For numerical values see Table 1050.)

820.1. $\text{ber } x + i \text{ bei } x = J_0(xi\sqrt{i}) = I_0(x\sqrt{i}) = \text{ber}_0 x + i \text{ bei}_0 x$.

820.2. $\text{ber}' x = \dfrac{d}{dx} \text{ber } x$, etc.

820.3. $\text{ber } x = 1 - \dfrac{(\frac{1}{2}x)^4}{(2!)^2} + \dfrac{(\frac{1}{2}x)^8}{(4!)^2} - \cdots$.

820.4. $\text{bei } x = \dfrac{(\frac{1}{2}x)^2}{(1!)^2} - \dfrac{(\frac{1}{2}x)^6}{(3!)^2} + \dfrac{(\frac{1}{2}x)^{10}}{(5!)^2} - \cdots$.

820.5. $\text{ber}' x = -\dfrac{(\frac{1}{2}x)^3}{1!2!} + \dfrac{(\frac{1}{2}x)^7}{3!4!} - \dfrac{(\frac{1}{2}x)^{11}}{5!6!} + \cdots$.

820.6. $\text{bei}' x = \frac{1}{2}x - \dfrac{(\frac{1}{2}x)^5}{2!3!} + \dfrac{(\frac{1}{2}x)^9}{4!5!} - \cdots$.

821.1.   For large values of x,
$$\text{ber } x \approx \frac{e^{x/\sqrt{2}}}{\sqrt{(2\pi x)}}\left[L_0(x) \cos\left(\frac{x}{\sqrt{2}} - \frac{\pi}{8}\right) - M_0(x) \sin\left(\frac{x}{\sqrt{2}} - \frac{\pi}{8}\right)\right],$$

821.2. $\text{bei } x \approx \dfrac{e^{x/\sqrt{2}}}{\sqrt{(2\pi x)}}\left[M_0(x) \cos\left(\dfrac{x}{\sqrt{2}} - \dfrac{\pi}{8}\right)\right.$
$$\left. + L_0(x) \sin\left(\frac{x}{\sqrt{2}} - \frac{\pi}{8}\right)\right],$$

where

821.3. $L_0(x) = 1 + \dfrac{1^2}{1!8x} \cos \dfrac{\pi}{4} + \dfrac{1^2 \cdot 3^2}{2!(8x)^2} \cos \dfrac{2\pi}{4}$

$$+ \dfrac{1^2 \cdot 3^2 \cdot 5^2}{3!(8x)^3} \cos \dfrac{3\pi}{4} + \cdots,$$

821.4. $M_0(x) = - \dfrac{1^2}{1!8x} \sin \dfrac{\pi}{4} - \dfrac{1^2 \cdot 3^2}{2!(8x)^2} \sin \dfrac{2\pi}{4}$

$$- \dfrac{1^2 \cdot 3^2 \cdot 5^2}{3!(8x)^3} \sin \dfrac{3\pi}{4} - \cdots.$$

821.5. $\operatorname{ber}' x \approx \dfrac{e^{x/\sqrt{2}}}{\sqrt{(2\pi x)}} \left[S_0(x) \cos \left(\dfrac{x}{\sqrt{2}} + \dfrac{\pi}{8} \right) \right.$

$$\left. - T_0(x) \sin \left(\dfrac{x}{\sqrt{2}} + \dfrac{\pi}{8} \right) \right],$$

821.6. $\operatorname{bei}' x \approx \dfrac{e^{x/\sqrt{2}}}{\sqrt{(2\pi x)}} \left[T_0(x) \cos \left(\dfrac{x}{\sqrt{2}} + \dfrac{\pi}{8} \right) \right.$

$$\left. + S_0(x) \sin \left(\dfrac{x}{\sqrt{2}} + \dfrac{\pi}{8} \right) \right],$$

where

821.7. $S_0(x) = 1 - \dfrac{1 \cdot 3}{1!8x} \cos \dfrac{\pi}{4} - \dfrac{1^2 \cdot 3 \cdot 5}{2!(8x)^2} \cos \dfrac{2\pi}{4}$

$$- \dfrac{1^2 \cdot 3^2 \cdot 5 \cdot 7}{3!(8x)^3} \cos \dfrac{3\pi}{4} - \dfrac{1^2 \cdot 3^2 \cdot 5^2 \cdot 7 \cdot 9}{4!(8x)^4} \cos \dfrac{4\pi}{4} - \cdots,$$

821.8. $T_0(x) = \dfrac{1 \cdot 3}{1!8x} \sin \dfrac{\pi}{4} + \dfrac{1^2 \cdot 3 \cdot 5}{2!(8x)^2} \sin \dfrac{2\pi}{4}$

$$+ \dfrac{1^2 \cdot 3^2 \cdot 5 \cdot 7}{3!(8x)^3} \sin \dfrac{3\pi}{4} + \dfrac{1^2 \cdot 3^2 \cdot 5^2 \cdot 7 \cdot 9}{4!(8x)^4} \sin \dfrac{4\pi}{4} + \cdots.$$

[Ref. 14.]

822.1.   When n is a positive integer,

$$\operatorname{ber}_n x + i \operatorname{bei}_n x = J_n(xi\sqrt{i}) = i^n I_n(x\sqrt{i}).$$

822.2. $\operatorname{ber}_n x = \displaystyle\sum_{p=0}^{\infty} \dfrac{(-1)^{n+p}(\frac{1}{2}x)^{n+2p}}{p!(n \mid p)!} \cos \dfrac{(n+2p)\pi}{4}$

where

$$p = 0, 1, 2, 3, \cdots.$$

822.3. $\text{bei}_n \, x = \sum\limits_{p=0}^{\infty} \dfrac{(-1)^{n+p+1}(\frac{1}{2}x)^{n+2p}}{p!(n+p)!} \sin \dfrac{(n+2p)\pi}{4}.$

822.4. $\text{ber}_n{}' \, x = \sum\limits_{p=0}^{\infty} \dfrac{(-1)^{n+p}\left(\dfrac{n}{2}+p\right)\left(\dfrac{1}{2}x\right)^{n+2p-1}}{p!(n+p)!} \cos \dfrac{(n+2p)\pi}{4}.$

822.5. $\text{bei}_n{}' \, x = \sum\limits_{p=0}^{\infty} \dfrac{(-1)^{n+p+1}\left(\dfrac{n}{2}+p\right)\left(\dfrac{1}{2}x\right)^{n+2p-1}}{p!(n+p)!} \sin \dfrac{(n+2p)\pi}{4}.$

823.1.   For large values of x, when n is a positive integer,

$$\text{ber}_n \, x \approx \frac{e^{x/\sqrt{2}}}{\sqrt{(2\pi x)}}\left[L_n(x) \cos\left(\frac{x}{\sqrt{2}} - \frac{\pi}{8} + \frac{n\pi}{2}\right) \right.$$
$$\left. - M_n(x) \sin\left(\frac{x}{\sqrt{2}} - \frac{\pi}{8} + \frac{n\pi}{2}\right) \right],$$

823.2. $\text{bei}_n \, x \approx \dfrac{e^{x/\sqrt{2}}}{\sqrt{(2\pi x)}}\left[M_n(x) \cos\left(\dfrac{x}{\sqrt{2}} - \dfrac{\pi}{8} + \dfrac{n\pi}{2}\right) \right.$
$$\left. + L_n(x) \sin\left(\frac{x}{\sqrt{2}} - \frac{\pi}{8} + \frac{n\pi}{2}\right) \right],$$

where

823.3. $L_n(x) = 1 - \dfrac{4n^2 - 1^2}{1!8x} \cos \dfrac{\pi}{4}$
$$+ \frac{(4n^2 - 1^2)(4n^2 - 3^2)}{2!(8x)^2} \cos \frac{2\pi}{4} - \cdots,$$

823.4. $M_n(x) = \dfrac{4n^2 - 1^2}{1!8x} \sin \dfrac{\pi}{4}$
$$- \frac{(4n^2 - 1^2)(4n^2 - 3^2)}{2!(8x)^2} \sin \frac{2\pi}{4} + \cdots.$$

823.5. $\text{ber}_n{}' \, x \approx \dfrac{e^{x/\sqrt{2}}}{\sqrt{(2\pi x)}}\left[S_n(x) \cos\left(\dfrac{x}{\sqrt{2}} + \dfrac{\pi}{8} + \dfrac{n\pi}{2}\right) \right.$
$$\left. - T_n(x) \sin\left(\frac{x}{\sqrt{2}} + \frac{\pi}{8} + \frac{n\pi}{2}\right) \right],$$

823.6. $\text{bei}_n' x \approx \dfrac{e^{x/\sqrt{2}}}{\sqrt{(2\pi x)}}\bigg[T_n(x) \cos \left(\dfrac{x}{\sqrt{2}} + \dfrac{\pi}{8} + \dfrac{n\pi}{2} \right),$

$$+ S_n(x) \sin \left(\dfrac{x}{\sqrt{2}} + \dfrac{\pi}{8} + \dfrac{n\pi}{2} \right) \bigg],$$

where

823.7. $S_n(x) = 1 - \dfrac{4n^2 + 1 \times 3}{1!8x} \cos \dfrac{\pi}{4}$

$$+ \dfrac{(4n^2 - 1^2)(4n^2 + 3 \times 5)}{2!(8x)^2} \cos \dfrac{2\pi}{4}$$

$$- \dfrac{(4n^2 - 1^2)(4n^2 - 3^2)(4n^2 \overset{\cdot}{+} 5 \times 7)}{3!(8x)^3} \cos \dfrac{3\pi}{4} + \cdots,$$

823.8. $T_n(x) = \dfrac{4n^2 + 1 \times 3}{1!8x} \sin \dfrac{\pi}{4}$

$$- \dfrac{(4n^2 - 1^2)(4n^2 + 3 \times 5)}{2!(8x)^2} \sin \dfrac{2\pi}{4}$$

$$+ \dfrac{(4n^2 - 1^2)(4n^2 - 3^2)(4n^2 + 5 \times 7)}{3!(8x)^3} \sin \dfrac{3\pi}{4} - \cdots.$$

Bessel Functions of Argument $xi\sqrt{i}$, of the Second Kind

(For numerical values see Table 1050.)

824.1. $\text{ker } x + i \text{ kei } x = K_0(x\sqrt{i}).$

824.2. $\text{ker}' x = \dfrac{d}{dx} \text{ker } x, \text{ etc.}$

824.3. $\text{ker } x = \left(\log \dfrac{2}{x} - \gamma \right) \text{ber } x + \dfrac{\pi}{4} \text{bei } x$

$$- (1 + \tfrac{1}{2}) \dfrac{(\tfrac{1}{2}x)^4}{(2!)^2} + (1 + \tfrac{1}{2} + \tfrac{1}{3} + \tfrac{1}{4}) \dfrac{(\tfrac{1}{2}x)^8}{(4!)^2}$$

$$- (1 + \tfrac{1}{2} + \tfrac{1}{3} + \tfrac{1}{4} + \tfrac{1}{5} + \tfrac{1}{6}) \dfrac{(\tfrac{1}{2}x)^{12}}{(6!)^2} + \cdots,$$

where

$$\gamma = 0.577\ 2157.$$

824.4. $\quad \text{kei } x = \left(\log \dfrac{2}{x} - \gamma\right) \text{bei } x - \dfrac{\pi}{4} \text{ ber } x$

$$+ \frac{(\frac{1}{2}x)^2}{(1!)^2} - (1 + \tfrac{1}{2} + \tfrac{1}{3}) \frac{(\frac{1}{2}x)^6}{(3!)^2}$$

$$+ (1 + \tfrac{1}{2} + \tfrac{1}{3} + \tfrac{1}{4} + \tfrac{1}{5}) \frac{(\frac{1}{2}x)^{10}}{(5!)^2} - \cdots .$$

824.5. $\quad \text{ker}' x = \left(\log \dfrac{2}{x} - \gamma\right) \text{ber}' x - \dfrac{1}{x} \text{ ber } x + \dfrac{\pi}{4} \text{ bei}' x$

$$- (1 + \tfrac{1}{2}) \frac{(\frac{1}{2}x)^3}{1!2!} + (1 + \tfrac{1}{2} + \tfrac{1}{3} + \tfrac{1}{4}) \frac{(\frac{1}{2}x)^7}{3!4!} - \cdots .$$

824.6. $\quad \text{kei}' x = \left(\log \dfrac{2}{x} - \gamma\right) \text{bei}' x - \dfrac{1}{x} \text{ bei } x - \dfrac{\pi}{4} \text{ ber}' x$

$$+ \tfrac{1}{2}x - (1 + \tfrac{1}{2} + \tfrac{1}{3}) \frac{(\frac{1}{2}x^5)}{2!3!}$$

$$+ (1 + \tfrac{1}{2} + \tfrac{1}{3} + \tfrac{1}{4} + \tfrac{1}{5}) \frac{(\frac{1}{2}x)^9}{4!5!} - \cdots .$$

825.1. $\quad$ For large values of x,

$$\text{ker } x \approx \left(\frac{\pi}{2x}\right)^{1/2} e^{-x/\sqrt{2}} \left[L_0(-x) \cos\left(\frac{x}{\sqrt{2}} + \frac{\pi}{8}\right)\right.$$

$$\left. + M_0(-x) \sin\left(\frac{x}{\sqrt{2}} + \frac{\pi}{8}\right)\right].$$

825.2. $\quad \text{kei } x \approx \left(\dfrac{\pi}{2x}\right)^{1/2} e^{-x/\sqrt{2}} \left[M_0(-x) \cos\left(\dfrac{x}{\sqrt{2}} + \dfrac{\pi}{8}\right)\right.$

$$\left. - L_0(-x) \sin\left(\frac{x}{\sqrt{2}} + \frac{\pi}{8}\right)\right].$$

See 821.3 and 821.4, changing x to $-x$.

825.3. $\quad \text{ker}' x \approx - \left(\dfrac{\pi}{2x}\right)^{1/2} e^{-x/\sqrt{2}} \left[S_0(-x) \cos\left(\dfrac{x}{\sqrt{2}} - \dfrac{\pi}{8}\right)\right.$

$$\left. + T_0(-x) \sin\left(\frac{x}{\sqrt{2}} - \frac{\pi}{8}\right)\right].$$

825.4. $\quad \text{kei}' x \approx - \left(\dfrac{\pi}{2x}\right)^{1/2} e^{-x/\sqrt{2}} \left[T_0(-x) \cos\left(\dfrac{x}{\sqrt{2}} - \dfrac{\pi}{8}\right)\right.$

$$\left. - S_0(-x) \sin\left(\frac{x}{\sqrt{2}} - \frac{\pi}{8}\right)\right].$$

See 821.7 and 821.8, changing x to $-x$.

826.1. When n is a positive integer,
$$\ker_n x + i \operatorname{kei}_n x = i^{-n} K_n(x\sqrt{i}). \qquad [\text{See } \mathbf{815.2.}]$$

826.2. $\ker_n x = \left(\log \dfrac{2}{x} - \gamma\right) \operatorname{ber}_n x + \dfrac{\pi}{4} \operatorname{bei}_n x$

$$+ \frac{1}{2} \sum_{p=0}^{n-1} \frac{(-1)^{n+p}(n-p-1)!}{p!} \left(\frac{x}{2}\right)^{2p-n} \cos \frac{(n+2p)\pi}{4}$$

$$+ \frac{1}{2} \sum_{p=0}^{\infty} \left(1 + \frac{1}{2} + \frac{1}{3} + \cdots + \frac{1}{p} + 1 + \frac{1}{2} + \frac{1}{3} + \cdots + \frac{1}{n+p}\right)$$
$$\frac{(-1)^{n+p}(\frac{1}{2}x)^{n+2p}}{p!(n+p)!} \cos \frac{(n+2p)\pi}{4}.$$

826.3. $\operatorname{kei}_n x = \left(\log \dfrac{2}{x} - \gamma\right) \operatorname{bei}_n x - \dfrac{\pi}{4} \operatorname{ber}_n x$

$$+ \frac{1}{2} \sum_{p=0}^{n-1} \frac{(-1)^{n+p}(n-p-1)!}{p!} \left(\frac{x}{2}\right)^{2p-n} \sin \frac{(n+2p)\pi}{4}$$

$$- \frac{1}{2} \sum_{p=0}^{\infty} \left(1 + \frac{1}{2} + \frac{1}{3} + \cdots + \frac{1}{p} + 1 + \frac{1}{2} + \frac{1}{3} + \cdots + \frac{1}{n+p}\right)$$
$$\frac{(-1)^{n+p}(\frac{1}{2}x)^{n+2p}}{p!(n+p)!} \sin \frac{(n+2p)\pi}{4}.$$

826.4. $\ker_n' x = \left(\log \dfrac{2}{x} - \gamma\right) \operatorname{ber}_n' x - \dfrac{\operatorname{ber}_n x}{x} + \dfrac{\pi}{4} \operatorname{bei}_n' x$

$$+ \frac{1}{4} \sum_{p=0}^{n-1} \frac{(-1)^{n+p}(2p-n)(n-p-1)!}{p!} \left(\frac{x}{2}\right)^{2p-n-1} \cos \frac{(n+2p)\pi}{4}$$

$$+ \frac{1}{4} \sum_{p=0}^{\infty} \left(1 + \frac{1}{2} + \frac{1}{3} + \cdots + \frac{1}{p} + 1 + \frac{1}{2} + \frac{1}{3} + \cdots + \frac{1}{n+p}\right)$$
$$\frac{(-1)^{n+p}(n+2p)(\frac{1}{2}x)^{n+2p-1}}{p!(n+p)!} \cos \frac{(n+2p)\pi}{4}.$$

826.5. $\operatorname{kei}_n' x = \left(\log \dfrac{2}{x} - \gamma\right) \operatorname{bei}_n' x - \dfrac{\operatorname{bei}_n x}{x} - \dfrac{\pi}{4} \operatorname{ber}_n' x$

$$+ \frac{1}{4} \sum_{p=0}^{n-1} \frac{(-1)^{n+p}(2p-n)(n-p-1)!}{p!} \left(\frac{x}{2}\right)^{2p-n-1} \sin \frac{(n+2p)\pi}{4}$$

$$- \frac{1}{4} \sum_{p=0}^{\infty} \left(1 + \frac{1}{2} + \frac{1}{3} + \cdots + \frac{1}{p} + 1 + \frac{1}{2} + \frac{1}{3} + \cdots + \frac{1}{n+p}\right)$$
$$\frac{(-1)^{n+p}(n+2p)(\frac{1}{2}x)^{n+2p-1}}{p!(n+p)!} \sin \frac{(n+2p)\pi}{4}.$$

827.1. For large values of x, when n is a positive integer,

$$\ker_n x \approx \left(\frac{\pi}{2x}\right)^{1/2} e^{-x/\sqrt{2}} \left[L_n(-x) \cos\left(\frac{x}{\sqrt{2}} + \frac{\pi}{8} + \frac{n\pi}{2}\right) \right.$$
$$\left. + M_n(-x) \sin\left(\frac{x}{\sqrt{2}} + \frac{\pi}{8} + \frac{n\pi}{2}\right) \right].$$

827.2. $$\ker_n x \approx \left(\frac{\pi}{2x}\right)^{1/2} e^{-x/\sqrt{2}} \left[M_n(-x) \cos\left(\frac{x}{\sqrt{2}} + \frac{\pi}{8} + \frac{n\pi}{2}\right) \right.$$
$$\left. - L_n(-x) \sin\left(\frac{x}{\sqrt{2}} + \frac{\pi}{8} + \frac{n\pi}{2}\right) \right]. \qquad \text{[See **823.3** and **823.4**.]}$$

827.3. $$\ker_n' x \approx -\left(\frac{\pi}{2x}\right)^{1/2} e^{-x/\sqrt{2}} \left[S_n(-x) \cos\left(\frac{x}{\sqrt{2}} - \frac{\pi}{8} + \frac{n\pi}{2}\right) \right.$$
$$\left. + T_n(-x) \sin\left(\frac{x}{\sqrt{2}} - \frac{\pi}{8} + \frac{n\pi}{2}\right) \right].$$

827.4. $$\ker_n' x \approx -\left(\frac{\pi}{2x}\right)^{1/2} e^{-x/\sqrt{2}} \left[T_n(-x) \cos\left(\frac{x}{\sqrt{2}} - \frac{\pi}{8} + \frac{n\pi}{2}\right) \right.$$
$$\left. - S_n(-x) \sin\left(\frac{x}{\sqrt{2}} - \frac{\pi}{8} + \frac{n\pi}{2}\right) \right]. \qquad \text{[See **823.7** and **823.8**.]}$$

Note that the series for large values of x are asymptotic expansions and there is a limit to the amount of precision which they will give.

Recurrence Formulas

828.1. $$\ber_1 x = \frac{1}{\sqrt{2}} (\ber' x - \bei' x).$$

828.2. $$\bei_1 x = \frac{1}{\sqrt{2}} (\ber' x + \bei' x).$$

828.3. $$\ber_2 x = \frac{2 \bei' x}{x} - \ber x.$$

828.4. $$\bei_2 x = -\frac{2 \ber' x}{x} - \bei x.$$

828.5. $$\ber_2' x = -\ber' x - \frac{2 \ber_2 x}{x}.$$

828.6. $$\bei_2' x = -\bei' x - \frac{2 \bei_2 x}{x}.$$

829.1. $\mathrm{ber}_{n+1}\, x = -\dfrac{n\sqrt{2}}{x}\,(\mathrm{ber}_n\, x - \mathrm{bei}_n\, x) - \mathrm{ber}_{n-1}\, x.$

829.2. $\mathrm{bei}_{n+1}\, x = -\dfrac{n\sqrt{2}}{x}\,(\mathrm{ber}_n\, x + \mathrm{bei}_n\, x) - \mathrm{bei}_{n-1}\, x.$

829.3. $\mathrm{ber}_n'\, x = -\dfrac{1}{\sqrt{2}}\,(\mathrm{ber}_{n-1}\, x + \mathrm{bei}_{n-1}\, x) - \dfrac{n\,\mathrm{ber}_n\, x}{x}.$

829.4. $\mathrm{bei}_n'\, x = \dfrac{1}{\sqrt{2}}\,(\mathrm{ber}_{n-1}\, x - \mathrm{bei}_{n-1}\, x) - \dfrac{n\,\mathrm{bei}_n\, x}{x}.$

830.   The formulas of **828–829** are applicable to Bessel functions of the second kind by changing ber to ker and bei to kei.

[Ref. 14, eq. (1)–(60).]

BESSEL FUNCTIONS—INTEGRALS

835.1. $\displaystyle\int x^n J_{n-1}(x)\,dx = x^n J_n(x).$

835.2. $\displaystyle\int x^{-n} J_{n+1}(x)\,dx = -x^{-n} J_n(x).$

835.3. $\displaystyle\int x^n I_{n-1}(x)\,dx = x^n I_n(x).$

835.4. $\displaystyle\int x^{-n} I_{n+1}(x)\,dx = x^{-n} I_n(x).$

835.5. $\displaystyle\int x^n K_{n-1}(x)\,dx = -x^n K_n(x).$

835.6. $\displaystyle\int x^{-n} K_{n+1}(x)\,dx = -x^{-n} K_n(x).$

836.1. $\displaystyle\int_0^x x\,\mathrm{ber}\, x\,dx = x\,\mathrm{bei}'\, x.$

836.2. $\displaystyle\int_0^x x\,\mathrm{bei}\, x\,dx = -x\,\mathrm{ber}'\, x.$

836.3. $\displaystyle\int_0^x x\,\mathrm{ker}\, x\,dx = x\,\mathrm{kei}'\, x.$

836.4. $\displaystyle\int_0^x x \text{ kei } x\, dx = - x \text{ ker}' x.$ [Ref. 12, p. 27.]

837.1. $\displaystyle\int x(\text{ber}_n{}^2 x + \text{bei}_n{}^2 x)dx = x(\text{ber}_n x \text{ bei}_n' x - \text{bei}_n x \text{ ber}_n' x).$

837.2. $\displaystyle\int x(\text{ber}_n'^2 x + \text{bei}_n'^2 x)dx = x (\text{bcr}_n x \text{ber}_n' x + \text{bei}_n x \text{ bei}_n' x).$

[Eq. 191 and 193, p. 170, Ref. 49.]

See also similar equations in $\text{ker}_n x$ and $\text{kei}_n x$, Ref. 49, p. 172, eq. 236 and 238.

⋆⋆ 11 ⋆⋆

SURFACE ZONAL HARMONICS

840. $P_0(\mu) = 1.$

$P_1(\mu) = \mu.$

$P_2(\mu) = \dfrac{1}{2}\,(3\mu^2 - 1).$

$P_3(\mu) = \dfrac{1}{2}\,(5\mu^3 - 3\mu).$

$P_4(\mu) = \dfrac{1}{2\cdot4}\,(5\cdot7\mu^4 - 2\cdot3\cdot5\mu^2 + 1\cdot3).$

$P_5(\mu) = \dfrac{1}{2\cdot4}\,(7\cdot9\mu^5 - 2\cdot5\cdot7\mu^3 + 3\cdot5\mu).$

$P_6(\mu) = \dfrac{1}{2\cdot4\cdot6}\,(7\cdot9\cdot11\mu^6 - 3\cdot5\cdot7\cdot9\mu^4 + 3\cdot3\cdot5\cdot7\mu^2 - 1\cdot3\cdot5).$

$P_7(\mu) = \dfrac{1}{2\cdot4\cdot6}\,(9\cdot11\cdot13\mu^7 - 3\cdot7\cdot9\cdot11\mu^5 + 3\cdot5\cdot7\cdot9\mu^3 - 3\cdot5\cdot7\mu).$

Note that the parentheses contain binomial coefficients as well as other factors.
[Ref. 25, p. 956.]

841. $P_m(\mu) = \dfrac{(2m-1)(2m-3)\cdots1}{m!}\left[\mu^m - \dfrac{m(m-1)}{2(2m-1)}\,\mu^{m-2}\right.$

$\left. + \dfrac{m(m-1)(m-2)(m-3)}{2\cdot4(2m-1)(2m-3)}\,\mu^{m-4} - \cdots\right].$

The series terminates with the term involving μ if m is odd and with the term independent of μ if m is even.   [Ref. 22, p. 145.]

842. $(m + 1)P_{m+1}(\mu) = (2m + 1)\mu P_m(\mu) - m P_{m-1}(\mu).$

[Ref. 22, p. 151.]

843. $(\mu^2 - 1)P_m'(\mu) = m\mu P_m(\mu) - m P_{m-1}(\mu).$

[Ref. 21, p. 137.]

844.   For large values of m,

$$P_m(\cos \theta) \approx \left(\frac{2}{m\pi \sin \theta}\right)^{1/2} \sin \left\{\left(m + \frac{1}{2}\right)\theta + \frac{\pi}{4}\right\}.$$

[Ref. 21, p. 137.]

844.1. $P_m(x) = \dfrac{1}{2^m m!} \dfrac{d^m}{dx^m} (x^2 - 1)^m.$ [Ref. 22, p. 160, eq. 1.]

844.2. $P_m(1) = 1.$

844.3. $P_{2m}(-x) = P_{2m}(x).$

844.4. $P_{2m+1}(-x) = -P_{2m+1}(x).$

[Ref. 22, p. 150, eq. 5–7.]

845.   First Derivatives, $P_m'(\mu) = \dfrac{d}{d\mu} P_m(\mu).$

$P_0'(\mu) = 0.$

$P_1'(\mu) = 1.$

$P_2'(\mu) = 3\mu.$

$P_3'(\mu) = \dfrac{1}{2} (3 \cdot 5\mu^2 - 1 \cdot 3).$

$P_4'(\mu) = \dfrac{1}{2} (5 \cdot 7\mu^3 - 3 \cdot 5\mu).$

$P_5'(\mu) = \dfrac{1}{2 \cdot 4} (5 \cdot 7 \cdot 9\mu^4 - 2 \cdot 3 \cdot 5 \cdot 7\mu^2 + 1 \cdot 3 \cdot 5).$

$P_6'(\mu) = \dfrac{1}{2 \cdot 4} (7 \cdot 9 \cdot 11\mu^5 - 2 \cdot 5 \cdot 7 \cdot 9\mu^3 + 3 \cdot 5 \cdot 7\mu).$

$P_7'(\mu) = \dfrac{1}{2 \cdot 4 \cdot 6} (7 \cdot 9 \cdot 11 \cdot 13\mu^6 - 3 \cdot 5 \cdot 7 \cdot 9 \cdot 11\mu^4 + 3 \cdot 3 \cdot 5 \cdot 7 \cdot 9\mu^2 - 1 \cdot 3 \cdot 5 \cdot 7).$

Note that the parentheses contain binomial coefficients as well as other factors.

[Ref. 25, p. 957.]

For tables of numerical values see Ref. 22, pp. 278–281, Ref. 45, pp. 188–197, and Refs. 52, 53, and 54.

⁂ 12 ⁂

DEFINITE INTEGRALS

850.1. $\displaystyle\int_0^\infty x^{n-1}e^{-x}dx = \int_0^1 \left(\log \frac{1}{x}\right)^{n-1} dx = \Gamma(n).$

[See Table **1005**.]

$\Gamma(n)$ is the **Gamma** function. The integral is finite when $n > 0$.

850.2. $\Gamma(n + 1) = n\Gamma(n).$

850.3. $\displaystyle\Gamma(n)\Gamma(1 - n) = \frac{\pi}{\sin n\pi},$ [n not an integer].

850.4. $\Gamma(n) = (n - 1)!$, when n is an integer > 0.

850.5. $\Gamma(1) = \Gamma(2) = 1.$

850.6. $\Gamma(\tfrac{1}{2}) = \sqrt{\pi}.$

850.7. $\displaystyle\Gamma(1\tfrac{1}{2}) = \frac{\sqrt{\pi}}{2}$

850.8. $\Gamma(n + \tfrac{1}{2}) = 1\cdot3\cdot5\cdots(2n - 3)(2n - 1)\sqrt{\pi}/2^n,$

[n an integer > 0]. [Ref. 10, p. 301.]

851.1. $\displaystyle\log \Gamma(1 + x) = -Cx + \frac{S_2 x^2}{2} - \frac{S_3 x^3}{3} + \frac{S_4 x^4}{4} - \cdots,$

[$x^2 < 1$],

where C is Euler's constant,

$$C = \lim_{p\to\infty} \left[-\log p + 1 + \frac{1}{2} + \frac{1}{3} + \cdots + \frac{1}{p}\right] = 0.577\,2157$$

and

$$S_p = 1 + \frac{1}{2^p} + \frac{1}{3^p} + \cdots = \zeta(p).$$

209

851.2. $\log \Gamma(1 + x) = \dfrac{1}{2} \log \dfrac{x\pi}{\sin x\pi} - Cx - \dfrac{S_3 x^3}{3} - \dfrac{S_5 x^5}{5} - \cdots.$

851.3. $\log \Gamma(1 + x) = \dfrac{1}{2} \log \dfrac{x\pi}{\sin x\pi} - \dfrac{1}{2} \log \dfrac{1 + x}{1 - x}$

$$+ (1 - C)x - (S_3 - 1)\dfrac{x^3}{3} - (S_5 - 1)\dfrac{x^5}{5} - \cdots.$$

Use **850.2** and **850.3** with these series for values of x greater than $^1/_2$.

[Ref. 7, par. 269–270 and Ref. 10, p. 303.]

851.4. $\Gamma(x + 1) \approx x^x e^{-x} \sqrt{(2\pi x)} \left[1 + \dfrac{1}{12x} + \dfrac{1}{288x^2} \right.$

$$\left. - \dfrac{139}{51,840x^3} - \dfrac{571}{2,488,320x^4} + \cdots \right],$$

where $\approx$ denotes approximate equality. This gives an asymptotic expression for $x!$ when x is a large integer.

[Ref. 44, vol. 1, p. 180.]   [See **11**.]

851.5. $\log \Gamma(x + 1) \approx \tfrac{1}{2} \log (2\pi) - x + (x + \tfrac{1}{2}) \log x$

$$+ \dfrac{B_1}{1 \cdot 2x} - \dfrac{B_2}{3 \cdot 4x^3} + \dfrac{B_3}{5 \cdot 6x^5} - \cdots.$$

[Stirling's series.]   [See **45** and **47.1**.]

This is an asymptotic series. The absolute value of the error is less than the absolute value of the first term neglected.

[Ref. 42, pp. 153–154.]

Note that $B_1 = 1/6$, $B_2 = 1/30$, $B_3 = 1/42$, etc., as in **45**.

852.1. $\displaystyle\int_0^\infty e^{-x} \log x \, dx = - C,$

where $C = 0.577\,2157$, as in **851.1**.

853.11. $\Pi(n) = \Gamma(n + 1).$ [See **850** and Table **1005**.]

$\Pi(n)$ is Gauss's function.

853.12. If n is a positive integer, $\Pi(n) = n!$.

853.13. $\Pi(0) = 1.$

853.21. $\int_0^1 x^{m-1}(1-x)^{n-1}dx = \dfrac{\Gamma(m)\,\Gamma(n)}{\Gamma(m+n)} = B(m,n)$, the Beta function,

$$[m, n > 0].$$

854.11. $\int_0^1 \dfrac{x^p dx}{1+x} = (-1)^p \left\{ \log 2 - 1 + \dfrac{1}{2} - \dfrac{1}{3} + \cdots + \dfrac{(-1)^p}{p} \right\},$

$$[p = 1, 2, \cdots].$$

854.12. $\int_0^1 \dfrac{x^{p-1}dx}{1+x^q} = \dfrac{1}{p} - \dfrac{1}{p+q} + \dfrac{1}{p+2q} - \dfrac{1}{p+3q} + \cdots$

$$[p, q > 0]. \quad \text{[See \textbf{35} and Ref. 34, p. 161.]}$$

854.21. $\int_0^1 \dfrac{dx}{1+x+x^2} = \dfrac{\pi}{3\sqrt{3}}.$

854.22. $\int_0^1 \dfrac{dx}{1-x+x^2} = \dfrac{2\pi}{3\sqrt{3}}.$

855.11. $\int_0^1 \dfrac{x^{p-1}dx}{(1-x)^p} = \dfrac{\pi}{\sin p\pi}$, $[0 < p < 1].$

855.12. $\int_0^1 \dfrac{x^p + x^{-p}}{1+x^2}\, dx = \dfrac{\pi}{2\cos\dfrac{p\pi}{2}}$, $[-1 < p < 1].$

$$\text{[Ref. 71, Art. 1101.]}$$

855.13. $\int_0^1 \dfrac{x^p + x^{-p}}{x^q + x^{-q}} \dfrac{dx}{x} = \dfrac{\pi}{2q\cos\left(\dfrac{p}{q}\dfrac{\pi}{2}\right)}$, $[-q < p < q].$

855.14. $\int_0^1 \dfrac{dx}{(1-x^q)^{1/q}} = \dfrac{\pi}{q\sin\dfrac{\pi}{q}}$, $[q > 1].$

855.15. $\int_0^1 \dfrac{x^{p-1}dx}{(1-x^q)^{p/q}} = \dfrac{\pi}{q\sin\dfrac{p\pi}{q}}$, $[0 < p < q].$

$$\text{[Ref. 71, Art. 1101, No. 6.]}$$

855.21. $\int_0^1 \dfrac{x^{m-1} + x^{n-1}}{(1+x)^{m+n}}\, dx = \dfrac{\Gamma(m)\Gamma(n)}{\Gamma(m+n)}$, $[m, n > 0].$

$$\text{[Ref. 6, Art. 122.]}$$

855.31. $\int_0^1 \dfrac{x^m dx}{\sqrt{(1-x^2)}} = \dfrac{2\cdot4\cdot6\,\cdots\,(m-1)}{3\cdot5\cdot7\,\cdots\,m},$

$$[m \text{ an odd integer} > 1],$$

$$= \dfrac{1\cdot3\cdot5\,\cdots\,(m-1)}{2\cdot4\cdot6\,\cdots\,m}\,\dfrac{\pi}{2},$$

$$[m \text{ an even, positive integer}],$$

$$= \dfrac{\sqrt{\pi}}{2}\,\dfrac{\Gamma\left(\dfrac{m+1}{2}\right)}{\Gamma\left(\dfrac{m}{2}+1\right)}, \qquad [m \text{ any value} > -1].$$

[Put sin $x = y$ in **858.44** or **858.45**.]

855.32. $\int_0^1 x^m \sqrt{(1-x^2)}\,dx = \dfrac{1}{m+2}\int_0^1 \dfrac{x^m dx}{\sqrt{(1-x^2)}},$

$$[m \text{ any value} > -1]. \quad [\text{See } \textbf{855.31.}]$$

855.33. $\int_0^1 \dfrac{dx}{\sqrt{(1-x^n)}} = \dfrac{\sqrt{\pi}}{n}\,\dfrac{\Gamma\left(\dfrac{1}{n}\right)}{\Gamma\left(\dfrac{1}{n}+\dfrac{1}{2}\right)}, \qquad [n > 0].$

855.34. $\int_0^1 \dfrac{x^m dx}{\sqrt{(1-x^n)}} = \dfrac{\sqrt{\pi}}{n}\,\dfrac{\Gamma\left(\dfrac{m+1}{n}\right)}{\Gamma\left(\dfrac{m+1}{n}+\dfrac{1}{2}\right)}, \qquad [m+1, n > 0].$

855.41. $\int_0^1 x^m(1-x^2)^p dx = \dfrac{\Gamma(p+1)\,\Gamma\left(\dfrac{m+1}{2}\right)}{2\,\Gamma\left(p+\dfrac{m+3}{2}\right)},$

$$[p+1, m+1 > 0]. \quad [\text{Ref. 7, Art. 272.}]$$

855.42. $\int_0^1 x^m(1-x^n)^p dx = \dfrac{\Gamma(p+1)\,\Gamma\left(\dfrac{m+1}{n}\right)}{n\Gamma\left(p+1+\dfrac{m+1}{n}\right)},$

$$[p+1, m+1, n > 0].$$

855.51. $\int_0^a x^{m-1}(a-x)^{n-1}dx = a^{m+n-1}\dfrac{\Gamma(m)\Gamma(n)}{\Gamma(m+n)},$ $[m, n > 0].$

856.01. $\int_0^\infty \dfrac{dx}{(1+x)x^q} = \dfrac{\pi}{\sin q\pi},$ $[0 < q < 1].$

Put $q = 1 - p$. Then,

856.02. $\int_0^\infty \dfrac{x^{p-1}dx}{1+x} = \dfrac{\pi}{\sin(\pi - p\pi)} = \dfrac{\pi}{\sin p\pi},$ $[0 < p < 1].$

[Ref. 71, Art. 871.]

856.03. $\int_0^\infty \dfrac{dx}{(1+x)\sqrt{x}} = \pi.$

856.04. $\int_0^\infty \dfrac{x^{p-1}dx}{a+x} = \dfrac{\pi\,a^{p-1}}{\sin p\pi},$ $[0 < p < 1].$

[Ref. 71, Art. 1101, eq. 11.]

856.05. $\int_0^\infty \dfrac{dx}{1+x^p} = \dfrac{\pi}{p \sin \dfrac{\pi}{p}},$ $[p > 1].$

856.06. $\int_0^\infty \dfrac{x^p\,dx}{(1+ax)^2} = \dfrac{p\pi}{a^{p+1}\sin p\pi},$ $[0 < p < 1].$

856.07. $\int_0^\infty \dfrac{x^p\,dx}{1+x^2} = \dfrac{\pi}{2\cos \dfrac{p\pi}{2}},$ $[-1 < p < 1].$

[Ref. 71, Art. 1101, eq. 4.]

856.08. $\int_0^\infty \dfrac{x^{p-1}\,dx}{1+x^q} = \dfrac{\pi}{q \sin \dfrac{p\pi}{q}},$ $[0 < p < q].$

[Ref. 71, Art. 1101, eq. 2.]

856.11. $\int_0^\infty \dfrac{x^{m-1}\,dx}{(1+x)^{m+n}} = \dfrac{\Gamma(m)\,\Gamma(n)}{\Gamma(m+n)},$ $[m, n > 0].$

856.12. $\int_0^\infty \dfrac{x^{m-1}\,dx}{(a+bx)^{m+n}} = \dfrac{\Gamma(m)\,\Gamma(n)}{a^n b^m\,\Gamma(m+n)},$ $[a, b, m, n > 0].$

[Ref. 6, Art. 122.]

856.21. $\int_0^\infty \dfrac{dx}{(a^2+x^2)^n} = \dfrac{1\cdot3\cdot5\,\cdots\,(2n-3)}{2\cdot4\cdot6\,\cdots\,(2n-2)}\dfrac{\pi}{2a^{2n-1}},$

$[a > 0; n = 2, 3\,\cdots].$

856.31. $\displaystyle\int_0^\infty \frac{dx}{(a^2 + x^2)(b^2 + x^2)} = \frac{\pi}{2ab(a + b)},$ $\qquad [a, b > 0].$

[Ref. 71, Art. 1042.]

856.32. $\displaystyle\int_0^\infty \frac{dx}{(a^2 + x^2)(a^n + x^n)} = \frac{\pi}{4a^{n+1}},$ $\qquad [a > 0].$

[Ref. 71, Art. 1092.]

856.33. $\displaystyle\int_0^\infty \left(\frac{1}{1 + x^2} - \frac{1}{1 + x}\right)\frac{dx}{x} = 0.$

857.01. $\displaystyle\int_0^\infty \frac{dx}{ax^2 + 2bx + c} = \frac{1}{\sqrt{(ac - b^2)}} \operatorname{ctn}^{-1}\frac{b}{\sqrt{(ac - b^2)}},$

$\qquad\qquad [a, ac - b^2 > 0].$

[Use principal values of ctn^{-1}, between 0 and π. See **500**.]

[Ref. 18, No. 67.]

857.02. $\displaystyle\int_0^\infty \frac{dx}{(ax^2 + 2bx + c)^{3/2}} = \frac{1}{b\sqrt{c} + c\sqrt{a}},$

$\qquad\qquad [a, c, b\sqrt{c} + c\sqrt{a} > 0].$

857.03. $\displaystyle\int_0^\infty \frac{x\,dx}{(ax^2 + 2bx + c)^{3/2}} = \frac{1}{a\sqrt{c} + b\sqrt{a}},$

$\qquad\qquad [a, c, a\sqrt{c} + b\sqrt{a} > 0].$

857.11. $\displaystyle\int_0^\infty \frac{dx}{ax^4 + 2bx^2 + c} = \frac{\pi}{2\sqrt{(ch)}},$

where $h = 2\{b + \sqrt{(ac)}\},$ $\qquad\qquad [a, c, h > 0].$

858.1. $\displaystyle\int_0^{\pi/2} \sin^2 x\,dx = \frac{\pi}{4}.$ $\qquad$ **858.2.** $\displaystyle\int_0^{\pi/2} \cos^2 x\,dx = \frac{\pi}{4}.$

858.3. $\displaystyle\int_0^\pi \sin^2 x\,dx = \int_0^\pi \cos^2 x\,dx = \frac{\pi}{2}.$

The average value of $\sin^2 x$ or $\cos^2 x$, taken over any multiple of $\pi/2$, is $1/2$. [Ref. 18, No. **61**.]

858.41. $\displaystyle\int_0^{\pi/2} \sin^2 mx\,dx = \int_0^{\pi/2} \cos^2 mx\,dx = \frac{\pi}{4},$ $\qquad [m = 1, 2, \cdots].$

858.42. $\displaystyle\int_0^\pi \sin^2 mx\,dx = \int_0^\pi \cos^2 m\,x dx = \frac{\pi}{2},$ $\qquad [m = 1, 2, \cdots].$

858.43. $\displaystyle\int_0^{2\pi} \sin^2 mx\, dx = \int_0^{2\pi} \cos^2 mx\, dx = \pi,$ $[m = 1, 2, \cdots].$

858.44. $\displaystyle\int_0^{\pi/2} \sin^p x\, dx = \int_0^{\pi/2} \cos^p x\, dx$

$$= \frac{2\cdot4\cdot6\cdots(p-1)}{3.5.7\cdots p},$$ $[p$ an odd integer $> 1],$

$$= \frac{1\cdot3\cdot5\cdots(p-1)}{2\cdot4\cdot6\cdots p}\frac{\pi}{2},$$

$[p$ an even, positive integer$],$

$$= \frac{\pi^{1/2}}{2}\frac{\Gamma\left(\dfrac{p+1}{2}\right)}{\Gamma\left(\dfrac{p}{2}+1\right)},$$

$[p$ any value $> -1].$

Alternatively, putting $m = 0$ in **858.502**, for the same numerical results as by **858.44**,

858.45. $\displaystyle\int_0^{\pi/2} \sin^p x\, dx = \int_0^{\pi/2} \cos^p x\, dx = \frac{\pi\Gamma(p+1)}{2^{p+1}\left\{\Gamma\left(\dfrac{p}{2}+1\right)\right\}^2},$

$[p > -1].$

858.46. $\displaystyle\int_0^{\pi} \sin^p x\, dx = \frac{\pi^{1/2}\,\Gamma\left(\dfrac{p+1}{2}\right)}{\Gamma\left(\dfrac{p}{2}+1\right)},$ $[p > -1],$

$$= 2\int_0^{\pi/2} \sin^p x\, dx.$$

[See **858.44**, or **858.45** may be used.]

858.47. $\displaystyle\int_0^{\pi} x \sin^p x\, dx = \frac{\pi^{3/2}}{2}\frac{\Gamma\left(\dfrac{p+1}{2}\right)}{\Gamma\left(\dfrac{p}{2}+1\right)},$ $[p > -1],$

$$= \pi\int_0^{\pi/2} \sin^p x\, dx.$$

[See **858.44**, or **858.45** may be used.]
[Ref. 71, Art. 992.]

858.48. $\displaystyle\int_0^{\pi/2} \tan^p x \, dx = \int_0^{\pi/2} \text{ctn}^p x \, dx = \frac{\pi}{2 \cos \dfrac{p\pi}{2}},$ $\qquad [p^2 < 1].$

858.491. $\displaystyle\int_0^{\pi/2} \frac{x \, dx}{\sin x} = 2\, G = 2(0.915\,9656).$ $\qquad$ [See **48.32.**]

[Ref. 69, vol. 2, pp. 2 and 123 (no. 32).]

858.492. $\displaystyle\int_0^{\pi/2} \frac{x \, dx}{\tan x} = \frac{\pi}{2} \log 2.$ $\qquad$ [Ref. 71, Art. 990, eq. 6.]

858.493. $\displaystyle\int_0^{\pi/2} \frac{x^2 \, dx}{\sin^2 x} = \pi \log 2.$ $\qquad$ [Ref. 71, Art. 990, eq. 7.]

858.501. $\displaystyle\int_0^{\pi/2} \cos^m x \cos mx \, dx = \frac{\pi}{2^{m+1}},$ $\qquad [m + 1 > 0].$

[Ref. 18, eq. 65.]

858.502. $\displaystyle\int_0^{\pi/2} \cos^p x \cos mx \, dx$

$$= \frac{\pi}{2^{p+1}} \frac{\Gamma(p + 1)}{\Gamma\left(\dfrac{p + m}{2} + 1\right) \Gamma\left(\dfrac{p - m}{2} + 1\right)},$$

$$[p + 1, \, p + m + 2, \, p - m + 2 > 0].$$

858.503. $\displaystyle\int_0^{\pi} \sin^m x \sin mx \, dx = \frac{\pi}{2^m} \sin \frac{m\pi}{2},$ $\qquad [m + 1 > 0].$

858.504. $\displaystyle\int_0^{\pi} \sin^p x \sin mx \, dx = \frac{\pi}{2^p} \frac{\sin \dfrac{m\pi}{2} \, \Gamma\left(p + 1\right)}{\Gamma\left(\dfrac{p + m}{2} + 1\right) \Gamma\left(\dfrac{p - m}{2} + 1\right)},$

$$[p + 1, \, p + m + 2, \, p - m + 2 > 0].$$

858.505. $\displaystyle\int_0^{\pi} \sin^m x \cos mx \, dx = \frac{\pi}{2^m} \cos \frac{m\pi}{2},$ $\qquad [m + 1 > 0].$

858.506. $\displaystyle\int_0^{\pi} \sin^p x \cos mx \, dx = \frac{\pi}{2^p} \frac{\cos \dfrac{m\pi}{2} \, \Gamma\left(p + 1\right)}{\Gamma\left(\dfrac{p + m}{2} + 1\right) \Gamma\left(\dfrac{p - m}{2} + 1\right)},$

$$[p + 1, \, p + m + 2, \, p - m + 2 > 0].$$

858.511. $\displaystyle\int_0^{\pi/2} \sin^{2a+1}x \cos^{2b+1} x \, dx = \frac{a!\,b!}{(a+b+1)!2}.$

858.512. $\displaystyle\int_0^{\pi/2} \sin^{2a+1}x \cos^{2b} x \, dx = \frac{2\cdot4\cdot6\cdots(2a)\cdot1\cdot3\cdot5\cdots(2b-1)}{1\cdot3\cdot5\cdots(2a+2b+1)}.$

858.513. $\displaystyle\int_0^{\pi/2} \sin^{2a}x \cos^{2b+1} x \, dx = \frac{1\cdot3\cdot5\cdots(2a-1)\cdot2\cdot4\cdot6\cdots(2b)}{1\cdot3\cdot5\cdots(2a+2b+1)}.$

858.514. $\displaystyle\int_0^{\pi/2} \sin^{2a}x \cos^{2b} x \, dx$

$$= \frac{\pi}{2}\frac{1\cdot3\cdot5\cdots(2a-1)\cdot1\cdot3\cdot5\cdots(2b-1)}{2\cdot4\cdot6\cdots(2a+2b)}.$$

In **858.511–.514**, a and b are integers, > 0.

858.515. $\displaystyle\int_0^{\pi/2} \sin^p x \cos^q x \, dx = \frac{\Gamma\left(\dfrac{p+1}{2}\right)\Gamma\left(\dfrac{q+1}{2}\right)}{2\Gamma\left(\dfrac{p+q}{2}+1\right)},$

[$p + 1, q + 1 > 0$, not necessarily integers.]   [Ref. 7, Art. 272.]

858.516. $\displaystyle\int_0^{\pi} \sin mx \sin nx \, dx = 0,$ $\qquad\qquad$ [$m \neq n$],

$\qquad\qquad\qquad\qquad\qquad = \dfrac{\pi}{2},$ $\qquad\qquad$ [$m = n$],

$\qquad\qquad\qquad\qquad\qquad\qquad\qquad$ [m and n, integers].

858.517. $\displaystyle\int_0^{\pi} \cos mx \cos nx \, dx = 0,$ $\qquad\qquad$ [$m \neq n$],

$\qquad\qquad\qquad\qquad\qquad = \dfrac{\pi}{2},$ $\qquad\qquad$ [$m = n$],

$\qquad\qquad\qquad\qquad\qquad\qquad\qquad$ [m and n, integers].

858.518. $\displaystyle\int_0^{\pi} \sin mx \cos nx \, dx = 0,$ $\qquad\qquad$ [$m = n$],

$\qquad\qquad\qquad\qquad\qquad = 0,$ $\qquad\qquad$ [$m \neq n$; $(m + n)$ even],

$\qquad\qquad\qquad\qquad\qquad = \dfrac{2m}{m^2 - n^2},$ $\qquad$ [$m \neq n$; $(m + n)$ odd],

$\qquad\qquad\qquad\qquad\qquad\qquad\qquad$ [m and n, integers].

[For **858.516–.518**, see Ref. 71, Art. 1121.]

858.520. $\displaystyle\int_0^{\pi/2} \frac{dx}{1 + a \sin x} = \int_0^{\pi/2} \frac{dx}{1 + a \cos x} = \dfrac{\dfrac{\pi}{2} - \sin^{-1} a}{\sqrt{(1 - a^2)}}$

$$= \frac{\cos^{-1} a}{\sqrt{(1 - a^2)}}.$$

In equations **858.520–.535**, $0 < a < 1$ and the angles $\sin^{-1} a$ and $\cos^{-1} a$ are in the first quadrant.

858.521. $\displaystyle\int_0^{\pi/2} \frac{dx}{1 - a \sin x} = \int_0^{\pi/2} \frac{dx}{1 - a \cos x} = \dfrac{\dfrac{\pi}{2} + \sin^{-1} a}{\sqrt{(1 - a^2)}}.$

[See note, **858.520.**]

858.522. $\displaystyle\int_0^{\pi} \frac{dx}{1 + a \sin x} = \frac{\pi - 2 \sin^{-1} a}{\sqrt{(1 - a^2)}} = \frac{2 \cos^{-1} a}{\sqrt{(1 - a^2)}}.$

[See note, **858.520.**]

858.523. $\displaystyle\int_0^{\pi} \frac{dx}{1 - a \sin x} = \frac{\pi + 2 \sin^{-1} a}{\sqrt{(1 - a^2)}}.$ [See note, **858.520.**]

858.524. $\displaystyle\int_0^{\pi} \frac{dx}{1 \pm a \cos x} = \frac{\pi}{\sqrt{(1 - a^2)}}.$ [See note, **858.520.**]

858.525. $\displaystyle\int_0^{2\pi} \frac{dx}{1 \pm a \sin x} = \int_0^{2\pi} \frac{dx}{1 \pm a \cos x} = \frac{2\pi}{\sqrt{(1 - a^2)}}.$

[See note, **858.520.**]

858.530. $\displaystyle\int_0^{\pi/2} \frac{dx}{(1 + a \sin x)^2} = \int_0^{\pi/2} \frac{dx}{(1 + a \cos x)^2}$

$$= \frac{\left(\dfrac{\pi}{2} - \sin^{-1} a\right)}{(1 - a^2)^{3/2}} - \frac{a}{1 - a^2}.$$

[See note, **858.520.**]

858.531. $\displaystyle\int_0^{\pi/2} \frac{dx}{(1 - a \sin x)^2} = \int_0^{\pi/2} \frac{dx}{(1 - a \cos x)^2}$

$$= \frac{\left(\dfrac{\pi}{2} + \sin^{-1} a\right)}{(1 - a^2)^{3/2}} + \frac{a}{1 - a^2}.$$

[See note, **858.520.**]

858.532. $\displaystyle\int_0^\pi \frac{dx}{(1 + a \sin x)^2} = \frac{\pi - 2 \sin^{-1} a}{(1 - a^2)^{3/2}} - \frac{2a}{1 - a^2}.$

[See note, **858.520.**]

858.533. $\displaystyle\int_0^\pi \frac{dx}{(1 - a \sin x)^2} = \frac{\pi + 2 \sin^{-1} a}{(1 - a^2)^{3/2}} + \frac{2a}{1 - a^2}.$

[See note, **858.520.**]

858.534. $\displaystyle\int_0^\pi \frac{dx}{(1 \pm a \cos x)^2} = \frac{\pi}{(1 - a^2)^{3/2}}.$ [See note, **858.520.**]

858.535. $\displaystyle\int_0^{2\pi} \frac{dx}{(1 \pm a \sin x)^2} = \int_0^{2\pi} \frac{dx}{(1 \pm a \cos x)^2} = \frac{2\pi}{(1 - a^2)^{3/2}}.$

[See note, **858.520.**]

858.536. $\displaystyle\int_0^\pi \frac{\cos mx \, dx}{1 + a \cos x} = \frac{\pi\{\sqrt{(1 - a^2)} - 1\}^m}{a^m \sqrt{(1 - a^2)}},$

$$[0 < a < 1; m = 0, 1, 2, \cdots].$$

858.537. $\displaystyle\int_0^{\pi/2} \frac{dx}{(a \sin x + b \cos x)^2} = \frac{1}{ab},$ $[a, b > 0].$

858.540. $\displaystyle\int_0^{\pi/2} \frac{dx}{1 + a^2 \sin^2 x} = \int_0^{\pi/2} \frac{dx}{1 + a^2 \cos^2 x} = \frac{\pi}{2\sqrt{(1 + a^2)}}.$

858.541. $\displaystyle\int_0^{\pi/2} \frac{dx}{1 - a^2 \sin^2 x} = \int_0^{\pi/2} \frac{dx}{1 - a^2 \cos^2 x} = \frac{\pi}{2\sqrt{(1 - a^2)}},$

$$[a^2 < 1].$$

858.542. $\displaystyle\int_0^{\pi/2} \frac{dx}{(1 + a^2 \sin^2 x)^2} = \int_0^{\pi/2} \frac{dx}{(1 + a^2 \cos^2 x)^2} = \frac{\pi}{4} \frac{(2 + a^2)}{(1 + a^2)^{3/2}}.$

858.543. $\displaystyle\int_0^{\pi/2} \frac{dx}{(1 - a^2 \sin^2 x)^2} = \int_0^{\pi/2} \frac{dx}{(1 - a^2 \cos^2 x)^2} = \frac{\pi}{4} \frac{(2 - a^2)}{(1 - a^2)^{3/2}},$

$$[a^2 < 1].$$

858.544. $\displaystyle\int_0^\pi \frac{x \sin mx \, dx}{1 + \cos^2 mx} = \frac{\pi^2}{4m^2},$ $[m > 0].$

858.545. $\displaystyle\int_0^\pi \frac{x \, dx}{1 + \cos \phi \sin x} = \frac{\pi\phi}{\sin \phi}.$ [Ref. 71, Art. 992, Ex. 2].

858.546. $\displaystyle\int_0^\pi \frac{\sin^2 x \, dx}{a + b \cos x} = \frac{\pi}{a + \sqrt{(a^2 - b^2)}},$ $[a, a^2 - b^2 > 0].$

858.550. $\displaystyle\int_0^{\pi/2} \frac{dx}{a^2 \sin^2 x + b^2 \cos^2 x} = \frac{\pi}{2ab},$ $[ab > 0]$.

858.551. $\displaystyle\int_0^{\pi} \frac{dx}{a^2 \sin^2 x + b^2 \cos^2 x} = \frac{\pi}{ab},$ $[ab > 0]$.

858.552. $\displaystyle\int_0^{\pi/2} \frac{\sin^2 x \, dx}{a^2 \sin^2 x + b^2 \cos^2 x} = \int_0^{\pi/2} \frac{dx}{a^2 + b^2 \operatorname{ctn}^2 x} = \frac{\pi}{2a(a+b)},$
$[a, b > 0]$.

858.553. $\displaystyle\int_0^{\pi/2} \frac{\cos^2 x \, dx}{a^2 \sin^2 x + b^2 \cos^2 x} = \int_0^{\pi/2} \frac{dx}{b^2 + a^2 \tan^2 x} = \frac{\pi}{2b(a+b)},$
$[a, b > 0]$.

858.554. $\displaystyle\int_0^{\pi/2} \frac{dx}{(a^2 \sin^2 x + b^2 \cos^2 x)^2} = \frac{\pi}{4} \frac{(a^2 + b^2)}{a^3 b^3},$ $[ab > 0]$.

858.555. $\displaystyle\int_0^{\pi/2} \frac{\sin^2 x \, dx}{(a^2 \sin^2 x + b^2 \cos^2 x)^2} = \frac{\pi}{4a^3 b},$ $[ab > 0]$.

858.556. $\displaystyle\int_0^{\pi/2} \frac{\cos^2 x \, dx}{(a^2 \sin^2 x + b^2 \cos^2 x)^2} = \frac{\pi}{4ab^3},$ $[ab > 0]$.

858.560. $\displaystyle\int_0^{\infty} \sin(a^2 x^2) \, dx = \int_0^{\infty} \cos(a^2 x^2) \, dx = \frac{\sqrt{\pi}}{2a\sqrt{2}},$ $[a > 0]$.
[Ref. 7, Art. 302.]

858.561. $\displaystyle\int_0^{\infty} \sin \frac{\pi x^2}{2} \, dx = \int_0^{\infty} \cos \frac{\pi x^2}{2} \, dx = \tfrac{1}{2}.$ [Fresnel's integrals.]

858.562. $\displaystyle\int_0^{\infty} \sin(x^p) \, dx = \Gamma\left(1 + \frac{1}{p}\right) \sin \frac{\pi}{2p},$ $[p > 1]$.

858.563. $\displaystyle\int_0^{\infty} \cos(x^p) \, dx = \Gamma\left(1 + \frac{1}{p}\right) \cos \frac{\pi}{2p},$. $[p > 1]$.

[For **858.562** and **.563**, see J. W. L. Glaisher, Quat. J. Math., vol. 13, 1875, p. 343.]

858.564. $\displaystyle\int_0^{\infty} \sin a^2 x^2 \cos mx \, dx = \frac{\sqrt{\pi}}{2a} \sin\left(\frac{\pi}{4} - \frac{m^2}{4a^2}\right),$ $[a > 0]$.

858.565. $\displaystyle\int_0^{\infty} \cos a^2 x^2 \cos mx \, dx = \frac{\sqrt{\pi}}{2a} \cos\left(\frac{\pi}{4} - \frac{m^2}{4a^2}\right),$ $[a > 0]$.

[For **858.564** and **.565**, see Ref. 71, p. 241, No. 49.]

858.601. $\int_0^\infty \dfrac{\sin mx}{x}\, dx = \dfrac{\pi}{2},\, 0,\ \text{or}\ -\dfrac{\pi}{2},\ \text{as}\ m\ \text{is positive, 0, or negative.}$

[Ref. 7, Art. 290.]

858.611. $\int_0^\infty \dfrac{\cos mx - \cos nx}{x}\, dx = \log \dfrac{n}{m},$ $\qquad\qquad [m, n > 0].$

[Ref. 71, Art. 1001, eq. 6.]

858.621. $\int_0^\infty \dfrac{\tan mx}{x}\, dx = \dfrac{\pi}{2},\, 0,\ \text{or}\ -\dfrac{\pi}{2},\ \text{as}\ m\ \text{is positive, 0, or negative.}$

[Ref. 71, Art. 1007.]

858.630. $\int_0^\infty \dfrac{\sin^2 mx}{x}\, dx = \int_0^\infty \dfrac{\cos^2 mx}{x}\, dx = \infty.$

858.631. $\int_0^\infty \dfrac{\sin^2 mx - \sin^2 nx}{x}\, dx = \dfrac{1}{2} \log \dfrac{m}{n},$ $\qquad [m, n \neq 0].$

858.632. $\int_0^\infty \dfrac{\cos^2 mx - \cos^2 nx}{x}\, dx = \dfrac{1}{2} \log \dfrac{n}{m},$ $\qquad [m, n \neq 0].$

858.641. $\int_0^\infty \dfrac{\sin^3 mx}{x}\, dx = \dfrac{\pi}{4},$ $\qquad [m > 0].$ $\qquad$ [Ref. 71, Art. 1023.]

858.649. $\int_0^\infty \dfrac{\sin^{2p+1} mx}{x}\, dx$

$$= \dfrac{1 \cdot 3 \cdot 5 \cdots (2p - 1)}{2 \cdot 4 \cdot 6 \cdots (2p)} \dfrac{\pi}{2}, \qquad [p = 1, 2, 3 \cdots ; m > 0].$$

[Ref. 71, Art. 1010.]

858.650. $\int_0^\infty \dfrac{\sin mx}{x^2}\, dx = \int_0^\infty \dfrac{\cos mx}{x^2}\, dx = \infty.$

[Ref. 16, Table 159, Nos. 1, 3.]

858.651. $\int_0^\infty \dfrac{\cos mx - \cos nx}{x^2}\, dx = (n - m) \dfrac{\pi}{2},$ $\qquad [n > m > 0].$

[Ref. 71, Art. 1020.]

858.652. $\int_0^\infty \dfrac{\sin^2 mx}{x^2}\, dx = |m| \dfrac{\pi}{2}.$ $\qquad$ [See **858.711.**]

858.653. $\int_0^\infty \dfrac{\sin^3 mx}{x^2}\, dx = \dfrac{3}{4} m \log 3.$ $\qquad$ [Ref. 71, Art. 1032.]

858.654. $\int_0^\infty \dfrac{\sin^4 mx}{x^2}\, dx = |m| \dfrac{\pi}{4}.$

858.659. $\int_0^\infty \dfrac{\sin^{2p} mx}{x^2}\, dx = \dfrac{1 \cdot 3 \cdot 5 \cdots (2p - 3)}{2 \cdot 4 \cdot 6 \cdots (2p - 2)} \dfrac{|m|\,\pi}{2},$

$$[p = 2, 3, 4, \ldots].$$
[Ref. 71, Art. 1026.]

858.661. $\int_0^\infty \dfrac{\sin^3 mx}{x^3}\, dx = \dfrac{3}{8} m^2\, \pi,$ \hspace{2cm} $[m > 0].$

858.701. $\int_0^\infty \dfrac{\sin mx \cos nx}{x}\, dx = \pi/2,$ \hspace{1.5cm} $[m > n > 0],$

$$= \pi/4, \hspace{2cm} [m = n > 0],$$
$$= 0, \hspace{2.2cm} [n > m > 0].$$
[Ref. 71, Art. 1011.]

858.702. $\int_0^\infty \dfrac{\sin mx \sin nx}{x}\, dx = \dfrac{1}{2} \log \dfrac{m + n}{m - n},$ \hspace{1cm} $[m > n > 0].$
[Ref. 71, Art. 1015.]

858.703. $\int_0^\infty \dfrac{\cos mx \cos nx}{x}\, dx = \infty.$ \hspace{1cm} [Ref. 71, Art. 1015.]

858.704. $\int_0^\infty \dfrac{\sin^2 ax \sin mx}{x}\, dx = \dfrac{\pi}{4},$ \hspace{1.5cm} $[2a > m > 0],$

$$= \dfrac{\pi}{8}, \hspace{2cm} [2a = m > 0],$$
$$= 0, \hspace{2.2cm} [m > 2a > 0].$$
[Ref. 67, p. 78, No. 8.]

858.711. $\int_0^\infty \dfrac{\sin mx \sin nx}{x^2}\, dx = \dfrac{\pi m}{2},$ \hspace{1.5cm} $[n \geqq m > 0],$

$$= \dfrac{\pi n}{2}, \hspace{2cm} [m \geqq n > 0].$$
[Ref. 71, Art. 1020.]

858.712. $\int_0^\infty \dfrac{\sin^2 ax \sin mx}{x^2}\, dx$

$$= \dfrac{m + 2a}{4} \log |m + 2a| + \dfrac{m - 2a}{4} \log |m - 2a| - \dfrac{m}{2} \log m,$$
$$[m > 0].$$
[Ref. 67, p. 78, No. 9.]

858.713. $\int_0^\infty \dfrac{\sin^2 ax \cos mx}{x^2}\, dx = \dfrac{\pi}{2}\left(a - \dfrac{m}{2}\right),$ $\qquad\left[a > \dfrac{m}{2} > 0\right],$

$$= 0, \qquad\qquad\qquad \left[\dfrac{m}{2} \geqq a \geqq 0\right].$$

[Ref. 73, vol. 2, p. 207, eq. 63.]

858.721. $\int_0^\infty \dfrac{1 - \cos mx}{x^2}\, dx = \dfrac{\pi |m|}{2}.$ $\qquad$ [Ref. 71, Art. 1016.]

858.731. $\int_0^\infty \dfrac{\sin^2 ax \sin mx}{x^3}\, dx = \dfrac{\pi\, am}{2} - \dfrac{\pi\, m^2}{8},$ $\qquad\left[a \geqq \dfrac{m}{2} > 0\right],$

$$= \dfrac{\pi\, a^2}{2}, \qquad\qquad \left[\dfrac{m}{2} \geqq a > 0\right].$$

858.801. $\int_0^\infty \dfrac{\sin mx}{\sqrt{x}}\, dx = \int_0^\infty \dfrac{\cos mx}{\sqrt{x}}\, dx = \dfrac{\sqrt{\pi}}{\sqrt{(2m)}},$ $\qquad [m > 0].$

[Ref. 7, Art. 302.]

858.802. $\int_0^\infty \dfrac{\sin mx}{x\sqrt{x}}\, dx = \sqrt{(2\pi m)},$ $\qquad\qquad [m > 0].$

[Ref. 67, p. 64, No. 9.]

858.811. $\int_0^\infty \dfrac{\sin mx}{x^p}\, dx = \dfrac{\pi m^{p-1}}{2 \sin\left(\dfrac{p\pi}{2}\right)\Gamma(p)},$ $\qquad [0 < p < 2; m > 0].$

[Ref. 75, p. 260.]

Alternatively, for the same numerical results as by the preceding,

858.812. $\int_0^\infty x^{q-1} \sin mx\, dx = \dfrac{\Gamma(q)}{m^q} \sin \dfrac{q\pi}{2},$

$$[- 1 < q < 0; 0 < q < 1; m > 0].$$

For $q = 0$ or near it, use **858.601** or **858.811**.

858.813. $\int_0^\infty \dfrac{\cos mx}{x^p}\, dx = \dfrac{\pi m^{p-1}}{2 \cos\left(\dfrac{p\pi}{2}\right)\Gamma(p)},$ $\qquad [0 < p < 1; m > 0].$

[Ref. 75, p. 260.]

Alternatively, for the same numerical results as by the preceding,

858.814. $\int_0^\infty x^{q-1} \cos mx\, dx = \dfrac{\Gamma(q)}{m^q} \cos \dfrac{q\pi}{2},$ $\qquad [0 < q < 1; m > 0].$

858.821. $\int_0^\infty \dfrac{\sin mx \cos nx}{\sqrt{x}} \, dx$

$$= \left\{ \dfrac{1}{\sqrt{(m+n)}} + \dfrac{1}{\sqrt{(m-n)}} \right\} \dfrac{\sqrt{\pi}}{2\sqrt{2}}, \qquad [m > n > 0],$$

$$= \left\{ \dfrac{1}{\sqrt{(m+n)}} - \dfrac{1}{\sqrt{(n-m)}} \right\} \dfrac{\sqrt{\pi}}{2\sqrt{2}}, \qquad [n > m > 0].$$

858.822. $\int_0^\infty \dfrac{\sin^2 mx}{x\sqrt{x}} \, dx = \sqrt{(m\pi)}, \qquad\qquad\qquad [m > 0].$

858.823. $\int_0^\infty \dfrac{\sin^3 mx}{\sqrt{x}} \, dx = \dfrac{(3\sqrt{3}-1)}{4} \sqrt{\left(\dfrac{\pi}{6m}\right)}, \qquad\qquad [m > 0].$

858.824. $\int_0^\infty \dfrac{\sin^3 mx}{x\sqrt{x}} \, dx = \dfrac{(3 - \sqrt{3})}{4} \sqrt{(2\,m\,\pi)}, \qquad\qquad [m > 0].$

858.831. $\int_0^\infty \dfrac{\tan^{-1} x \, dx}{(1+x)\sqrt{x}} = \dfrac{\pi^2}{4}.$

858.832. $\int_0^\infty \tan^{-1}\dfrac{a}{x} \sin mx \, dx = \dfrac{\pi}{2m}(1 - e^{-am}), \qquad [a, m > 0].$

[Ref. 67, p. 87, No. 5.]

858.841. $\int_0^\infty \dfrac{\sin x \, dx}{x\sqrt{(1 - k^2 \sin^2 x)}} = K(k), \qquad\qquad [0 < k < 1].$

858.842. $\int_0^\infty \dfrac{\sin x \, dx}{x\sqrt{(1 - k^2 \cos^2 x)}} = K(k), \qquad\qquad [0 < k < 1].$

858.843. $\int_0^\infty \dfrac{\sin x \cos x \, dx}{x\sqrt{(1 - k^2 \sin^2 x)}}$

$$= \dfrac{1}{k^2}\{E(k) - (1 - k^2)K(k)\}, \qquad [0 < k < 1].$$

For **858.841–.843** see **773.1, 774.1,** and Tables 1040, 1041.

859.001. $\int_0^\infty \dfrac{\cos mx}{a^2 + x^2} \, dx = \dfrac{\pi}{2a} e^{-ma}, \qquad\qquad [a > 0; m \geqq 0].$

[Ref. 71, Art. 1048.]

859.002. $\int_0^\infty \dfrac{\sin^2 mx}{a^2 + x^2} \, dx = \dfrac{\pi}{4a}(1 - e^{-2ma}), \qquad\qquad [a > 0; m \geqq 0].$

859.003. $\int_0^\infty \dfrac{\cos^2 mx}{a^2 + x^2}\, dx = \dfrac{\pi}{4a}\, (1 + e^{-2ma}),$ $\qquad\qquad [a > 0; m \geqq 0].$

859.004. $\int_0^\infty \dfrac{x \sin mx}{a^2 + x^2}\, dx = \dfrac{\pi}{2}\, e^{-ma},$ $\qquad\qquad\qquad [a \geqq 0; m > 0].$

$\qquad\qquad\qquad\qquad\qquad\qquad\qquad\qquad\qquad$ [Ref. 71, Art. 1050.]

859.005. $\int_0^\infty \dfrac{\sin mx}{x(a^2 + x^2)}\, dx = \dfrac{\pi}{2a^2}\, (1 - e^{-ma}),$ $\qquad [a > 0; m \geqq 0].$

$\qquad\qquad\qquad\qquad\qquad\qquad\qquad\qquad\qquad$ [Ref. 71, Art. 1051.]

859.006. $\int_0^\infty \dfrac{\sin mx \sin nx}{a^2 + x^2}\, dx = \dfrac{\pi}{2a}\, e^{-ma} \sinh na,$

$\qquad\qquad\qquad\qquad\qquad\qquad\qquad [a > 0; m \geqq n \geqq 0],$

$\qquad\qquad\qquad\qquad\quad = \dfrac{\pi}{2a}\, e^{-na} \sinh ma,$

$\qquad\qquad\qquad\qquad\qquad\qquad\qquad [a > 0; n \geqq m \geqq 0].$

859.007. $\int_0^\infty \dfrac{\cos mx \cos nx}{a^2 + x^2}\, dx = \dfrac{\pi}{2a}\, e^{-ma} \cosh na,$

$\qquad\qquad\qquad\qquad\qquad\qquad\qquad [a > 0; m \geqq n \geqq 0],$

$\qquad\qquad\qquad\qquad\quad = \dfrac{\pi}{2a}\, e^{-na} \cosh ma,$

$\qquad\qquad\qquad\qquad\qquad\qquad\qquad [a > 0; n \geqq m \geqq 0].$

859.008. $\int_0^\infty \dfrac{x \sin mx \cos nx}{a^2 + x^2}\, dx = \dfrac{\pi}{2}\, e^{-ma} \cosh na,$

$\qquad\qquad\qquad\qquad\qquad\qquad\qquad [a > 0; m > n > 0],$

$\qquad\qquad\qquad\qquad\quad = -\dfrac{\pi}{2}\, e^{-na} \sinh ma,$

$\qquad\qquad\qquad\qquad\qquad\qquad\qquad [a > 0; n > m > 0].$

859.011. $\int_0^\infty \dfrac{\cos mx}{(a^2 + x^2)^2}\, dx = \dfrac{\pi}{4a^3}\, (1 + ma)\, e^{-ma},$ $\qquad [a, m > 0].$

859.012. $\int_0^\infty \dfrac{x \sin mx}{(a^2 + x^2)^2}\, dx = \dfrac{\pi m}{4a}\, e^{-ma},$ $\qquad\qquad\qquad [a, m > 0].$

859.013. $\int_0^\infty \dfrac{x^2 \cos mx}{(a^2 + x^2)^2}\, dx = \dfrac{\pi}{4a}\, (1 - ma)\, e^{-ma},$ $\qquad [a, m > 0].$

$\qquad\qquad\qquad$ [For **859.011, .012** and **.013**, see Ref. 74, p. 221.]

859.014. $\int_0^\infty \dfrac{\sin mx}{x(a^2 + x^2)^2} \, dx = \dfrac{\pi}{2a^4}\left(1 - \dfrac{2 + ma}{2} e^{-ma}\right),$ $\quad [a, m > 0]$.

[Ref. 74, p. 225, No. 9.]

859.021. $\int_0^\infty \dfrac{\cos mx \, dx}{(a^2 + x^2)(b^2 + x^2)} = \left(\dfrac{e^{-mb}}{b} - \dfrac{e^{-ma}}{a}\right)\dfrac{\pi}{2(a^2 - b^2)},$

$[a, b, m > 0; a \neq b]$. [Ref. 74, p. 220, No. 3.]

859.022. $\int_0^\infty \dfrac{x \sin mx \, dx}{(a^2 + x^2)(b^2 + x^2)} = \left(\dfrac{e^{-mb} - e^{-ma}}{a^2 - b^2}\right)\dfrac{\pi}{2},$

$[a, b, m > 0; a \neq b]$. [Ref. 74, p. 221, No. 4.]

859.031. $\int_0^\infty \dfrac{\cos mx}{x^4 + 4a^4} \, dx = \dfrac{\pi e^{-ma}}{8a^3}(\sin ma + \cos ma)$.

859.032. $\int_0^\infty \dfrac{\sin mx}{x(x^4 + 4a^4)} \, dx = \dfrac{\pi}{8a^4}(1 - e^{-ma}\cos ma)$.

859.033. $\int_0^\infty \dfrac{x \sin mx}{x^4 + 4a^4} \, dx = \dfrac{\pi}{4a^2} e^{-ma} \sin ma$.

859.034. $\int_0^\infty \dfrac{x^2 \cos mx}{x^4 + 4a^4} \, dx = \dfrac{\pi}{4a} e^{-ma}(\cos ma - \sin ma)$.

859.035. $\int_0^\infty \dfrac{x^3 \sin mx}{x^4 + 4a^4} \, dx = \dfrac{\pi}{2} e^{-ma} \cos ma$.

[For **859.031–.035**, $a, m > 0$. See Ref. 71, Art. 1059.]

859.041. $\int_0^\infty \dfrac{\cos mx}{\sqrt{(a^2 + x^2)}} \, dx = K_0(ma)$, as in **815.1**, $\quad [ma > 0]$.

[Ref. 13, p. 183. For numerical values, see Refs. 50, 12, and 45.]

859.042. $\int_0^1 \dfrac{\cos mx}{\sqrt{(1 - x^2)}} \, dx = \dfrac{\pi}{2} J_0(m)$.

[For numerical values, see Refs. 50, 63, 17 and 45.]

859.043. $\int_0^\pi \dfrac{dx}{\sqrt{(a \pm b \cos x)}} = \dfrac{2}{\sqrt{(a + b)}} K\left(\sqrt{\dfrac{2b}{a + b}}\right),$

$[0 < b < a]$.

$\left[\dfrac{2b}{a + b} = k^2 = \sin^2 \theta. \text{ See } \mathbf{773.1} \text{ and Table 1040.}\right]$

859.100. $\displaystyle\int_0^{\pi/2} \frac{dx}{1 + 2a \sin x + a^2} = \int_0^{\pi/2} \frac{dx}{1 + 2a \cos x + a^2}$

$$= \frac{\dfrac{\pi}{2} - \sin^{-1}\dfrac{2a}{1+a^2}}{|1-a^2|} = \frac{\cos^{-1}\dfrac{2a}{1+a^2}}{|1-a^2|}.$$

[See note, **859.112.**]

859.101. $\displaystyle\int_0^{\pi/2} \frac{dx}{1 - 2a \sin x + a^2} = \int_0^{\pi/2} \frac{dx}{1 - 2a \cos x + a^2}$

$$= \frac{\dfrac{\pi}{2} + \sin^{-1}\dfrac{2a}{1+a^2}}{|1-a^2|}.$$

[See note, **859.112.**]

859.111. $\displaystyle\int_0^{\pi} \frac{dx}{1 + 2a \sin x + a^2} = \frac{\pi - 2\sin^{-1}\dfrac{2a}{1+a^2}}{|1-a^2|}$

$$= \frac{2\cos^{-1}\dfrac{2a}{1+a^2}}{|1-a^2|}.$$

[See note, **859.112.**]

859.112. $\displaystyle\int_0^{\pi} \frac{dx}{1 - 2a \sin x + a^2} = \frac{\pi + 2\sin^{-1}\dfrac{2a}{1+a^2}}{|1-a^2|}.$

For **859.100–.112,** $a > 0$; $a \neq 1$; $\sin^{-1}$ and $\cos^{-1}$ denote angles in the first quadrant.

859.113. $\displaystyle\int_0^{\pi} \frac{dx}{a^2 \pm 2ab \cos x + b^2} = \frac{\pi}{|a^2 - b^2|},$ $\qquad [a^2 \neq b^2].$

[Ref. 7, Art. 46.]

859.121. $\displaystyle\int_0^{\pi} \frac{\sin x \, dx}{1 - 2a \cos x + a^2} = \frac{2}{a}\tanh^{-1}a = \frac{1}{a}\log\frac{1+a}{1-a},$ $\quad [a^2 < 1],$

$$= \frac{2}{a}\coth^{-1}a = \frac{1}{a}\log\frac{a+1}{a-1}, \quad [a^2 > 1].$$

[Ref. 71, Art. 1151.]

859.122. $\displaystyle\int_0^\pi \frac{\cos mx\, dx}{1 - 2a \cos x + a^2} = \frac{\pi a^m}{1 - a^2},$ $[a^2 < 1; m = 0, 1, 2, \cdots],$

$$= \frac{\pi}{a^m(a^2 - 1)},$$

$$[a^2 > 1; m = 0, 1, 2, \cdots]. \qquad \text{[Ref. 7, Art. 296.]}$$

859.123. $\displaystyle\int_0^\pi \frac{x \sin x\, dx}{1 - 2a \cos x + a^2} = \frac{\pi}{a} \log (1 + a),$ $\qquad\qquad [a^2 \leqq 1],$

$$= \frac{\pi}{a} \log \left(1 + \frac{1}{a}\right), \qquad\qquad [a^2 \geqq 1].$$

859.124. $\displaystyle\int_0^\pi \frac{(a - b \cos x)\, dx}{a^2 - 2ab \cos x + b^2} = \frac{\pi}{a},$ $\qquad\qquad [a > b > 0],$

$$= 0, \qquad\qquad [b > a > 0].$$

$$\text{[Ref. 7, Art. 46.]}$$

859.131. $\displaystyle\int_0^\pi \frac{\sin^2 x\, dx}{a^2 - 2ab \cos x + b^2} = \frac{\pi}{2a^2},$ $\qquad\qquad [a > b > 0],$

$$= \frac{\pi}{2b^2}, \qquad\qquad [b > a > 0].$$

859.132. $\displaystyle\int_0^\pi \frac{\cos^2 x\, dx}{1 - 2a \cos x + a^2} = \frac{\pi}{2} \left(\frac{1 + a^2}{1 - a^2}\right),$ $\qquad\qquad [a^2 < 1],$

$$= \frac{\pi}{2a^2}\left(\frac{a^2 + 1}{a^2 - 1}\right), \qquad\qquad [a^2 > 1].$$

859.141. $\displaystyle\int_0^\pi \frac{\sin x \sin mx\, dx}{1 - 2a \cos x + a^2} = \frac{\pi a^{m-1}}{2},$ $[a^2 < 1; m = 1, 2, 3, \cdots],$

$$= \frac{\pi}{2a^{m+1}}, \quad [a^2 > 1; m = 1, 2, 3, \cdots].$$

859.142. $\displaystyle\int_0^\pi \frac{\cos x \cos mx\, dx}{1 - 2a \cos x + a^2} = \frac{\pi a^{m-1}}{2} \left(\frac{1 + a^2}{1 - a^2}\right),$

$$[a^2 < 1; m = 1, 2, 3, \cdots],$$

$$= \frac{\pi}{2a^{m+1}} \left(\frac{a^2 + 1}{a^2 - 1}\right),$$

$$[a^2 > 1; m = 1, 2, 3, \cdots].$$

859.151. $\int_0^\pi \dfrac{\sin x \, dx}{\sqrt{(a^2 - 2ab \cos x + b^2)}} = \dfrac{2}{a},$ $[a > b > 0],$

$$= \dfrac{2}{b},$$ $[b > a > 0].$

859.161. $\int_0^1 \dfrac{dx}{1 + 2x \cos \phi + x^2} = \dfrac{\phi}{2 \sin \phi},$ $[-\pi < \phi < \pi].$

When $\phi = 0,$

859.162. $\int_0^1 \dfrac{dx}{1 + 2x + x^2} = \dfrac{1}{2}.$ [See **90.2.**]

859.163. $\int_0^\infty \dfrac{dx}{1 + 2x \cos \phi + x^2} = \dfrac{\phi}{\sin \phi},$ $[-\pi < \phi < \pi].$

When $\phi = 0,$

859.164. $\int_0^\infty \dfrac{dx}{1 + 2x + x^2} = 1.$ [See **90.2.**]

859.165. $\int_0^\infty \dfrac{dx}{1 - 2x^2 \cos \phi + x^4} = \dfrac{\pi}{4 \sin \dfrac{\phi}{2}},$ $[0 < \phi < \pi].$

859.166. $\int_0^\infty \dfrac{x^p dx}{1 + 2x \cos \phi + x^2} = \dfrac{\pi \sin p\phi}{\sin p\pi \sin \phi},$

$[0 < p < 1; 0 < \phi < \pi].$

860.01. $\int_0^\infty e^{-ax} \, dx = \dfrac{1}{a},$ $[a > 0].$

860.02. $\int_0^\infty x \, e^{-ax} \, dx = \dfrac{1}{a^2},$ $[a > 0].$

860.03. $\int_0^\infty x^2 \, e^{-ax} \, dx = \dfrac{2}{a^3},$ $[a > 0].$

860.04. $\int_0^\infty x^{1/2} \, e^{-ax} \, dx = \dfrac{\sqrt{\pi}}{2a\sqrt{a}},$ $[a > 0].$

860.05. $\int_0^\infty x^{-1/2} \, e^{-ax} \, dx = \dfrac{\sqrt{\pi}}{\sqrt{a}},$ $[a > 0].$

860.06. $\int_0^\infty x^{p-\frac{1}{2}} \, e^{-ax} \, dx = \dfrac{1 \cdot 3 \cdot 5 \cdots (2p - 1)}{2^p} \dfrac{\sqrt{\pi}}{a^{p+\frac{1}{2}}},$

$[a > 0; p = 1, 2, \cdots].$ [See **860.07**, 2nd solution.]

860.07. $\displaystyle\int_0^\infty x^n e^{-ax} dx = \frac{n!}{a^{n+1}},$ $\qquad [a > 0; n = 1, 2, \cdots],$

$\qquad\qquad\qquad = \dfrac{\Gamma(n+1)}{a^{n+1}},$ $\qquad [a > 0; n + 1 > 0].$

[Ref. 7, Art. 260.]

860.11. $\displaystyle\int_0^\infty e^{-r^2 x^2} dx = \frac{\sqrt{\pi}}{2r},$ $\qquad [r > 0].$ [Ref. 7, Art. 272.]

860.12. $\displaystyle\int_0^\infty x e^{-r^2 x^2} dx = \frac{1}{2r^2}.$

860.13. $\displaystyle\int_0^\infty x^2 e^{-r^2 x^2} dx = \frac{\sqrt{\pi}}{4r^3},$ $\qquad\qquad\qquad\qquad [r > 0].$

860.15. $\displaystyle\int_0^\infty x^{2a+1} e^{-r^2 x^2} dx = \frac{a!}{2r^{2a+2}},$ $\qquad [r > 0; a = 1, 2, \cdots].$

860.16. $\displaystyle\int_0^\infty x^{2a} e^{-r^2 x^2} dx = \frac{1 \cdot 3 \cdot 5 \cdots (2a-1)}{2^{a+1} r^{2a+1}} \sqrt{\pi},$

$\qquad\qquad\qquad\qquad\qquad [r > 0; a = 1, 2, \cdots].$

860.17. $\displaystyle\int_0^\infty x^n e^{-r^2 x^2} dx = \frac{\Gamma\left(\dfrac{n+1}{2}\right)}{2r^{n+1}},$ $\qquad [n + 1, r > 0],$

(putting $m = 2$ in **860.19.**)

860.18. $\displaystyle\int_0^\infty e^{-(rx)^m} dx = \frac{1}{mr} \Gamma\left(\frac{1}{m}\right),$ $\qquad [r, m > 0],$

(Putting $n=0$ in **860.19.**)

860.19. $\displaystyle\int_0^\infty x^n e^{-(rx)^m} dx = \frac{1}{mr^{n+1}} \Gamma\left(\frac{n+1}{m}\right),$ $\qquad [n + 1, r, m > 0].$

[Ref. 39, Sec. 100.]

By assigning values to n, r, and m, the preceding equations in **860** can be obtained.

860.21. $\displaystyle\int_0^\infty \frac{e^{-ax}}{x} dx = \infty.$ $\qquad\qquad$ [Ref. 7, Art. 288.]

860.22. $\displaystyle\int_0^\infty \frac{e^{-ax} - e^{-bx}}{x} dx = \log\frac{b}{a},$ $\qquad\qquad [a, b > 0].$

[Ref. 7, Art. 288.]

860.23. $\displaystyle\int_0^\infty \frac{e^{-ax^c} - e^{-bx^c}}{x}\, dx = \frac{1}{c}\log\frac{b}{a},$ $\qquad\qquad [a, b, c > 0].$

860.24. $\displaystyle\int_0^\infty \frac{1 - e^{-ax^2}}{x^2}\, dx = \sqrt{(a\pi)},$ $\qquad\qquad [a > 0].$

860.25. $\displaystyle\int_0^\infty e^{-a^2x^2 - \frac{b^2}{x^2}}\, dx = \frac{\sqrt{\pi}}{2a}\, e^{-2ab},$ $\qquad\qquad [a, b > 0].$

860.30. $\displaystyle\int_0^\infty \frac{dx}{e^{ax} - 1} = \infty,$ $\qquad\qquad [a > 0].$

860.31. $\displaystyle\int_0^\infty \frac{x\, dx}{e^{ax} - 1} = \frac{\pi^2}{6a^2},$ $\qquad\qquad [a > 0].$

860.32. $\displaystyle\int_0^\infty \frac{x^2\, dx}{e^{ax} - 1} = \frac{2\zeta(3)}{a^3}, \; = \frac{2(1.202057)}{a^3},$ $\qquad\qquad [a > 0].$
$\qquad\qquad\qquad\qquad\qquad\qquad\qquad\qquad\qquad$ [See **48.003.**]

860.33. $\displaystyle\int_0^\infty \frac{x^3\, dx}{e^{ax} - 1} = \frac{\pi^4}{15a^4},$ $\qquad\qquad [a > 0].$

860.37. $\displaystyle\int_0^\infty \frac{x^{2n-1}\, dx}{e^{ax} - 1} = \frac{(2n-1)!}{a^{2n}}\, \zeta(2n) = \frac{2^{2n-2}\, \pi^{2n}}{na^{2n}}\, B_n,$
$\qquad\qquad\qquad\quad [a > 0; n = 1, 2, \cdots].$ $\qquad$ [See **45.**]

860.38. $\displaystyle\int_0^\infty \frac{x^{2n}\, dx}{e^{ax} - 1}.$ $\qquad$ Use **860.39.**

860.39. $\displaystyle\int_0^\infty \frac{x^{p-1}\, dx}{e^{ax} - 1} = \frac{\Gamma(p)}{a^p}\left[1 + \frac{1}{2^p} + \frac{1}{3^p} + \cdots\right] = \frac{\Gamma(p)}{a^p}\, \zeta(p),$
$\qquad\quad [a, p > 0; p \text{ not necessarily an integer}].$ $\qquad$ [See **48.09.**]

For numerical values of $\zeta(p)$, including decimal values of p, see table of the Riemann Zeta function, Ref. 45.

860.40. $\displaystyle\int_0^\infty \frac{dx}{e^{ax} + 1} = \frac{\log 2}{a},$ $\qquad\qquad [a > 0].$ $\quad$ [See **601.01.**]

860.41. $\displaystyle\int_0^\infty \frac{x\, dx}{e^{ax} + 1} = \frac{\pi^2}{12a^2},$ $\qquad\qquad [a > 0].$ $\quad$ [See **48.22.**]

860.42. $\displaystyle\int_0^\infty \frac{x^2\, dx}{e^{ax} + 1} = \frac{2!}{a^3}\frac{3}{4}\, \zeta(3) = \frac{3(1.202\,057)}{2a^3},$ $\qquad\qquad [a > 0].$

$\qquad\qquad\qquad\qquad\qquad\qquad\qquad\qquad\qquad$ [See **48.003** and **48.23.**]

860.43. $\int_0^\infty \dfrac{x^3\,dx}{e^{ax}+1} = \dfrac{7}{120}\dfrac{\pi^4}{a^4},$ $\qquad\qquad\qquad\qquad$ $[a > 0].$

860.47. $\int_0^\infty \dfrac{x^{2n-1}\,dx}{e^{ax}+1} = \dfrac{2^{2n-1}-1}{2n}\dfrac{\pi^{2n}}{a^{2n}}\,B_n,$ $\qquad$ $[a > 0; n = 1, 2, \cdots].$

$\qquad\qquad\qquad\qquad\qquad\qquad\qquad\qquad\qquad$ [See **45** and **48.28**.]

860.48. $\int_0^\infty \dfrac{x^{2n}\,dx}{e^{ax}+1}\cdot$ $\qquad$ Use **860.49**.

860.49. $\int_0^\infty \dfrac{x^{p-1}\,dx}{e^{ax}+1} = \dfrac{\Gamma(p)}{a^p}\left(1 - \dfrac{2}{2^p}\right)\zeta(p),$

$\qquad\qquad\qquad\qquad$ $[a > 0; p > 1; p$ not necessarily an integer].

$\qquad$ [See **48.29** and Ref. 45.] $\qquad$ [For $p = 1$, see **860.40**.]

860.500. $\int_0^\infty \dfrac{dx}{\sinh ax} = \int_0^\infty \dfrac{2\,dx}{e^{ax}-e^{-ax}} = \infty.$

860.501. $\int_0^\infty \dfrac{x\,dx}{\sinh ax} = \dfrac{\pi^2}{4a^2},$ $\qquad\qquad\qquad\qquad\qquad$ $[a > 0].$

860.502. $\int_0^\infty \dfrac{x^2\,dx}{\sinh ax} = 2(2!)\left(1 + \dfrac{1}{3^3} + \dfrac{1}{5^3} + \dfrac{1}{7^3} + \cdots\right)\Big/a^3,$

$\qquad\qquad = \dfrac{7}{2a^3}\,\zeta(3), = 4(1.05180)/a^3.$ $\qquad$ [See **48.13**.]

860.503. $\int_0^\infty \dfrac{x^3\,dx}{\sinh ax} = \dfrac{\pi^4}{8a^4},$ $\qquad\qquad\qquad\qquad\qquad$ $[a > 0].$

860.504. $\int_0^\infty \dfrac{x^4\,dx}{\sinh ax} = \dfrac{93}{2}\,(1.03693)/a^5.$

860.507. $\int_0^\infty \dfrac{x^{2n-1}\,dx}{\sinh ax} = \dfrac{\pi^{2n}(2^{2n}-1)}{2n\,a^{2n}}\,B_n,$ $\qquad$ $[a > 0; n = 1, 2, \cdots].$

$\qquad\qquad\qquad\qquad\qquad\qquad\qquad\qquad\qquad$ [See **45** and **48.18**.]

860.508. $\int_0^\infty \dfrac{x^{2n}\,dx}{\sinh ax}\cdot$ $\qquad$ Use **860.509**.

860.509. $\int_0^\infty \dfrac{x^{p-1}\,dx}{\sinh ax} = \dfrac{2\Gamma(p)}{a^p}\left(1 - \dfrac{1}{2^p}\right)\zeta(p),$

$\qquad\qquad\qquad\qquad$ $[a > 0; p > 1; p$ not necessarily an integer].

$\qquad\qquad\qquad\qquad\qquad\qquad$ [See **48.19** and Ref. 45.]

860.511. $\int_0^\infty \dfrac{x \, dx}{\sinh^2 ax} = \infty.$

860.512. $\int_0^\infty \dfrac{x^2 \, dx}{\sinh^2 ax} = \dfrac{\pi^2}{6a^3},$ $\qquad\qquad [a > 0].$

860.513. $\int_0^\infty \dfrac{x^3 \, dx}{\sinh^2 ax} = \dfrac{3}{2} \, (1.202057)/a^4.$

860.514. $\int_0^\infty \dfrac{x^4 \, dx}{\sinh^2 ax} = \dfrac{\pi^4}{30a^5},$ $\qquad\qquad [a > 0].$

860.518. $\int_0^\infty \dfrac{x^{2n} \, dx}{\sinh^2 ax} = \dfrac{\pi^{2n}}{a^{2n+1}} \, B_n,$ $\qquad [a > 0].$ $\quad$ [See **45**.]

860.519. $\int_0^\infty \dfrac{x^p \, dx}{\sinh^2 ax} = \dfrac{\Gamma(p+1)}{2^{p-1} \, a^{p+1}} \left[1 + \dfrac{1}{2^p} + \dfrac{1}{3^p} + \dfrac{1}{4^p} + \cdots \right],$

$\qquad\qquad\qquad = \dfrac{\Gamma(p+1)}{2^{p-1} \, a^{p+1}} \, \zeta(p),$

$\qquad [a > 0; \; p > 1; \; p \text{ not necessarily an integer.}]$ $\qquad$ [See **48.09**.]

860.530. $\int_0^\infty \dfrac{dx}{\cosh ax} = \int_0^\infty \dfrac{2 \, dx}{e^{ax} + e^{-ax}} = \dfrac{\pi}{2a},$ $\quad [a > 0].$ $\quad$ [See **679.10**.]

860.531. $\int_0^\infty \dfrac{x \, dx}{\cosh ax} = 2G/a^2 = 2(0.9159656)/a^2.$ $\qquad$ [See **48.32**.]

860.532. $\int_0^\infty \dfrac{x^2 \, dx}{\cosh ax} = \dfrac{\pi^3}{8a^3},$ $\qquad\qquad [a > 0].$

860.533. $\int_0^\infty \dfrac{x^3 \, dx}{\cosh ax} = 12(0.98894455)/a^4.$ $\qquad$ [See **48.34**.]

860.534. $\int_0^\infty \dfrac{x^4 \, dx}{\cosh ax} = \dfrac{5}{32} \dfrac{\pi^5}{a^5},$ $\qquad\qquad [a > 0].$

860.538. $\int_0^\infty \dfrac{x^{2n} \, dx}{\cosh ax} = \dfrac{\pi^{2n+1}}{(2a)^{2n+1}} \, E_n,$ $\qquad [a > 0; \, n = 0, 1, 2, \cdots].$

$\qquad\qquad\qquad\qquad\qquad\qquad\qquad\qquad\qquad$ [See **45**.]

860.539. $\int_0^\infty \dfrac{x^{p-1} \, dx}{\cosh ax} = \dfrac{2\Gamma(p)}{a^p} \left(1 - \dfrac{1}{3^p} + \dfrac{1}{5^p} - \dfrac{1}{7^p} + \cdots \right),$

$\qquad\qquad\qquad\qquad\qquad\qquad\qquad\qquad\qquad [a, p > 0].$

For integral values of p, see **48.31**–**.39**. For non-integral values of p, the numerical sum of the series is to be computed.

860.541. $\displaystyle\int_0^\infty \frac{x\,dx}{\cosh^2 ax} = \frac{\log 2}{a^2}.$

860.542. $\displaystyle\int_0^\infty \frac{x^2\,dx}{\cosh^2 ax} = \frac{\pi^2}{12\,a^3},$ $\qquad\qquad [a > 0].$

860.543. $\displaystyle\int_0^\infty \frac{x^3\,dx}{\cosh^2 ax} = \frac{9}{8}\,(1.202057)/a^4 = \frac{9}{8}\,\frac{\zeta(3)}{a^4}.$

860.544. $\displaystyle\int_0^\infty \frac{x^4\,dx}{\cosh^2 ax} = \frac{7}{240}\,\frac{\pi^4}{a^5},$ $\qquad\qquad [a > 0].$

860.548. $\displaystyle\int_0^\infty \frac{x^{2n}\,dx}{\cosh^2 ax} = \frac{(2^{2n-1} - 1)}{2^{2n-1}\,a^{2n+1}}\,\pi^{2n}B_n,$ $\qquad [a > 0;\, n = 1,\, 2,\, \cdots].$

860.549. $\displaystyle\int_0^\infty \frac{x^p\,dx}{\cosh^2 ax} = \frac{\left(1 - \dfrac{2}{2^p}\right)}{2^{p-1}}\,\frac{\Gamma(p + 1)}{a^{p+1}}\,\zeta(p),$ $\qquad [a > 0;\, p > 1].$

$\qquad\qquad\qquad\qquad\qquad\qquad\qquad\qquad\qquad\qquad\qquad$ [See **48.29** and Ref. **45.**]

When $p = 1$, use **860.541.**

860.80. $\displaystyle\int_0^\infty e^{-ax}\sin mx\,dx = \frac{m}{a^2 + m^2},$ $\qquad\qquad [a > 0].$

$\qquad\qquad\qquad\qquad\qquad\qquad\qquad\qquad\qquad\qquad\qquad\qquad$ [Ref. 7, Art. 291.]

860.81. $\displaystyle\int_0^\infty x\,e^{-ax}\sin mx\,dx = \frac{2am}{(a^2 + m^2)^2},$ $\qquad\qquad [a > 0].$

860.82. $\displaystyle\int_0^\infty x^2\,e^{-ax}\sin mx\,dx = \frac{2m(3a^2 - m^2)}{(a^2 + m^2)^3},$ $\qquad\qquad [a > 0].$

860.89. $\displaystyle\int_0^\infty x^{p-1}\,e^{-ax}\sin mx\,dx = \frac{\Gamma(p)\sin p\theta}{(a^2 + m^2)^{p/2}},$ $\qquad\qquad [p,\, a,\, m > 0].$

where $\sin\theta = m/r,\ \cos\theta = a/r,\ r = (a^2 + m^2)^{1/2}.$

860.90. $\displaystyle\int_0^\infty e^{-ax}\cos mx\,dx = \frac{a}{a^2 + m^2},$ $\qquad\qquad [a > 0].$

$\qquad\qquad\qquad\qquad\qquad\qquad\qquad\qquad\qquad\qquad\qquad\qquad$ [Ref. 7, Art. 291.]

860.91. $\displaystyle\int_0^\infty x\,e^{-ax}\cos mx\,dx = \frac{a^2 - m^2}{(a^2 + m^2)^2},$ $\qquad\qquad [a > 0].$

860.92. $\displaystyle\int_0^\infty x^2\,e^{-ax}\cos mx\,dx = \frac{2a(a^2 - 3m^2)}{(a^2 + m^2)^3},$ $\qquad\qquad [a > 0].$

860.99. $\displaystyle\int_0^\infty x^{p-1} e^{-ax} \cos mx \, dx = \frac{\Gamma(p) \cos p\theta}{(a^2 + m^2)^{p/2}},$

[$a, p > 0$], where θ is as in **860.89**.

[For **860.89** and **.99**, see Ref. 6, Art. 124, and Ref. 7, Art. 301.]

861.01. $\displaystyle\int_0^\infty \frac{e^{-ax}}{x} \sin mx \, dx = \tan^{-1} \frac{m}{a},$ $\qquad\qquad$ [$a > 0$].

[Ref. 7, Art. 285.]

861.02. $\displaystyle\int_0^\infty \frac{e^{-ax}}{x} \cos mx \, dx = \infty.$ $\qquad$ [Ref. 16, Table 365, No. 3.]

861.03. $\displaystyle\int_0^\infty \frac{e^{-ax}}{x} (1 - \cos mx) \, dx = \frac{1}{2} \log \left(1 + \frac{m^2}{a^2}\right),$ $\qquad$ [$a > 0$].

[Ref. 71, Art. 1036.]

861.04. $\displaystyle\int_0^\infty \frac{e^{-ax}}{x} (\cos mx - \cos nx) \, dx = \frac{1}{2} \log \frac{a^2+n^2}{a^2+m^2},$ $\qquad$ [$a > 0$].

[Ref. 71, Art. 1001.]

861.05. $\displaystyle\int_0^\infty \frac{e^{-ax} - e^{-bx}}{x} \cos mx \, dx = \frac{1}{2} \log \frac{b^2 + m^2}{a^2 + m^2},$ $\qquad$ [$a, b > 0$].

[Ref. 71, Art. 1035.]

861.06. $\displaystyle\int_0^\infty e^{-ax} \cos^2 mx \, dx = \frac{a^2 + 2m^2}{a(a^2 + 4m^2)},$ $\qquad\qquad$ [$a > 0$].

[Ref. 67, p. 155, No. 45.]

861.10. $\displaystyle\int_0^\infty e^{-ax} \sin^2 mx \, dx = \frac{2m^2}{a(a^2 + 4m^2)},$ $\qquad\qquad$ [$a > 0$].

[Ref. 67, p. 150, No. 3.]

861.11. $\displaystyle\int_0^\infty \frac{e^{-ax}}{x} \sin^2 mx \, dx = \frac{1}{4} \log \left(1 + \frac{4m^2}{a^2}\right),$ $\qquad$ [$a > 0$].

861.12. $\displaystyle\int_0^\infty \frac{e^{-ax}}{x^2} \sin^2 mx \, dx = m \tan^{-1} \frac{2m}{a} - \frac{a}{4} \log \left(1 + \frac{4m^2}{a^2}\right),$

[$a > 0$].

861.13. $\displaystyle\int_0^\infty e^{-ax} \sin mx \sin nx \, dx = \frac{2amn}{\{a^2 + (m-n)^2\}\{a^2 + (m+n)^2\}},$

[$a > 0$].

861.14. $\int_0^\infty e^{-ax} \sin mx \cos nx \, dx = \dfrac{m(a^2 + m^2 - n^2)}{\{a^2 + (m - n)^2\}\{a^2 + (m + n)^2\}}$,

$$[a > 0].$$

861.15. $\int_0^\infty e^{-ax} \cos mx \cos nx \, dx = \dfrac{a(a^2 + m^2 + n^2)}{\{a^2 + (m - n)^2\}\{a^2 + (m + n)^2\}}$,

$$[a > 0].$$

861.16. $\int_0^\infty \dfrac{e^{-ax}}{x} \sin mx \sin nx \, dx = \dfrac{1}{4} \log \dfrac{a^2 + (m + n)^2}{a^2 + (m - n)^2}$, $[a > 0]$.

861.20. $\int_0^\infty e^{-a^2x^2} \cos mx \, dx = \dfrac{\sqrt{\pi}}{2a} e^{-m^2/(4a^2)}$, $[a > 0]$.

$$[\text{Ref. 7, Art. 283.}]$$

861.21. $\int_0^\infty x \, e^{-a^2x^2} \sin mx \, dx = \dfrac{m\sqrt{\pi}}{4a^3} e^{-m^2/(4a^2)}$, $[a > 0]$.

861.22. $\int_0^\infty \dfrac{e^{-a^2x^2}}{x} \sin mx \, dx = \dfrac{\pi}{2} \operatorname{erf}\left(\dfrac{m}{2a}\right)$, $[a > 0]$.

See **590.** For numerical values, see Ref. 55e or Ref. 45.

861.31. $\int_0^\infty \dfrac{e^{-ax}}{\sqrt{x}} \cos mx \, dx = \dfrac{\{a + \sqrt{(a^2 + m^2)}\}^{1/2}\sqrt{\pi}}{(a^2 + m^2)^{1/2} \sqrt{2}}$, $[a > 0]$.

$$[\text{Ref. 67, p. 14, No. 4.}]$$

861.32. $\int_0^\infty e^{-ax} \sin \sqrt{(mx)} \, dx = \dfrac{\sqrt{(\pi m)}}{2a\sqrt{a}} e^{-m/(4a)}$, $[a, m > 0]$.

861.33. $\int_0^\infty \dfrac{e^{-ax}}{\sqrt{x}} \cos \sqrt{(mx)} \, dx = \dfrac{\sqrt{\pi}}{\sqrt{a}} e^{-m/(4a)}$, $[a, m > 0]$.

861.41. $\int_0^\infty e^{-ax} \sin (px + q) \, dx = \dfrac{a \sin q + p \cos q}{a^2 + p^2}$, $[a > 0]$.

861.42. $\int_0^\infty e^{-ax} \cos (px + q) \, dx = \dfrac{a \cos q - p \sin q}{a^2 + p^2}$, $[a > 0]$.

861.51. $\int_0^{\pi/2} \dfrac{\sinh x}{\sin x} \, dx = \dfrac{\pi}{2}$. $[\text{Ref. 72, p. 241.}]$

861.61. $\int_0^\infty \dfrac{\sin mx}{\sinh ax} \, dx = \int_0^\infty \dfrac{2 \sin mx}{e^{ax} - e^{-ax}} \, dx = \dfrac{\pi}{2a} \tanh \dfrac{\pi m}{2a}$, $[a > 0]$.

861.62. $\displaystyle\int_0^\infty \frac{\cos mx}{\cosh ax}\,dx = \frac{\pi}{2a\cosh \dfrac{\pi m}{2a}},$ $\qquad [a > 0]$.

861.63. $\displaystyle\int_0^\infty \frac{\sinh px}{\sinh qx}\,dx = \frac{\pi}{2q}\tan\frac{\pi p}{2q},$ $\qquad [-q < p < q; q > 0.]$

861.64. $\displaystyle\int_0^\infty \frac{\cosh px}{\cosh qx}\,dx = \frac{\pi}{2q\cos\dfrac{\pi p}{2q}},$ $\qquad [-q < p < q; q > 0]$.

861.65. $\displaystyle\int_0^\infty \tanh qx \sin mx\,dx = \frac{\pi}{2q\sinh\dfrac{\pi m}{2q}},$ $\qquad [q > 0]$.

861.66. $\displaystyle\int_0^\infty \frac{\sin mx}{\tanh qx}\,dx = \frac{\pi}{2q\tanh\dfrac{\pi m}{2q}},$ $\qquad [q > 0]$.

[For **861.61–.66,** see Ref. 71, Art. 1106.]

861.71. $\displaystyle\int_0^\infty \frac{\sin mx \cos nx}{\sinh ax}\,dx = \frac{\pi\sinh\dfrac{m\pi}{a}}{2a\left(\cosh\dfrac{m\pi}{a} + \cosh\dfrac{n\pi}{a}\right)},$ $\qquad [a > 0]$.

861.72. $\displaystyle\int_0^\infty \frac{\cos mx \cos nx}{\cosh ax}\,dx = \frac{\pi\cosh\dfrac{m\pi}{2a}\cosh\dfrac{n\pi}{2a}}{a\left(\cosh\dfrac{m\pi}{a} + \cosh\dfrac{n\pi}{a}\right)},$ $\qquad [a > 0]$.

[For **861.71–.72,** see Ref. 67, p. 36.]

861.73. $\displaystyle\int_0^\infty \frac{\cos mx}{\cosh^2 ax}\,dx = \frac{\pi m}{2a^2\sinh\dfrac{\pi m}{2a}},$ $\qquad [a > 0]$.

[Ref. 67, p. 30, No. 2.]

861.81. $\displaystyle\int_0^\infty \frac{x\sin mx}{\cosh ax}\,dx = \frac{\pi^2}{4a^2}\frac{\sinh\left(\dfrac{\pi m}{2a}\right)}{\cosh^2\left(\dfrac{\pi m}{2a}\right)},$ $\qquad [a > 0]$.

[Ref. 67, p. 89, No. 12.]

861.82. $\displaystyle\int_0^\infty \frac{x \cos mx}{\sinh ax}\,dx = \frac{\pi^2}{4a^2 \cosh^2\left(\dfrac{\pi m}{2a}\right)},$ $\qquad [a > 0].$

[Ref. 67, p. 31, No. 18.]

861.83. $\displaystyle\int_0^\infty \frac{\tanh ax}{x} \cos mx\,dx = \log \operatorname{ctnh}\left(\frac{\pi m}{4a}\right),$ $\qquad [a, m > 0].$

[Ref. 67, p. 33, No. 28.]

For a large collection of definite integrals involving cos mx, sin mx, and e^{-ax}, see *Tables of Integral Transforms*, by Erdélyi, Magnus, Oberhettinger, and Tricomi, Ref. 67.

862.01. $\displaystyle\int_0^\infty e^{-ax} \sinh bx\,dx = \frac{b}{a^2 - b^2},$ $\qquad [a > b \geqq 0].$

862.02. $\displaystyle\int_0^\infty e^{-ax} \cosh bx\,dx = \frac{a}{a^2 - b^2},$ $\qquad [a > b \geqq 0].$

862.03. $\displaystyle\int_0^\infty xe^{-a^2x^2} \sinh bx\,dx = \frac{b\sqrt{\pi}}{4a^3} e^{b^2/(4a^2)},$ $\qquad [a > 0].$

862.04. $\displaystyle\int_0^\infty e^{-ax} \sinh(b\sqrt{x})\,dx = \frac{b\sqrt{\pi}}{2a\sqrt{a}} e^{b^2/(4a)},$ $\qquad [a > 0].$

862.11. $\displaystyle\int_0^\infty \frac{\sin mx}{e^{ax} + 1}\,dx = \frac{1}{2m} - \frac{\pi}{2a \sinh(\pi m/a)},$ $\qquad [a > m > 0].$

862.12. $\displaystyle\int_0^\infty \frac{\sin mx}{e^{ax} - 1}\,dx = \frac{\pi}{2a \tanh(\pi m/a)} - \frac{1}{2m},$ $\qquad [a > m > 0].$

862.21. $\displaystyle\int_0^\infty \frac{\sinh px}{e^{ax} + 1}\,dx = \frac{\pi}{2a \sin(\pi p/a)} - \frac{1}{2p},$ $\qquad [a > p > 0].$

862.22. $\displaystyle\int_0^\infty \frac{\sinh px}{e^{ax} - 1}\,dx = \frac{1}{2p} - \frac{\pi}{2a \tan(\pi p/a)},$ $\qquad [a > p > 0].$

862.31. $\displaystyle\int_0^\infty \frac{\sin mx\,dx}{(1 + x^2) \sinh \pi x} = -\frac{m}{2e^m} + (\sinh m) \log(1 + e^{-m}),$

$\qquad [m > 0].$

862.32. $\displaystyle\int_0^\infty \frac{\tanh \dfrac{\pi x}{2}}{1 + x^2} \sin mx\,dx = \frac{m}{e^m} - (\sinh m) \log(1 - e^{-2m}),$

$\qquad [m > 0].$

862.33. $\displaystyle\int_0^\infty \frac{\sin mx\, dx}{(1 + x^2)\tanh\left(\dfrac{\pi x}{2}\right)} = -(\sinh m)\log\tanh\frac{m}{2},\quad [m > 0].$

862.41. $\displaystyle\int_0^\infty \frac{\sinh px}{\sinh qx}\cos mx\, dx = \frac{\pi}{2q}\,\frac{\sin\dfrac{p\pi}{q}}{\cos\dfrac{p\pi}{q} + \cosh\dfrac{m\pi}{q}},$

$$[-q \leqq p \leqq q; q > 0].$$

862.42. $\displaystyle\int_0^\infty \frac{\cosh px}{\sinh qx}\sin mx\, dx = \frac{\pi}{2q}\,\frac{\sinh\dfrac{m\pi}{q}}{\cos\dfrac{p\pi}{q} + \cosh\dfrac{m\pi}{q}},$

$$[-q \leqq p \leqq q; q > 0].$$

862.43. $\displaystyle\int_0^\infty \frac{\sinh px}{\cosh qx}\sin mx\, dx = \frac{\pi}{q}\,\frac{\sin\dfrac{p\pi}{2q}\sinh\dfrac{m\pi}{2q}}{\cos\dfrac{p\pi}{q} + \cosh\dfrac{m\pi}{q}},$

$$[-q \leqq p \leqq q; q > 0].$$

862.44. $\displaystyle\int_0^\infty \frac{\cosh px}{\cosh qx}\cos mx\, dx = \frac{\pi}{q}\,\frac{\cos\dfrac{p\pi}{2q}\cosh\dfrac{m\pi}{2q}}{\cos\dfrac{p\pi}{q} + \cosh\dfrac{m\pi}{q}},$

$$[-q \leqq p \leqq q; q > 0].$$

[For **862.41–44**, see Ref. 71, Art.1105.]

863.01. $\displaystyle\int_0^1 \left(\log\frac{1}{x}\right)^q dx = \Gamma(q + 1),\quad [q + 1 > 0].\qquad$ [See **850.1.**]

863.02. $\displaystyle\int_0^1 x^p \log\frac{1}{x}\, dx = \frac{1}{(p + 1)^2},\qquad\qquad [p + 1 > 0].$

863.03. $\displaystyle\int_0^1 x^p \left(\log\frac{1}{x}\right)^2 dx = \frac{2}{(p + 1)^3},\qquad\qquad [p + 1 > 0].$

863.04. $\displaystyle\int_0^1 x^p \left(\log\frac{1}{x}\right)^q dx = \frac{\Gamma(q + 1)}{(p + 1)^{q+1}},\qquad [p + 1, q + 1 > 0].$

863.05. $\displaystyle\int_0^1 \left(\log\frac{1}{x}\right)^{1/2} dx = \frac{\sqrt{\pi}}{2}.$

863.06. $\displaystyle\int_0^1 \left(\log\frac{1}{x}\right)^{-1/2} dx = \sqrt{\pi}.$

863.10. $\displaystyle\int_0^1 \frac{\log\dfrac{1}{x}}{1-x} dx = \frac{\pi^2}{6}.$

863.11. $\displaystyle\int_0^1 \frac{x\log\dfrac{1}{x}}{1-x} dx = \frac{\pi^2}{6} - \frac{1}{1^2}.$ \qquad [Ref. 71, Art. 1077, No. 8.]

863.12. $\displaystyle\int_0^1 \frac{x^2\log\dfrac{1}{x}}{1-x} dx = \frac{\pi^2}{6} - \frac{1}{1^2} - \frac{1}{2^2}.$

[Ref. 69, vol. 2, p. 74, Nos. 4, 4a.]

863.20. $\displaystyle\int_0^1 \frac{\log\dfrac{1}{x}}{1+x} dx = \frac{\pi^2}{12}.$

863.21. $\displaystyle\int_0^1 \frac{x\log\dfrac{1}{x}}{1+x} dx = 1 - \frac{\pi^2}{12}.$

863.22. $\displaystyle\int_0^1 \frac{x^2\log\dfrac{1}{x}}{1+x} dx = \frac{\pi^2}{12} - \frac{3}{4}.$

[See Ref. 69, vol. 2, pp. 74, 75, Nos. 4, 6, 7.]

863.30. $\displaystyle\int_0^1 \frac{1+x}{1-x}\log\frac{1}{x} dx = \frac{\pi^2}{3} - 1.$

863.31. $\displaystyle\int_0^1 \frac{\log\dfrac{1}{x}}{1-x^2} dx = \frac{\pi^2}{8}.$

863.32. $\displaystyle\int_0^1 \frac{x\log\dfrac{1}{x}}{1-x^2} dx = \frac{\pi^2}{24}.$

863.33. $\displaystyle\int_0^1 \frac{x^2 \log \dfrac{1}{x}}{1 - x^2}\, dx = \frac{\pi^2}{8} - 1.$

863.34. $\displaystyle\int_0^1 \frac{\log \dfrac{1}{x}}{1 + x^2}\, dx = G = 0.915\,9656.$ [See **48.32.**]

[For **863.31–.34,** see Ref. 71, Art. 1079, Nos. 1, 8, 9, 1.]

863.35. $\displaystyle\int_0^1 \frac{\log \dfrac{1}{x}}{(1 + x)^2}\, dx = \log 2.$

863.36. $\displaystyle\int_0^1 \frac{1 - x}{1 + x^3} \log \frac{1}{x}\, dx = \frac{2\pi^2}{27}.$ [Ref. 71, Art. 1083.]

863.37. $\displaystyle\int_0^1 \frac{1 + x}{1 - x^3} \log \frac{1}{x}\, dx = \frac{4\pi^2}{27}.$ [Ref. 71, Art. 1083.]

863.41. $\displaystyle\int_0^1 \frac{\log \dfrac{1}{x}}{\sqrt{(1 - x^2)}}\, dx = \frac{\pi}{2} \log 2.$ [Ref. 71, Arts. 990, 1081.]

863.42. $\displaystyle\int_0^1 \frac{x \log \dfrac{1}{x}}{\sqrt{(1 - x^2)}}\, dx = 1 - \log 2.$

863.43. $\displaystyle\int_0^1 \left(\log \frac{1}{x}\right) \sqrt{(1 - x^2)}\, dx = \frac{\pi}{4}\left(\log 2 + \frac{1}{2}\right).$

863.51. $\displaystyle\int_0^1 \frac{x^p\, dx}{\log x} = \infty.$

863.52. $\displaystyle\int_0^1 \frac{x^p - x^q}{\log x}\, dx = \log \frac{p + 1}{q + 1},$ $[p + 1, q + 1 > 0].$

[Ref. 71, Art. 1099.]

863.53. $\displaystyle\int_0^1 \frac{x^p - x^q}{\log x} x^r\, dx = \log \frac{p + r + 1}{q + r + 1},$

$[p + r + 1, q + r + 1 > 0].$

863.54. $\displaystyle\int_0^1 \frac{(1 - x^p)(1 - x^q)}{\log x}\, dx = \log \frac{(p + q + 1)}{(p + 1)(q + 1)},$

$$[p + 1, q + 1, p + q + 1 > 0].$$

863.55. $\displaystyle\int_0^1 \left(\frac{1 - x}{1 + x}\right) \frac{dx}{\log x} = \log \frac{2}{\pi}.$

863.61. $\displaystyle\int_0^1 \frac{(\log x)^2}{1 + x^2}\, dx = \frac{\pi^3}{16}.$ [Ref. 71, p. 269.]

863.71. $\displaystyle\int_0^1 \log(1 - x)\, dx = -1.$

863.72. $\displaystyle\int_0^1 x \log(1 - x)\, dx = -\frac{3}{4}.$

863.73. $\displaystyle\int_0^1 x^p \log(1 - x)\, dx = -\frac{1}{p + 1} \sum_{n=1}^{p+1} \frac{1}{n},$ $[p = 0, 1, 2, \cdots].$

863.74. $\displaystyle\int_0^1 \frac{\log(1 - x)}{x}\, dx = -\frac{\pi^2}{6}.$

863.81. $\displaystyle\int_0^1 \log(1 + x)\, dx = 2 \log 2 - 1.$

863.82. $\displaystyle\int_0^1 x \log(1 + x)\, dx = \frac{1}{4}.$

863.83. $\displaystyle\int_0^1 x^{2p} \log(1 + x)\, dx = \frac{1}{2p + 1} \left\{ 2 \log 2 - \sum_{n=1}^{2p+1} \frac{(-1)^{n-1}}{n} \right\},$

$$[p = 0, 1, 2 \cdots].$$

863.84. $\displaystyle\int_0^1 x^{2p-1} \log(1 + x)\, dx = \frac{1}{2p} \sum_{n=1}^{2p} \frac{(-1)^{n-1}}{n},$ $[p = 1, 2, \cdots].$

863.85. $\displaystyle\int_0^1 \frac{\log(1 + x)}{x}\, dx = \frac{\pi^2}{12}.$

863.91. $\displaystyle\int_0^1 \log x \log(1 - x)\, dx = 2 - \frac{\pi^2}{6}.$

863.92. $\displaystyle\int_0^1 \log x \log(1 + x)\, dx = 2 - 2 \log 2 - \frac{\pi^2}{12}.$

863.93. $\displaystyle\int_0^1 x \log x \log(1 - x)\, dx = 1 - \frac{\pi^2}{12}.$

863.94. $\displaystyle\int_0^1 x \log x \log (1 + x)\, dx = \frac{\pi^2}{24} - \frac{1}{2}.$

864.01. $\displaystyle\int_0^1 \frac{\log (1 - x)}{1 + x^2}\, dx = \frac{\pi}{8} \log 2 - \mathrm{G} = \frac{\pi}{8} \log 2 - 0.915\ 9656.$

[See **48.32.**]

864.02. $\displaystyle\int_0^1 \frac{\log (1 + x)}{1 + x^2}\, dx = \frac{\pi}{8} \log 2 \cdot$ [Ref. 7, Art. 51.]

864.03. $\displaystyle\int_0^1 \frac{1}{1 + x^2} \log \frac{1 + x}{x}\, dx = \frac{\pi}{8} \log 2 + \mathrm{G}.$ [See **48.32.**]

864.04. $\displaystyle\int_0^1 \frac{\log (1 + x^2)}{1 + x^2}\, dx = \frac{\pi}{2} \log 2 - \mathrm{G}.$ [See **48.32.**]

864.11. $\displaystyle\int_0^1 \log \left(\frac{1 + x}{1 - x}\right) dx = 2 \log 2.$

864.12. $\displaystyle\int_0^1 \frac{1}{x} \log \left(\frac{1 + x}{1 - x}\right) dx = \frac{\pi^2}{4}.$

864.21. $\displaystyle\int_0^1 \frac{\log (1 - x^p)}{x}\, dx = -\frac{\pi^2}{6p},$ $[p > 0].$

864.22. $\displaystyle\int_0^1 \frac{\log (1 + x^p)}{x}\, dx = \frac{\pi^2}{12p},$ $[p > 0].$

864.31. $\displaystyle\int_0^1 \frac{\log (1 - x)}{\sqrt{(1 - x^2)}}\, dx = -2\,\mathrm{G} - \frac{\pi}{2} \log 2.$ [See **48.32.**]

864.32. $\displaystyle\int_0^1 \frac{\log (1 + x)}{\sqrt{(1 - x^2)}}\, dx = 2\mathrm{G} - \frac{\pi}{2} \log 2.$ [See **48.32.**]

864.33. $\displaystyle\int_0^1 \frac{(\log x)^2}{\sqrt{(1 - x^2)}}\, dx = \frac{\pi}{2} \left\{\frac{\pi^2}{12} + (\log 2)^2\right\}\cdot$

864.41. $\displaystyle\int_0^1 (1 - x)\, e^{-x} \log \frac{1}{x}\, dx = 1 - \frac{1}{e}\cdot$

864.51. $\displaystyle\int_0^\infty \frac{x^{p-1}}{1 - x} \log \frac{1}{x}\, dx = \frac{\pi^2}{\sin^2 p\pi},$ $[0 < p < 1].$

864.52. $\displaystyle\int_0^\infty \frac{x^{p-1}}{1 + x} \log \frac{1}{x}\, dx = \frac{\pi^2 \cos p\pi}{\sin^2 p\pi},$ $[0 < p < 1].$

[Ref. 71, Art. 1101, eq. (18).]

864.53. $\displaystyle\int_0^\infty \frac{\log x}{x^2 - 1}\, dx = \frac{\pi^2}{4}.$ [Ref. 71, Art. 1084, No. 1.]

864.54. $\displaystyle\int_0^\infty \frac{\log x}{x^2 + a^2}\, dx = \frac{\pi}{2a} \log a,$ $[a > 0].$

864.55. $\displaystyle\int_0^\infty \frac{(\log x)^2}{x^2 + 1}\, dx = \frac{\pi^3}{8}.$ [Ref. 71, p. 269.]

864.61. $\displaystyle\int_0^\infty \frac{\log (1 + x)}{1 + x^2}\, dx = \frac{\pi}{4} \log 2 + G.$ [See **48.32.**]

864.62. $\displaystyle\int_0^\infty \frac{\log (1 + a^2 x^2)}{b^2 + x^2}\, dx = \frac{\pi}{b} \log (1 + ab),$ $[a, b > 0].$

864.63. $\displaystyle\int_0^\infty \frac{\log (a^2 + x^2)}{b^2 + x^2}\, dx = \frac{\pi}{b} \log (a + b),$ $[a, b > 0].$

[For **864.62** and **.63**, see Ref. 71, Art. 1044.]

864.71. $\displaystyle\int_0^\infty \frac{\log (1 + x)}{x^{2-p}}\, dx = \frac{\pi}{(1 - p) \sin p\pi},$ $[0 < p < 1].$

864.72. $\displaystyle\int_0^\infty \frac{\log (1 + x)}{x^{1+q}}\, dx = \frac{\pi}{q \sin \pi q},$ $[0 < q < 1].$

864.73. $\displaystyle\int_0^\infty \frac{\log (1 + x^p)}{x^q}\, dx = \frac{\pi}{(q - 1) \sin \dfrac{\pi(q - 1)}{p}},$ $[0 < q - 1 < p].$

864.74. $\displaystyle\int_0^\infty \frac{\log |x^p - 1|}{x^q}\, dx = \frac{\pi}{(q - 1) \tan \dfrac{\pi(q - 1)}{p}},$

$[0 < q - 1 < p].$

865.01. $\displaystyle\int_0^{\pi/4} \log \sin x\, dx = -\frac{\pi}{4} \log 2 - \frac{G}{2}.$ [See **48.32.**]

865.02. $\displaystyle\int_0^{\pi/4} \log \cos x\, dx = -\frac{\pi}{4} \log 2 + \frac{G}{2}.$ [See **48.32.**]

865.03. $\displaystyle\int_0^{\pi/4} \log \tan x\, dx = -G.$ [See **48.32.**]

865.04. $\displaystyle\int_0^{\pi/4} \log (1 + \tan x)\, dx = \frac{\pi}{8} \log 2.$ [Ref. 7, Art. 51.]

865.05. $\displaystyle\int_0^{\pi/4} \log\,(1 - \tan x)\,dx = \frac{\pi}{8}\log 2 - G.$ [See **48.32.**]

865.11. $\displaystyle\int_0^{\pi/2} \log \sin x\,dx = \int_0^{\pi/2} \log \cos x\,dx = -\frac{\pi}{2}\log 2.$

                                            [Ref. 7, Art. 51.]

865.12. $\displaystyle\int_0^{\pi/2} \log \tan x\,dx = 0.$

865.21. $\displaystyle\int_0^{\pi/2} (\sin x)\log \sin x\,dx = \log 2 - 1.$

865.22. $\displaystyle\int_0^{\pi/2} (\cos x)\log \cos x\,dx = \log 2 - 1.$

865.23. $\displaystyle\int_0^{\pi/2} (\cos x)\log \sin x\,dx = \int_0^{\pi/2} (\sin x)\log \cos x\,dx = -1.$

865.24. $\displaystyle\int_0^{\pi/2} (\tan x)\log \sin x\,dx = -\frac{\pi^2}{24}.$

865.25. $\displaystyle\int_0^{\pi/2} (\sin^2 x)\log \sin x\,dx = \frac{\pi}{8}(1 - 2\log 2).$

865.26. $\displaystyle\int_0^{\pi/2} (\cos 2nx)\log \sin x\,dx = -\frac{\pi}{4n},$ $[n > 0].$

                                  [Ref. 71, Art. 1081.]

865.27. $\displaystyle\int_0^{\pi/2} \frac{\log \cos x}{\sin x}\,dx = -\frac{\pi^2}{8}.$

865.31. $\displaystyle\int_0^{\pi/2} \log\,(1 + \cos x)\,dx = 2G - \frac{\pi \log 2}{2}.$ [See **48.32.**]

865.32. $\displaystyle\int_0^{\pi/2} \log\,(1 - \cos x)\,dx = -2G - \frac{\pi \log 2}{2}.$ [See **48.32.**]

865.33. $\displaystyle\int_0^{\pi/2} \log\,(1 + \tan x)\,dx = G + \frac{\pi \log 2}{4}.$ [See **48.32.**]

865.34. $\displaystyle\int_0^{\pi/2} \log\,(1 + p\sin^2 x)\,dx$

$$= \int_0^{\pi/2} \log\,(1 + p\cos^2 x)\,dx = \pi \log \frac{1 + \sqrt{(1 + p)}}{2},$$

$$[(1 + p) \geqq 0].$$

865.35. $\displaystyle\int_0^{\pi/2} \log(a^2 \sin^2 x + b^2 \cos^2 x)\, dx = \pi \log \frac{a+b}{2},$

$$[a, b > 0].\quad [\text{Ref. 71, Art. 1044.}]$$

865.36. $\displaystyle\int_0^{\pi/2} \log(a^2 + b^2 \tan^2 x)\, dx$

$$= \int_0^{\pi/2} \log(a^2 + b^2 \cot^2 x)\, dx = \pi \log(a+b),$$

$$[a, b > 0].$$

865.37. $\displaystyle\int_0^{\pi/2} \log\left(\frac{a + b \sin x}{a - b \sin x}\right) \frac{dx}{\sin x}$

$$= \int_0^{\pi/2} \log\left(\frac{a + b \cos x}{a - b \cos x}\right) \frac{dx}{\cos x} = \pi \sin^{-1}\left(\frac{b}{a}\right),$$

$$[b^2 \leqq a^2].$$

865.41. $\displaystyle\int_0^{\pi} \log \sin x\, dx = -\pi \log 2.$

865.42. $\displaystyle\int_0^{\pi} x \log \sin x\, dx = -\frac{\pi^2 \log 2}{2}.$

865.43. $\displaystyle\int_0^{\pi} \log(1 \pm \cos x)\, dx = -\pi \log 2.$

865.44. $\displaystyle\int_0^{\pi} \log(a \pm b \cos x)\, dx = \pi \log \left\{\frac{a + \sqrt{(a^2 - b^2)}}{2}\right\},$

$$[a \geqq b > 0].$$

865.45. $\displaystyle\int_0^{\pi} \frac{\log(1 + a \cos x)}{\cos x}\, dx = \pi \sin^{-1} a,\qquad [0 < a < 1].$

865.51. $\displaystyle\int_0^1 (\cos mx) \log x\, dx = -\frac{Si(m)}{m}.$ [See **431.11.**]

For numerical values see Ref. 55f, 4, 17, or 45.

865.52. $\displaystyle\int_0^1 x^{p-1} \sin(q \log x)\, dx = \frac{-q}{p^2 + q^2},\qquad [p > 0].$

865.53. $\displaystyle\int_0^1 x^{p-1} \cos(q \log x)\, dx = \frac{p}{p^2 + q^2},\qquad [p > 0].$

865.54. $\displaystyle\int_0^1 \frac{x^p \sin(q \log x)}{\log x}\, dx = \tan^{-1}\frac{q}{p+1},\qquad [p + 1 > 0].$

865.61. $\displaystyle\int_0^\infty \log\left(1 + \frac{a^2}{x^2}\right) \cos mx\, dx = \frac{\pi}{m}\,(1 - e^{-am})$, $[a, m > 0]$.

865.62. $\displaystyle\int_0^\infty \log\left(\frac{a^2 + x^2}{b^2 + x^2}\right) \cos mx\, dx = \frac{\pi}{m}\,(e^{-bm} - e^{-am})$,

$$[a, b, m > 0].$$

[For **865.61** and **.62** see Ref. 67, p. 18.]

865.63. $\displaystyle\int_0^\infty \frac{\sin mx}{x}\log\frac{1}{x}\,dx = \frac{\pi}{2}\,(\log m + C)$, $[m > 0]$.

[See **851.1.**]

865.64. $\displaystyle\int_0^\infty \frac{\log(\sin^2 mx)}{a^2 + x^2}\,dx = \frac{\pi}{a}\log\left(\frac{\sinh ma}{e^{ma}}\right)$, $[m, a > 0]$,

$$\text{or} = \frac{\pi}{a}\log\left(\frac{1 - e^{-2ma}}{2}\right), \qquad [m, a > 0].$$

865.65. $\displaystyle\int_0^\infty \frac{\log(\cos^2 mx)}{a^2 + x^2}\,dx = \frac{\pi}{a}\log\left(\frac{\cosh ma}{e^{ma}}\right)$, $[m, a > 0]$.

865.66. $\displaystyle\int_0^\infty \frac{\log(\tan^2 mx)}{a^2 + x^2}\,dx = \frac{\pi}{a}\log\tanh ma$, $[m, a > 0]$.

[For **865.64–.66,** see Ref. 73, Art. 193.]

865.71. $\displaystyle\int_0^\pi \log(1 \pm 2a\cos x + a^2)\,dx = 2\pi\log a$, $[a > 1]$,

$$= 0, \qquad\qquad [a^2 \le 1].$$

865.72. $\displaystyle\int_0^\pi \log(a^2 \pm 2ab\cos x + b^2)\,dx = 2\pi\log a$, $[a \ge b > 0]$,

$$= 2\pi\log b, \quad\quad [b \ge a > 0],$$
$$= 0, \qquad [1 = b \ge a > 0].$$

[Ref. 7, Art. 292.]

865.73. $\displaystyle\int_0^{2\pi} \log(a^2 \pm 2ab\cos x + b^2)\,dx = 4\pi\log a$, $[a \ge b > 0]$,

$$= 4\pi\log b, \quad\quad [b \ge a > 0],$$
$$= 0, \qquad [1 = b \ge a > 0].$$

865.74. $\displaystyle\int_0^\pi \log(1 - 2a\cos x + a^2)\cos mx\, dx = -\frac{\pi}{m}\,a^m$,

$$[0 < a < 1; m = 1, 2, \cdots].$$

865.75. $\displaystyle\int_0^\infty \frac{\log\left(1 \pm 2p \cos mx + p^2\right)}{a^2 + x^2}\, dx = \frac{\pi}{a} \log\left(1 \pm pe^{-ma}\right),$

$$[0 < p \leqq 1;\, m,\, a > 0],$$

$$= \frac{\pi}{a} \log\left(p \pm e^{-ma}\right),$$

$$[p \geqq 1;\, m,\, a > 0].$$

865.81. $\displaystyle\int_0^\infty \log\left(1 + e^{-x}\right) dx = \frac{\pi^2}{12}.$

865.82. $\displaystyle\int_0^\infty \log\left(1 - e^{-x}\right) dx = -\frac{\pi^2}{6}.$

865.83. $\displaystyle\int_0^\infty \log\left(\frac{1 - e^{-x}}{1 + e^{-x}}\right) dx = \int_0^\infty \log \tanh \frac{x}{2}\, dx = -\frac{\pi^2}{4}.$

[Ref. 40, Art. 91.]

865.901. $\displaystyle\int_0^\infty e^{-ax} \log \frac{1}{x}\, dx = \frac{1}{a}\left(\log a + C\right),$ $[a > 0].$ [See **852.1**.]

For the constant C, see **851.1**. $C = 0.5772157$

865.902. $\displaystyle\int_0^\infty x e^{-ax} \log \frac{1}{x}\, dx = \frac{1}{a^2}\left(\log a - 1 + C\right),$ $[a > 0].$

865.903. $\displaystyle\int_0^\infty x^2 e^{-ax} \log \frac{1}{x}\, dx = \frac{2}{a^3}\left(\log a - \frac{3}{2} + C\right),$ $[a > 0].$

865.904. $\displaystyle\int_0^\infty \frac{e^{-ax} \log \dfrac{1}{x}}{\sqrt{x}}\, dx = \frac{\sqrt{\pi}}{\sqrt{a}}\left(\log a + 2 \log 2 + C\right),$ $[a > 0].$

865.905. $\displaystyle\int_0^\infty \frac{1}{x}\left(\frac{1}{1 + px} - e^{-ax}\right) dx = \log \frac{a}{p} + C,$ $[a,\, p > 0].$

865.906. $\displaystyle\int_0^\infty \frac{1}{x}\left(\frac{1}{1 + p^2 x^2} - e^{-ax}\right) dx = \log \frac{a}{p} + C,$ $[a,\, p > 0].$

865.907. $\displaystyle\int_0^\infty \left(\frac{1}{e^x - 1} - \frac{1}{x e^x}\right) dx = C.$

865.908. $\displaystyle\int_0^1 \log\left(\log \frac{1}{x}\right) dx = -C.$

865.909. $\displaystyle\int_0^1 \left(\frac{1}{\log x} + \frac{1}{1 - x}\right) dx = C.$

865.911. $\displaystyle\int_0^\infty \left(\log \frac{1}{x}\right) e^{-ax} \sin mx \, dx$

$$= \frac{1}{(a^2 + m^2)} \left\{\frac{m}{2} \log (a^2 + m^2) - a \tan^{-1} \frac{m}{a} + m\, C\right\},$$

$$[m, a > 0]. \quad [\text{See } \mathbf{851.1}.]$$

865.912. $\displaystyle\int_0^\infty \left(\log \frac{1}{x}\right) e^{-ax} \cos mx \, dx$

$$= \frac{1}{(a^2 + m^2)} \left\{\frac{a}{2} \log (a^2 + m^2) + m \tan^{-1} \frac{m}{a} + a C\right\},$$

$$[m, a > 0]. \quad [\text{See } \mathbf{851.1}.]$$

865.913. $\displaystyle\int_0^\infty \frac{\left(\log \dfrac{1}{x}\right) e^{-ax} \sin mx}{x} \, dx = \left\{\frac{1}{2} \log (a^2 + m^2) + C\right\} \tan^{-1} \frac{m}{a},$

$$[m, a > 0]. \quad [\text{See } \mathbf{851.1}.]$$

A very large proportion of the definite integrals from **850** to **865** may be found in *Nouvelles Tables d'Intégrales Définies*, by B. de Haan, Ref. 16. See also Ref. 76.

866.01. $\displaystyle\int_0^\pi \cos (x \sin \phi) \, d\phi = \int_0^\pi \cos (x \cos \phi) \, d\phi = \pi J_0(x).$

866.02. $\displaystyle\int_0^\pi \cos (n\phi - x \sin \phi) \, d\phi = \pi J_n(x),$

where n is zero or any positive integer.

$$[\text{Bessel's Integral. Ref. 12, p. 32, eq. (9).}]$$

866.03. $\displaystyle\int_0^\pi e^{p \cos x} \, dx = \pi I_0(p), \quad \text{as in } \mathbf{813.1.} \qquad [\text{Ref. 13, p. 181.}]$

For numerical values see Ref. 50, 55g, 17, or 45.

880. Simpson's Rule. When there are a number of values of $y = f(x)$ for values of x at equal intervals, h, apart, an approximate numerical integration is given by

$$\int_{x=a}^{b} f(x) \, dx \approx \frac{h}{3} \left[y_0 + 4y_1 + 2y_2 + 4y_3 + 2y_4 + \ldots \right.$$
$$\left. + 4y_{2n-1} + y_{2n} \right]$$

where $h = x_1 - x_0 =$ the constant interval of x, so that $2nh = b - a$. The coefficients are alternately 4 and 2 as indicated. The approximation is in general more accurate as n is larger. In this way, a numerical result can often be obtained when the algebraic expression cannot be integrated in suitable form. This computation can be performed as one continuous operation on a manual calculating machine, using a table of $f(x)$.

881. An estimate of the error in the above approximate formula is

$$\frac{nh^5 f^{iv}(x)}{90} = \frac{(b - a) h^4 f^{iv}(x)}{180}$$

where the largest entry found in the fourth column of differences in the table of $f(x)$, in the range between a and b, may be used for the numerical value of $h^4 f^{iv}(x)$. See also pages 184–5 of *Methods of Advanced Calculus*, by Philip Franklin (Ref. 39).

882. The following alternative formula is more accurate, with many functions, than No. 880. It also can be computed in one continuous operation on a manual calculating machine.

$$\int_{x=a}^{b} f(x) \, dx \approx \frac{h}{4.5} \left[1.4\, y_0 + 6.4\, y_1 + 2.4\, y_2 + 6.4\, y_3 + 2.8\, y_4 \right.$$
$$+ 6.4\, y_5 + 2.4\, y_6 + 6.4\, y_7 + 2.8\, y_8$$
$$\ldots\ldots\ldots\ldots\ldots\ldots\ldots\ldots\ldots$$
$$\left. + 6.4\, y_{4n-3} + 2.4\, y_{4n-2} + 6.4\, y_{4n-1} + 1.4\, y_{4n} \right]$$

where $4nh = b - a$.

883. Trapezoidal Rule.

$$\int_{x=a}^{b} f(x) \, dx \approx h \left[\frac{1}{2} y_0 + y_1 + y_2 + \ldots + y_{n-1} + \frac{1}{2} y_n \right]$$

where $b - a = nh$ and where n may be odd or even.

·· 13 ··

DIFFERENTIAL EQUATIONS

890.1. Separation of the variables. If the equation can be put in the form $f_1(x)dx = f_2(y)dy$, each term may be integrated.

890.2. Separation of the variables by a substitution—Homogeneous equations. If the equation is of the form

$$f_1(x, y)dx + f_2(x, y)dy = 0,$$

where the functions are homogeneous in x and y and are of the same degree, let $y = ux$. Then

$$\frac{dx}{x} = -\frac{f_2(1, u)du}{f_1(1, u) + uf_2(1, u)}.$$

If more convenient let $x = uy$.

890.3. Separation of the variables by a substitution, for equations of the form

$$f_1(xy)y\, dx + f_2(xy)x\, dy = 0,$$

where f_1 and f_2 are any functions. Let $y = u/x$. Then

$$\frac{dx}{x} = \frac{f_2(u)du}{u\{f_2(u) - f_1(u)\}}.$$

890.4. An equation of the form

$$(ax + by + c)dx + (fx + gy + h)dy = 0$$

can be made homogeneous by putting $x = x' + m$ and $y = y' + n$. The quantities m and n can be found by solving the two simultaneous

251

equations in m and n required to make the original equation homogeneous. This method does not apply if

$$\frac{ax + by}{fx + gy} = \text{a constant,}$$

but we can then solve by substituting $ax + by = u$ and eliminating y or x.

890.5.   Exact differential equations. If $M\,dx + N\,dy = 0$ is an equation in which

$$\frac{\partial M}{\partial y} = \frac{\partial N}{\partial x},$$

it is an exact differential equation.

Integrate $\int M\,dx$, regarding y as a constant and adding an unknown function of y, say $f(y)$; differentiate the result with respect to y and equate the new result to N; from the resulting equation determine the unknown function of y. The solution is then

$$\int M\,dx + f(y) + c = 0.$$

If more convenient, interchange M and N and also x and y in the above rule.

[See Ref. 32, *A Course in Mathematics*, by F. S. Woods and F. H. Bailey, vol. 2, ed. of 1909, p. 270.]

891.1.   Linear equations of the first order. A differential equation is linear when it has only the first power of the function and of its derivatives. The linear equation of the first order is of the form

$$\frac{dy}{dx} + Py = Q \quad \text{or} \quad dy + Py\,dx = Q\,dx,$$

where P and Q are independent of y but may involve x.

Insert $e^{\int P\,dx}$ as an integrating factor. The solution is

$$y = e^{-\int P\,dx}\left[\int e^{\int P\,dx}Q\,dx + c\right].$$

891.2. Bernoulli's equation. If the equation is of the form

$$\frac{dy}{dx} + Py = Qy^n,$$

where P and Q do not involve y, it can be made linear by substituting $1/y^{n-1} = u$. Divide the equation by y^n before making the substitution.

892. Equations of the first order but not of the first degree.

Let

$$\frac{dy}{dx} = p.$$

If possible, solve the resulting equation for p. The equations given by putting p equal to the values so found may often be integrated, thus furnishing solutions of the given equation.

893.1. Equations of the second order, not containing y directly. Let $dy/dx = p$. The equation will become one of the first order in p and x. It may be possible to solve this by one of the methods of the preceding paragraphs.

893.2. Equations of the second order, not containing x directly.

Let

$$\frac{dy}{dx} = p$$

Then

$$\frac{d^2y}{dx^2} = \frac{dp}{dy}\frac{dy}{dx} = p\frac{dp}{dy}.$$

The resulting equation is of the first order in p and y and it may be possible to solve it by one of the methods of the preceding paragraphs.

894. To solve

$$\frac{d^2y}{dx^2} + A\frac{dy}{dx} + By = 0,$$

where A and B are constants, find the roots of the auxiliary equation $p^2 + Ap + B = 0$. If the roots are real and unequal quantities a and b, the solution is $y = he^{ax} + ke^{bx}$, where h and k are constants.

If the roots are complex quantities $m + in$ and $m - in$,

$$y = e^{mx}(h \cos nx + k \sin nx).$$

If the roots are equal and are a, a,

$$y = e^{ax}(hx + k).$$

895. Equations of the nth order of the form

$$\frac{d^n y}{dx^n} + A \frac{d^{n-1} y}{dx^{n-1}} + B \frac{d^{n-2} y}{dx^{n-2}} + \cdots + Ky = 0,$$

where $A, B, \cdots, K$ are constants. This is a linear differential equation. For each distinct real root a of the auxiliary equation

$$p^n + Ap^{n-1} + Bp^{n-2} + \cdots + K = 0,$$

there is a term he^{ax} in the solution. The terms of the solution are to be added together.

When a occurs twice among the n roots of the auxiliary equation, the corresponding term is $e^{ax}(hx + k)$.

When a occurs three times, the corresponding term is

$$e^{ax}(hx^2 + kx + l),$$

and so forth.

When there is a pair of imaginary roots $m + in$ and $m - in$, there is a term in the solution

$$e^{mx}(h \cos nx + k \sin nx).$$

When the same pair occurs twice, the corresponding term in the solution is

$$e^{mx}\{(hx + k) \cos nx + (sx + t) \sin nx\}$$

and so forth.

896. Linear differential equations with constant coefficients.

$$\frac{d^n y}{dx^n} + A \frac{d^{n-1} y}{dx^{n-1}} + B \frac{d^{n-2} y}{dx^{n-2}} + \cdots + Ky = X$$

where X may involve x.

First solve the equation obtained by putting $X = 0$, as in **894** or **895**. Add to this solution a **particular integral** which satisfies the original equation and which need not contain constants of integration since n such constants have already been put in the solution.

897.   The "homogeneous linear equation" of the second order,

$$x^2 \frac{d^2y}{dx^2} + Ax \frac{dy}{dx} + By = f(x)$$

becomes a linear equation with constant coefficients

$$\frac{d^2y}{dv^2} + (A - 1) \frac{dy}{dv} + By = f(e^v)$$

by substituting $x = e^v$.

[See Ref. 8, *Elements of the Infinitesimal Calculus*, Chaps. 44–45, by G. H. Chandler, or other textbooks.]

898.   Linear partial differential equation of the first order,

$$P \frac{\partial z}{\partial x} + Q \frac{\partial z}{\partial y} = R.$$

To solve this, first solve the equations

$$\frac{dx}{P} = \frac{dy}{Q} = \frac{dz}{R},$$

and place the solution in the form $u = c_1$, $v = c_2$. Then

$$\phi(u, v) = 0,$$

where ϕ is an arbitrary function, is the solution required.

[Ref. 11, p. 292.]

APPENDIX

A. Tables of Numerical Values

B. References

Table 1000—$\sqrt{(a^2 + b^2)}/a$

b/a	$\dfrac{\sqrt{(a^2 + b^2)}}{a}$	b/a	$\dfrac{\sqrt{(a^2 + b^2)}}{a}$	b/a	$\dfrac{\sqrt{(a^2 + b^2)}}{a}$	b/a	$\dfrac{\sqrt{(a^2 + b^2)}}{a}$	b/a	$\dfrac{\sqrt{(a^2 + b^2)}}{a}$	b/a	$\dfrac{\sqrt{(a^2 + b^2)}}{a}$	b/a	$\dfrac{\sqrt{(a^2 + b^2)}}{a}$	b/a	$\dfrac{\sqrt{(a^2 + b^2)}}{a}$
0	1.000	.175	1.015	.300	1.044	.350	1.059	.400	1.077	.450	1.097	.500	1.118	.550	1.141
.010	1.000	.180	1.016	.302	1.045	.352	1.060	.402	1.078	.452	1.097	.502	1.119	.552	1.142
.020	1.000	.185	1.017	.304	1.045	.354	1.061	.404	1.079	.454	1.098	.504	1.120	.554	1.143
.030	1.000	.190	1.018	.306	1.046	.356	1.061	.406	1.079	.456	1.099	.506	1.121	.556	1.144
.040	1.001	.195	1.019	.308	1.046	.358	1.062	.408	1.080	.458	1.100	.508	1.122	.558	1.145
.050	1.001	.200	1.020	.310	1.047	.360	1.063	.410	1.081	.460	1.101	.510	1.123	.560	1.146
.060	1.002	.205	1.021	.312	1.048	.362	1.064	.412	1.082	.462	1.102	.512	1.123	.562	1.147
.070	1.002	.210	1.022	.314	1.048	.364	1.064	.414	1.082	.464	1.102	.514	1.124	.564	1.148
.080	1.003	.215	1.023	.316	1.049	.366	1.065	.416	1.083	.466	1.103	.516	1.125	.566	1.149
.090	1.004	.220	1.024	.318	1.049	.368	1.066	.418	1.084	.468	1.104	.518	1.126	.568	1.150
.100	1.005	.225	1.025	.320	1.050	.370	1.066	.420	1.085	.470	1.105	.520	1.127	.570	1.151
.105	1.005	.230	1.026	.322	1.051	.372	1.067	.422	1.085	.472	1.106	.522	1.128	.572	1.152
.110	1.006	.235	1.027	.324	1.051	.374	1.068	.424	1.086	.474	1.107	.524	1.129	.574	1.153
.115	1.007	.240	1.028	.326	1.052	.376	1.068	.426	1.087	.476	1.108	.526	1.130	.576	1.154
.120	1.007	.245	1.030	.328	1.052	.378	1.069	.428	1.088	.478	1.108	.528	1.131	.578	1.155
.125	1.008	.250	1.031	.330	1.053	.380	1.070	.430	1.089	.480	1.109	.530	1.132	.580	1.156
.130	1.008	.255	1.032	.332	1.054	.382	1.070	.432	1.089	.482	1.110	.532	1.133	.582	1.157
.135	1.009	.260	1.033	.334	1.054	.384	1.071	.434	1.090	.484	1.111	.534	1.134	.584	1.158
.140	1.010	.265	1.035	.336	1.055	.386	1.072	.436	1.091	.486	1.112	.536	1.135	.586	1.159
.145	1.010	.270	1.036	.338	1.056	.388	1.073	.438	1.092	.488	1.113	.538	1.136	.588	1.160
.150	1.011	.275	1.037	.340	1.056	.390	1.073	.440	1.093	.490	1.114	.540	1.136	.590	1.161
.155	1.012	.280	1.038	.342	1.057	.392	1.074	.442	1.093	.492	1.114	.542	1.137	.592	1.162
.160	1.013	.285	1.040	.344	1.058	.394	1.075	.444	1.094	.494	1.115	.544	1.138	.594	1.163
.165	1.014	.290	1.041	.346	1.058	.396	1.076	.446	1.095	.496	1.116	.546	1.139	.596	1.164
.170	1.014	.295	1.043	.348	1.059	.398	1.076	.448	1.096	.498	1.117	.548	1.140	.598	1.165

Table 1000 (continued)—$\sqrt{(a^2+b^2)}/a$

b/a	$\frac{\sqrt{(a^2+b^2)}}{a}$	b/a	$\frac{\sqrt{(a^2+b^2)}}{a}$	b/a	$\frac{\sqrt{(a^2+b^2)}}{a}$	b/a	$\frac{\sqrt{(a^2+b^2)}}{a}$	b/a	$\frac{\sqrt{(a^2+b^2)}}{a}$	b/a	$\frac{\sqrt{(a^2+b^2)}}{a}$	b/a	$\frac{\sqrt{(a^2+b^2)}}{a}$	b/a	$\frac{\sqrt{(a^2+b^2)}}{a}$
.600	1.166	.650	1.193	.700	1.221	.750	1.250	.800	1.281	.850	1.312	.900	1.345	.950	1.379
.602	1.167	.652	1.194	.702	1.222	.752	1.251	.802	1.282	.852	1.314	.902	1.347	.952	1.381
.604	1.168	.654	1.195	.704	1.223	.754	1.252	.804	1.283	.854	1.315	.904	1.348	.954	1.382
.606	1.169	.656	1.196	.706	1.224	.756	1.254	.806	1.284	.856	1.316	.906	1.349	.956	1.383
.608	1.170	.658	1.197	.708	1.225	.758	1.255	.808	1.286	.858	1.318	.908	1.351	.958	1.385
.610	1.171	.660	1.198	.710	1.226	.760	1.256	.810	1.287	.860	1.319	.910	1.352	.960	1.386
.612	1.172	.662	1.199	.712	1.228	.762	1.257	.812	1.288	.862	1.320	.912	1.353	.962	1.388
.614	1.173	.664	1.200	.714	1.229	.764	1.258	.814	1.289	.864	1.322	.914	1.355	.964	1.389
.616	1.175	.666	1.201	.716	1.230	.766	1.260	.816	1.291	.866	1.323	.916	1.356	.966	1.390
.618	1.176	.668	1.203	.718	1.231	.768	1.261	.818	1.292	.868	1.324	.918	1.357	.968	1.392
.620	1.177	.670	1.204	.720	1.232	.770	1.262	.820	1.293	.870	1.325	.920	1.359	.970	1.393
.622	1.178	.672	1.205	.722	1.233	.772	1.263	.822	1.294	.872	1.327	.922	1.360	.972	1.395
.624	1.179	.674	1.206	.724	1.235	.774	1.265	.824	1.296	.874	1.328	.924	1.362	.974	1.396
.626	1.180	.676	1.207	.726	1.236	.776	1.266	.826	1.297	.876	1.329	.926	1.363	.976	1.397
.628	1.181	.678	1.208	.728	1.237	.778	1.267	.828	1.298	.878	1.331	.928	1.364	.978	1.399
.630	1.182	.680	1.209	.730	1.238	.780	1.268	.830	1.300	.880	1.332	.930	1.366	.980	1.400
.632	1.183	.682	1.210	.732	1.239	.782	1.269	.832	1.301	.882	1.333	.932	1.367	.982	1.402
.634	1.184	.684	1.212	.734	1.240	.784	1.271	.834	1.302	.884	1.335	.934	1.368	.984	1.403
.636	1.185	.686	1.213	.736	1.242	.786	1.272	.836	1.303	.886	1.336	.936	1.370	.986	1.404
.638	1.186	.688	1.214	.738	1.243	.788	1.273	.838	1.305	.888	1.337	.938	1.371	.988	1.406
.640	1.187	.690	1.215	.740	1.244	.790	1.274	.840	1.306	.890	1.339	.940	1.372	.990	1.407
.642	1.188	.692	1.216	.742	1.245	.792	1.276	.842	1.307	.892	1.340	.942	1.374	.992	1.409
.644	1.189	.694	1.217	.744	1.246	.794	1.277	.844	1.309	.894	1.341	.944	1.375	.994	1.410
.646	1.191	.696	1.218	.746	1.248	.796	1.278	.846	1.310	.896	1.343	.946	1.377	.996	1.411
.648	1.192	.698	1.220	.748	1.249	.798	1.279	.848	1.311	.898	1.344	.948	1.378	.998	1.413
														1.000	1.414

$$\sqrt{(a^2+b^2)} = a + \frac{b^2}{2a} - \frac{b^4}{8a^3} + \cdots [b^2 < a^2].$$ The approximation $\sqrt{(a^2+b^2)} = a + \frac{b^2}{2a}$ is correct within 1/1000 when $b/a < 0.3$.

Table 1005—Gamma Function [See 850]

n	Γ(n)	Diff.	n	Γ(n)	Diff.	n	Γ(n)	Diff.	n	Γ(n)	Diff.
1.00	1.000 00	−567	1.25	.906 40	−200	1.50	.886 23	36	1.75	.919 06	231
1.01	.994 33	−549	1.26	.904 40	−190	1.51	.886 59	45	1.76	.921 37	239
1.02	.988 84	−529	1.27	.902 50	−178	1.52	.887 04	53	1.77	.923 76	247
1.03	.983 55	−511	1.28	.900 72	−168	1.53	.887 57	61	1.78	.926 23	254
1.04	.978 44	−494	1.29	.899 04	−157	1.54	.888 18	69	1.79	.928 77	261
1.05	.973 50	−476	1.30	.897 47	−147	1.55	.888 87	77	1.80	.931 38	270
1.06	.968 74	−459	1.31	.896 00	−136	1.56	.889 64	85	1.81	.934 08	277
1.07	.964 15	−442	1.32	.894 64	−126	1.57	.890 49	93	1.82	.936 85	284
1.08	.959 73	−427	1.33	.893 38	−116	1.58	.891 42	101	1.83	.939 69	292
1.09	.955 46	−411	1.34	.892 22	−107	1.59	.892 43	109	1.84	.942 61	300
1.10	.951 35	−395	1.35	.891 15	−97	1.60	.893 52	116	1.85	.945 61	308
1.11	.947 40	−381	1.36	.890 18	−87	1.61	.894 68	124	1.86	.948 69	315
1.12	.943 59	−366	1.37	.889 31	−77	1.62	.895 92	132	1.87	.951 84	323
1.13	.939 93	−351	1.38	.888 54	−69	1.63	.897 24	140	1.88	.955 07	331
1.14	.936 42	−338	1.39	.887 85	−59	1.64	.898 64	148	1.89	.958 38	339
1.15	.933 04	−324	1.40	.887 26	−50	1.65	.900 12	155	1.90	.961 77	346
1.16	.929 80	−310	1.41	.886 76	−40	1.66	.901 67	163	1.91	.965 23	354
1.17	.926 70	−297	1.42	.886 36	−32	1.67	.903 30	170	1.92	.968 77	363
1.18	.923 73	−284	1.43	.886 04	−23	1.68	.905 00	178	1.93	.972 40	370
1.19	.920 89	−272	1.44	.885 81	−15	1.69	.906 78	186	1.94	.976 10	378
1.20	.918 17	−259	1.45	.885 66	−6	1.70	.908 64	193	1.95	.979 88	386
1.21	.915 58	−247	1.46	.885 60	3	1.71	.910 57	201	1.96	.983 74	394
1.22	.913 11	−236	1.47	.885 63	12	1.72	.912 58	209	1.97	.987 68	403
1.23	.910 75	−223	1.48	.885 75	20	1.73	.914 67	216	1.98	.991 71	410
1.24	.908 52	−212	1.49	.885 95	28	1.74	.916 83	223	1.99	.995 81	419
									2.00	1.000 00	

For other values of n, use this table and make successive applications of the following equation:

$$\Gamma(n + 1) = n\Gamma(n).$$

For more complete tables, see Ref. 6, p. 169, Ref. 44, v. 1, pp. 196–273, and Ref. 45.

Table 1010—Trigonometric Functions

[Characteristics of Logarithms omitted — determine by the usual rule from the value]

Radians	Degrees	Sine Value	Log₁₀	Tangent Value	Log₁₀	Cotangent Value	Log₁₀	Cosine Value	Log₁₀		
.0000	0° 00′	.0000	——	.0000	——	——	——	1.0000	.0000	90° 00′	1.5708
.0029	10	.0029	.4637	.0029	.4637	343.77	.5363	1.0000	.0000	50	1.5679
.0058	20	.0058	.7648	.0058	.7648	171.89	.2352	1.0000	.0000	40	1.5650
.0087	30	.0087	.9408	.0087	.9409	114.59	.0591	1.0000	.0000	30	1.5621
.0116	40	.0116	.0658	.0116	.0658	85.940	.9342	.9999	.0000	20	1.5592
.0145	50	.0145	.1627	.0145	.1627	68.750	.8373	.9999	.0000	10	1.5563
.0175	1° 00′	.0175	.2419	.0175	.2419	57.290	.7581	.9998	.9999	89° 00′	1.5533
.0204	10	.0204	.3088	.0204	.3089	49.104	.6911	.9998	.9999	50	1.5504
.0233	20	.0233	.3668	.0233	.3669	42.964	.6331	.9997	.9999	40	1.5475
.0262	30	.0262	.4179	.0262	.4181	38.188	.5819	.9997	.9999	30	1.5446
.0291	40	.0291	.4637	.0291	.4638	34.368	.5362	.9996	.9998	20	1.5417
.0320	50	.0320	.5050	.0320	.5053	31.242	.4947	.9995	.9998	10	1.5388
.0349	2° 00′	.0349	.5428	.0349	.5431	28.636	.4569	.9994	.9997	88° 00′	1.5359
.0378	10	.0378	.5776	.0378	.5779	26.432	.4221	.9993	.9997	50	1.5330
.0407	20	.0407	.6097	.0407	.6101	24.542	.3899	.9992	.9996	40	1.5301
.0436	30	.0436	.6397	.0437	.6401	22.904	.3599	.9990	.9996	30	1.5272
.0465	40	.0465	.6677	.0466	.6682	21.470	.3318	.9989	.9995	20	1.5243
.0495	50	.0494	.6940	.0495	.6945	20.206	.3055	.9988	.9995	10	1.5213
.0524	3° 00′	.0523	.7188	.0524	.7194	19.081	.2806	.9986	.9994	87° 00′	1.5184
.0553	10	.0552	.7423	.0553	.7429	18.075	.2571	.9985	.9993	50	1.5155
.0582	20	.0581	.7645	.0582	.7652	17.169	.2348	.9983	.9993	40	1.5126
.0611	30	.0610	.7857	.0612	.7865	16.350	.2135	.9981	.9992	30	1.5097
.0640	40	.0640	.8059	.0641	.8067	15.605	.1933	.9980	.9991	20	1.5068
.0669	50	.0669	.8251	.0670	.8261	14.924	.1739	.9978	.9990	10	1.5039
.0698	4° 00′	.0698	.8436	.0699	.8446	14.301	.1554	.9976	.9989	86° 00′	1.5010
.0727	10	.0727	.8613	.0729	.8624	13.727	.1376	.9974	.9989	50	1.4981
.0756	20	.0756	.8783	.0758	.8795	13.197	.1205	.9971	.9988	40	1.4952
.0785	30	.0785	.8946	.0787	.8960	12.706	.1040	.9969	.9987	30	1.4923
.0814	40	.0814	.9104	.0816	.9118	12.251	.0882	.9967	.9986	20	1.4893
.0844	50	.0843	.9256	.0846	.9272	11.826	.0728	.9964	.9985	10	1.4864
.0873	5° 00′	.0872	.9403	.0875	.9420	11.430	.0580	.9962	.9983	85° 00′	1.4835
.0902	10	.0901	.9545	.0904	.9563	11.059	.0437	.9959	.9982	50	1.4806
.0931	20	.0929	.9682	.0934	.9701	10.712	.0299	.9957	.9981	40	1.4777
.0960	30	.0958	.9816	.0963	.9836	10.385	.0164	.9954	.9980	30	1.4748
.0989	40	.0987	.9945	.0992	.9966	10.078	.0034	.9951	.9979	20	1.4719
.1018	50	.1016	.0070	.1022	.0093	9.7882	.9907	.9948	.9977	10	1.4690
.1047	6° 00′	.1045	.0192	.1051	.0216	9.5144	.9784	.9945	.9976	84° 00′	1.4661
.1076	10	.1074	.0311	.1080	.0336	9.2553	.9664	.9942	.9975	50	1.4632
.1105	20	.1103	.0426	.1110	.0453	9.0098	.9547	.9939	.9973	40	1.4603
.1134	30	.1132	.0539	.1139	.0567	8.7769	.9433	.9936	.9972	30	1.4573
.1164	40	.1161	.0648	.1169	.0678	8.5555	.9322	.9932	.9971	20	1.4544
.1193	50	.1190	.0755	.1198	.0786	8.3450	.9214	.9929	.9969	10	1.4515
.1222	7° 00′	.1219	.0859	.1228	.0891	8.1443	.9109	.9925	.9968	83° 00′	1.4486
.1251	10	.1248	.0961	.1257	.0995	7.9530	.9005	.9922	.9966	50	1.4457
.1280	20	.1276	.1060	.1287	.1096	7.7704	.8904	.9918	.9964	40	1.4428
.1309	30	.1305	.1157	.1317	.1194	7.5958	.8806	.9914	.9963	30	1.4399
.1338	40	.1334	.1252	.1346	.1291	7.4287	.8709	.9911	.9961	20	1.4370
.1367	50	.1363	.1345	.1376	.1385	7.2687	.8615	.9907	.9959	10	1.4341
.1396	8° 00′	.1392	.1436	.1405	.1478	7.1154	.8522	.9903	.9958	82° 00′	1.4312
.1425	10	.1421	.1525	.1435	.1569	6.9682	.8431	.9899	.9956	50	1.4283
.1454	20	.1449	.1612	.1465	.1658	6.8269	.8342	.9894	.9954	40	1.4254
.1484	30	.1478	.1697	.1495	.1745	6.6912	.8255	.9890	.9952	30	1.4224
.1513	40	.1507	.1781	.1524	.1831	6.5606	.8169	.9886	.9950	20	1.4195
.1542	50	.1536	.1863	.1554	.1915	6.4348	.8085	.9881	.9948	10	1.4166
.1571	9° 00′	.1564	.1943	.1584	.1997	6.3138	.8003	.9877	.9946	81° 00′	1.4137
		Value Cosine	Log₁₀	Value Cotangent	Log₁₀	Value Tangent	Log₁₀	Value Sine	Log₁₀	Degrees	Radians

261

Table 1010 (*continued*)—Trigonometric Functions

[Characteristics of Logarithms omitted — determine by the usual rule from the value]

RADIANS	DEGREES	SINE Value	SINE Log₁₀	TANGENT Value	TANGENT Log₁₀	COTANGENT Value	COTANGENT Log₁₀	COSINE Value	COSINE Log₁₀		
.1571	9° 00′	.1564	.1943	.1584	.1997	6.3138	.8003	.9877	.9946	81° 00′	1.4137
.1600	10	.1593	.2022	.1614	.2078	6.1970	.7922	.9872	.9944	50	1.4108
.1629	20	.1622	.2100	.1644	.2158	6.0844	.7842	.9868	.9942	40	1.4079
.1658	30	.1650	.2176	.1673	.2236	5.9758	.7764	.9863	.9940	30	1.4050
.1687	40	.1679	.2251	.1703	.2313	5.8708	.7687	.9858	.9938	20	1.4021
.1716	50	.1708	.2324	.1733	.2389	5.7694	.7611	.9853	.9936	10	1.3992
.1745	10° 00′	.1736	.2397	.1763	.2463	5.6713	.7537	.9848	.9934	80° 00′	1.3963
.1774	10	.1765	.2468	.1793	.2536	5.5764	.7464	.9843	.9931	50	1.3934
.1804	20	.1794	.2538	.1823	.2609	5.4845	.7391	.9838	.9929	40	1.3904
.1833	30	.1822	.2606	.1853	.2680	5.3955	.7320	.9833	.9927	30	1.3875
.1862	40	.1851	.2674	.1883	.2750	5.3093	.7250	.9827	.9924	20	1.3846
.1891	50	.1880	.2740	.1914	.2819	5.2257	.7181	.9822	.9922	10	1.3817
.1920	11° 00′	.1908	.2806	.1944	.2887	5.1446	.7113	.9816	.9919	79° 00′	1.3788
.1949	10	.1937	.2870	.1974	.2953	5.0658	.7047	.9811	.9917	50	1.3759
.1978	20	.1965	.2931	.2004	.3020	4.9894	.6980	.9805	.9914	40	1.3730
.2007	30	.1994	.2997	.2035	.3085	4.9152	.6915	.9799	.9912	30	1.3701
.2036	40	.2022	.3058	.2065	.3149	4.8430	.6851	.9793	.9909	20	1.3672
.2065	50	.2051	.3119	.2095	.3212	4.7729	.6788	.9787	.9907	10	1.3643
.2094	12° 00′	.2079	.3179	.2126	.3275	4.7046	.6725	.9781	.9904	78° 00′	1.3614
.2123	10	.2108	.3238	.2156	.3336	4.6382	.6664	.9775	.9901	50	1.3584
.2153	20	.2136	.3296	.2186	.3397	4.5736	.6603	.9769	.9899	40	1.3555
.2182	30	.2164	.3353	.2217	.3458	4.5107	.6542	.9763	.9896	30	1.3526
.2211	40	.2193	.3410	.2247	.3517	4.4494	.6483	.9757	.9893	20	1.3497
.2240	50	.2221	.3466	.2278	.3576	4.3897	.6424	.9750	.9890	10	1.3468
.2269	13° 00′	.2250	.3521	.2309	.3634	4.3315	.6366	.9744	.9887	77° 00′	1.3439
.2298	10	.2278	.3575	.2339	.3691	4.2747	.6309	.9737	.9884	50	1.3410
.2327	20	.2306	.3629	.2370	.3748	4.2193	.6252	.9730	.9881	40	1.3381
.2356	30	.2334	.3682	.2401	.3804	4.1653	.6196	.9724	.9878	30	1.3352
.2385	40	.2363	.3734	.2432	.3859	4.1126	.6141	.9717	.9875	20	1.3323
.2414	50	.2391	.3786	.2462	.3914	4.0611	.6086	.9710	.9872	10	1.3294
.2443	14° 00′	.2419	.3837	.2493	.3968	4.0108	.6032	.9703	.9869	76° 00′	1.3265
.2473	10	.2447	.3887	.2524	.4021	3.9617	.5979	.9696	.9866	50	1.3235
.2502	20	.2476	.3937	.2555	.4074	3.9136	.5926	.9689	.9863	40	1.3206
.2531	30	.2504	.3986	.2586	.4127	3.8667	.5873	.9681	.9859	30	1.3177
.2560	40	.2532	.4035	.2617	.4178	3.8208	.5822	.9674	.9856	20	1.3148
.2589	50	.2560	.4083	.2648	.4230	3.7760	.5770	.9667	.9853	10	1.3119
.2618	15° 00′	.2588	.4130	.2679	.4281	3.7321	.5719	.9659	.9849	75° 00′	1.3090
.2647	10	.2616	.4177	.2711	.4331	3.6891	.5669	.9652	.9846	50	1.3061
.2676	20	.2644	.4223	.2742	.4381	3.6470	.5619	.9644	.9843	40	1.3032
.2705	30	.2672	.4269	.2773	.4430	3.6059	.5570	.9636	.9839	30	1.3003
.2734	40	.2700	.4314	.2805	.4479	3.5656	.5521	.9628	.9836	20	1.2974
.2763	50	.2728	.4359	.2836	.4527	3.5261	.5473	.9621	.9832	10	1.2945
.2793	16° 00′	.2756	.4403	.2867	.4575	3.4874	.5425	.9613	.9828	74° 00′	1.2915
.2822	10	.2784	.4447	.2899	.4622	3.4495	.5378	.9605	.9825	50	1.2886
.2851	20	.2812	.4491	.2931	.4669	3.4124	.5331	.9596	.9821	40	1.2857
.2880	30	.2840	.4533	.2962	.4716	3.3759	.5284	.9588	.9817	30	1.2828
.2909	40	.2868	.4576	.2994	.4762	3.3402	.5238	.9580	.9814	20	1.2799
.2938	50	.2896	.4618	.3026	.4808	3.3052	.5192	.9572	.9810	10	1.2770
.2967	17° 00′	.2924	.4659	.3057	.4853	3.2709	.5147	.9563	.9806	73° 00′	1.2741
.2996	10	.2952	.4700	.3089	.4898	3.2371	.5102	.9555	.9802	50	1.2712
.3025	20	.2979	.4741	.3121	.4943	3.2041	.5057	.9546	.9798	40	1.2683
.3054	30	.3007	.4781	.3153	.4987	3.1716	.5013	.9537	.9794	30	1.2654
.3083	40	.3035	.4821	.3185	.5031	3.1397	.4969	.9528	.9790	20	1.2625
.3113	50	.3062	.4861	.3217	.5075	3.1084	.4925	.9520	.9786	10	1.2595
.3142	18° 00′	.3090	.4900	.3249	.5118	3.0777	.4882	.9511	.9782	72° 00′	1.2566
		Value COSINE	Log₁₀	Value COTANGENT	Log₁₀	Value TANGENT	Log₁₀	Value SINE	Log₁₀	DEGREES	RADIANS

Table 1010 (*continued*)—**Trigonometric Functions**

[Characteristics of Logarithms omitted — determine by the usual rule from the value]

Radians	Degrees	Sine Value	Log₁₀	Tangent Value	Log₁₀	Cotangent Value	Log₁₀	Cosine Value	Log₁₀		
.3142	**18° 00′**	.3090	.4900	.3249	.5118	3.0777	.4882	.9511	.9782	**72° 00′**	1.2566
.3171	10	.3118	.4939	.3281	.5161	3.0475	.4839	.9502	.9778	50	1.2537
.3200	20	.3145	.4977	.3314	.5203	3.0178	.4797	.9492	.9774	40	1.2508
.3229	30	.3173	.5015	.3346	.5245	2.9887	.4755	.9483	.9770	30	1.2479
.3258	40	.3201	.5052	.3378	.5287	2.9600	.4713	.9474	.9765	20	1.2450
.3287	50	.3228	.5090	.3411	.5329	2.9319	.4671	.9465	.9761	10	1.2421
.3316	**19° 00′**	.3256	.5126	.3443	.5370	2.9042	.4630	.9455	.9757	**71° 00′**	1.2392
.3345	10	.3283	.5163	.3476	.5411	2.8770	.4589	.9446	.9752	50	1.2363
.3374	20	.3311	.5199	.3508	.5451	2.8502	.4549	.9436	.9748	40	1.2334
.3403	30	.3338	.5235	.3541	.5491	2.8239	.4509	.9426	.9743	30	1.2305
.3432	40	.3365	.5270	.3574	.5531	2.7980	.4469	.9417	.9739	20	1.2275
.3462	50	.3393	.5306	.3607	.5571	2.7725	.4429	.9407	.9734	10	1.2246
.3491	**20° 00′**	.3420	.5341	.3640	.5611	2.7475	.4389	.9397	.9730	**70° 00′**	1.2217
.3520	10	.3448	.5375	.3673	.5650	2.7228	.4350	.9387	.9725	50	1.2188
.3549	20	.3475	.5409	.3706	.5689	2.6985	.4311	.9377	.9721	40	1.2159
.3578	30	.3502	.5443	.3739	.5727	2.6746	.4273	.9367	.9716	30	1.2130
.3607	40	.3529	.5477	.3772	.5766	2.6511	.4234	.9356	.9711	20	1.2101
.3636	50	.3557	.5510	.3805	.5804	2.6279	.4196	.9346	.9706	10	1.2072
.3665	**21° 00′**	.3584	.5543	.3839	.5842	2.6051	.4158	.9336	.9702	**69° 00′**	1.2043
.3694	10	.3611	.5576	.3872	.5879	2.5826	.4121	.9325	.9697	50	1.2014
.3723	20	.3638	.5609	.3906	.5917	2.5605	.4083	.9315	.9692	40	1.1985
.3752	30	.3665	.5641	.3939	.5954	2.5386	.4046	.9304	.9687	30	1.1956
.3782	40	.3692	.5673	.3973	.5991	2.5172	.4009	.9293	.9682	20	1.1926
.3811	50	.3719	.5704	.4006	.6028	2.4960	.3972	.9283	.9677	10	1.1897
.3840	**22° 00′**	.3746	.5736	.4040	.6064	2.4751	.3936	.9272	.9672	**68° 00′**	1.1868
.3869	10	.3773	.5767	.4074	.6100	2.4545	.3900	.9261	.9667	50	1.1839
.3898	20	.3800	.5798	.4108	.6136	2.4342	.3864	.9250	.9661	40	1.1810
.3927	30	.3827	.5828	.4142	.6172	2.4142	.3828	.9239	.9656	30	1.1781
.3956	40	.3854	.5859	.4176	.6208	2.3945	.3792	.9228	.9651	20	1.1752
.3985	50	.3881	.5889	.4210	.6243	2.3750	.3757	.9216	.9646	10	1.1723
.4014	**23° 00′**	.3907	.5919	.4245	.6279	2.3559	.3721	.9205	.9640	**67° 00′**	1.1694
.4043	10	.3934	.5948	.4279	.6314	2.3369	.3686	.9194	.9635	50	1.1665
.4072	20	.3961	.5978	.4314	.6348	2.3183	.3652	.9182	.9629	40	1.1636
.4102	30	.3987	.6007	.4348	.6383	2.2998	.3617	.9171	.9624	30	1.1606
.4131	40	.4014	.6036	.4383	.6417	2.2817	.3583	.9159	.9618	20	1.1577
.4160	50	.4041	.6065	.4417	.6452	2.2637	.3548	.9147	.9613	10	1.1548
.4189	**24° 00′**	.4067	.6093	.4452	.6486	2.2460	.3514	.9135	.9607	**66° 00′**	1.1519
.4218	10	.4094	.6121	.4487	.6520	2.2286	.3480	.9124	.9602	50	1.1490
.4247	20	.4120	.6149	.4522	.6553	2.2113	.3447	.9112	.9596	40	1.1461
.4276	30	.4147	.6177	.4557	.6587	2.1943	.3413	.9100	.9590	30	1.1432
.4305	40	.4173	.6205	.4592	.6620	2.1775	.3380	.9088	.9584	20	1.1403
.4334	50	.4200	.6232	.4628	.6654	2.1609	.3346	.9075	.9579	10	1.1374
.4363	**25° 00′**	.4226	.6259	.4663	.6687	2.1445	.3313	.9063	.9573	**65° 00′**	1.1345
.4392	10	.4253	.6286	.4699	.6720	2.1283	.3280	.9051	.9567	50	1.1316
.4422	20	.4279	.6313	.4734	.6752	2.1123	.3248	.9038	.9561	40	1.1286
.4451	30	.4305	.6340	.4770	.6785	2.0965	.3215	.9026	.9555	30	1.1257
.4480	40	.4331	.6366	.4806	.6817	2.0809	.3183	.9013	.9549	20	1.1228
.4509	50	.4358	.6392	.4841	.6850	2.0655	.3150	.9001	.9543	10	1.1199
.4538	**26° 00′**	.4384	.6418	.4877	.6882	2.0503	.3118	.8988	.9537	**64° 00′**	1.1170
.4567	10	.4410	.6444	.4913	.6914	2.0353	.3086	.8975	.9530	50	1.1141
.4596	20	.4436	.6470	.4950	.6946	2.0204	.3054	.8962	.9524	40	1.1112
.4625	30	.4462	.6495	.4986	.6977	2.0057	.3023	.8949	.9518	30	1.1083
.4654	40	.4488	.6521	.5022	.7009	1.9912	.2991	.8936	.9512	20	1.1054
.4683	50	.4514	.6546	.5059	.7040	1.9768	.2960	.8923	.9505	10	1.1025
.4712	**27° 00′**	.4540	.6570	.5095	.7072	1.9626	.2928	.8910	.9499	**63° 00′**	1.0996
		Value Cosine	Log₁₀	Value Cotangent	Log₁₀	Value Tangent	Log₁₀	Value Sine	Log₁₀	Degrees	Radians

Table 1010 (*continued*)—Trigonometric Functions

[Characteristics of Logarithms omitted — determine by the usual rule from the value]

RADIANS	DEGREES	SINE Value	Log₁₀	TANGENT Value	Log₁₀	COTANGENT Value	Log₁₀	COSINE Value	Log₁₀		
.4712	27°00'	.4540	.6570	.5095	.7072	1.9626	.2928	.8910	.9499	63°00'	1.0996
.4741	10	.4566	.6595	.5132	.7103	1.9486	.2897	.8897	.9492	50	1.0966
.4771	20	.4592	.6620	.5169	.7134	1.9347	.2866	.8884	.9486	40	1.0937
.4800	30	.4617	.6644	.5206	.7165	1.9210	.2835	.8870	.9479	30	1.0908
.4829	40	.4643	.6668	.5243	.7196	1.9074	.2804	.8857	.9473	20	1.0879
.4858	50	.4669	.6692	.5280	.7226	1.8940	.2774	.8843	.9466	10	1.0850
.4887	28°00'	.4695	.6716	.5317	.7257	1.8807	.2743	.8829	.9459	62°00'	1.0821
.4916	10	.4720	.6740	.5354	.7287	1.8676	.2713	.8816	.9453	50	1.0792
.4945	20	.4746	.6763	.5392	.7317	1.8546	.2683	.8802	.9446	40	1.0763
.4974	30	.4772	.6787	.5430	.7348	1.8418	.2652	.8788	.9439	30	1.0734
.5003	40	.4797	.6810	.5467	.7378	1.8291	.2622	.8774	.9432	20	1.0705
.5032	50	.4823	.6833	.5505	.7408	1.8165	.2592	.8760	.9425	10	1.0676
.5061	29°00'	.4848	.6856	.5543	.7438	1.8040	.2562	.8746	.9418	61°00'	1.0647
.5091	10	.4874	.6878	.5581	.7467	1.7917	.2533	.8732	.9411	50	1.0617
.5120	20	.4899	.6901	.5619	.7497	1.7796	.2503	.8718	.9404	40	1.0588
.5149	30	.4924	.6923	.5658	.7526	1.7675	.2474	.8704	.9397	30	1.0559
.5178	40	.4950	.6946	.5696	.7556	1.7556	.2444	.8689	.9390	20	1.0530
.5207	50	.4975	.6968	.5735	.7585	1.7437	.2415	.8675	.9383	10	1.0501
.5236	30°00'	.5000	.6990	.5774	.7614	1.7321	.2386	.8660	.9375	60°00'	1.0472
.5265	10	.5025	.7012	.5812	.7644	1.7205	.2356	.8646	.9368	50	1.0443
.5294	20	.5050	.7033	.5851	.7673	1.7090	.2327	.8631	.9361	40	1.0414
.5323	30	.5075	.7055	.5890	.7701	1.6977	.2299	.8616	.9353	30	1.0385
.5352	40	.5100	.7076	.5930	.7730	1.6864	.2270	.8601	.9346	20	1.0356
.5381	50	.5125	.7097	.5969	.7759	1.6753	.2241	.8587	.9338	10	1.0327
.5411	31°00'	.5150	.7118	.6009	.7788	1.6643	.2212	.8572	.9331	59°00'	1.0297
.5440	10	.5175	.7139	.6048	.7816	1.6534	.2184	.8557	.9323	50	1.0268
.5469	20	.5200	.7160	.6088	.7845	1.6426	.2155	.8542	.9315	40	1.0239
.5498	30	.5225	.7181	.6128	.7873	1.6319	.2127	.8526	.9308	30	1.0210
.5527	40	.5250	.7201	.6168	.7902	1.6212	.2098	.8511	.9300	20	1.0181
.5556	50	.5275	.7222	.6208	.7930	1.6107	.2070	.8496	.9292	10	1.0152
.5585	32°00'	.5299	.7242	.6249	.7958	1.6003	.2042	.8480	.9284	58°00'	1.0123
.5614	10	.5324	.7262	.6289	.7986	1.5900	.2014	.8465	.9276	50	1.0094
.5643	20	.5348	.7282	.6330	.8014	1.5798	.1986	.8450	.9268	40	1.0065
.5672	30	.5373	.7302	.6371	.8042	1.5697	.1958	.8434	.9260	30	1.0036
.5701	40	.5398	.7322	.6412	.8070	1.5597	.1930	.8418	.9252	20	1.0007
.5730	50	.5422	.7342	.6453	.8097	1.5497	.1903	.8403	.9244	10	.9977
.5760	33°00'	.5446	.7361	.6494	.8125	1.5399	.1875	.8387	.9236	57°00'	.9948
.5789	10	.5471	.7380	.6536	.8153	1.5301	.1847	.8371	.9228	50	.9919
.5818	20	.5495	.7400	.6577	.8180	1.5204	.1820	.8355	.9219	40	.9890
.5847	30	.5519	.7419	.6619	.8208	1.5108	.1792	.8339	.9211	30	.9861
.5876	40	.5544	.7438	.6661	.8235	1.5013	.1765	.8323	.9203	20	.9832
.5905	50	.5568	.7457	.6703	.8263	1.4919	.1737	.8307	.9194	10	.9803
.5934	34°00'	.5592	.7476	.6745	.8290	1.4826	.1710	.8290	.9186	56°00'	.9774
.5963	10	.5616	.7494	.6787	.8317	1.4733	.1683	.8274	.9177	50	.9745
.5992	20	.5640	.7513	.6830	.8344	1.4641	.1656	.8258	.9169	40	.9716
.6021	30	.5664	.7531	.6873	.8371	1.4550	.1629	.8241	.9160	30	.9687
.6050	40	.5688	.7550	.6916	.8398	1.4460	.1602	.8225	.9151	20	.9657
.6080	50	.5712	.7568	.6959	.8425	1.4370	.1575	.8208	.9142	10	.9628
.6109	35°00'	.5736	.7586	.7002	.8452	1.4281	.1548	.8192	.9134	55°00'	.9599
.6138	10	.5760	.7604	.7046	.8479	1.4193	.1521	.8175	.9125	50	.9570
.6167	20	.5783	.7622	.7089	.8506	1.4106	.1494	.8158	.9116	40	.9541
.6196	30	.5807	.7640	.7133	.8533	1.4019	.1467	.8141	.9107	30	.9512
.6225	40	.5831	.7657	.7177	.8559	1.3934	.1441	.8124	.9098	20	.9483
.6254	50	.5854	.7675	.7221	.8586	1.3848	.1414	.8107	.9089	10	.9454
.6283	36°00'	.5878	.7692	.7265	.8613	1.3764	.1387	.8090	.9080	54°00'	.9425
		Value COSINE	Log₁₀	Value COTANGENT	Log₁₀	Value TANGENT	Log₁₀	Value SINE	Log₁₀	DEGREES	RADIANS

Table 1010 (*continued*)—**Trigonometric Functions**

[Characteristics of Logarithms omitted — determine by the usual rule from the value]

RADIANS	DEGREES	SINE Value	SINE Log₁₀	TANGENT Value	TANGENT Log₁₀	COTANGENT Value	COTANGENT Log₁₀	COSINE Value	COSINE Log₁₀		
.6283	36° 00′	.5878	.7692	.7265	.8613	1.3764	.1387	.8090	.9080	54° 00′	.9425
.6312	10	.5901	.7710	.7310	.8639	1.3680	.1361	.8073	.9070	50	.9396
.6341	20	.5925	.7727	.7355	.8666	1.3597	.1334	.8056	.9061	40	.9367
.6370	30	.5948	.7744	.7400	.8692	1.3514	.1308	.8039	.9052	30	.9338
.6400	40	.5972	.7761	.7445	.8718	1.3432	.1282	.8021	.9042	20	.9308
.6429	50	.5995	.7778	.7490	.8745	1.3351	.1255	.8004	.9033	10	.9279
.6458	37° 00′	.6018	.7795	.7536	.8771	1.3270	.1229	.7986	.9023	53° 00′	.9250
.6487	10	.6041	.7811	.7581	.8797	1.3190	.1203	.7969	.9014	50	.9221
.6516	20	.6065	.7828	.7627	.8824	1.3111	.1176	.7951	.9004	40	.9192
.6545	30	.6088	.7844	.7673	.8850	1.3032	.1150	.7934	.8995	30	.9163
.6574	40	.6111	.7861	.7720	.8876	1.2954	.1124	.7916	.8985	20	.9134
.6603	50	.6134	.7877	.7766	.8902	1.2876	.1098	.7898	.8975	10	.9105
.6632	38° 00′	.6157	.7893	.7813	.8928	1.2799	.1072	.7880	.8965	52° 00′	.9076
.6661	10	.6180	.7910	.7860	.8954	1.2723	.1046	.7862	.8955	50	.9047
.6690	20	.6202	.7926	.7907	.8980	1.2647	.1020	.7844	.8945	40	.9018
.6720	30	.6225	.7941	.7954	.9006	1.2572	.0994	.7826	.8935	30	.8988
.6749	40	.6248	.7957	.8002	.9032	1.2497	.0968	.7808	.8925	20	.8959
.6778	50	.6271	.7973	.8050	.9058	1.2423	.0942	.7790	.8915	10	.8930
.6807	39° 00′	.6293	.7989	.8098	.9084	1.2349	.0916	.7771	.8905	51° 00′	.8901
.6836	10	.6316	.8004	.8146	.9110	1.2276	.0890	.7753	.8895	50	.8872
.6865	20	.6338	.8020	.8195	.9135	1.2203	.0865	.7735	.8884	40	.8843
.6894	30	.6361	.8035	.8243	.9161	1.2131	.0839	.7716	.8874	30	.8814
.6923	40	.6383	.8050	.8292	.9187	1.2059	.0813	.7698	.8864	20	.8785
.6952	50	.6406	.8066	.8342	.9212	1.1988	.0788	.7679	.8853	10	.8756
.6981	40° 00′	.6428	.8081	.8391	.9238	1.1918	.0762	.7660	.8843	50° 00′	.8727
.7010	10	.6450	.8096	.8441	.9264	1.1847	.0736	.7642	.8832	50	.8698
.7039	20	.6472	.8111	.8491	.9289	1.1778	.0711	.7623	.8821	40	.8668
.7069	30	.6494	.8125	.8541	.9315	1.1708	.0685	.7604	.8810	30	.8639
.7098	40	.6517	.8140	.8591	.9341	1.1640	.0659	.7585	.8300	20	.8610
.7127	50	.6539	.8155	.8642	.9366	1.1571	.0634	.7566	.8789	10	.8581
.7156	41° 00′	.6561	.8169	.8693	.9392	1.1504	.0608	.7547	.8778	49° 00′	.8552
.7185	10	.6583	.8184	.8744	.9417	1.1436	.0583	.7528	.8767	50	.8523
.7214	20	.6604	.8198	.8796	.9443	1.1369	.0557	.7509	.8756	40	.8494
.7243	30	.6626	.8213	.8847	.9468	1.1303	.0532	.7490	.8745	30	.8465
.7272	40	.6648	.8227	.8899	.9494	1.1237	.0506	.7470	.8733	20	.8436
.7301	50	.6670	.8241	.8952	.9519	1.1171	.0481	.7451	.8722	10	.8407
.7330	42° 00′	.6691	.8255	.9004	.9544	1.1106	.0456	.7431	.8711	48° 00′	.8378
.7359	10	.6713	.8269	.9057	.9570	1.1041	.0430	.7412	.8699	50	.8348
.7389	20	.6734	.8283	.9110	.9595	1.0977	.0405	.7392	.8688	40	.8319
.7418	30	.6756	.8297	.9163	.9621	1.0913	.0379	.7373	.8676	30	.8290
.7447	40	.6777	.8311	.9217	.9646	1.0850	.0354	.7353	.8665	20	.8261
.7476	50	.6799	.8324	.9271	.9671	1.0786	.0329	.7333	.8653	10	.8232
.7505	43° 00′	.6820	.8338	.9325	.9697	1.0724	.0303	.7314	.8641	47° 00′	.8203
.7534	10	.6841	.8351	.9380	.9722	1.0661	.0278	.7294	.8629	50	.8174
.7563	20	.6862	.8365	.9435	.9747	1.0599	.0253	.7274	.8618	40	.8145
.7592	30	.6884	.8378	.9490	.9772	1.0538	.0228	.7254	.8606	30	.8116
.7621	40	.6905	.8391	.9545	.9798	1.0477	.0202	.7234	.8594	20	.8087
.7650	50	.6926	.8405	.9601	.9823	1.0416	.0177	.7214	.8582	10	.8058
.7679	44° 00′	.6947	.8418	.9657	.9848	1.0355	.0152	.7193	.8569	46° 00′	.8029
.7709	10	.6967	.8431	.9713	.9874	1.0295	.0126	.7173	.8557	50	.7999
.7738	20	.6988	.8444	.9770	.9899	1.0235	.0101	.7153	.8545	40	.7970
.7767	30	.7009	.8457	.9827	.9924	1.0176	.0076	.7133	.8532	30	.7941
.7796	40	.7030	.8469	.9884	.9949	1.0117	.0051	.7112	.8520	20	.7912
.7825	50	.7050	.8482	.9942	.9975	1.0058	.0025	.7092	.8507	10	.7883
.7854	45° 00′	.7071	.8495	1.0000	.0000	1.0000	.0000	.7071	.8495	45° 00′	.7854
		Value COSINE	Log₁₀	Value COTANGENT	Log₁₀	Value TANGENT	Log₁₀	Value SINE	Log₁₀	DEGREES	RADIANS

Table 1011—Degrees, Minutes, and Seconds to Radians

Degrees						Minutes		Seconds	
0°	0.00000 00	60°	1.04719 76	120°	2.09439 51	0′	0.00000 00	0″	0.00000 00
1	0.01745 33	61	1.06465 08	121	2.11184 84	1	0.0002909	1	0.00000 48
2	0.03490 66	62	1.08210 41	122	2.12930 17	2	0.0005818	2	0.00000 97
3	0.05235 99	63	1.09955 74	123	2.14675 50	3	0.0008727	3	0.00001 45
4	0.06981 32	64	1.11701 07	124	2.16420 83	4	0.0011636	4	0.00001 94
5	0.08726 65	65	1.13446 40	125	2.18166 16	5	0.0014544	5	0.00002 42
6	0.10471 98	66	1.15191 73	126	2.19911 49	6	0.0017453	6	0.00002 91
7	0.12217 30	67	1.16937 06	127	2.21656 82	7	0.0020362	7	0.00003 39
8	0.13962 63	68	1.18682 39	128	2.23402 14	8	0.0023271	8	0.00003 88
9	0.15707 96	69	1.20427 72	129	2.25147 47	9	0.0026180	9	0.00004 36
10	0.17453 29	70	1.22173 05	130	2.26892 80	10	0.0029089	10	0.00004 85
11	0.19192 62	71	1.23918 38	131	2.28658 13	11	0.0031998	11	0.00005 33
12	0.20943 95	72	1.25663 71	132	2.30383 46	12	0.0034907	12	0.00005 82
13	0.22689 28	73	1.27409 04	133	2.32128 79	13	0.0037815	13	0.00006 30
14	0.24434 61	74	1.29154 36	134	2.33874 12	14	0.0040724	14	0.00006 79
15	0.26179 94	75	1.30899 69	135	2.35619 45	15	0.0043633	15	0.00007 27
16	0.27925 27	76	1.32645 02	136	2.37364 78	16	0.0046542	16	0.00007 76
17	0.29670 60	77	1.34390 35	137	2.39110 11	17	0.0049451	17	0.00008 24
18	0.31415 93	78	1.36135 68	138	2.40855 44	18	0.0052360	18	0.00008 73
19	0.33161 26	79	1.37881 01	139	2.42600 77	19	0.0055269	19	0.00009 21
20	0.34906 59	80	1.39626 34	140	2.44346 10	20	0.0058178	20	0.00009 70
21	0.36651 91	81	1.41371 67	141	2.46091 42	21	0.0061087	21	0.00010 18
22	0.38397 24	82	1.43117 00	142	2.47836 75	22	0.0063995	22	0.00010 67
23	0.40142 57	83	1.44862 33	143	2.49582 08	23	0.0066904	23	0.00011 15
24	0.41887 90	84	1.46607 66	144	2.51327 41	24	0.0069813	24	0.00011 64
25	0.43633 23	85	1.48352 99	145	2.53072 74	25	0.0072722	25	0.00012 12
26	0.45378 56	86	1.50098 32	146	2.54818 07	26	0.0075631	26	0.00012 61
27	0.47123 89	87	1.51843 64	147	2.56563 40	27	0.0078540	27	0.00013 09
28	0.48869 22	88	1.53588 97	148	2.58308 73	28	0.0081449	28	0.00013 57
29	0.50614 55	89	1.55334 30	149	2.60054 06	29	0.0084358	29	0.00014 06
30	0.52359 88	90	1.57079 63	150	2.61799 39	30	0.0087266	30	0.00014 54
31	0.54105 21	91	1.58824 96	151	2.63544 72	31	0.0090175	31	0.00015 03
32	0.55850 54	92	1.60570 29	152	2.65290 05	32	0.0093084	32	0.00015 51
33	0.57595 87	93	1.62315 62	153	2.67035 38	33	0.0095993	33	0.00016 00
34	0.59341 19	94	1.64060 95	154	2.68780 70	34	0.0098902	34	0.00016 48
35	0.61086 52	95	1.65806 28	155	2.70526 03	35	0.0101811	35	0.00016 97
36	0.62831 85	96	1.67551 61	156	2.72271 36	36	0.0104720	36	0.00017 45
37	0.64577 18	97	1.69296 94	157	2.74016 69	37	0.0107629	37	0.00017 94
38	0.66322 51	98	1.71042 27	158	2.75762 02	38	0.0110538	38	0.00018 42
39	0.68067 84	99	1.72787 60	159	2.77507 35	39	0.0113446	39	0.00018 91
40	0.69813 17	100	1.74532 93	160	2.79252 68	40	0.0116355	40	0.00019 39
41	0.71558 50	101	1.76278 25	161	2.80998 01	41	0.0119264	41	0.00019 88
42	0.73303 83	102	1.78023 58	162	2.82743 34	42	0.0122173	42	0.00020 36
43	0.75049 16	103	1.79768 91	163	2.84488 67	43	0.0125082	43	0.00020 85
44	0.76794 49	104	1.81514 24	164	2.86234 00	44	0.0127991	44	0.00021 33
45	0.78539 82	105	1.83259 57	165	2.87979 33	45	0.0130900	45	0.00021 82
46	0.80285 15	106	1.85004 90	166	2.89724 66	46	0.0133809	46	0.00022 30
47	0.82030 47	107	1.86750 23	167	2.91469 99	47	0.0136717	47	0.00022 79
48	0.83775 80	108	1.88495 56	168	2.93215 31	48	0.0139626	48	0.00023 27
49	0.85521 13	109	1.90240 89	169	2.94960 64	49	0.0142535	49	0.00023 76
50	0.87266 46	110	1.91986 22	170	2.96705 97	50	0.0145444	50	0.00024 24
51	0.89011 79	111	1.93731 55	171	2.98451 30	51	0.0148353	51	0.00024 73
52	0.90757 12	112	1.95476 88	172	3.00196 63	52	0.0151262	52	0.00025 21
53	0.92502 45	113	1.97222 21	173	3.01941 96	53	0.0154171	53	0.00025 70
54	0.94247 78	114	1.98967 53	174	3.03687 29	54	0.0157080	54	0.00026 18
55	0.95993 11	115	2.00712 86	175	3.05432 62	55	0.0159989	55	0.00026 66
56	0.97738 44	116	2.02458 19	176	3.07177 95	56	0.0162897	56	0.00027 15
57	0.99483 77	117	2.04203 52	177	3.08923 28	57	0.0165806	57	0.00027 63
58	1.01229 10	118	2.05948 85	178	3.10668 61	58	0.0168715	58	0.00028 12
59	1.02974 43	119	2.07694 18	179	3.12413 94	59	0.0171624	59	0.00028 60
60	1.04719 76	120	2.09439 51	180	3.14159 27	60	0.0174533	60	0.00029 09

Table 1012—Radians to Degrees, Minutes, and Seconds

	RADIANS	TENTHS	HUNDREDTHS	THOUSANDTHS	TEN-THOUSANDTHS
1	57°17'44".8	5°43'46".5	0°34'22".6	0° 3'26".3	0° 0'20".6
2	114°35'29".6	11°27'33".0	1° 8'45".3	0° 6'52".5	0° 0'41".3
3	171°53'14".4	17°11'19".4	1°43'07".9	0°10'18".8	0° 1'01".9
4	229°10'59".2	22°55'05".9	2°17'30".6	0°13'45".1	0° 1'22".5
5	286°28'44".0	28°38'52".4	2°51'53".2	0°17'11".3	0° 1'43".1
6	343°46'28".8	34°22'38".9	3°26'15".9	0°20'37".6	0° 2'03".8
7	401° 4'13".6	40° 6'25".4	4° 0'38".5	0°24'03".9	0° 2'24".4
8	458°21'58".4	45°50'11".8	4°35'01".2	0°27'30".1	0° 2'45".0
9	515°39'43".3	51°33'58".3	5° 9'23".8	0°30'56".4	0° 3'05".6

In decimals,

$$1 \text{ radian} = 180/\pi = 57.295\ 77951 \text{ degrees}$$
$$1 \text{ degree} = \pi/180 = 0.017453\ 29252 \text{ radians.}$$

Trigonometric tables such as Tables 1015 and 1016 on the pages following often may be used advantageously by first converting the angles of a problem to decimals of degrees.

In these tables, where the name of the function is given at the top of the page, the degrees for that function are to be read from the left-hand column and the top line. The degrees for the function named at the bottom of the page are to be read from the right-hand column and the bottom line.

Table 1015—Trigonometric Functions
SIN

deg.	0	1	2	3	4	5	6	7	8	9	(10)	
0.0	.00 000	017	035	052	070	087	105	122	140	157	175	.9
.1	175	192	209	227	244	262	279	297	314	332	349	.8
.2	349	367	384	401	419	436	454	471	489	506	524	.7
.3	524	541	559	576	593	611	628	646	663	681	698	.6
.4	698	716	733	750	768	785	803	820	838	855	873	.5
.5	873	890	908	925	942	960	977	995	012	030	047	.4
.6	.01 047	065	082	100	117	134	152	169	187	204	222	.3
.7	222	239	257	274	292	309	326	344	361	379	396	.2
.8	396	414	431	449	466	483	501	518	536	553	571	.1
.9	571	588	606	623	641	658	675	693	710	728	745	89.0
1.0	745	763	780	798	815	832	850	867	885	902	920	.9
.1	920	937	955	972	990	007	024	042	059	077	094	.8
.2	.02 094	112	129	147	164	181	199	216	234	251	269	.7
.3	269	286	304	321	339	356	373	391	408	426	443	.6
.4	443	461	478	496	513	530	548	565	583	600	618	.5
.5	618	635	653	670	687	705	722	740	757	775	792	.4
.6	792	810	827	845	862	879	897	914	932	949	967	.3
.7	967	984	002	019	036	054	071	089	106	124	141	.2
.8	.03 141	159	176	193	211	228	246	263	281	298	316	.1
.9	316	333	350	368	385	403	420	438	455	473	490	88.0
2.0	490	507	525	542	560	577	595	612	629	647	664	.9
.1	664	682	699	717	734	752	769	786	804	821	839	.8
.2	839	856	874	891	909	926	943	961	978	996	013	.7
.3	.04 013	031	048	065	083	100	118	135	153	170	188	.6
.4	188	205	222	240	257	275	292	310	327	345	362	.5 Diff.
.5	362	379	397	414	432	449	467	484	501	519	536	.4
.6	536	554	571	589	606	623	641	658	676	693	711	.3 17–18
.7	711	728	746	763	780	798	815	833	850	868	885	.2
.8	885	902	920	937	955	972	990	007	024	042	059	.1
.9	.05 059	077	094	112	129	146	164	181	199	216	234	87.0
3.0	234	251	268	286	303	321	338	356	373	390	408	.9
.1	408	425	443	460	478	495	512	530	547	565	582	.8
.2	582	600	617	634	652	669	687	704	722	739	756	.7
.3	756	774	791	809	826	844	861	878	896	913	931	.6
.4	931	948	965	983	000	018	035	053	070	087	105	.5
.5	.06 105	122	140	157	175	192	209	227	244	262	279	.4
.6	279	296	314	331	349	366	384	401	418	436	453	.3
.7	453	471	488	505	523	540	558	575	593	610	627	.2
.8	627	645	662	680	697	714	732	749	767	784	802	.1
.9	802	819	836	854	871	889	906	923	941	958	976	86.0
4.0	976	993	010	028	045	063	080	098	115	132	150	.9
.1	.07 150	167	185	202	219	237	254	272	289	306	324	.8
.2	324	341	359	376	393	411	428	446	463	480	498	.7
.3	498	515	533	550	567	585	602	620	637	655	672	.6
.4	672	689	707	724	742	759	776	794	811	829	846	.5
.5	846	863	881	898	916	933	950	968	985	002	020	.4
.6	.08 020	037	055	072	089	107	124	142	159	176	194	.3
.7	194	211	229	246	263	281	298	316	333	350	368	.2
.8	368	385	403	420	437	455	472	490	507	524	542	.1
.9	542	559	576	594	611	629	646	663	681	698	716	85.0 deg.
(10)	9	8	7	6	5	4	3	2	1	0		

COS

Table 1015 (continued)—Sin and Cos of Hundredths of Degrees

SIN

	0	1	2	3	4	5	6	7	8	9	(10)	
deg. 5.0	.08 716	733	750	768	785	803	820	837	855	872	889	.9
.1	889	907	924	942	959	976	994	011	028	046	063	.8
.2	.09 063	081	098	115	133	150	168	185	202	220	237	.7
.3	237	254	272	289	307	324	341	359	376	393	411	.6
.4	411	428	446	463	480	498	515	532	550	567	585	.5
.5	585	602	619	637	654	671	689	706	724	741	758	.4
.6	758	776	793	810	828	845	863	880	897	915	932	.3
.7	932	949	967	984	001	019	036	054	071	088	106	.2
.8	.10 106	123	140	158	175	192	210	227	245	262	279	.1
.9	279	297	314	331	349	366	383	401	418	435	453	84.0
6.0	453	470	488	505	522	540	557	574	592	609	626	.9
.1	626	644	661	678	696	713	731	748	765	783	800	.8
.2	800	817	835	852	869	887	904	921	939	956	973	.7
.3	973	991	008	025	043	060	078	095	112	130	147	.6
.4	.11 147	164	182	199	216	234	251	268	286	303	320	.5
.5	320	338	355	372	390	407	424	442	459	476	494	.4
.6	494	511	528	546	563	580	598	615	632	650	667	.3
.7	667	684	702	719	736	754	771	788	806	823	840	.2
.8	840	858	875	892	910	927	944	962	979	996	014	.1
.9	.12 014	031	048	066	083	100	118	135	152	170	187	83.0
7.0	187	204	222	239	256	274	291	308	326	343	360	.9
.1	360	377	395	412	429	447	464	481	499	516	533	.8
.2	533	551	568	585	603	620	637	655	672	689	706	.7
.3	706	724	741	758	776	793	810	828	845	862	880	.6
.4	880	897	914	931	949	966	983	001	018	035	053	.5 Diff.
.5	.13 053	070	087	105	122	139	156	174	191	208	226	.4
.6	226	243	260	278	295	312	329	347	364	381	399	.3 17–18
.7	399	416	433	451	468	485	502	520	537	554	572	.2
.8	572	589	606	623	641	658	675	693	710	727	744	.1
.9	744	762	779	796	814	831	848	865	883	900	917	82.0
8.0	917	935	952	969	986	004	021	038	056	073	090	.9
.1	.14 090	107	125	142	159	177	194	211	228	246	263	.8
.2	263	280	297	315	332	349	367	384	401	418	436	.7
.3	436	453	470	487	505	522	539	557	574	591	608	.6
.4	608	626	643	660	677	695	712	729	746	764	781	.5
.5	781	798	815	833	850	867	885	902	919	936	954	.4
.6	954	971	988	005	023	040	057	074	092	109	126	.3
.7	.15 126	143	161	178	195	212	230	247	264	281	299	.2
.8	299	316	333	350	368	385	402	419	437	454	471	.1
.9	471	488	506	523	540	557	574	592	609	626	643	81.0
9.0	643	661	678	695	712	730	747	764	781	799	816	.9
.1	816	833	850	868	885	902	919	936	954	971	988	.8
.2	988	005	023	040	057	074	091	109	126	143	160	.7
.3	.16 160	178	195	212	229	246	264	281	298	315	333	.6
.4	333	350	367	384	401	419	436	453	470	488	505	.5
.5	505	522	539	556	574	591	608	625	642	660	677	.4
.6	677	694	711	728	746	763	780	797	815	832	849	.3
.7	849	866	883	901	918	935	952	969	987	004	021	.2
.8	.17 021	038	055	073	090	107	124	141	159	176	193	.1
.9	193	210	227	244	262	279	296	313	330	348	365	80.0 deg.
	(10)	9	8	7	6	5	4	3	2	1	0	

COS

269

Table 1015 (*continued*)—Trigonometric Functions
SIN

deg.	0	1	2	3	4	5	6	7	8	9	(10)	
10.0	.17 365	382	399	416	434	451	468	485	502	519	537	.9
.1	537	554	571	588	605	623	640	657	674	691	708	.8
.2	708	726	743	760	777	794	812	829	846	863	880	.7
.3	880	897	915	932	949	966	983	000	018	035	052	.6
.4	.18 052	069	086	103	121	138	155	172	189	206	224	.5
.5	224	241	258	275	292	309	327	344	361	378	395	.4
.6	395	412	429	447	464	481	498	515	532	550	567	.3
.7	567	584	601	618	635	652	670	687	704	721	738	.2
.8	738	755	772	790	807	824	841	858	875	892	910	.1
.9	910	927	944	961	978	995	012	029	047	064	081	79.0
11.0	.19 081	098	115	132	149	167	184	201	218	235	252	.9
.1	252	269	286	304	321	338	355	372	389	406	423	.8
.2	423	441	458	475	492	509	526	543	560	577	595	.7
.3	595	612	629	646	663	680	697	714	732	749	766	.6
.4	766	783	800	817	834	851	868	885	903	920	937	.5
.5	937	954	971	988	005	022	039	056	074	091	108	.4
.6	.20 108	125	142	159	176	193	210	227	245	262	279	.3
.7	279	296	313	330	347	364	381	398	415	433	450	.2
.8	450	467	484	501	518	535	552	569	586	603	620	.1
.9	620	637	655	672	689	706	723	740	757	774	791	78.0
12.0	791	808	825	842	859	877	894	911	928	945	962	.9
.1	962	979	996	013	030	047	064	081	098	115	132	.8
.2	.21 132	150	167	184	201	218	235	252	269	286	303	.7
.3	303	320	337	354	371	388	405	422	439	456	474	.6
.4	474	491	508	525	542	559	576	593	610	627	644	.5 Diff.
.5	644	661	678	695	712	729	746	763	780	797	814	.4
.6	814	831	848	865	882	899	917	934	951	968	985	.3 16–18
.7	985	002	019	036	053	070	087	104	121	138	155	.2
.8	.22 155	172	189	206	223	240	257	274	291	308	325	.1
.9	325	342	359	376	393	410	427	444	461	478	495	77.0
13.0	495	512	529	546	563	580	597	614	631	648	665	.9
.1	665	682	699	716	733	750	767	784	801	818	835	.8
.2	835	852	869	886	903	920	937	954	971	988	005	.7
.3	.23 005	022	039	056	073	090	107	124	141	158	175	.6
.4	175	192	209	226	243	260	277	294	311	328	345	.5
.5	345	362	378	395	412	429	446	463	480	497	514	.4
.6	514	531	548	565	582	599	616	633	650	667	684	.3
.7	684	701	718	735	752	769	786	802	819	836	853	.2
.8	853	870	887	904	921	938	955	972	989	006	023	.1
.9	.24 023	040	057	074	091	108	124	141	158	175	192	76.0
14.0	192	209	226	243	260	277	294	311	328	345	362	.9
.1	362	378	395	412	429	446	463	480	497	514	531	.8
.2	531	548	565	581	598	615	632	649	666	683	700	.7
.3	700	717	734	751	768	784	801	818	835	852	869	.6
.4	869	886	903	920	937	954	970	987	004	021	038	.5
.5	.25 038	055	072	089	106	122	139	156	173	190	207	.4
.6	207	224	241	258	274	291	308	325	342	359	376	.3
.7	376	393	410	426	443	460	477	494	511	528	545	.2
.8	545	561	578	595	612	629	646	663	680	696	713	.1
.9	713	730	747	764	781	798	814	831	848	865	882	75.0 deg.
	(10)	9	8	7	6	5	4	3	2	1	0	

COS

Table 1015 (*continued*)—**Sin and Cos of Hundredths of Degrees**

SIN

deg.	0	1	2	3	4	5	6	7	8	9	(10)	
15.0	.25 882	899	916	932	949	966	983	000	017	034	050	.9
.1	.26 050	067	084	101	118	135	152	168	185	202	219	.8
.2	219	236	253	269	286	303	320	337	354	370	387	.7
.3	387	404	421	438	455	471	488	505	522	539	556	.6
.4	556	572	589	606	623	640	657	673	690	707	724	.5
.5	724	741	757	774	791	808	825	842	858	875	892	.4
.6	892	909	926	942	959	976	993	010	026	043	060	.3
.7	.27 060	077	094	110	127	144	161	178	194	211	228	.2
.8	228	245	262	278	295	312	329	346	362	379	396	.1
.9	396	413	429	446	463	480	497	513	530	547	564	74.0
16.0	564	581	597	614	631	648	664	681	698	715	731	.9
.1	731	748	765	782	799	815	832	849	866	882	899	.8
.2	899	916	933	949	966	983	000	016	033	050	067	.7
.3	.28 067	083	100	117	134	150	167	184	201	217	234	.6
.4	234	251	268	284	301	318	335	351	368	385	402	.5
.5	402	418	435	452	468	485	502	519	535	552	569	.4
.6	569	586	602	619	636	652	669	686	703	719	736	.3
.7	736	753	769	786	803	820	836	853	870	886	903	.2
.8	903	920	937	953	970	987	003	020	037	054	070	.1
.9	.29 070	087	104	120	137	154	170	187	204	220	237	73.0
17.0	237	254	271	287	304	321	337	354	371	387	404	.9
.1	404	421	437	454	471	487	504	521	537	554	571	.8
.2	571	587	604	621	637	654	671	687	704	721	737	.7
.3	737	754	771	787	804	821	837	854	871	887	904	.6
.4	904	921	937	954	971	987	004	021	037	054	071	.5 Diff.
.5	.30 071	087	104	121	137	154	170	187	204	220	237	.4
.6	237	254	270	287	304	320	337	353	370	387	403	.3 16–17
.7	403	420	437	453	470	486	503	520	536	553	570	.2
.8	570	586	603	619	636	653	669	686	702	719	736	.1
.9	736	752	769	785	802	819	835	852	868	885	902	72.0
18.0	902	918	935	951	968	985	001	018	034	051	068	.9
.1	.31 068	084	101	117	134	151	167	184	200	217	233	.8
.2	233	250	267	283	300	316	333	350	366	383	399	.7
.3	399	416	432	449	466	482	499	515	532	548	565	.6
.4	565	581	598	615	631	648	664	681	697	714	730	.5
.5	730	747	764	780	797	813	830	846	863	879	896	.4
.6	896	912	929	946	962	979	995	012	028	045	061	.3
.7	.32 061	078	094	111	127	144	160	177	194	210	227	.2
.8	227	243	260	276	293	309	326	342	359	375	392	.1
.9	392	408	425	441	458	474	491	507	524	540	557	71.0
19.0	557	573	590	606	623	639	656	672	689	705	722	.9
.1	722	738	755	771	788	804	821	837	854	870	887	.8
.2	887	903	920	936	953	969	986	002	018	035	051	.7
.3	.33 051	068	084	101	117	134	150	167	183	200	216	.6
.4	216	233	249	265	282	298	315	331	348	364	381	.5
.5	381	397	414	430	446	463	479	496	512	529	545	.4
.6	545	562	578	594	611	627	644	660	677	693	710	.3
.7	710	726	742	759	775	792	808	825	841	857	874	.2
.8	874	890	907	923	939	956	972	989	005	022	038	.1
.9	.34 038	054	071	087	104	120	136	153	169	186	202	70.0 deg.
	(10)	9	8	7	6	5	4	3	2	1	0	

COS

271

Table 1015 (*continued*)—Trigonometric Functions
SIN

	0	1	2	3	4	5	6	7	8	9	(10)	
deg.												
20.0	.34 202	218	235	251	268	284	300	317	333	350	366	.9
.1	366	382	399	415	432	448	464	481	497	513	530	.8
.2	530	546	563	579	595	612	628	644	661	677	694	.7
.3	694	710	726	743	759	775	792	808	824	841	857	.6
.4	857	874	890	906	923	939	955	972	988	004	021	.5
.5	.35 021	037	053	070	086	102	119	135	151	168	184	.4
.6	184	201	217	233	250	266	282	298	315	331	347	.3
.7	347	364	380	396	413	429	445	462	478	494	511	.2
.8	511	527	543	560	576	592	609	625	641	657	674	.1
.9	674	690	706	723	739	755	772	788	804	821	837	69.0
21.0	837	853	869	886	902	918	935	951	967	983	000	.9
.1	.36 000	016	032	049	065	081	097	114	130	146	162	.8
.2	162	179	195	211	228	244	260	276	293	309	325	.7
.3	325	341	358	374	390	406	423	439	455	471	488	.6
.4	488	504	520	536	553	569	585	601	618	634	650	.5
.5	650	666	683	699	715	731	748	764	780	796	812	.4
.6	812	829	845	861	877	894	910	926	942	958	975	.3
.7	975	991	007	023	040	056	072	088	104	121	137	.2
.8	.37 137	153	169	185	202	218	234	250	266	283	299	.1
.9	299	315	331	347	364	380	396	412	428	444	461	68.0
22.0	461	477	493	509	525	542	558	574	590	606	622	.9
.1	622	639	655	671	687	703	719	736	752	768	784	.8
.2	784	800	816	833	849	865	881	897	913	929	946	.7
.3	946	962	978	994	010	026	042	059	075	091	107	.6
.4	.38 107	123	139	155	172	188	204	220	236	252	268	.5 Diff.
.5	268	284	301	317	333	349	365	381	397	413	430	.4
.6	430	446	462	478	494	510	526	542	558	575	591	.3 15–17
.7	591	607	623	639	655	671	687	703	719	735	752	.2
.8	752	768	784	800	816	832	848	864	880	896	912	.1
.9	912	928	945	961	977	993	009	025	041	057	073	67.0
23.0	.39 073	089	105	121	137	153	169	186	202	218	234	.9
.1	234	250	266	282	298	314	330	346	362	378	394	.8
.2	394	410	426	442	458	474	490	506	522	539	555	.7
.3	555	571	587	603	619	635	651	667	683	699	715	.6
.4	715	731	747	763	779	795	811	827	843	859	875	.5
.5	875	891	907	923	939	955	971	987	003	019	035	.4
.6	.40 035	051	067	083	099	115	131	147	163	179	195	.3
.7	195	211	227	243	259	275	291	307	323	339	355	.2
.8	355	370	386	402	418	434	450	466	482	498	514	.1
.9	514	530	546	562	578	594	610	626	642	658	674	66.0
24.0	674	690	706	721	737	753	769	785	801	817	833	.9
.1	833	849	865	881	897	913	929	945	960	976	992	.8
.2	992	008	024	040	056	072	088	104	120	136	151	.7
.3	.41 151	167	183	199	215	231	247	263	279	295	310	.6
.4	310	326	342	358	374	390	406	422	438	453	469	.5
.5	469	485	501	517	533	549	565	580	596	612	628	.4
.6	628	644	660	676	692	707	723	739	755	771	787	.3
.7	787	803	818	834	850	866	882	898	914	929	945	.2
.8	945	961	977	993	009	024	040	056	072	088	104	.1
.9	.42 104	119	135	151	167	183	199	214	230	246	262	65.0 deg.
	(10)	9	8	7	6	5	4	3	2	1	0	

COS

Table 1015 (*continued*)—Sin and Cos of Hundredths of Degrees

SIN

deg.	0	1	2	3	4	5	6	7	8	9	(10)	
25.0	.42 262	278	293	309	325	341	357	373	388	404	420	.9
.1	420	436	452	467	483	499	515	531	546	562	578	.8
.2	578	594	610	625	641	657	673	688	704	720	736	.7
.3	736	752	767	783	799	815	830	846	862	878	894	.6
.4	894	909	925	941	957	972	988	004	020	035	051	.5
.5	.43 051	067	083	098	114	130	146	161	177	193	209	.4
.6	209	224	240	256	272	287	303	319	334	350	366	.3
.7	366	382	397	413	429	445	460	476	492	507	523	.2
.8	523	539	555	570	586	602	617	633	649	664	680	.1
.9	680	696	712	727	743	759	774	790	806	821	837	64.0
26.0	837	853	868	884	900	916	931	947	963	978	994	.9
.1	994	010	025	041	057	072	088	104	119	135	151	.8
.2	.44 151	166	182	198	213	229	245	260	276	291	307	.7
.3	307	323	338	354	370	385	401	417	432	448	464	.6
.4	464	479	495	510	526	542	557	573	589	604	620	.5
.5	620	635	651	667	682	698	713	729	745	760	776	.4
.6	776	792	807	823	838	854	870	885	901	916	932	.3
.7	932	947	963	979	994	010	025	041	057	072	088	.2
.8	.45 088	103	119	134	150	166	181	197	212	228	243	.1
.9	243	259	275	290	306	321	337	352	368	383	399	63.0
27.0	399	415	430	446	461	477	492	508	523	539	554	.9
.1	554	570	586	601	617	632	648	663	679	694	710	.8
.2	710	725	741	756	772	787	803	818	834	849	865	.7
.3	865	880	896	911	927	942	958	973	989	004	020	.6
.4	.46 020	035	051	066	082	097	113	128	144	159	175	.5 Diff.
.5	175	190	206	221	237	252	268	283	299	314	330	.4
.6	330	345	361	376	391	407	422	438	453	469	484	.3 15–16
.7	484	500	515	531	546	561	577	592	608	623	639	.2
.8	639	654	670	685	700	716	731	747	762	778	793	.1
.9	793	808	824	839	855	870	886	901	916	932	947	62.0
28.0	947	963	978	993	009	024	040	055	070	086	101	.9
.1	.47 101	117	132	147	163	178	194	209	224	240	255	.8
.2	255	270	286	301	317	332	347	363	378	393	409	.7
.3	409	424	440	455	470	486	501	516	532	547	562	.6
.4	562	578	593	608	624	639	655	670	685	701	716	.5
.5	716	731	747	762	777	793	808	823	839	854	869	.4
.6	869	885	900	915	930	946	961	976	992	007	022	.3
.7	.48 022	038	053	068	084	099	114	129	145	160	175	.2
.8	175	191	206	221	237	252	267	282	298	313	328	.1
.9	328	344	359	374	389	405	420	435	450	466	481	61.0
29.0	481	496	511	527	542	557	573	588	603	618	634	.9
.1	634	649	664	679	695	710	725	740	755	771	786	.8
.2	786	801	816	832	847	862	877	893	908	923	938	.7
.3	938	953	969	984	999	014	030	045	060	075	090	.6
.4	.49 090	106	121	136	151	166	182	197	212	227	242	.5
.5	242	258	273	288	303	318	333	349	364	379	394	.4
.6	394	409	425	440	455	470	485	500	516	531	546	.3
.7	546	561	576	591	606	622	637	652	667	682	697	.2
.8	697	713	728	743	758	773	788	803	819	834	849	.1
.9	849	864	879	894	909	924	940	955	970	985	000	60.0
.50												deg.
	(10)	9	8	7	6	5	4	3	2	1	0	

COS
273

Table 1015 (*continued*)—Trigonometric Functions
SIN

deg.	0	1	2	3	4	5	6	7	8	9	(10)	
30.0	.50 000	015	030	045	060	076	091	106	121	136	151	.9
.1	151	166	181	196	211	227	242	257	272	287	302	.8
.2	302	317	332	347	362	377	392	408	423	438	453	.7
.3	453	468	483	498	513	528	543	558	573	588	603	.6
.4	603	618	633	649	664	679	694	709	724	739	754	.5
.5	754	769	784	799	814	829	844	859	874	889	904	.4
.6	904	919	934	949	964	979	994	009	024	039	054	.3
.7	.51 054	069	084	099	114	129	144	159	174	189	204	.2
.8	204	219	234	249	264	279	294	309	324	339	354	.1
.9	354	369	384	399	414	429	444	459	474	489	504	59.0
31.0	504	519	534	549	564	579	594	608	623	638	653	.9
.1	653	668	683	698	713	728	743	758	773	788	803	.8
.2	803	818	833	847	862	877	892	907	922	937	952	.7
.3	952	967	982	997	012	026	041	056	071	086	101	.6
.4	.52 101	116	131	146	161	175	190	205	220	235	250	.5
.5	250	265	280	294	309	324	339	354	369	384	399	.4
.6	399	413	428	443	458	473	488	503	517	532	547	.3
.7	547	562	577	592	607	621	636	651	666	681	696	.2
.8	696	710	725	740	755	770	785	799	814	829	844	.1
.9	844	859	873	888	903	918	933	948	962	977	992	58.0
32.0	992	007	022	036	051	066	081	095	110	125	140	.9
.1	.53 140	155	169	184	199	214	229	243	258	273	288	.8
.2	288	302	317	332	347	361	376	391	406	420	435	.7
.3	435	450	465	479	494	509	524	538	553	568	583	.6
.4	583	597	612	627	642	656	671	686	701	715	730	.5 Diff.
.5	730	745	759	774	789	804	818	833	848	862	877	.4
.6	877	892	906	921	936	951	965	980	995	009	024	.3 14–15
.7	.54 024	039	053	068	083	097	112	127	141	156	171	.2
.8	171	185	200	215	229	244	259	273	288	303	317	.1
.9	317	332	347	361	376	391	405	420	435	449	464	57.0
33.0	464	479	493	508	522	537	552	566	581	596	610	.9
.1	610	625	639	654	669	683	698	713	727	742	756	.8
.2	756	771	786	800	815	829	844	859	873	888	902	.7
.3	902	917	931	946	961	975	990	004	019	034	048	.6
.4	.55 048	063	077	092	106	121	135	150	165	179	194	.5
.5	194	208	223	237	252	266	281	296	310	325	339	.4
.6	339	354	368	383	397	412	426	441	455	470	484	.3
.7	484	499	513	528	543	557	572	586	601	615	630	.2
.8	630	644	659	673	688	702	717	731	746	760	775	.1
.9	775	789	803	818	832	847	861	876	890	905	919	56.0
34.0	919	934	948	963	977	992	006	021	035	049	064	.9
.1	.56 064	078	093	107	122	136	151	165	179	194	208	.8
.2	208	223	237	252	266	280	295	309	324	338	353	.7
.3	353	367	381	396	410	425	439	453	468	482	497	.6
.4	497	511	525	540	554	569	583	597	612	626	641	.5
.5	641	655	669	684	698	713	727	741	756	770	784	.4
.6	784	799	813	827	842	856	871	885	899	914	928	.3
.7	928	942	957	971	985	000	014	028	043	057	071	.2
.8	.57 071	086	100	114	129	143	157	172	186	200	215	.1
.9	215	229	243	258	272	286	300	315	329	343	358	55.0 deg.
(10)		9	8	7	6	5	4	3	2	1	0	

Table 1015 *(continued)*—**Sin and Cos of Hundredths of Degrees**

SIN

deg.	0	1	2	3	4	5	6	7	8	9	(10)	
35.0	.57 358	372	386	401	415	429	443	458	472	486	501	.9
.1	501	515	529	543	558	572	586	600	615	629	643	.8
.2	643	657	672	686	700	715	729	743	757	772	786	.7
.3	786	800	814	828	843	857	871	885	900	914	928	.6
.4	928	942	957	971	985	999	013	028	042	056	070	.5
.5	.58 070	085	099	113	127	141	156	170	184	198	212	.4
.6	212	226	241	255	269	283	297	312	326	340	354	.3
.7	354	368	382	397	411	425	439	453	467	482	496	.2
.8	496	510	524	538	552	567	581	595	609	623	637	.1
.9	637	651	666	680	694	708	722	736	750	764	779	54.0
36.0	779	793	807	821	835	849	863	877	891	906	920	.9
.1	920	934	948	962	976	990	004	018	032	046	061	.8
.2	.59 061	075	089	103	117	131	145	159	173	187	201	.7
.3	201	215	229	244	258	272	286	300	314	328	342	.6
.4	342	356	370	384	398	412	426	440	454	468	482	.5
.5	482	496	510	524	538	552	566	580	594	608	622	.4
.6	622	636	651	665	679	693	707	721	735	749	763	.3
.7	763	777	790	804	818	832	846	860	874	888	902	.2
.8	902	916	930	944	958	972	986	000	014	028	042	.1
.9	.60 042	056	070	084	098	112	126	140	154	168	182	53.0
37.0	182	195	209	223	237	251	265	279	293	307	321	.9
.1	321	335	349	363	376	390	404	418	432	446	460	.8
.2	460	474	488	502	516	529	543	557	571	585	599	.7
.3	599	613	627	640	654	668	682	696	710	724	738	.6
.4	738	751	765	779	793	807	821	835	848	862	876	.5 Diff.
.5	876	890	904	918	932	945	959	973	987	001	015	.4
.6	.61 015	028	042	056	070	084	097	111	125	139	153	.3 13–15
.7	153	167	180	194	208	222	236	249	263	277	291	.2
.8	291	304	318	332	346	360	373	387	401	415	429	.1
.9	429	442	456	470	484	497	511	525	539	552	566	52.0
38.0	566	580	594	607	621	635	649	662	676	690	704	.9
.1	704	717	731	745	759	772	786	800	813	827	841	.8
.2	841	855	868	882	896	909	923	937	951	964	978	.7
.3	978	992	005	019	033	046	060	074	087	101	115	.6
.4	.62 115	128	142	156	169	183	197	210	224	238	251	.5
.5	251	265	279	292	306	320	333	347	361	374	388	.4
.6	388	402	415	429	443	456	470	483	497	511	524	.3
.7	524	538	552	565	579	592	606	620	633	647	660	.2
.8	660	674	688	701	715	728	742	756	769	783	796	.1
.9	796	810	823	837	851	864	878	891	905	918	932	51.0
39.0	932	946	959	973	986	000	013	027	040	054	068	.9
.1	.63 068	081	095	108	122	135	149	162	176	189	203	.8
.2	203	216	230	243	257	271	284	298	311	325	338	.7
.3	338	352	365	379	392	406	419	433	446	460	473	.6
.4	473	487	500	514	527	540	554	567	581	594	608	.5
.5	608	621	635	648	662	675	689	702	715	729	742	.4
.6	742	756	769	783	796	810	823	836	850	863	877	.3
.7	877	890	904	917	930	944	957	971	984	998	011	.2
.8	.64 011	024	038	051	065	078	091	105	118	132	145	.1
.9	145	158	172	185	199	212	225	239	252	265	279	50.0 deg.
	(10)	9	8	7	6	5	4	3	2	1	0	

COS

Table 1015 *(continued)*—**Trigonometric Functions**
SIN

deg.	0	1	2	3	4	5	6	7	8	9	(10)	
40.0	.64 279	292	305	319	332	346	359	372	386	399	412	.9
.1	412	426	439	452	466	479	492	506	519	532	546	.8
.2	546	559	572	586	599	612	626	639	652	666	679	.7
.3	679	692	706	719	732	746	759	772	785	799	812	.6
.4	812	825	839	852	865	878	892	905	918	932	945	.5
.5	945	958	971	985	998	011	024	038	051	064	077	.4
.6	.65 077	091	104	117	130	144	157	170	183	197	210	.3
.7	210	223	236	250	263	276	289	302	316	329	342	.2
.8	342	355	368	382	395	408	421	434	448	461	474	.1
.9	474	487	500	514	527	540	553	566	580	593	606	49.0
41.0	606	619	632	645	659	672	685	698	711	724	738	.9
.1	738	751	764	777	790	803	816	830	843	856	869	.8
.2	869	882	895	908	921	935	948	961	974	987	000	.7
.3	.66 000	013	026	039	053	066	079	092	105	118	131	.6
.4	131	144	157	170	184	197	210	223	236	249	262	.5
.5	262	275	288	301	314	327	340	353	367	380	393	.4
.6	393	406	419	432	445	458	471	484	497	510	523	.3
.7	523	536	549	562	575	588	601	614	627	640	653	.2
.8	653	666	679	692	705	718	731	744	757	770	783	.1
.9	783	796	809	822	835	848	861	874	887	900	913	48.0
42.0	913	926	939	952	965	978	991	004	017	030	043	.9
.1	.67 043	056	069	082	094	107	120	133	146	159	172	.8
.2	172	185	198	211	224	237	250	263	275	288	301	.7
.3	301	314	327	340	353	366	379	392	404	417	430	.6
.4	430	443	456	469	482	495	508	520	533	546	559	.5 Diff.
.5	559	572	585	598	610	623	636	649	662	675	688	.4
.6	688	700	713	726	739	752	765	777	790	803	816	.3 12–14
.7	816	829	842	854	867	880	893	906	919	931	944	.2
.8	944	957	970	983	995	008	021	034	047	059	072	.1
.9	.68 072	085	098	110	123	136	149	162	174	187	200	47.0
43.0	200	213	225	238	251	264	276	289	302	315	327	.9
.1	327	340	353	366	378	391	404	417	429	442	455	.8
.2	455	467	480	493	506	518	531	544	556	569	582	.7
.3	582	595	607	620	633	645	658	671	683	696	709	.6
.4	709	721	734	747	759	772	785	797	810	823	835	.5
.5	835	848	861	873	886	899	911	924	937	949	962	.4
.6	962	975	987	000	012	025	038	050	063	076	088	.3
.7	.69 088	101	113	126	139	151	164	177	189	202	214	.2
.8	214	227	240	252	265	277	290	302	315	328	340	.1
.9	340	353	365	378	390	403	416	428	441	453	466	46.0
44.0	466	478	491	503	516	529	541	554	566	579	591	.9
.1	591	604	616	629	641	654	666	679	691	704	717	.8
.2	717	729	742	754	767	779	792	804	817	829	842	.7
.3	842	854	867	879	891	904	916	929	941	954	966	.6
.4	966	979	991	004	016	029	041	054	066	078	091	.5
.5	.70 091	103	116	128	141	153	166	178	190	203	215	.4
.6	215	228	240	253	265	277	290	302	315	327	339	.3
.7	339	352	364	377	389	401	414	426	439	451	463	.2
.8	463	476	488	501	513	525	538	550	562	575	587	.1
.9	587	600	612	624	637	649	661	674	686	698	711	45.0 deg.
	(10)	9	8	7	6	5	4	3	2	1	0	

COS

Table 1015 (*continued*)—Sin and Cos of Hundredths of Degrees

SIN

deg.	0	1	2	3	4	5	6	7	8	9	(10)	
45.0	.70 711	723	735	748	760	772	785	797	809	822	834	.9
.1	834	846	859	871	883	896	908	920	932	945	957	.8
.2	957	969	982	994	006	019	031	043	055	068	080	.7
.3	.71 080	092	104	117	129	141	154	166	178	190	203	.6
.4	203	215	227	239	252	264	276	288	301	313	325	.5
.5	325	337	350	362	374	386	398	411	423	435	447	.4
.6	447	459	472	484	496	508	520	533	545	557	569	.3
.7	569	581	594	606	618	630	642	655	667	679	691	.2
.8	691	703	715	728	740	752	764	776	788	800	813	.1
.9	813	825	837	849	861	873	885	898	910	922	934	44.0
46.0	934	946	958	970	982	995	007	019	031	043	055	.9
.1	.72 055	067	079	091	104	116	128	140	152	164	176	.8
.2	176	188	200	212	224	236	248	261	273	285	297	.7
.3	297	309	321	333	345	357	369	381	393	405	417	.6
.4	417	429	441	453	465	477	489	501	513	525	537	.5
.5	537	549	561	573	585	597	609	621	633	645	657	.4
.6	657	669	681	693	705	717	729	741	753	765	777	.3
.7	777	789	801	813	825	837	849	861	873	885	897	.2
.8	897	909	921	933	945	957	969	980	992	004	016	.1
.9	.73 016	028	040	052	064	076	088	100	112	123	135	43.0
47.0	135	147	159	171	183	195	207	219	231	242	254	.9
.1	254	266	278	290	302	314	326	337	349	361	373	.8
.2	373	385	397	409	420	432	444	456	468	480	491	.7
.3	491	503	515	527	539	551	562	574	586	598	610	.6
.4	610	622	633	645	657	669	681	692	704	716	728	.5 Diff.
.5	728	740	751	763	775	787	798	810	822	834	846	.4
.6	846	857	869	881	893	904	916	928	940	951	963	.3 11–13
.7	963	975	987	998	010	022	034	045	057	069	080	.2
.8	.74 080	092	104	116	127	139	151	162	174	186	198	.1
.9	198	209	221	233	244	256	268	279	291	303	314	42.0
48.0	314	326	338	350	361	373	385	396	408	419	431	.9
.1	431	443	454	466	478	489	501	513	524	536	548	.8
.2	548	559	571	582	594	606	617	629	641	652	664	.7
.3	664	675	687	699	710	722	733	745	757	768	780	.6
.4	780	791	803	815	826	838	849	861	872	884	896	.5
.5	896	907	919	930	942	953	965	976	988	000	011	.4
.6	.75 011	023	034	046	057	069	080	092	103	115	126	.3
.7	126	138	149	161	172	184	195	207	218	230	241	.2
.8	241	253	264	276	287	299	310	322	333	345	356	.1
.9	356	368	379	391	402	414	425	437	448	460	471	41.0
49.0	471	482	494	505	517	528	540	551	562	574	585	.9
.1	585	597	608	620	631	642	654	665	677	688	700	.8
.2	700	711	722	734	745	756	768	779	791	802	813	.7
.3	813	825	836	848	859	870	882	893	904	916	927	.6
.4	927	938	950	961	973	984	995	007	018	029	041	.5
.5	.76 041	052	063	075	086	097	109	120	131	143	154	.4
.6	154	165	176	188	199	210	222	233	244	256	267	.3
.7	267	278	289	301	312	323	335	346	357	368	380	.2
.8	380	391	402	413	425	436	447	458	470	481	492	.1
.9	492	503	515	526	537	548	560	571	582	593	604	40.0 deg.
(10)	9	8	7	6	5	4	3	2	1	0		

COS

Table 1015 (*continued*)—Trigonometric Functions
SIN

deg.	0	1	2	3	4	5	6	7	8	9	(10)	
50.0	.76 604	616	627	638	649	661	672	683	694	705	717	.9
.1	717	728	739	750	761	772	784	795	806	817	828	.8
.2	828	840	851	862	873	884	895	906	918	929	940	.7
.3	940	951	962	973	985	996	007	018	029	040	051	.6
.4	.77 051	062	074	085	096	107	118	129	140	151	162	.5
.5	162	174	185	196	207	218	229	240	251	262	273	.4
.6	273	284	296	307	318	329	340	351	362	373	384	.3
.7	384	395	406	417	428	439	450	461	472	483	494	.2
.8	494	505	517	528	539	550	561	572	583	594	605	.1
.9	605	616	627	638	649	660	671	682	693	704	715	39.0
51.0	715	726	737	748	759	769	780	791	802	813	824	.9
.1	824	835	846	857	868	879	890	901	912	923	934	.8
.2	934	945	956	967	978	988	999	010	021	032	043	.7
.3	.78 043	054	065	076	087	098	108	119	130	141	152	.6
.4	152	163	174	185	196	206	217	228	239	250	261	.5
.5	261	272	283	293	304	315	326	337	348	359	369	.4
.6	369	380	391	402	413	424	434	445	456	467	478	.3
.7	478	488	499	510	521	532	542	553	564	575	586	.2
.8	586	596	607	618	629	640	650	661	672	683	694	.1
.9	694	704	715	726	737	747	758	769	780	790	801	38.0
52.0	801	812	823	833	844	855	866	876	887	898	908	.9
.1	908	919	930	941	951	962	973	983	994	005	016	.8
.2	.79 016	026	037	048	058	069	080	090	101	112	122	.7
.3	122	133	144	154	165	176	186	197	208	218	229	.6
.4	229	240	250	261	272	282	293	303	314	325	335	.5 Diff.
.5	335	346	357	367	378	388	399	410	420	431	441	.4
.6	441	452	463	473	484	494	505	516	526	537	547	.3 10–12
.7	547	558	568	579	590	600	611	621	632	642	653	.2
.8	653	664	674	685	695	706	716	727	737	748	758	.1
.9	758	769	779	790	800	811	822	832	843	853	864	37.0
53.0	864	874	885	895	906	916	927	937	948	958	968	.9
.1	968	979	989	000	010	021	031	042	052	063	073	.8
.2	.80 073	084	094	104	115	125	136	146	157	167	178	.7
.3	178	188	198	209	219	230	240	251	261	271	282	.6
.4	282	292	303	313	323	334	344	355	365	375	386	.5
.5	386	396	406	417	427	438	448	458	469	479	489	.4
.6	489	500	510	520	531	541	551	562	572	582	593	.3
.7	593	603	613	624	634	644	655	665	675	686	696	.2
.8	696	706	717	727	737	748	758	768	778	789	799	.1
.9	799	809	820	830	840	850	861	871	881	891	902	36.0
54.0	902	912	922	932	943	953	963	973	984	994	004	.9
.1	.81 004	014	025	035	045	055	066	076	086	096	106	.8
.2	106	117	127	137	147	157	168	178	188	198	208	.7
.3	208	219	229	239	249	259	269	280	290	300	310	.6
.4	310	320	330	341	351	361	371	381	391	401	412	.5
.5	412	422	432	442	452	462	472	482	493	503	513	.4
.6	513	523	533	543	553	563	573	583	594	604	614	.3
.7	614	624	634	644	654	664	674	684	694	704	714	.2
.8	714	725	735	745	755	765	775	785	795	805	815	.1
.9	815	825	835	845	855	865	875	885	895	905	915	35.0 deg.
	(10)	9	8	7	6	5	4	3	2	1	0	

COS

Table 1015 (*continued*)—Sin and Cos of Hundredths of Degrees

SIN

deg.	0	1	2	3	4	5	6	7	8	9	(10)	
55.0	.81 915	925	935	945	955	965	975	985	995	005̄	015̄	.9
.1	.82 015	025	035	045	055	065	075	085	095	105	115	.8
.2	115	125	135	145	155	165	175	185	195	204	214	.7
.3	214	224	234	244	254	264	274	284	294	304	314	.6
.4	314	324	333	343	353	363	373	383	393	403	413	.5
.5	413	423	432	442	452	462	472	482	492	501	511	.4
.6	511	521	531	541	551	561	570	580	590	600	610	.3
.7	610	620	629	639	649	659	669	679	688	698	708	.2
.8	708	718	728	737	747	757	767	777	786	796	806	.1
.9	806	816	826	835	845	855	865	874	884	894	904	34.0
56.0	904	914	923	933	943	953	962	972	982	991	001̄	.9
.1	.83 001	011	021	030	040	050	060	069	079	089	098	.8
.2	098	108	118	128	137	147	157	166	176	186	195	.7
.3	195	205	215	224	234	244	253	263	273	282	292	.6
.4	292	302	311	321	331	340	350	360	369	379	389	.5
.5	389	398	408	417	427	437	446	456	466	475	485	.4
.6	485	494	504	514	523	533	542	552	562	571	581	.3
.7	581	590	600	609	619	629	638	648	657	667	676	.2
.8	676	686	696	705	715	724	734	743	753	762	772	.1
.9	772	781	791	800	810	819	829	839	848	858	867	33.0
57.0	867	877	886	896	905	915	924	934	943	953	962	.9
.1	962	971	981	990	000̄	009̄	019̄	028̄	038̄	047̄	057̄	.8
.2	.84 057	066	076	085	094	104	113	123	132	142	151	.7
.3	151	161	170	179	189	198	208	217	226	236	245	.6
.4	245	255	264	273	283	292	302	311	320	330	339	.5 Diff.
.5	339	349	358	367	377	386	395	405	414	423	433	.4
.6	433	442	451	461	470	480	489	498	508	517	526	.3 8–10
.7	526	536	545	554	563	573	582	591	601	610	619	.2
.8	619	629	638	647	656	666	675	684	694	703	712	.1
.9	712	721	731	740	749	759	768	777	786	796	805	32.0
58.0	805	814	823	833	842	851	860	869	879	888	897	.9
.1	897	906	916	925	934	943	952	962	971	980	989	.8
.2	989	998	008̄	017̄	026̄	035̄	044̄	054̄	063̄	072̄	081̄	.7
.3	.85 081	090	099	109	118	127	136	145	154	164	173	.6
.4	173	182	191	200	209	218	228	237	246	255	264	.5
.5	264	273	282	291	300	310	319	328	337	346	355	.4
.6	355	364	373	382	391	401	410	419	428	437	446	.3
.7	446	455	464	473	482	491	500	509	518	527	536	.2
.8	536	545	555	564	573	582	591	600	609	618	627	.1
.9	627	636	645	654	663	672	681	690	699	708	717	31.0
59.0	717	726	735	744	753	762	771	780	789	798	806	.9
.1	806	815	824	833	842	851	860	869	878	887	896	.8
.2	896	905	914	923	932	941	950	958	967	976	985	.7
.3	985	994	003̄	012̄	021̄	030̄	039̄	048̄	056̄	065̄	074̄	.6
.4	.86 074	083	092	101	110	119	127	136	145	154	163	.5
.5	163	172	181	189	198	207	216	225	234	243	251	.4
.6	251	260	269	278	287	295	304	313	322	331	340	.3
.7	340	348	357	366	375	384	392	401	410	419	427	.2
.8	427	436	445	454	463	471	480	489	498	506	515	.1
.9	515	524	533	541	550	559	568	576	585	594	603	30.0
												deg.
	(10)	9	8	7	6	5	4	3	2	1	0	

COS

Table 1015 (*continued*)—**Trigonometric Functions**

SIN

deg.	0	1	2	3	4	5	6	7	8	9	(10).	
60.0	.86 603	611	620	629	637	646	655	664	672	681	690	.9
.1	690	698	707	716	724	733	742	751	759	768	777	.8
.2	777	785	794	803	811	820	829	837	846	855	863	.7
.3	863	872	880	889	898	906	915	924	932	941	949	.6
.4	949	958	967	975	984	993	001	010	018	027	036	.5
.5	.87 036	044	053	061	070	079	087	096	104	113	121	.4
.6	121	130	139	147	156	164	173	181	190	198	207	.3
.7	207	215	224	233	241	250	258	267	275	284	292	.2
.8	292	301	309	318	326	335	343	352	360	369	377	.1
.9	377	386	394	403	411	420	428	437	445	454	462	29.0
61.0	462	470	479	487	496	504	513	521	530	538	546	.9
.1	546	555	563	572	580	589	597	605	614	622	631	.8
.2	631	639	647	656	664	673	681	689	698	706	715	.7
.3	715	723	731	740	748	756	765	773	782	790	798	.6
.4	798	807	815	823	832	840	848	857	865	873	882	.5
.5	882	890	898	907	915	923	932	940	948	957	965	.4
.6	965	973	981	990	998	006	015	023	031	039	048	.3
.7	.88 048	056	064	073	081	089	097	106	114	122	130	.2
.8	130	139	147	155	163	172	180	188	196	204	213	.1
.9	213	221	229	237	246	254	262	270	278	287	295	28.0
62.0	295	303	311	319	328	336	344	352	360	368	377	.9
.1	377	385	393	401	409	417	426	434	442	450	458	.8
.2	458	466	474	483	491	499	507	515	523	531	539	.7
.3	539	547	556	564	572	580	588	596	604	612	620	.6
.4	620	628	637	645	653	661	669	677	685	693	701	.5 Diff.
.5	701	709	717	725	733	741	749	757	765	774	782	.4
.6	782	790	798	806	814	822	830	838	846	854	862	.3 7–9
.7	862	870	878	886	894	902	910	918	926	934	942	.2
.8	942	950	958	966	974	981	989	997	005	013	021	.1
.9	.89 021	029	037	045	053	061	069	077	085	093	101	27.0
63.0	101	109	116	124	132	140	148	156	164	172	180	.9
.1	180	188	196	203	211	219	227	235	243	251	259	.8
.2	259	266	274	282	290	298	306	314	321	329	337	.7
.3	337	345	353	361	368	376	384	392	400	408	415	.6
.4	415	423	431	439	447	454	462	470	478	486	493	.5
.5	493	501	509	517	525	532	540	548	556	563	571	.4
.6	571	579	587	594	602	610	618	625	633	641	649	.3
.7	649	656	664	672	680	687	695	703	710	718	726	.2
.8	726	734	741	749	757	764	772	780	787	795	803	.1
.9	803	810	818	826	833	841	849	856	864	872	879	26.0
64.0	879	887	895	902	910	918	925	933	941	948	956	.9
.1	956	963	971	979	986	994	001	009	017	024	032	.8
.2	.90 032	039	047	055	062	070	077	085	093	100	108	.7
.3	108	115	123	130	138	146	153	161	168	176	183	.6
.4	183	191	198	206	213	221	228	236	243	251	259	.5
.5	259	266	274	281	289	296	304	311	319	326	334	.4
.6	334	341	348	356	363	371	378	386	393	401	408	.3
.7	408	416	423	431	438	446	453	460	468	475	483	.2
.8	483	490	498	505	512	520	527	535	542	549	557	.1
.9	557	564	572	579	586	594	601	609	616	623	631	25.0 deg.
(10)	9	8	7	6	5	4	3	2	1		0	

COS

Table 1015 (*continued*)—Sin and Cos of Hundredths of Degrees

SIN

deg.		0	1	2	3	4	5	6	7	8	9	(10)		Diff.
65.0	.90	631	638	646	653	660	668	675	682	690	697	704	.9	7–8
.1		704	712	719	726	734	741	748	756	763	770	778	.8	
.2		778	785	792	800	807	814	822	829	836	844	851	.7	
.3		851	858	865	873	880	887	895	902	909	916	924	.6	
.4		924	931	938	945	953	960	967	974	982	989	996	.5	
.5		996	003	011	018	025	032	040	047	054	061	068	.4	
.6	.91	068	076	083	090	097	104	112	119	126	133	140	.3	
.7		140	148	155	162	169	176	183	191	198	205	212	.2	
.8		212	219	226	233	241	248	255	262	269	276	283	.1	
.9		283	291	298	305	312	319	326	333	340	347	355	24.0	
66.0		355	362	369	376	383	390	397	404	411	418	425	.9	
.1		425	432	440	447	454	461	468	475	482	489	496	.8	
.2		496	503	510	517	524	531	538	545	552	559	566	.7	
.3		566	573	580	587	594	601	608	615	622	629	636	.6	7
.4		636	643	650	657	664	671	678	685	692	699	706	.5	
.5		706	713	720	727	734	741	748	755	762	769	775	.4	
.6		775	782	789	796	803	810	817	824	831	838	845	.3	
.7		845	852	858	865	872	879	886	893	900	907	914	.2	
.8		914	920	927	934	941	948	955	962	968	975	982	.1	
.9		982	989	996	003	010	016	023	030	037	044	050	23.0	
67.0	.92	050	057	064	071	078	085	091	098	105	112	119	.9	
.1		119	125	132	139	146	152	159	166	173	180	186	.8	
.2		186	193	200	207	213	220	227	234	240	247	254	.7	
.3		254	261	267	274	281	287	294	301	308	314	321	.6	
.4		321	328	334	341	348	355	361	368	375	381	388	.5	
.5		388	395	401	408	415	421	428	435	441	448	455	.4	
.6		455	461	468	475	481	488	494	501	508	514	521	.3	
.7		521	528	534	541	547	554	561	567	574	580	587	.2	
.8		587	594	600	607	613	620	627	633	640	646	653	.1	
.9		653	659	666	673	679	686	692	699	705	712	718	22.0	
68.0		718	725	731	738	745	751	758	764	771	777	784	.9	
.1		784	790	797	803	810	816	823	829	836	842	849	.8	6–7
.2		849	855	862	868	874	881	887	894	900	907	913	.7	
.3		913	920	926	933	939	945	952	958	965	971	978	.6	
.4		978	984	990	997	003	010	016	023	029	035	042	.5	
.5	.93	042	048	055	061	067	074	080	086	093	099	106	.4	
.6		106	112	118	125	131	137	144	150	156	163	169	.3	
.7		169	175	182	188	194	201	207	213	220	226	232	.2	
.8		232	239	245	251	258	264	270	276	283	289	295	.1	
.9		295	302	308	314	320	327	333	339	346	352	358	21.0	
69.0		358	364	371	377	383	389	396	402	408	414	420	.9	
.1		420	427	433	439	445	452	458	464	470	476	483	.8	
.2		483	489	495	501	507	514	520	526	532	538	544	.7	
.3		544	551	557	563	569	575	581	588	594	600	606	.6	
.4		606	612	618	624	630	637	643	649	655	661	667	.5	
.5		667	673	679	686	692	698	704	710	716	722	728	.4	
.6		728	734	740	746	753	759	765	771	777	783	789	.3	
.7		789	795	801	807	813	819	825	831	837	843	849	.2	
.8		849	855	861	867	873	879	885	891	897	903	909	.1	6
.9		909	915	921	927	933	939	945	951	957	963	969	20.0 deg.	
		(10)	9	8	7	6	5	4	3	2	1	0		

COS

Table 1015 (*continued*)—Trigonometric Functions
SIN

deg.	0	1	2	3	4	5	6	7	8	9	(10)		Diff.
70.0	.93 969	975	981	987	993	999	005	011	017	023	029	.9	6
.1	.94 029	035	041	047	053	058	064	070	076	082	088	.8	
.2	088	094	100	106	112	118	123	129	135	141	147	.7	
.3	147	153	159	165	171	176	182	188	194	200	206	.6	
.4	206	212	217	223	229	235	241	247	252	258	264	.5	
.5	264	270	276	282	287	293	299	305	311	316	322	.4	
.6	322	328	334	340	345	351	357	363	369	374	380	.3	
.7	380	386	392	397	403	409	415	420	426	432	438	.2	
.8	438	443	449	455	461	466	472	478	483	489	495	.1	
.9	495	501	506	512	518	523	529	535	540	546	552	19.0	
71.0	552	558	563	569	575	580	586	592	597	603	609	.9	
.1	609	614	620	625	631	637	642	648	654	659	665	.8	
.2	665	671	676	682	687	693	699	704	710	715	721	.7	
.3	721	727	732	738	743	749	755	760	766	771	777	.6	
.4	777	782	788	794	799	805	810	816	821	827	832	.5	
.5	832	838	843	849	854	860	866	871	877	882	888	.4	5–6
.6	888	893	899	904	910	915	921	926	932	937	943	.3	
.7	943	948	954	959	964	970	975	981	986	992	997	.2	
.8	997	003	008	014	019	024	030	035	041	046	052	.1	
.9	.95 052	057	062	068	073	079	084	089	095	100	106	18.0	
72.0	106	111	116	122	127	133	138	143	149	154	159	.9	
.1	159	165	170	176	181	186	192	197	202	208	213	.8	
.2	213	218	224	229	234	240	245	250	256	261	266	.7	
.3	266	271	277	282	287	293	298	303	309	314	319	.6	
.4	319	324	330	335	340	345	351	356	361	366	372	.5	
.5	372	377	382	387	393	398	403	408	414	419	424	.4	
.6	424	429	434	440	445	450	455	460	466	471	476	.3	
.7	476	481	486	492	497	502	507	512	518	523	528	.2	
.8	528	533	538	543	548	554	559	564	569	574	579	.1	
.9	579	584	590	595	600	605	610	615	620	625	630	17.0	
73.0	630	636	641	646	651	656	661	666	671	676	681	.9	
.1	681	686	691	697	702	707	712	717	722	727	732	.8	
.2	732	737	742	747	752	757	762	767	772	777	782	.7	
.3	782	787	792	797	802	807	812	817	822	827	832	.6	5
.4	832	837	842	847	852	857	862	867	872	877	882	.5	
.5	882	887	892	897	902	907	912	917	922	926	931	.4	
.6	931	936	941	946	951	956	961	966	971	976	981	.3	
.7	981	985	990	995	000	005	010	015	020	024	029	.2	
.8	.96 029	034	039	044	049	054	059	063	068	073	078	.1	
.9	078	083	088	092	097	102	107	112	117	121	126	16.0	
74.0	126	131	136	141	145	150	155	160	165	169	174	.9	
.1	174	179	184	188	193	198	203	208	212	217	222	.8	
.2	222	227	231	236	241	246	250	255	260	264	269	.7	
.3	269	274	279	283	288	293	297	302	307	312	316	.6	
.4	316	321	326	330	335	340	344	349	354	358	363	.5	
.5	363	368	372	377	382	386	391	396	400	405	410	.4	
.6	410	414	419	423	428	433	437	442	447	451	456	.3	
.7	456	460	465	470	474	479	483	488	492	497	502	.2	
.8	502	506	511	515	520	524	529	534	538	543	547	.1	
.9	547	552	556	561	565	570	574	579	584	588	593	15.0	4–5
	(10)	9	8	7	6	5	4	3	2	1	0	deg.	

COS

Table 1015 (*continued*)—**Sin and Cos of Hundredths of Degrees**

SIN

deg.	0	1	2	3	4	5	6	7	8	9	(10)		Diff.
75.0	.96 593	597	602	606	611	615	620	624	629	633	638	.9	4–5
.1	638	642	647	651	656	660	664	669	673	678	682	.8	
.2	682	687	691	696	700	705	709	713	718	722	727	.7	
.3	727	731	736	740	744	749	753	758	762	767	771	.6	
.4	771	775	780	784	788	793	797	802	806	810	815	.5	
.5	815	819	823	828	832	837	841	845	850	854	858	.4	
.6	858	863	867	871	876	880	884	889	893	897	902	.3	
.7	902	906	910	914	919	923	927	932	936	940	945	.2	
.8	945	949	953	957	962	966	970	974	979	983	987	.1	
.9	987	991	996	000	004	008	013	017	021	025	030	14.0	
76.0	.97 030	034	038	042	046	051	055	059	063	067	072	.9	
.1	072	076	080	084	088	093	097	101	105	109	113	.8	
.2	113	118	122	126	130	134	138	142	147	151	155	.7	
.3	155	159	163	167	171	176	180	184	188	192	196	.6	
.4	196	200	204	208	212	217	221	225	229	233	237	.5	
.5	237	241	245	249	253	257	261	265	269	274	278	.4	
.6	278	282	286	290	294	298	302	306	310	314	318	.3	
.7	318	322	326	330	334	338	342	346	350	354	358	.2	4
.8	358	362	366	370	374	378	382	386	390	394	398	.1	
.9	398	402	406	409	413	417	421	425	429	433	437	13.0	
77.0	437	441	445	449	453	457	461	464	468	472	476	.9	
.1	476	480	484	488	492	496	499	503	507	511	515	.8	
.2	515	519	523	527	530	534	538	542	546	550	553	.7	
.3	553	557	561	565	569	573	576	580	584	588	592	.6	
.4	592	595	599	603	607	611	614	618	622	626	630	.5	
.5	630	633	637	641	645	648	652	656	660	663	667	.4	
.6	667	671	675	678	682	686	690	693	697	701	705	.3	
.7	705	708	712	716	719	723	727	731	734	738	742	.2	
.8	742	745	749	753	756	760	764	767	771	775	778	.1	
.9	778	782	786	789	793	797	800	804	807	811	815	12.0	
78.0	815	818	822	826	829	833	836	840	844	847	851	.9	
.1	851	854	858	862	865	869	872	876	880	883	887	.8	
.2	887	890	894	897	901	905	908	912	915	919	922	.7	
.3	922	926	929	933	936	940	943	947	951	954	958	.6	
.4	958	961	965	968	972	975	979	982	986	989	992	.5	3–4
.5	992	996	999	003	006	010	013	017	020	024	027	.4	
.6	.98 027	031	034	037	041	044	048	051	055	058	061	.3	
.7	061	065	068	072	075	079	082	085	089	092	096	.2	
.8	096	099	102	106	109	112	116	119	123	126	129	.1	
.9	129	133	136	139	143	146	149	153	156	159	163	11.0	
79.0	163	166	169	173	176	179	183	186	189	193	196	.9	
.1	196	199	202	206	209	212	216	219	222	225	229	.8	
.2	229	232	235	239	242	245	248	252	255	258	261	.7	
.3	261	265	268	271	274	277	281	284	287	290	294	.6	
.4	294	297	300	303	306	310	313	316	319	322	325	.5	
.5	325	329	332	335	338	341	345	348	351	354	357	.4	
.6	357	360	363	367	370	373	376	379	382	385	389	.3	
.7	389	392	395	398	401	404	407	410	413	416	420	.2	
.8	420	423	426	429	432	435	438	441	444	447	450	.1	
.9	450	453	456	459	463	466	469	472	475	478	481	10.0	3
												deg.	
(10)	9	8	7	6	5	4	3	2	1	0			

COS

Table 1015 (*continued*)—Trigonometric Functions
SIN

deg.	0	1	2	3	4	5	6	7	8	9	(10)		Diff.
80.0	.98 481	484	487	490	493	496	499	502	505	508	511	.9	
.1	511	514	517	520	523	526	529	532	535	538	541	.8	3
.2	541	544	547	550	553	556	559	562	564	567	570	.7	
.3	570	573	576	579	582	585	588	591	594	597	600	.6	
.4	600	603	605	608	611	614	617	620	623	626	629	.5	
.5	629	631	634	637	640	643	646	649	652	654	657	.4	
.6	657	660	663	666	669	671	674	677	680	683	686	.3	
.7	686	688	691	694	697	700	702	705	708	711	714	.2	
.8	714	716	719	722	725	728	730	733	736	739	741	.1	
.9	741	744	747	750	752	755	758	761	763	766	769	9.0	
81.0	769	772	774	777	780	782	785	788	791	793	796	.9	
.1	796	799	801	804	807	809	812	815	817	820	823	.8	
.2	823	826	828	831	833	836	839	841	844	847	849	.7	
.3	849	852	855	857	860	863	865	868	870	873	876	.6	
.4	876	878	881	883	886	889	891	894	896	899	902	.5	
.5	902	904	907	909	912	914	917	920	922	925	927	.4	
.6	927	930	932	935	937	940	942	945	948	950	953	.3	
.7	953	955	958	960	963	965	968	970	973	975	978	.2	2–3
.8	978	980	983	985	988	990	993	995	997	000	002	.1	
.9	.99 002	005	007	010	012	015	017	020	022	024	027	8.0	
82.0	027	029	032	034	036	039	041	044	046	049	051	.9	
.1	051	053	056	058	061	063	065	068	070	072	075	.8	
.2	075	077	080	082	084	087	089	091	094	096	098	.7	
.3	098	101	103	105	108	110	112	115	117	119	122	.6	
.4	122	124	126	128	131	133	135	138	140	142	144	.5	
.5	144	147	149	151	154	156	158	160	163	165	167	.4	
.6	167	169	172	174	176	178	181	183	185	187	189	.3	
.7	189	192	194	196	198	200	203	205	207	209	211	.2	
.8	211	214	216	218	220	222	225	227	229	231	233	.1	
.9	233	235	238	240	242	244	246	248	250	252	255	7.0	
83.0	255	257	259	261	263	265	267	269	272	274	276	.9	
.1	276	278	280	282	284	286	288	290	292	294	297	.8	
.2	297	299	301	303	305	307	309	311	313	315	317	.7	
.3	317	319	321	323	325	327	329	331	333	335	337	.6	
.4	337	339	341	343	345	347	349	351	353	355	357	.5	2
.5	357	359	361	363	365	367	369	371	373	375	377	.4	
.6	377	379	381	383	385	386	388	390	392	394	396	.3	
.7	396	398	400	402	404	406	408	409	411	413	415	.2	
.8	415	417	419	421	423	424	426	428	430	432	434	.1	
.9	434	436	437	439	441	443	445	447	449	450	452	6.0	
84.0	452	454	456	458	459	461	463	465	457	468	470	.9	
.1	470	472	474	476	477	479	481	483	485	486	488	.8	
.2	488	490	492	493	495	497	499	500	502	504	506	.7	
.3	506	507	509	511	512	514	516	518	519	521	523	.6	
.4	523	524	526	528	530	531	533	535	536	538	540	.5	
.5	540	541	543	545	546	548	550	551	553	555	556	.4	
.6	556	558	559	561	563	564	566	568	569	571	572	.3	
.7	572	574	576	577	579	580	582	584	585	587	588	.2	
.8	588	590	592	593	595	596	598	599	601	603	604	.1	
.9	604	606	607	609	610	612	613	615	616	618	619	5.0	1–2
	(10)	9	8	7	6	5	4	3	2	1	0	deg.	

COS

Table 1015 (continued)—Sin and Cos of Hundredths of Degrees

SIN

deg.	0	1	2	3	4	5	6	7	8	9	(10)		Diff.
85.0	.99 619	621	623	624	626	627	629	630	632	633	635	.9	1–2
.1	635	636	638	639	640	642	643	645	646	648	649	.8	
.2	649	651	652	654	655	657	658	659	661	662	664	.7	
.3	664	665	667	668	669	671	672	674	675	676	678	.6	
.4	678	679	681	682	683	685	686	688	689	690	692	.5	
.5	692	693	694	696	697	699	700	701	703	704	705	.4	
.6	705	707	708	709	711	712	713	715	716	717	719	.3	
.7	719	720	721	722	724	725	726	728	729	730	731	.2	
.8	731	733	734	735	737	738	739	740	742	743	744	.1	
.9	744	745	747	748	749	750	752	753	754	755	756	4.0	
86.0	.99 756	758	759	760	761	762	764	765	766	767	768	.9	
.1	768	770	771	772	773	774	775	777	778	779	780	.8	
.2	780	781	782	784	785	786	787	788	789	790	792	.7	
.3	792	793	794	795	796	797	798	799	800	802	803	.6	
.4	803	804	805	806	807	808	809	810	811	812	813	.5	
.5	813	815	816	817	818	819	820	821	822	823	824	.4	
.6	824	825	826	827	828	829	830	831	832	833	834	.3	
.7	834	835	836	837	838	839	840	841	842	843	844	.2	1
.8	844	845	846	847	848	849	850	851	852	853	854	.1	
.9	854	855	856	856	857	858	859	860	861	862	863	3.0	
87.0	.99 863	864	865	866	867	867	868	869	870	871	872	.9	
.1	872	873	874	875	875	876	877	878	879	880	881	.8	
.2	881	881	882	883	884	885	886	887	887	888	889	.7	
.3	889	890	891	891	892	893	894	895	895	896	897	.6	
.4	897	898	899	899	900	901	902	903	903	904	905	.5	
.5	905	906	906	907	908	909	909	910	911	912	912	.4	
.6	912	913	914	914	915	916	917	917	918	919	919	.3	
.7	919	920	921	922	922	923	924	924	925	926	926	.2	
.8	926	927	928	928	929	930	930	931	932	932	933	.1	
.9	933	933	934	935	935	936	937	937	938	938	939	2.0	
88.0	.99 939	940	940	941	941	942	943	943	944	944	945	.9	
.1	945	946	946	947	947	948	948	949	950	950	951	.8	
.2	951	951	952	952	953	953	954	954	955	955	956	.7	
.3	956	957	957	958	958	959	959	960	960	961	961	.6	
.4	961	961	962	962	963	963	964	964	965	965	966	.5	
.5	966	966	967	967	968	968	968	969	969	970	970	.4	
.6	970	971	971	971	972	972	973	973	973	974	974	.3	
.7	974	975	975	975	976	976	977	977	977	978	978	.2	
.8	978	978	979	979	980	980	980	981	981	981	982	.1	
.9	982	982	982	983	983	983	984	984	984	984	985	1.0	
89.0	.99 985	985	985	986	986	986	987	987	987	987	988	.9	
.1	988	988	988	988	989	989	989	990	990	990	990	.8	
.2	900	990	991	991	991	991	992	992	992	992	993	.7	
.3	993	993	993	993	993	994	994	994	994	994	995	.6	
.4	995	995	995	995	995	995	996	996	996	996	996	.5	
.5	996	996	996	997	997	997	997	997	997	997	998	.4	
.6	998	998	998	998	998	998	998	998	998	999	999	.3	
.7	999	999	999	999	999	999	999	999	999	999	999	.2	
.8	999	999	000	000	000	000	000	000	000	000	000	.1	
.9	1.00 000	000	000	000	000	000	000	000	000	000	000	0.0	0
												deg.	
90.0	1.00 000												
(10)	9	8	7	6	5	4	3	2	1	0			

COS

285

Table 1016—Trigonometric Functions
TAN

	0	1	2	3	4	5	6	7	8	9	(10)	
deg.												
0.0	.00 000	017	035	052	070	087	105	122	140	157	175	.9
.1	175	192	209	227	244	262	279	297	314	332	349	.8
.2	349	367	384	401	419	436	454	471	489	506	524	.7
.3	524	541	559	576	593	611	628	646	663	681	698	.6
.4	698	716	733	751	768	785	803	820	838	855	873	.5
.5	873	890	908	925	943	960	977	995	012	030	047	.4
.6	.01 047	065	082	100	117	135	152	169	187	204	222	.3
.7	222	239	257	274	292	309	327	344	361	379	396	.2
.8	396	414	431	449	466	484	501	519	536	553	571	.1
.9	571	588	606	623	641	658	676	693	711	728	746	89.0
1.0	746	763	780	798	815	833	850	868	885	903	920	.9
.1	920	938	955	972	990	007	025	042	060	077	095	.8
.2	.02 095	112	130	147	165	182	199	217	234	252	269	.7
.3	269	287	304	322	339	357	374	392	409	426	444	.6
.4	444	461	479	496	514	531	549	566	584	601	619	.5
.5	619	636	654	671	688	706	723	741	758	776	793	.4
.6	793	811	828	846	863	881	898	916	933	950	968	.3
.7	968	985	003	020	038	055	073	090	108	125	143	.2
.8	.03 143	160	178	195	213	230	247	265	282	300	317	.1
.9	317	335	352	370	387	405	422	440	457	475	492	88.0
2.0	492	510	527	545	562	579	597	614	632	649	667	.9
.1	667	684	702	719	737	754	772	789	807	824	842	.8
.2	842	859	877	894	912	929	946	964	981	999	016	.7
.3	.04 016	034	051	069	086	104	121	139	156	174	191	.6
.4	191	209	226	244	261	279	296	314	331	349	366	.5 Diff.
.5	366	384	401	419	436	454	471	489	506	523	541	.4
.6	541	558	576	593	611	628	646	663	681	698	716	.3 17–18
.7	716	733	751	768	786	803	821	838	856	873	891	.2
.8	891	908	926	943	961	978	996	013	031	048	066	.1
.9	.05 066	083	101	118	136	153	171	188	206	223	241	87.0
3.0	241	258	276	293	311	328	346	363	381	398	416	.9
.1	416	433	451	468	486	503	521	538	556	573	591	.8
.2	591	608	626	643	661	678	696	713	731	748	766	.7
.3	766	783	801	818	836	854	871	889	906	924	941	.6
.4	941	959	976	994	011	029	046	064	081	099	116	.5
.5	.06 116	134	151	169	186	204	221	239	256	274	291	.4
.6	291	309	327	344	362	379	397	414	432	449	467	.3
.7	467	484	502	519	537	554	572	589	607	624	642	.2
.8	642	660	677	695	712	730	747	765	782	800	817	.1
.9	817	835	852	870	887	905	923	940	958	975	993	86.0
4.0	993	010	028	045	063	080	098	115	133	151	168	.9
.1	.07 168	186	203	221	238	256	273	291	308	326	344	.8
.2	344	361	379	396	414	431	449	466	484	501	519	.7
.3	519	537	554	572	589	607	624	642	659	677	695	.6
.4	695	712	730	747	765	782	800	817	835	853	870	.5
.5	870	888	905	923	940	958	976	993	011	028	046	.4
.6	.08 046	063	081	099	116	134	151	169	186	204	221	.3
.7	221	239	257	274	292	309	327	345	362	380	397	.2
.8	397	415	432	450	468	485	503	520	538	555	573	.1
.9	573	591	608	626	643	661	679	696	714	731	749	85.0 deg.
	(10)	9	8	7	6	5	4	3	2	1	0	

COT

286

Table 1016 (continued)—Tan and Cot of Hundredths of Degrees

TAN

deg.		0	1	2	3	4	5	6	7	8	9	(10)	
5.0	.08	749	766	784	802	819	837	854	872	890	907	925	.9
.1		925	942	960	978	995	013	030	048	066	083	101	.8
.2	.09	101	118	136	154	171	189	206	224	242	259	277	.7
.3		277	294	312	330	347	365	382	400	418	435	453	.6
.4		453	470	488	506	523	541	558	576	594	611	629	.5
.5		629	647	664	682	699	717	735	752	770	787	805	.4
.6		805	823	840	858	876	893	911	928	946	964	981	.3
.7		981	999	017	034	052	069	087	105	122	140	158	.2
.8	.10	158	175	193	211	228	246	263	281	299	316	334	.1
.9		334	352	369	387	405	422	440	457	475	493	510	84.0
6.0		510	528	546	563	581	599	616	634	652	669	687	.9
.1		687	705	722	740	758	775	793	811	828	846	863	.8
.2		863	881	899	916	934	952	969	987	005	022	040	.7
.3	.11	040	058	075	093	111	128	146	164	181	199	217	.6
.4		217	234	252	270	287	305	323	341	358	376	394	.5
.5		394	411	429	447	464	482	500	517	535	553	570	.4
.6		570	588	606	623	641	659	677	694	712	730	747	.3
.7		747	765	783	800	818	836	853	871	889	907	924	.2
.8		924	942	960	977	995	013	031	048	066	084	101	.1
.9	.12	101	119	137	154	172	190	208	225	243	261	278	83.0
7.0		278	296	314	332	349	367	385	402	420	438	456	.9
.1		456	473	491	509	527	544	562	580	597	615	633	.8
.2		633	651	668	686	704	722	739	757	775	793	810	.7
.3		810	828	846	864	881	899	917	934	952	970	988	.6
.4		988	005	023	041	059	076	094	112	130	147	165	.5 Diff.
.5	.13	165	183	201	219	236	254	272	290	307	325	343	.4
.6		343	361	378	396	414	432	449	467	485	503	521	.3 17–18
.7		521	538	556	574	592	609	627	645	663	681	698	.2
.8		698	716	734	752	769	787	805	823	841	858	876	.1
.9		876	894	912	930	947	965	983	001	018	036	054	82.0
8.0	.14	054	072	090	107	125	143	161	179	196	214	232	.9
.1		232	250	268	286	303	321	339	357	375	392	410	.8
.2		410	428	446	464	481	499	517	535	553	571	588	.7
.3		588	606	624	642	660	678	695	713	731	749	767	.6
.4		767	785	802	820	838	856	874	892	909	927	945	.5
.5		945	963	981	999	016	034	052	070	088	106	124	.4
.6	.15	124	141	159	177	195	213	231	249	266	284	302	.3
.7		302	320	338	356	374	391	409	427	445	463	481	.2
.8		481	499	517	534	552	570	588	606	624	642	660	.1
.9		660	677	695	713	731	749	767	785	803	821	838	81.0
9.0		838	856	874	892	910	928	946	964	982	000	017	.9
.1	.16	017	035	053	071	089	107	125	143	161	179	196	.8
.2		196	214	232	250	268	286	304	322	340	358	376	.7
.3		376	394	411	429	447	465	483	501	519	537	555	.6
.4		555	573	591	609	627	645	663	680	698	716	734	.5
.5		734	752	770	788	806	824	842	860	878	896	914	.4
.6		914	932	950	968	986	004	021	039	057	075	093	.3
.7	.17	093	111	129	147	165	183	201	219	237	255	273	.2
.8		273	291	309	327	345	363	381	399	417	435	453	.1
.9		453	471	489	507	525	543	561	579	597	615	633	80.0 deg.
	(10)	9	8	7	6	5	4	3	2	1	0		

COT

Table 1016 (*continued*)—Trigonometric Functions
TAN

deg.	0	1	2	3	4	5	6	7	8	9	(10)	
10.0	.17 633	651	669	687	705	723	741	759	777	795	813	.9
.1	813	831	849	867	885	903	921	939	957	975	993	.8
.2	993	011	029	047	065	083	101	119	137	155	173	.7
.3	.18 173	191	209	227	245	263	281	299	317	335	353	.6
.4	353	371	390	408	426	444	462	480	498	516	534	.5
.5	534	552	570	588	606	624	642	660	678	696	714	.4
.6	714	733	751	769	787	805	823	841	859	877	895	.3
.7	895	913	931	949	968	986	004	022	040	058	076	.2
.8	.19 076	094	112	130	148	166	185	203	221	239	257	.1
.9	257	275	293	311	329	347	366	384	402	420	438	79.0
11.0	438	456	474	492	510	529	547	565	583	601	619	.9
.1	619	637	655	674	692	710	728	746	764	782	801	.8
.2	801	819	837	855	873	891	909	928	946	964	982	.7
.3	982	000	018	036	055	073	091	109	127	145	164	.6
.4	.20 164	182	200	218	236	254	273	291	309	327	345	.5
.5	345	363	382	400	418	436	454	472	491	509	527	.4
.6	527	545	563	582	600	618	636	654	673	691	709	.3
.7	709	727	745	764	782	800	818	836	855	873	891	.2
.8	891	909	928	946	964	982	000	019	037	055	073	.1
.9	.21 073	092	110	128	146	164	183	201	219	237	256	78.0
12.0	256	274	292	310	329	347	365	383	402	420	438	.9
.1	438	456	475	493	511	529	548	566	584	602	621	.8
.2	621	639	657	676	694	712	730	749	767	785	804	.7
.3	804	822	840	858	877	895	913	932	950	968	986	.6
.4	986	005	023	041	060	078	096	115	133	151	169	.5 Diff.
.5	.22 169	188	206	224	243	261	279	298	316	334	353	.4
.6	353	371	389	408	426	444	463	481	499	518	536	.3 18–19
.7	536	554	573	591	609	628	646	664	683	701	719	.2
.8	719	738	756	775	793	811	830	848	866	885	903	.1
.9	903	921	940	958	977	995	013	032	050	068	087	77.0
13.0	.23 087	105	124	142	160	179	197	216	234	252	271	.9
.1	271	289	308	326	344	363	381	400	418	436	455	.8
.2	455	473	492	510	528	547	565	584	602	621	639	.7
.3	639	657	676	694	713	731	750	768	786	805	823	.6
.4	823	842	860	879	897	916	934	953	971	989	008	.5
.5	.24 008	026	045	063	082	100	119	137	156	174	193	.4
.6	193	211	229	248	266	285	303	322	340	359	377	.3
.7	377	396	414	433	451	470	488	507	525	544	562	.2
.8	562	581	599	618	636	655	673	692	710	729	748	.1
.9	747	766	785	803	822	840	859	877	896	914	933	76.0
14.0	933	951	970	988	007	026	044	063	081	100	118	.9
.1	.25 118	137	155	174	192	211	230	248	267	285	304	.8
.2	304	322	341	360	378	397	415	434	453	471	490	.7
.3	490	508	527	545	564	583	601	620	638	657	676	.6
.4	676	694	713	731	750	769	787	806	825	843	862	.5
.5	862	880	899	918	936	955	974	992	011	029	048	.4
.6	.26 048	067	085	104	123	141	160	179	197	216	235	.3
.7	235	253	272	290	309	328	346	365	384	402	421	.2
.8	421	440	458	477	496	515	533	552	571	589	608	.1
.9	608	627	645	664	683	701	720	739	758	776	795	75.0
	(10)	9	8	7	6	5	4	3	2	1	0	deg.

COT

288

Table 1016 (continued)—Tan and Cot of Hundredths of Degrees

TAN

deg.	0	1	2	3	4	5	6	7	8	9	(10)		Diff.
15.0	.26 795	814	832	851	870	888	907	926	945	963	982	.9	18–19
.1	982	001	020	038	057	076	094	113	132	151	169	.8	
.2	.27 169	188	207	226	244	263	282	301	319	338	357	.7	
.3	357	376	394	413	432	451	469	488	507	526	545	.6	
.4	545	563	582	601	620	638	657	676	695	714	732	.5	
.5	732	751	770	789	808	826	845	864	883	902	921	.4	
.6	921	939	958	977	996	015	033	052	071	090	109	.3	
.7	.28 109	128	146	165	184	203	222	241	259	278	297	.2	
.8	297	316	335	354	373	391	410	429	448	467	486	.1	
.9	486	505	523	542	561	580	599	618	637	656	675	74.0	
16.0	675	693	712	731	750	769	788	807	826	845	864	.9	
.1	864	882	901	920	939	958	977	996	015	034	053	.8	
.2	.29 053	072	091	109	128	147	166	185	204	223	242	.7	
.3	242	261	280	299	318	337	356	375	394	413	432	.6	
.4	432	451	470	489	507	526	545	564	583	602	621	.5	
.5	621	640	659	678	697	716	735	754	773	792	811	.4	19
.6	811	830	849	868	887	906	925	944	963	982	001	.3	
.7	.30 001	020	039	059	078	097	116	135	154	173	192	.2	
.8	192	211	230	249	268	287	306	325	344	363	382	.1	
.9	382	401	420	440	459	478	497	516	535	554	573	73.0	
17.0	573	592	611	630	649	669	688	707	726	745	764	.9	
.1	764	783	802	821	840	860	879	898	917	936	955	.8	
.2	955	974	993	013	032	051	070	089	108	127	147	.7	
.3	.31 147	166	185	204	223	242	261	281	300	319	338	.6	
.4	338	357	376	396	415	434	453	472	492	511	530	.5	
.5	530	549	568	587	607	626	645	664	683	703	722	.4	
.6	722	741	760	780	799	818	837	856	876	895	914	.3	
.7	914	933	953	972	991	010	029	049	068	087	106	.2	
.8	.32 106	126	145	164	184	203	222	241	261	280	299	.1	
.9	299	318	338	357	376	396	415	434	453	473	492	72.0	
18.0	492	511	531	550	569	588	608	627	646	666	685	.9	
.1	685	704	724	743	762	782	801	820	840	859	878	.8	
.2	878	898	917	936	956	975	994	014	033	052	072	.7	
.3	.33 072	091	111	130	149	169	188	207	227	246	266	.6	
.4	266	285	304	324	343	363	382	401	421	440	460	.5	
.5	460	479	498	518	537	557	576	595	615	634	654	.4	
.6	654	673	693	712	731	751	770	790	809	829	848	.3	
.7	848	868	887	907	926	945	965	984	004	023	043	.2	
.8	.34 043	062	082	101	121	140	160	179	199	218	238	.1	
.9	238	257	277	296	316	335	355	374	394	413	433	71.0	
19.0	433	452	472	491	511	530	550	569	589	609	628	.9	
.1	628	648	667	687	706	726	745	765	785	804	824	.8	
.2	824	843	863	882	902	922	941	961	980	000	020	.7	
.3	.35 020	039	059	078	098	118	137	157	176	196	216	.6	
.4	216	235	255	274	294	314	333	353	373	392	412	.5	
.5	412	432	451	471	490	510	530	549	569	589	608	.4	
.6	608	628	648	667	687	707	726	746	766	785	805	.3	
.7	805	825	845	864	884	904	923	943	963	983	002	.2	
.8	.36 002	022	042	061	081	101	121	140	160	180	199	.1	
.9	199	219	239	259	278	298	318	338	357	377	397	70.0	20
	(10)	9	8	7	6	5	4	3	2	1	0	deg.	

COT

Table 1016 (*continued*)—Trigonometric Functions
TAN

deg.	0	1	2	3	4	5	6	7	8	9	(10)	Diff.
20.0	.36 397	417	437	456	476	496	516	535	555	575	595	.9
.1	595	615	634	654	674	694	714	733	753	773	793	.8
.2	793	813	832	852	872	892	912	932	951	971	991	.7
.3	991	011	031	051	071	090	110	130	150	170	190	.6
.4	.37 190	210	229	249	269	289	309	329	349	369	388	.5
.5	388	408	428	448	468	488	508	528	548	568	588	.4
.6	588	607	627	647	667	687	707	727	747	767	787	.3
.7	787	807	827	847	867	887	907	927	946	966	986	.2
.8	986	006	026	046	066	086	106	126	146	166	186	.1 20
.9	.38 186	206	226	246	266	286	306	326	346	366	386	69.0
21.0	386	406	426	446	467	487	507	527	547	567	587	.9
.1	587	607	627	647	667	687	707	727	747	767	787	.8
.2	787	808	828	848	868	888	908	928	948	968	988	.7
.3	988	008	029	049	069	089	109	129	149	169	190	.6
.4	.39 190	210	230	250	270	290	310	331	351	371	391	.5
.5	391	411	431	452	472	492	512	532	552	573	593	.4
.6	593	613	633	653	674	694	714	734	754	775	795	.3
.7	795	815	835	856	876	896	916	936	957	977	997	.2
.8	997	017	038	058	078	098	119	139	159	179	200	.1
.9	.40 200	220	240	261	281	301	321	342	362	382	403	68.0
22.0	403	423	443	464	484	504	524	545	565	585	606	.9
.1	606	626	646	667	687	707	728	748	769	789	809	.8
.2	809	830	850	870	891	911	931	952	972	993	013	.7
.3	.41 013	033	054	074	095	115	135	156	176	197	217	.6
.4	217	237	258	278	299	319	340	360	380	401	421	.5
.5	421	442	462	483	503	524	544	565	585	606	626	.4
.6	626	646	667	687	708	728	749	769	790	810	831	.3
.7	831	851	872	892	913	933	954	975	995	016	036	.2
.8	.42 036	057	077	098	118	139	159	180	201	221	242	.1
.9	242	262	283	303	324	345	365	386	406	427	447	67.0
23.0	447	468	489	509	530	551	571	592	612	633	654	.9
.1	654	674	695	716	736	757	777	798	819	839	860	.8
.2	860	881	901	922	943	963	984	005	025	046	067	.7
.3	.43 067	087	108	129	150	170	191	212	232	253	274	.6
.4	274	295	315	336	357	378	398	419	440	460	481	.5
.5	481	502	523	544	564	585	606	627	647	668	689	.4
.6	689	710	731	751	772	793	814	834	855	876	897	.3
.7	897	918	939	959	980	001	022	043	064	084	105	.2
.8	.44 105	126	147	168	189	210	230	251	272	293	314	.1
.9	314	335	356	377	397	418	439	460	481	502	523	66.0
24.0	523	544	565	586	607	627	648	669	690	711	732	.9
.1	732	753	774	795	816	837	858	879	900	921	942	.8
.2	942	963	984	005	026	047	068	089	110	131	152	.7 21
.3	.45 152	173	194	215	236	257	278	299	320	341	362	.6
.4	362	383	404	425	446	467	488	509	530	552	573	.5
.5	573	594	615	636	657	678	699	720	741	762	784	.4
.6	784	805	826	847	868	889	910	931	953	974	995	.3
.7	995	016	037	058	079	101	122	143	164	185	206	.2
.8	.46 206	228	249	270	291	312	334	355	376	397	418	.1
.9	418	440	461	482	503	525	546	567	588	610	631	65.0
												deg.
(10)	9	8	7	6	5	4	3	2	1	0		

Table 1016 (*continued*)—Tan and Cot of Hundredths of Degrees

TAN

deg.	0	1	2	3	4	5	6	7	8	9	(10)		Diff.
25.0	.46 631	652	673	695	716	737	758	780	801	822	843	.9	
.1	843	865	886	907	929	950	971	992	014	035	056	.8	
.2	.47 056	078	099	120	142	163	184	206	227	248	270	.7	
.3	270	291	312	334	355	377	398	419	441	462	483	.6	
.4	483	505	526	548	569	590	612	633	655	676	698	.5	
.5	698	719	740	762	783	805	826	848	869	891	912	.4	
.6	912	933	955	976	998	019	041	062	084	105	127	.3	
.7	.48 127	148	170	191	213	234	256	277	299	320	342	.2	
.8	342	363	385	407	428	450	471	493	514	536	557	.1	
.9	557	579	601	622	644	665	687	708	730	752	773	64.0	
26.0	773	795	816	838	860	881	903	925	946	968	989	.9	
.1	989	011	033	054	076	098	119	141	163	184	206	.8	
.2	.49 206	228	249	271	293	315	336	358	380	401	423	.7	
.3	423	445	467	488	510	532	553	575	597	619	640	.6	
.4	640	662	684	706	727	749	771	793	815	836	858	.5	
.5	858	880	902	924	945	967	989	011	033	054	076	.4	
.6	.50 076	098	120	142	164	185	207	229	251	273	295	.3	
.7	295	317	339	360	382	404	426	448	470	492	514	.2	
.8	514	536	557	579	601	623	645	667	689	711	733	.1	
.9	733	755	777	799	821	843	865	887	909	931	953	63.0	
27.0	953	975	997	019	041	063	085	107	129	151	173	.9	22
.1	.51 173	195	217	239	261	283	305	327	349	371	393	.8	
.2	393	415	437	459	481	503	525	548	570	592	614	.7	
.3	614	636	658	680	702	724	747	769	791	813	835	.6	
.4	835	857	879	902	924	946	968	990	012	035	057	.5	
.5	.52 057	079	101	123	145	168	190	212	234	257	279	.4	
.6	279	301	323	345	368	390	412	434	457	479	501	.3	
.7	501	523	546	568	590	613	635	657	679	702	724	.2	
.8	724	746	769	791	813	836	858	880	903	925	947	.1	
.9	947	970	992	014	037	059	081	104	126	149	171	62.0	
28.0	.53 171	193	216	238	261	283	305	328	350	373	395	.9	
.1	395	417	440	462	485	507	530	552	575	597	620	.8	
.2	620	642	664	687	709	732	754	777	799	822	844	.7	
.3	844	867	889	912	935	957	980	002	025	047	070	.6	
.4	.54 070	092	115	137	160	183	205	228	250	273	296	.5	
.5	296	318	341	363	386	409	431	454	476	499	522	.4	
.6	522	544	567	590	612	635	658	680	703	726	748	.3	
.7	748	771	794	816	839	862	885	907	930	953	975	.2	
.8	975	998	021	044	066	089	112	135	157	180	203	.1	
.9	.55 203	226	249	271	294	317	340	362	385	408	431	61.0	
29.0	431	454	477	499	522	545	568	591	614	636	659	.9	
.1	659	682	705	728	751	774	797	819	842	865	888	.8	
.2	888	911	934	957	980	003	026	049	071	094	117	.7	
.3	.56 117	140	163	186	209	232	255	278	301	324	347	.6	
.4	347	370	393	416	439	462	485	508	531	554	577	.5	
.5	577	600	623	646	669	693	716	739	762	785	808	.4	23
.6	808	831	854	877	900	923	947	970	993	016	039	.3	
.7	.57 039	062	085	108	132	155	178	201	224	247	271	.2	
.8	271	294	317	340	363	386	410	433	456	479	503	.1	
.9	503	526	549	572	595	619	642	665	688	712	735	60.0 deg.	
	(10)	9	8	7	6	5	4	3	2	1	0		

COT

Table 1016 (*continued*)—Trigonometric Functions

TAN

deg.	0	1	2	3	4	5	6	7	8	9	(10)	Diff.
30.0	.57 735	758	782	805	828	851	875	898	921	945	968	.9 23–24
.1	968	991	015	038	061	085	108	131	155	178	201	.8
.2	.58 201	225	248	272	295	318	342	365	388	412	435	.7
.3	435	459	482	506	529	552	576	599	623	646	670	.6
.4	670	693	717	740	764	787	811	834	857	881	905	.5
.5	905	928	952	975	999	022	046	069	093	116	140	.4
.6	.59 140	163	187	211	234	258	281	305	328	352	376	.3
.7	376	399	423	446	470	494	517	541	565	588	612	.2
.8	612	636	659	683	707	730	754	778	801	825	849	.1
.9	849	872	896	920	944	967	991	015	039	062	086	59.0
31.0	.60 086	110	134	157	181	205	229	252	276	300	324	.9
.1	324	348	371	395	419	443	467	491	514	538	562	.8
.2	562	586	610	634	658	681	705	729	753	777	801	.7
.3	801	825	849	873	897	921	944	968	992	016	040	.6
.4	.61 040	064	088	112	136	160	184	208	232	256	280	.5 24
.5	280	304	328	352	376	400	424	448	472	496	520	.4
.6	520	544	569	593	617	641	665	689	713	737	761	.3
.7	761	785	809	834	848	882	906	930	954	978	003	.2
.8	.62 003	027	051	075	099	124	148	172	196	220	245	.1
.9	245	269	293	317	341	366	390	414	438	463	487	58.0
32.0	487	511	535	560	584	608	633	657	681	706	730	.9
.1	730	754	779	803	827	852	876	900	925	949	973	.8
.2	973	998	022	047	071	095	120	144	169	193	217	.7
.3	.63 217	242	266	291	315	340	364	389	413	437	462	.6
.4	462	486	511	535	560	584	609	633	658	682	707	.5
.5	707	732	756	781	805	830	854	879	903	928	953	.4
.6	953	977	002	026	051	076	100	125	150	174	199	.3
.7	.64 199	224	248	273	297	322	347	372	396	421	446	.2
.8	446	470	495	520	544	569	594	619	643	668	693	.1
.9	693	718	742	767	792	817	842	866	891	916	941	57.0
33.0	941	966	990	015	040	065	090	115	139	164	189	.9
.1	.65 189	214	239	264	289	314	339	363	388	413	438	.8
.2	438	463	488	513	538	563	588	613	638	663	688	.7
.3	688	713	738	763	788	813	838	863	888	913	938	.6 25
.4	938	963	988	013	038	063	088	113	138	163	189	.5
.5	.66 189	214	239	264	289	314	339	364	390	415	440	.4
.6	440	465	490	515	541	566	591	616	641	666	692	.3
.7	692	717	742	767	793	818	843	868	894	919	944	.2
.8	944	969	995	020	045	071	096	121	147	172	197	.1
.9	.67 197	223	248	273	299	324	349	375	400	425	451	56.0
34.0	451	476	502	527	552	578	603	629	654	680	705	.9
.1	705	731	756	781	807	832	858	883	909	934	960	.8
.2	960	985	011	036	062	088	113	139	164	190	215	.7
.3	.68 215	241	267	292	318	343	369	395	420	446	471	.6
.4	471	497	523	548	574	600	625	651	677	702	728	.5
.5	728	754	780	805	831	857	882	908	934	960	985	.4
.6	985	011	037	063	088	114	140	166	192	217	243	.3
.7	.69 243	269	295	321	347	372	398	424	450	476	502	.2
.8	502	528	554	579	605	631	657	683	709	735	761	.1
.9	761	787	813	839	865	891	917	943	969	995	021	55.0 26
.70												deg.
(10)	9	8	7	6	5	4	3	2	1	0		

COT

Table 1016 *(continued)*—Tan and Cot of Hundredths of Degrees

TAN

	0	1	2	3	4	5	6	7	8	9	(10)		Diff.
deg.													
35.0	.70 021	047	073	099	125	151	177	203	229	255	281	.9	26
.1	281	307	333	359	386	412	438	464	490	516	542	.8	
.2	542	568	595	621	647	673	699	725	752	778	804	.7	
.3	804	830	856	883	909	935	961	988	014	040	066	.6	
.4	.71 066	093	119	145	171	198	224	250	277	303	329	.5	
.5	329	356	382	408	435	461	487	514	540	567	593	.4	
.6	593	619	646	672	699	725	751	778	804	831	857	.3	
.7	857	884	910	937	963	990	016	043	069	096	122	.2	
.8	.72 122	149	175	202	228	255	282	308	335	361	388	.1	
.9	388	415	441	468	494	521	548	574	601	628	654	54.0	
36.0	654	681	708	734	761	788	814	841	868	895	921	.9	
.1	921	948	975	001	028	055	082	109	135	162	189	.8	
.2	.73 189	216	243	269	296	323	350	377	404	430	457	.7	
.3	457	484	511	538	565	592	619	646	672	699	726	.6	
.4	726	753	780	807	834	861	888	915	942	969	996	.5	27
.5	996	023	050	077	104	131	158	185	212	239	267	.4	
.6	.74 267	294	321	348	375	402	429	456	483	511	538	.3	
.7	538	565	592	619	646	674	701	728	755	782	810	.2	
.8	810	837	864	891	918	946	973	000	028	055	082	.1	
.9	.75 082	109	137	164	191	219	246	273	301	328	355	53.0	
37.0	355	383	410	438	465	492	520	547	575	602	629	.9	
.1	629	657	684	712	739	767	794	822	849	877	904	.8	
.2	904	932	959	987	014	042	069	097	124	152	180	.7	
.3	.76 180	207	235	262	290	318	345	373	400	428	456	.6	
.4	456	483	511	539	566	594	622	650	677	705	733	.5	
.5	733	760	788	816	844	871	899	927	955	983	010	.4	
.6	.77 010	038	066	094	122	149	177	205	233	261	289	.3	
.7	289	317	345	372	400	428	456	484	512	540	568	.2	
.8	568	596	624	652	680	708	736	764	792	820	848	.1	28
.9	848	876	904	932	960	988	016	044	072	100	129	52.0	
38.0	.78 129	157	185	213	241	269	297	325	354	382	410	.9	
.1	410	438	466	495	523	551	579	607	636	664	692	.8	
.2	692	721	749	777	805	834	862	890	919	947	975	.7	
.3	975	004	032	060	089	117	145	174	202	231	259	.6	
.4	.79 259	287	316	344	373	401	430	458	487	515	544	.5	
.5	544	572	601	629	658	686	715	743	772	800	829	.4	
.6	829	858	886	915	943	972	001	029	058	086	115	.3	
.7	.80 115	144	172	201	230	258	287	316	345	373	402	.2	
.8	402	431	460	488	517	546	575	603	632	661	690	.1	
.9	690	719	747	776	805	834	863	892	921	950	978	51.0	
39.0	978	007	036	065	094	123	152	181	210	239	268	.9	
.1	.81 268	297	326	355	384	413	442	471	500	529	558	.8	29
.2	558	587	616	645	674	703	733	762	791	820	849	.7	
.3	849	878	907	937	966	995	024	053	082	112	141	.6	
.4	.82 141	170	199	229	258	287	316	346	375	404	434	.5	
.5	434	463	492	522	551	580	610	639	668	698	727	.4	
.6	727	757	786	815	845	874	904	933	963	992	022	.3	
.7	.83 022	051	081	110	140	169	199	228	258	287	317	.2	
.8	317	346	376	406	435	465	494	524	554	583	613	.1	
.9	613	643	672	702	732	761	791	821	850	880	910	50.0 deg.	29–30
(10)	9	8	7	6	5	4	3	2	1		0		

COT

Table 1016 (*continued*)—Trigonometric Functions
TAN

deg.	0	1	2	3	4	5	6	7	8	9	(10)		Diff.
40.0	.83 910	940	969	999	029	059	089	118	148	178	208	.9	
.1	.84 208	238	267	297	327	357	387	417	447	477	507	.8	
.2	507	536	566	596	626	656	686	716	746	776	806	.7	30
.3	806	836	866	896	926	956	986	016	046	077	107	.6	
.4	.85 107	137	167	197	227	257	287	318	348	378	408	.5	
.5	408	438	468	499	529	559	589	620	650	680	710	.4	
.6	710	741	771	801	832	862	892	923	953	983	014	.3	
.7	.86 014	044	074	105	135	166	196	226	257	287	318	.2	
.8	318	348	379	409	440	470	501	531	562	592	623	.1	
.9	623	653	684	714	745	776	806	837	867	898	929	49.0	
41.0	929	959	990	021	051	082	113	143	174	205	236	.9	
.1	.87 236	266	297	328	359	389	420	451	482	513	543	.8	
.2	543	574	605	636	667	698	729	759	790	821	852	.7	
.3	852	883	914	945	976	007	038	069	100	131	162	.6	31
.4	.88 162	193	224	255	286	317	348	379	410	441	473	.5	
.5	473	504	535	566	597	628	659	691	722	753	784	.4	
.6	784	815	847	878	909	940	972	003	034	065	097	.3	
.7	.89 097	128	159	191	222	253	285	316	348	379	410	.2	
.8	410	442	473	505	536	567	599	630	662	693	725	.1	
.9	725	756	788	819	851	883	914	946	977	009	040	48.0	
42.0	.90 040	072	104	135	167	199	230	262	294	325	357	.9	
.1	357	389	420	452	484	516	547	579	611	643	674	.8	
.2	674	706	738	770	802	834	865	897	929	961	993	.7	
.3	993	025	057	089	121	153	185	217	249	281	313	.6	32
.4	.91 313	345	377	409	441	473	505	537	569	601	633	.5	
.5	633	665	697	729	762	794	826	858	890	923	955	.4	
.6	955	987	019	051	084	116	148	180	213	245	277	.3	
.7	.92 277	310	342	374	407	439	471	504	536	569	601	.2	
.8	601	633	666	698	731	763	796	828	861	893	926	.1	
.9	926	958	991	023	056	088	121	154	186	219	252	47.0	
43.0	.93 252	284	317	349	382	415	447	480	513	546	578	.9	
.1	578	611	644	677	709	742	775	808	841	873	906	.8	
.2	906	939	972	005	038	071	104	136	169	202	235	.7	
.3	.94 235	268	301	334	367	400	433	466	499	532	565	.6	33
.4	565	598	631	665	698	731	764	797	830	863	896	.5	
.5	896	930	963	996	029	062	096	129	162	195	229	.4	
.6	.95 229	262	295	329	362	395	429	462	495	529	562	.3	
.7	562	595	629	662	696	729	763	796	830	863	897	.2	
.8	897	930	964	997	031	064	098	131	165	199	232	.1	
.9	.96 232	266	299	333	367	400	434	468	501	535	569	46.0	
44.0	569	603	636	670	704	738	771	805	839	873	907	.9	
.1	907	941	974	008	042	076	110	144	178	212	246	.8	
.2	.97 246	280	314	348	382	416	450	484	518	552	586	.7	34
.3	586	620	654	688	722	756	791	825	859	893	927	.6	
.4	927	961	996	030	064	098	133	167	201	235	270	.5	
.5	.98 270	304	338	373	407	441	476	510	545	579	613	.4	
.6	613	648	682	717	751	786	820	855	889	924	958	.3	
.7	958	993	027	062	097	131	166	200	235	270	304	.2	
.8	.99 304	339	374	408	443	478	512	547	582	617	652	.1	
.9	652	686	721	756	791	826	860	895	930	965	000	45.0	34–35
1.00												deg.	
	(10)	9	8	7	6	5	4	3	2	1	0		

COT

Table 1016 (*continued*)—Tan and Cot of Hundredths of Degrees

TAN

		0	1	2	3	4	5	6	7	8	9	(10)		Diff.
deg.														
45.0	1.0	000	003	007	010	014	017	021	024	028	031	035	.9	3–4
.1		035	038	042	045	049	052	056	060	063	067	070	.8	
.2		070	074	077	081	084	088	091	095	098	102	105	.7	
.3		105	109	112	116	119	123	126	130	134	137	141	.6	
.4		141	144	148	151	155	158	162	165	169	173	176	.5	
.5		176	180	183	187	190	194	197	201	205	208	212	.4	
.6		212	215	219	222	226	230	233	237	240	244	247	.3	
.7		247	251	255	258	262	265	269	272	276	280	283	.2	
.8		283	287	290	294	298	301	305	308	312	316	319	.1	
.9		319	323	326	330	334	337	341	344	348	352	355	44.0	
46.0		355	359	363	366	370	373	377	381	384	388	392	.9	
.1		392	395	399	402	406	410	413	417	421	424	428	.8	
.2		428	432	435	439	442	446	450	453	457	461	464	.7	
.3		464	468	472	475	479	483	486	490	494	497	501	.6	
.4		501	505	508	512	516	519	523	527	530	534	538	.5	
.5		538	541	545	549	553	556	560	564	567	571	575	.4	
.6		575	578	582	586	590	593	597	601	604	608	612	.3	
.7		612	615	619	623	627	630	634	638	641	645	649	.2	
.8		649	653	656	660	664	668	671	675	679	682	686	.1	
.9		686	690	694	697	701	705	709	712	716	720	724	43.0	
47.0		724	727	731	735	739	742	746	750	754	758	761	.9	
.1		761	765	769	773	776	780	784	788	791	795	799	.8	
.2		799	803	807	810	814	818	822	826	829	833	837	.7	
.3		837	841	844	848	852	856	860	863	867	871	875	.6	
.4		875	879	883	886	890	894	898	902	905	909	913	.5	
.5		913	917	921	925	928	932	936	940	944	948	951	.4	
.6		951	955	959	963	967	971	974	978	982	986	990	.3	
.7		990	994	998	001	005	009	013	017	021	025	028	.2	
.8	1.1	028	032	036	040	044	048	052	056	059	063	067	.1	
.9		067	071	075	079	083	087	091	094	098	102	106	42.0	
48.0		106	110	114	118	122	126	130	133	137	141	145	.9	
.1		145	149	153	157	161	165	169	173	177	180	184	.8	
.2		184	188	192	196	200	204	208	212	216	220	224	.7	
.3		224	228	232	236	240	243	247	251	255	259	263	.6	
.4		263	267	271	275	279	283	287	291	295	299	303	.5	
.5		303	307	311	315	319	323	327	331	335	339	343	.4	
.6		343	347	351	355	359	363	367	371	375	379	383	.3	4
.7		383	387	391	395	399	403	407	411	415	419	423	.2	
.8		423	427	431	435	439	443	447	451	455	459	463	.1	
.9		463	467	471	475	479	483	487	492	496	500	504	41.0	
49.0		504	508	512	516	520	524	528	532	536	540	544	.9	
.1		544	548	552	557	561	565	569	573	577	581	585	.8	
.2		585	589	593	597	601	606	610	614	618	622	626	.7	
.3		626	630	634	638	643	647	651	655	659	663	667	.6	
.4		667	671	675	680	684	688	692	696	700	704	708	.5	
.5		708	713	717	721	725	729	733	738	742	746	750	.4	
.6		750	754	758	762	767	771	775	779	783	787	792	.3	
.7		792	796	800	804	808	812	817	821	825	829	833	.2	
.8		833	838	842	846	850	854	859	863	867	871	875	.1	
.9		875	880	884	888	892	896	901	905	909	913	918	40.0	4–5
													deg.	
	(10)		9	8	7	6	5	4	3	2	1	0		

COT

Table 1016 (*continued*)—Trigonometric Functions

TAN

deg.	0	1	2	3	4	5	6	7	8	9	(10)	Diff.
50.0	1.1 918	922	926	930	934	939	943	947	951	956	960	.9
.1	960	964	968	973	977	981	985	990	994	998	002	.8
.2	1.2 002	007	011	015	019	024	028	032	037	041	045	.7
.3	045	049	054	058	062	066	071	075	079	084	088	.6
.4	088	092	097	101	105	109	114	118	122	127	131	.5
.5	131	135	140	144	148	153	157	161	166	170	174	.4
.6	174	179	183	187	192	196	200	205	209	213	218	.3
.7	218	222	226	231	235	239	244	248	252	257	261	.2
.8	261	266	270	274	279	283	287	292	296	301	305	.1
.9	305	309	314	318	323	327	331	336	340	345	349	39.0
51.0	349	353	358	362	367	371	375	380	384	389	393	.9
.1	393	398	402	406	411	415	420	424	429	433	437	.8
.2	437	442	446	451	455	460	464	469	473	478	482	.7
.3	482	487	491	495	500	504	509	513	518	522	527	.6
.4	527	531	536	540	545	549	554	558	563	567	572	.5 4–5
.5	572	576	581	585	590	594	599	603	608	612	617	.4
.6	617	621	626	630	635	640	644	649	653	658	662	.3
.7	662	667	671	676	680	685	689	694	699	703	708	.2
.8	708	712	717	721	726	731	735	740	744	749	753	.1
.9	753	758	763	767	772	776	781	786	790	795	799	38.0
52.0	799	804	809	813	818	822	827	832	836	841	846	.9
.1	846	850	855	859	864	869	873	878	883	887	892	.8
.2	892	897	901	906	911	915	920	924	929	934	938	.7
.3	938	943	948	952	957	962	967	971	976	981	985	.6
.4	985	990	995	999	004	009	013	018	023	028	032	.5
.5	1.3 032	037	042	046	051	056	061	065	070	075	079	.4
.6	079	084	089	094	098	103	108	113	117	122	127	.3
.7	127	132	136	141	146	151	155	160	165	170	175	.2
.8	175	179	184	189	194	198	203	208	213	218	222	.1
.9	222	227	232	237	242	246	251	256	261	266	270	37.0
53.0	270	275	280	285	290	295	299	304	309	314	319	.9
.1	319	324	328	333	338	343	348	353	358	362	367	.8
.2	367	372	377	382	387	392	397	401	406	411	416	.7
.3	416	421	426	431	436	440	445	450	455	460	465	.6
.4	465	470	475	480	485	490	495	499	504	509	514	.5
.5	514	519	524	529	534	539	544	549	554	559	564	.4
.6	564	569	574	579	584	588	593	598	603	608	613	.3
.7	613	618	623	628	633	638	643	648	653	658	663	.2 5
.8	663	668	673	678	683	688	693	698	703	708	713	.1
.9	713	718	723	729	734	739	744	749	754	759	764	36.0
54.0	764	769	774	779	784	789	794	799	804	809	814	.9
.1	814	820	825	830	835	840	845	850	855	860	865	.8
.2	865	870	876	881	886	891	896	901	906	911	916	.7
.3	916	922	927	932	937	942	947	952	958	963	968	.6
.4	968	973	978	983	988	994	999	004	009	014	019	.5
.5	1.4 019	025	030	035	040	045	051	056	061	066	071	.4
.6	071	077	082	087	092	097	103	108	113	118	124	.3
.7	124	129	134	139	144	150	155	160	165	171	176	.2
.8	176	181	186	192	197	202	207	213	218	223	229	.1
.9	229	234	239	244	250	255	260	266	271	276	281	35.0 5–6 deg.
	(10)	9	8	7	6	5	4	3	2	1	0	

COT

296

Table 1016 (*continued*)—Tan and Cot of Hundredths of Degrees

TAN

deg.	0	1	2	3	4	5	6	7	8	9	(10)	Diff.	
55.0	1.4 281	287	292	297	303	308	313	319	324	329	335	.9	
.1	335	340	345	351	356	361	367	372	377	383	388	.8	
.2	388	393	399	404	410	415	420	426	431	436	442	.7	
.3	442	447	453	458	463	469	474	480	485	490	496	.6	
.4	496	501	507	512	517	523	528	534	539	545	550	.5	
.5	550	556	561	566	572	577	583	588	594	599	605	.4	
.6	605	610	616	621	627	632	637	643	648	654	659	.3	
.7	659	665	670	676	681	687	692	698	704	709	715	.2	5–6
.8	715	720	726	731	737	742	748	753	759	764	770	.1	
.9	770	775	781	787	792	798	803	809	814	820	826	34.0	
56.0	826	831	837	842	848	854	859	865	870	876	882	.9	
.1	882	887	893	898	904	910	915	921	927	932	938	.8	
.2	938	943	949	955	960	966	972	977	983	989	994	.7	
.3	994	000	006	011	017	023	028	034	040	046	051	.6	
.4	1.5 051	057	063	068	074	080	085	091	097	103	108	.5	
.5	108	114	120	126	131	137	143	149	154	160	166	.4	
.6	166	172	177	183	189	195	200	206	212	218	224	.3	
.7	224	229	235	241	247	253	258	264	270	276	282	.2	
.8	282	287	293	299	305	311	317	322	328	334	340	.1	
.9	340	346	352	358	363	369	375	381	387	393	399	33.0	
57.0	399	405	410	416	422	428	434	440	446	452	458	.9	
.1	458	464	469	475	481	487	493	499	505	511	517	.8	
.2	517	523	529	535	541	547	553	559	565	571	577	.7	
.3	577	583	589	595	601	607	613	619	625	631	637	.6	6
.4	637	643	649	655	661	667	673	679	685	691	697	.5	
.5	697	703	709	715	721	727	733	739	745	751	757	.4	
.6	757	764	770	776	782	788	794	800	806	812	818	.3	
.7	818	825	831	837	843	849	855	861	867	874	880	.2	
.8	880	886	892	898	904	911	917	923	929	935	941	.1	
.9	941	948	954	960	966	972	979	985	991	997	003	32.0	
58.0	1.6 003	010	016	022	028	034	041	047	053	059	066	.9	
.1	066	072	078	084	091	097	103	110	116	122	128	.8	
.2	128	135	141	147	154	160	166	172	179	185	191	.7	
.3	191	198	204	210	217	223	229	236	242	248	255	.6	
.4	255	261	267	274	280	287	293	299	306	312	319	.5	
.5	319	325	331	338	344	351	357	363	370	376	383	.4	
.6	383	389	395	402	408	415	421	428	434	441	447	.3	
.7	447	454	460	467	473	479	486	492	499	505	512	.2	6–7
.8	512	518	525	531	538	545	551	558	564	571	577	.1	
.9	577	584	590	597	603	610	617	623	630	636	643	31.0	
59.0	643	649	656	663	669	676	682	689	696	702	709	.9	
.1	709	715	722	729	735	742	749	755	762	769	775	.8	
.2	775	782	788	795	802	808	815	822	829	835	842	.7	
.3	842	849	855	862	869	875	882	889	896	902	909	.6	
.4	909	916	923	929	936	943	950	956	963	970	977	.5	
.5	977	983	990	997	004	011	017	024	031	038	045	.4	
.6	1.7 045	051	058	065	072	079	086	092	099	106	113	.3	
.7	113	120	127	134	140	147	154	161	168	175	182	.2	
.8	182	189	196	202	209	216	223	230	237	244	251	.1	
.9	251	258	265	272	279	286	293	300	307	314	321	30.0	7
deg.	(10)	9	8	7	6	5	4	3	2	1	0		

COT

Table 1016 (*continued*)—**Trigonometric Functions**
TAN

deg.	0	1	2	3	4	5	6	7	8	9	(10)		Diff.
60.0	1.7 321	327	334	341	348	355	362	369	376	384	391	.9	7
.1	391	398	405	412	419	426	433	440	447	454	461	.8	
.2	461	468	475	482	489	496	503	511	518	525	532	.7	
.3	532	539	546	553	560	567	575	582	589	596	603	.6	
.4	603	610	617	625	632	639	646	653	661	668	675	.5	
.5	675	682	689	697	704	711	718	725	733	740	747	.4	
.6	747	754	762	769	776	783	791	798	805	813	820	.3	
.7	820	827	834	842	849	856	864	871	878	886	893	.2	
.8	893	900	908	915	922	930	937	944	952	959	966	.1	
.9	966	974	981	989	996	003	011	018	026	033	040	29.0	
61.0	1.8 040	048	055	063	070	078	085	093	100	107	115	.9	
.1	115	122	130	137	145	152	160	167	175	182	190	.8	
.2	190	197	205	213	220	228	235	243	250	258	265	.7	
.3	265	273	281	288	296	303	311	318	326	334	341	.6	
.4	341	349	357	364	372	379	387	395	402	410	418	.5	
.5	418	425	433	441	448	456	464	471	479	487	495	.4	
.6	495	502	510	518	526	533	541	549	556	564	572	.3	
.7	572	580	588	595	603	611	619	626	634	642	650	.2	
.8	650	658	666	673	681	689	697	705	713	720	728,	.1	
.9	728	736	744	752	760	768	776	784	791	799	807	28.0	
62.0	807	815	823	831	839	847	855	863	871	879	887	.9	
.1	887	895	903	911	919	927	935	943	951	959	967	.8	8
.2	967	975	983	991	999	007	015	023	031	039	047	.7	
.3	1.9 047	055	063	071	080	088	096	104	112	120	128	.6	
.4	128	136	145	153	161	169	177	185	193	202	210	.5	
.5	210	218	226	234	243	251	259	267	275	284	292	.4	
.6	292	300	308	317	325	333	342	350	358	366	375	.3	
.7	375	383	391	400	408	416	425	433	441	450	458	.2	
.8	458	466	475	483	491	500	508	517	525	533	542	.1	
.9	542	550	559	567	575	584	592	601	609	618	626	27.0	
63.0	626	635	643	652	660	669	677	686	694	703	711	.9	
.1	711	720	728	737	745	754	762	771	779	788	797	.8	
.2	797	805	814	822	831	840	848	857	866	874	883	.7	
.3	883	891	900	909	917	926	935	943	952	961	970	.6	
.4	970	978	987	996	004	013	022	031	039	048	057	.5	
.5	2.0 057	066	074	083	092	101	110	118	127	136	145	.4	
.6	145	154	163	171	180	189	198	207	216	225	233	.3	
.7	233	242	251	260	269	278	287	296	305	314	323	.2	
.8	323	332	341	350	359	368	377	386	395	404	413	.1	9
.9	413	422	431	440	449	458	467	476	485	494	503	26.0	
64.0	503	512	521	530	539	549	558	567	576	585	594	.9	
.1	594	603	612	622	631	640	649	658	668	677	686	.8	
.2	686	695	704	714	723	732	741	751	760	769	778	.7	
.3	778	788	797	806	816	825	834	844	853	862	872	.6	
.4	872	881	890	900	909	918	928	937	947	956	965	.5	
.5	965	975	984	994	003	013	022	032	041	050	060	.4	
.6	2.1 060	069	079	088	098	107	117	127	136	146	155	.3	
.7	155	165	174	184	193	203	213	222	232	241	251	.2	
.8	251	261	270	280	290	299	309	319	328	338	348	.1	
.9	348	357	367	377	387	396	406	416	426	435	445	25.0	9–10 deg.
	(10)	9	8	7	6	5	4	3	2	1	0		

COT

Table 1016 (*continued*)—Tan and Cot of Hundredths of Degrees

TAN

	0	1	2	3	4	5	6	7	8	9	(10)		Diff.
deg.													
65.0	2.1 445	455	465	474	484	494	504	514	523	533	543	.9	
.1	543	553	563	573	583	592	602	612	622	632	642	.8	
.2	642	652	662	672	682	692	702	712	722	732	742	.7	
.3	742	752	762	772	782	792	802	812	822	832	842	.6	10
.4	842	852	862	872	882	892	902	913	923	933	943	.5	
.5	943	953	963	973	984	994	004	014	024	035	045	.4	
.6	2.2 045	055	065	076	086	096	106	117	127	137	148	.3	
.7	148	158	168	179	189	199	210	220	230	241	251	.2	
.8	251	261	272	282	293	303	313	324	334	345	355	.1	
.9	355	366	376	387	397	408	418	429	439	450	460	24.0	
66.0	460	471	481	492	503	513	524	534	545	556	566	.9	
.1	566	577	588	598	609	620	630	641	652	662	673	.8	
.2	673	684	694	705	716	727	737	748	759	770	781	.7	
.3	781	791	802	813	824	835	846	856	867	878	889	.6	
.4	889	900	911	922	933	944	955	966	976	987	998	.5	
.5	998	009	020	031	042	053	064	075	087	098	109	.4	11
.6	2.3 109	120	131	142	153	164	175	186	197	209	220	.3	
.7	220	231	242	253	264	276	287	298	309	321	332	.2	
.8	332	343	354	366	377	388	399	411	422	433	445	.1	
.9	445	456	467	479	490	501	513	524	536	547	559	23.0	
67.0	559	570	581	593	604	616	627	639	650	662	673	.9	
.1	673	685	696	708	719	731	743	754	766	777	789	.8	
.2	789	801	812	824	836	847	859	871	882	894	906	.7	
.3	906	917	929	941	953	964	976	988	000	012	023	.6	
.4	2.4 023	035	047	059	071	083	095	106	118	130	142	.5	
.5	142	154	166	178	190	202	214	226	238	250	262	.4	12
.6	262	274	286	298	310	322	334	346	358	370	383	.3	
.7	383	395	407	419	431	443	455	468	480	492	504	.2	
.8	504	516	529	541	553	566	578	590	602	615	627	.1	
.9	627	639	652	664	676	689	701	714	726	738	751	22.0	
68.0	751	763	776	788	801	813	826	838	851	863	876	.9	
.1	876	888	901	913	926	939	951	964	976	989	002	.8	
.2	2.5 002	014	027	040	052	065	078	091	103	116	129	.7	
.3	129	142	154	167	180	193	206	219	231	244	257	.6	
.4	257	270	283	296	309	322	335	348	361	373	386	.5	
.5	386	399	412	426	439	452	465	478	491	504	517	.4	13
.6	517	530	543	556	570	583	596	609	622	635	649	.3	
.7	649	662	675	688	702	715	728	742	755	768	782	.2	
.8	782	795	808	822	835	848	862	875	889	902	916	.1	
.9	916	929	943	956	970	983	997	010	024	037	051	21.0	
69.0	2.6 051	064	078	092	105	119	133	146	160	174	187	.9	
.1	187	201	215	229	242	256	270	284	298	311	325	.8	
.2	325	339	353	367	381	395	408	422	436	450	464	.7	
.3	464	478	492	506	520	534	548	562	576	590	605	.6	14
.4	605	619	633	647	661	675	689	704	718	732	746	.5	
.5	746	760	775	789	803	818	832	846	860	875	889	.4	
.6	889	904	918	932	947	961	976	990	005	019	034	.3	
.7	2.7 034	048	063	077	092	106	121	135	150	165	179	.2	
.8	179	194	209	223	238	253	267	282	297	312	326	.1	
.9	326	341	356	371	386	400	415	430	445	460	475	20.0 deg.	15
	(10)	9	8	7	6	5	4	3	2	1	0		

COT

Table 1016 (*continued*)—**Trigonometric Functions**
TAN

deg.		0	1	2	3	4	5	6	7	8	9	(10)		Diff.
70.0	2.7	475	490	505	520	535	550	565	580	595	610	625	.9	15
.1		625	640	655	670	685	700	715	731	746	761	776	.8	
.2		776	791	807	822	837	852	868	883	898	914	929	.7	
.3		929	944	960	975	990	006	021	037	052	068	083	.6	
.4	2.8	083	099	114	130	145	161	177	192	208	223	239	.5	
.5		239	255	270	286	302	318	333	349	365	381	397	.4	
.6		397	412	428	444	460	476	492	508	524	540	556	.3	
.7		556	572	588	604	620	636	652	668	684	700	716	.2	16
.8		716	732	748	765	781	797	813	829	846	862	878	.1	
.9		878	895	911	927	944	960	976	993	009	026	042	19.0	
71.0	2.9	042	059	075	092	108	125	141	158	174	191	208	.9	
.1		208	224	241	258	274	291	308	324	341	358	375	.8	
.2		375	392	408	425	442	459	476	493	510	527	544	.7	
.3		544	561	578	595	612	629	646	663	680	697	714	.6	17
.4		714	732	749	766	783	800	818	835	852	870	887	.5	
.5		887	904	922	939	956	974	991	009	026	044	061	.4	
.6	3.0	061	079	096	114	131	149	167	184	202	220	237	.3	
.7		237	255	273	290	308	326	344	362	379	397	415	.2	
.8		415	433	451	469	487	505	523	541	559	577	595	.1	18
.9		595	613	631	649	668	686	704	722	740	759	777	18.0	
72.0		777	795	813	832	850	868	887	905	924	942	961	.9	
.1		961	979	998	016	035	053	072	090	109	128	146	.8	
.2	3.1	146	165	184	202	221	240	259	278	296	315	334	.7	
.3		334	353	372	391	410	429	448	467	486	505	524	.6	19
.4		524	543	562	581	601	620	639	658	677	697	716	.5	
.5		716	735	755	774	793	813	832	852	871	891	910	.4	
.6		910	930	949	969	988	008	028	047	067	087	106	.3	
.7	3.2	106	126	146	166	185	205	225	245	265	285	305	.2	
.8		305	325	345	365	385	405	425	445	465	485	506	.1	20
.9		506	526	546	566	586	607	627	647	668	688	709	17.0	
73.0		709	729	749	770	790	811	831	852	873	893	914	.9	
.1		914	935	955	976	997	017	038	059	080	101	122	.8	
.2	3.3	122	143	163	184	205	226	247	268	290	311	332	.7	21
.3		332	353	374	395	416	438	459	480	502	523	544	.6	
.4		544	566	587	609	630	652	673	695	716	738	759	.5	
.5		759	781	803	824	846	868	890	912	933	955	977	.4	
.6		977	999	021	043	065	087	109	131	153	175	197	.3	22
.7	3.4	197	220	242	264	286	308	331	353	375	398	420	.2	
.8		420	443	465	488	510	533	555	578	600	623	646	.1	
.9		646	669	691	714	737	760	782	805	828	851	874	16.0	
74.0		874	897	920	943	966	989	012	036	059	082	105	.9	23
.1	3.5	105	129	152	175	199	222	245	269	292	316	339	.8	
.2		339	363	386	410	434	457	481	505	529	552	576	.7	
.3		576	600	624	648	672	696	720	744	768	792	816	.6	24
.4		816	840	864	889	913	937	961	986	010	034	059	.5	
.5	3.6	059	083	108	132	157	181	206	231	255	280	305	.4	
.6		305	330	354	379	404	429	454	479	504	529	554	.3	25
.7		554	579	604	629	654	680	705	730	755	781	806	.2	
.8		806	832	857	882	908	933	959	985	010	036	062	.1	
.9	3.7	062	087	113	139	165	191	217	242	268	294	321	15.0	26
													deg.	
	(10)	9	8	7	6	5	4	3	2	1	0			

COT

Table 1016 (*continued*)—Tan and Cot of Hundredths of Degrees

TAN

deg.	0	1	2	3	4	5	6	7	8	9	(10)		Diff.
75.0	3.7 321	347	373	399	425	451	477	504	530	556	583	.9	26
.1	583	609	636	662	689	715	742	768	795	822	848	.8	
.2	848	875	902	929	956	983	010	037	064	091	118	.7	27
.3	3.8 118	145	172	199	226	254	281	308	336	363	391	.6	
.4	391	418	446	473	501	528	556	584	612	639	667	.5	
.5	667	695	723	751	779	807	835	863	891	919	947	.4	28
.6	947	976	004	032	061	089	117	146	174	203	232	.3	
.7	3.9 232	260	289	318	346	375	404	433	462	491	520	.2	29
.8	520	549	578	607	636	665	694	724	753	782	812	.1	
.9	812	841	871	900	930	959	989	019	048	078	108	14.0	
76.0	4.0 108	138	168	197	227	257	287	318	348	378	408	.9	30
.1	408	438	469	499	529	560	590	621	651	682	713	.8	
.2	713	743	774	805	836	867	898	929	960	991	022	.7	31
.3	4.1 022	053	084	115	146	178	209	241	272	304	335	.6	
.4	335	367	398	430	462	493	525	557	589	621	653	.5	32
.5	653	685	717	749	781	814	846	878	911	943	976	.4	
.6	976	008	041	073	106	139	171	204	237	270	303	.3	33
.7	4.2 303	336	369	402	435	468	502	535	568	602	635	.2	
.8	635	669	702	736	770	803	837	871	905	938	972	.1	
.9	972	006	040	075	109	143	177	212	246	280	315	13.0	34
77.0	4.3 315	349	384	418	453	488	523	557	592	627	662	.9	35
.1	662	697	732	768	803	838	873	909	944	980	015	.8	
.2	4.4 015	051	086	122	158	194	230	265	301	337	373	.7	36
.3	373	410	446	482	518	555	591	628	664	701	737	.6	
.4	737	774	811	848	885	922	959	996	033	070	107	.5	37
.5	4.5 107	144	182	219	257	294	332	369	407	445	483	.4	
.6	483	520	558	596	634	673	711	749	787	826	864	.3	38
.7	864	903	941	980	018	057	096	135	174	213	252	.2	39
.8	4.6 252	291	330	369	409	448	487	527	567	606	646	.1	
.9	646	686	725	765	805	845	885	925	966	006	046	12.0	40
78.0	4.7 046	087	127	168	208	249	290	331	371	412	453	.9	41
.1	453	494	536	577	618	659	701	742	784	826	867	.8	
.2	867	909	951	993	035	077	119	161	203	246	288	.7	42
.3	4.8 288	331	373	416	459	501	544	587	630	673	716	.6	43
.4	716	759	803	846	889	933	977	020	064	108	152	.5	
.5	4.9 152	196	240	284	328	372	416	461	505	550	594	.4	44
.6	594	639	684	729	774	819	864	909	954	000	045	.3	45
.7	5.0 045	091	136	182	228	273	319	365	411	457	504	.2	46
.8	504	550	596	643	689	736	783	830	876	923	970	.1	47
.9	970	018	065	112	159	207	254	302	350	398	446	11.0	
79.0	5.1 446	494	542	590	638	686	735	783	832	880	929	.9	48
.1	929	978	027	076	125	174	224	273	323	372	422	.8	49
.2	5.2 422	472	521	571	621	672	722	772	822	873	924	.7	50
.3	924	974	025	076	127	178	229	280	332	383	435	.6	51
.4	5.3 435	486	538	590	642	694	746	798	850	903	955	.5	52
.5	955	008	060	113	166	219	272	325	379	432	486	.4	53
.6	5.4 486	539	593	647	701	755	809	863	917	972	026	.3	54
.7	5.5 026	081	136	191	246	301	356	411	467	522	578	.2	55
.8	578	633	689	745	801	857	914	970	026	083	140	.1	56
.9	5.6 140	196	253	310	368	425	482	540	597	655	713	10.0	57
												deg.	
	(10)	9	8	7	6	5	4	3	2	1	0		

COT

Table 1016 (*continued*)—Trigonometric Functions
TAN

deg.	0	1	2	3	4	5	6	7	8	9	(10)	
80.0	5.6 713	771	829	887	945	004	062	121	180	238	297	.9
.1	5.7 297	357	416	475	535	594	654	714	774	834	894	.8
.2	894	954	015	075	136	197	257	319	380	441	502	.7
.3	5.8 502	564	626	687	749	811	874	936	998	061	124	.6
.4	5.9 124	186	249	312	376	439	502	566	630	694	758	.5
.5	758	822	886	950	015	080	144	209	275	340	405	.4
.6	6.0 405	471	536	602	668	734	800	867	933	000	066	.3
.7	6.1 066	133	200	267	335	402	470	538	606	674	742	.2
.8	742	810	879	947	016	085	154	223	293	362	432	.1
.9	6.2 432	502	572	642	712	783	853	924	995	066	138	9.0
81.0	6.3 138	209	280	352	424	496	568	641	713	786	859	.9
.1	859	932	005	078	152	225	299	373	447	522	596	.8
.2	6.4 596	671	746	821	896	971	047	122	198	274	350	.7
.3	6.5 350	427	503	580	657	734	811	889	966	044	122	.6
.4	6.6 122	200	278	357	436	514	594	673	752	832	912	.5
.5	912	992	072	152	233	313	394	475	557	638	720	.4
.6	6.7 720	802	884	966	049	131	214	297	380	464	548	.3
.7	6.8 548	631	715	800	884	969	054	139	224	310	395	.2
.8	6.9 395	481	567	654	740	827	914	001	088	176	264	.1
.9	7.0 264	352	440	528	617	706	795	884	974	064	154	8.0
82.0	7.1 154	244	334	425	516	607	698	790	882	974	066	.9
.1	7.2 066	159	251	344	438	531	625	719	813	907	002	.8
.2	7.3 002	097	192	287	383	479	575	671	768	865	962	.7
.3	962	059	157	254	352	451	549	648	747	847	947	.6
.4	7.4 947	046	147	247	348	449	550	651	753	855	958	.5
.5	7.5 958	060	163	266	369	473	577	681	786	891	996	.4
.6	7.6 996											
	7.7	101	207	313	419	525	632	739	847	954	062	.3
.7	7.8 062	170	279	388	497	606	716	826	937	047	158	.2
.8	7.9 158	269	381	493	605	718	830	944	057	171	285	.1
.9	8.0 285	399	514	629	744	860	976					
	8.1							092	209	326	443	7.0 Diff.
83.0	8. 144	156	168	180	192	204	215	227	239	251	264	.9 11–13
.1	264	276	288	300	312	324	337	349	361	374	386	.8 12–13
.2	386	399	411	424	436	449	462	474	487	500	513	.7 12–13
.3	513	525	538	551	564	577	590	603	616	630	643	.6 12–14
.4	643	656	669	683	696	709	723	736	750	763	777	.5 13–14
.5	777	791	804	818	832	846	859	873	887	901	915	.4 13–14
.6	915	929	943	958	972	986	000	015	029	043	058	.3 14–15
.7	9. 058	072	087	102	116	131	146	160	175	190	205	.2 14–15
.8	205	220	235	250	265	281	296	311	326	342	357	.1 15–16
.9	357	373	388	404	419	435	451	467	483	498	514	6.0 15–16
84.0	514	530	546	563	579	595	611	627	644	660	677	.9 16–17
.1	677	693	710	727	743	760	777	794	811	828	845	.8 16–17
.2	845	862	879	896	914	931	948	966	983	001	019	.7 17–18
.3	10. 019	036	054	072	090	108	126	144	162	180	199	.6 17–19
.4	199	217	236	254	273	291	310	329	348	366	385	.5 18–19
.5	385	404	424	443	462	481	501	520	540	559	579	.4 19–20
.6	579	599	618	638	658	678	698	719	739	759	780	.3 19–21
.7	780	800	821	841	862	883	904	925	946	967	988	.2 20–21
.8	988	009	031	052	074	095	117	139	161	183	205	.1 21–22
.9	11. 205	227	249	271	294	316	339	362	384	407	430	5.0 22–23
												deg.
	(10)	9	8	7	6	5	4	3	2	1	0	

COT

302

Table 1016 (continued)—Tan and Cot of Hundredths of Degrees

TAN

deg.		0	1	2	3	4	5	6	7	8	9	(10)	Diff.
85.0	11.	430	453	476	499	523	546	570	593	617	641	664	.9 23–24
.1		664	688	713	737	761	785	810	834	859	884	909	.8 24–25
.2		909	934	959	984	009	035	060	086	111	137	163	.7 25–26
.3	12.	163	189	215	242	268	295	321	348	375	402	429	.6 26–27
.4		429	456	483	511	538	566	594	622	650	678	706	.5 27–28
.5		706	735	763	792	821	850	879	908	937	967	996	.4 28–30
.6		996	026	056	086	116	146	177	207	238	269	300	.3 30–31
.7	13.	300	331	362	393	425	457	488	520	553	585	617	.2 31–33
.8		617	650	683	716	749	782	815	849	883	917	951	.1 33–34
.9		951	985	019	054	089	124	159	194	229	265	301	4.0 34–36
86.0	14.	301	337	373	409	446	482	519	556	593	631	669	.9 36–38
.1		669	706	744	783	821	860	898	937	977	016	056	.8 37–40
.2	15.	056	096	136	176	216	257	298	339	380	422	464	.7 40–42
.3		464	506	548	591	633	676	719	763	806	850	895	.6 42–45
.4		895	939	984	028	074	119	165	211	257	303	350	.5 44–47
.5	16.	350	397	444	492	539	587	636	684	733	782	832	.4 47–50
.6		832	882	932	982	033	084	135	187	238	291	343	.3 50–53
.7	17.	343	396	449	503	556	611	665	720	775	830	886	.2 53–56
.8		886	942	999	056	113	171	229	287	346	405	464	.1 56–59
.9	18.	464	524	585	645	706	768	830	892	955	018	081	3.0 60–63
87.0	19.	081	145	209	274	339	405	471	538	605	672	740	.9
.1		740	809	878	947	017	087	158	229	301	374	446	.8
.2	20.	446	520	594	668	743	819	895	972	049	127	205	.7
.3	21.	205	284	363	444	524	606	688	770	853	937	022	.6
.4	22.	022	107	193	279	366	454	543	632	722	812	904	.5
.5		904	996	089	182	277	372	468	564	662	760	859	.4
.6	23.	859	959	060	162	264	368	472	577	683	790	898	.3
.7	24.	898											
	25.		007	116	227	339	452	565	680	796	913	031	.2
.8	26.	031	150	270	391	513	637	761	887	014	142	271	.1
.9	27.	271	402	534	667	801	937						
	28.							074	213	352	494	636	2.0 deg.
	(10)	9	8	7	6	5	4	3	2	1	0		

COT

For 88° and 89° see the following two pages.

For a more extended table of trigonometric functions of decimals of degrees, see Reference 58.

NOTE—Tables 1015 and 1016 are from *Mathematical Tables*, Reference 45.

Table 1016 *(continued)*—Trigonometric Functions
TAN | | | | TAN

deg.			deg.		
88.00	28.636	2.00	88.50	38.188	1.50
.01	28.780	1.99	.51	38.445	.49
.02	28.926	.98	.52	38.705	.48
.03	29.073	.97	.53	38.968	.47
.04	29.221	.96	.54	39.235	.46
.05	29.371	.95	.55	39.506	.45
.06	29.523	.94	.56	39.780	.44
.07	29.676	.93	.57	40.059	.43
.08	29.830	.92	.58	40.341	.42
.09	29.987	.91	.59	40.627	.41
88.10	30.145	1.90	88.60	40.917	1.40
.11	30.304	.89	.61	41.212	.39
.12	30.466	.88	.62	41.511	.38
.13	30.629	.87	.63	41.814	.37
.14	30.793	.86	.64	42.121	.36
.15	30.960	.85	.65	42.433	.35
.16	31.128	.84	.66	42.750	.34
.17	31.299	.83	.67	43.072	.33
.18	31.471	.82	.68	43.398	.32
.19	31.645	.81	.69	43.730	.31
88.20	31.821	1.80	88.70	44.066	1.30
.21	31.998	.79	.71	44.408	.29
.22	32.178	.78	.72	44.755	.28
.23	32.360	.77	.73	45.107	.27
.24	32.544	.76	.74	45.466	.26
.25	32.730	.75	.75	45.829	.25
.26	32.918	.74	.76	46.199	.24
.27	33.109	.73	.77	46.575	.23
.28	33.301	.72	.78	46.957	.22
.29	33.496	.71	.79	47.345	.21
88.30	33.694	1.70	88.80	47.740	1.20
.31	33.893	.69	.81	48.141	.19
.32	34.095	.68	.82	48.549	.18
.33	34.299	.67	.83	48.964	.17
.34	34.506	.66	.84	49.386	.16
.35	34.715	.65	.85	49.816	.15
.36	34.927	.64	.86	50.253	.14
.37	35.141	.63	.87	50.698	.13
.38	35.358	.62	.88	51.150	.12
.39	35.578	.61	.89	51.611	.11
88.40	35.801	1.60	88.90	52.081	1.10
.41	36.026	.59	.91	52.559	.09
.42	36.254	.58	.92	53.045	.08
.43	36.485	.57	.93	53.541	.07
.44	36.719	.56	.94	54.046	.06
.45	36.956	.55	.95	54.561	.05
.46	37.196	.54	.96	55.086	.04
.47	37.439	.53	.97	55.621	.03
.48	37.686	.52	.98	56.166	.02
.49	37.935	1.51	.99	56.723	1.01
		deg.			deg.

COT | | | | COT

Table 1016 *(continued)*—Tan and Cot of Hundredths of Degrees

	TAN			TAN	
deg.			deg.		
89.00	57.290	1.00	89.50	114.589	0.50
.01	57.869	0.99	.51	116.927	.49
.02	58.459	.98	.52	119.363	.48
.03	59.062	.97	.53	121 903	.47
.04	59.678	.96	.54	124.553	.46
.05	60.306	.95	.55	127.321	.45
.06	60.947	.94	.56	130.215	.44
.07	61.603	.93	.57	133.243	.43
.08	62.273	.92	.58	136.416	.42
.09	62.957	.91	.59	139.743	.41
89.10	63.657	0.90	89.60	143.237	0.40
.11	64.372	.89	.61	146.910	.39
.12	65.104	.88	.62	150.776	.38
.13	65.852	.87	.63	154.851	.37
.14	66.618	.86	.64	159.153	.36
.15	67.402	.85	.65	163.700	.35
.16	68.204	.84	.66	168.515	.34
.17	69.026	.83	.67	173.622	.33
.18	69.868	.82	.68	179.047	.32
.19	70.731	.81	.69	184.823	.31
89.20	71.615	0.80	89.70	190.984	0.30
.21	72.522	.79	.71	197.570	.29
.22	73.452	.78	.72	204.626	.28
.23	74.406	.77	.73	212.205	.27
.24	75.385	.76	.74	220.367	.26
.25	76.390	.75	.75	229.182	.25
.26	77.422	.74	.76	238.731	.24
.27	78.483	.73	.77	249.111	.23
.28	79.573	.72	.78	260.434	.22
.29	80.694	.71	.79	272.836	.21
89.30	81.847	0.70	89.80	286.478	0.20
.31	83.033	.69	.81	301.56	.19
.32	84.255	.68	.82	318.31	.18
.33	85.512	.67	.83	337.03	.17
.34	86.808	.66	.84	358.10	.16
.35	88.144	.65	.85	381.97	.15
.36	89.521	.64	.86	409.25	.14
.37	90.942	.63	.87	440.74	.13
.38	92.409	.62	.88	477.46	.12
.39	93.924	.61	.89	520.87	.11
89.40	95.489	0.60	89.90	572.96	0.10
.41	97.108	.59	.91	636.62	.09
.42	98.782	.58	.92	716.20	.08
.43	100.516	.57	.93	818.51	.07
.44	102.311	.56	.94	954.93	.06
.45	104.171	.55	.95	1145.92	.05
.46	106.100	.54	.96	1432.4	.04
.47	108.102	.53	.97	1909.9	.03
.48	110.181	.52	.98	2864.8	.02
.49	112.342	0.51	.99	5729.6	.01
			90.00	Infin.	0.00
		deg.			deg.

	COT			COT	

Table 1020—Logarithms to Base 10

N	0	1	2	3	4	5	6	7	8	9	1 2 3	4 5 6
10	0000	0043	0086	0128	0170	0212	0253	0294	0334	0374	4 8 12	17
11	0414	0453	0492	0531	0569	0607	0645	0682	0719	0755	4 8 11	15 19
12	0792	0828	0864	0899	0934	0969	1004	1038	1072	1106	3 7 10	14 17 2
13	1139	1173	1206	1239	1271	1303	1335	1367	1399	1430	3 6 10	13 16 19
14	1461	1492	1523	1553	1584	1614	1644	1673	1703	1732	3 6 9	12 15 18
15	1761	1790	1818	1847	1875	1903	1931	1959	1987	2014	3 6 8	11 14 17	20
16	2041	2068	2095	2122	2148	2175	2201	2227	2253	2279	3 5 8	11 13 16	18 21
17	2304	2330	2355	2380	2405	2430	2455	2480	2504	2529	2 5 7	10 12 15	17 20 22
18	2553	2577	2601	2625	2648	2672	2695	2718	2742	2765	2 5 7	9 12 14	16 19 21
19	2788	2810	2833	2856	2878	2900	2923	2945	2967	2989	2 4 7	9 11 13	16 18 20
20	3010	3032	3054	3075	3096	3118	3139	3160	3181	3201	2 4 6	8 11 13	15 17 19
21	3222	3243	3263	3284	3304	3324	3345	3365	3385	3404	2 4 6	8 10 12	14 16 18
22	3424	3444	3464	3483	3502	3522	3541	3560	3579	3598	2 4 6	8 10 12	14 16 17
23	3617	3636	3655	3674	3692	3711	3729	3747	3766	3784	2 4 6	7 9 11	13 15 17
24	3802	3820	3838	3856	3874	3892	3909	3927	3945	3962	2 4 5	7 9 11	12 14 16
25	3979	3997	4014	4031	4048	4065	4082	4099	4116	4133	2 4 5	7 9 10	12 14 16
26	4150	4166	4183	4200	4216	4232	4249	4265	4281	4298	2 3 5	7 8 10	11 13 15
27	4314	4330	4346	4362	4378	4393	4409	4425	4440	4456	2 3 5	6 8 9	11 12 14
28	4472	4487	4502	4518	4533	4548	4564	4579	4594	4609	2 3 5	6 8 9	11 12 14
29	4624	4639	4654	4669	4683	4698	4713	4728	4742	4757	1 3 4	6 7 9	10 12 13
30	4771	4786	4800	4814	4829	4843	4857	4871	4886	4900	1 3 4	6 7 9	10 11 13
31	4914	4928	4942	4955	4969	4983	4997	5011	5024	5038	1 3 4	5 7 8	10 11 12
32	5051	5065	5079	5092	5105	5119	5132	5145	5159	5172	1 3 4	5 7 8	9 11 12
33	5185	5198	5211	5224	5237	5250	5263	5276	5289	5302	1 3 4	5 7 8	9 11 12
34	5315	5328	5340	5353	5366	5378	5391	5403	5416	5428	1 2 4	5 6 8	9 10 11
35	5441	5453	5465	5478	5490	5502	5514	5527	5539	5551	1 2 4	5 6 7	9 10 11
36	5563	5575	5587	5599	5611	5623	5635	5647	5658	5670	1 2 4	5 6 7	8 10 11
37	5682	5694	5705	5717	5729	5740	5752	5763	5775	5786	1 2 4	5 6 7	8 9 11
38	5798	5809	5821	5832	5843	5855	5866	5877	5888	5899	1 2 3	5 6 7	8 9 10
39	5911	5922	5933	5944	5955	5966	5977	5988	5999	6010	1 2 3	4 5 7	8 9 10
40	6021	6031	6042	6053	6064	6075	6085	6096	6107	6117	1 2 3	4 5 6	8 9 10
41	6128	6138	6149	6160	6170	6180	6191	6201	6212	6222	1 2 3	4 5 6	7 8 9
42	6232	6243	6253	6263	6274	6284	6294	6304	6314	6325	1 2 3	4 5 6	7 8 9
43	6335	6345	6355	6365	6375	6385	6395	6405	6415	6425	1 2 3	4 5 6	7 8 9
44	6435	6444	6454	6464	6474	6484	6493	6503	6513	6522	1 2 3	4 5 6	7 8 9
45	6532	6542	6551	6561	6571	6580	6590	6599	6609	6618	1 2 3	4 5 6	7 8 9
46	6628	6637	6646	6656	6665	6675	6684	6693	6702	6712	1 2 3	4 5 6	7 7 8
47	6721	6730	6739	6749	6758	6767	6776	6785	6794	6803	1 2 3	4 5 6	7 7 8
48	6812	6821	6830	6839	6848	6857	6866	6875	6884	6893	1 2 3	4 5 6	7 7 8
49	6902	6911	6920	6928	6937	6946	6955	6964	6972	6981	1 2 3	4 4 5	6 7 8
50	6990	6998	7007	7016	7024	7033	7042	7050	7059	7067	1 2 3	3 4 5	6 7 8
51	7076	7084	7093	7101	7110	7118	7126	7135	7143	7152	1 2 3	3 4 5	6 7 8
52	7160	7168	7177	7185	7193	7202	7210	7218	7226	7235	1 2 3	3 4 5	6 7 7
53	7243	7251	7259	7267	7275	7284	7292	7300	7308	7316	1 2 2	3 4 5	6 6 7
54	7324	7332	7340	7348	7356	7364	7372	7380	7388	7396	1 2 2	3 4 5	6 6 7
N	0	1	2	3	4	5	6	7	8	9	1 2 2	4 5 6	7 8 9

The proportional parts are stated in full for every tenth at the right-hand side. The logarithm of any number of four significant figures can be read directly by add-

306

Table 1020 (*continued*)—Logarithms to Base 10

N	0	1	2	3	4	5	6	7	8	9	1	2	3	4	5	6	7	8	9
55	7404	7412	7419	7427	7435	7443	7451	7459	7466	7474	1	2	2	3	4	5	5	6	7
56	7482	7490	7497	7505	7513	7520	7528	7536	7543	7551	1	2	2	3	4	5	5	6	7
57	7559	7566	7574	7582	7589	7597	7604	7612	7619	7627	1	1	2	3	4	5	5	6	7
58	7634	7642	7649	7657	7664	7672	7679	7686	7694	7701	1	1	2	3	4	4	5	6	7
59	7709	7716	7723	7731	7738	7745	7752	7760	7767	7774	1	1	2	3	4	4	5	6	7
60	7782	7789	7796	7803	7810	7818	7825	7832	7839	7846	1	1	2	3	4	4	5	6	6
61	7853	7860	7868	7875	7882	7889	7896	7903	7910	7917	1	1	2	3	3	4	5	6	6
62	7924	7931	7938	7945	7952	7959	7966	7973	7980	7987	1	1	2	3	3	4	5	5	6
63	7993	8000	8007	8014	8021	8028	8035	8041	8048	8055	1	1	2	3	3	4	5	5	6
64	8062	8069	8075	8082	8089	8096	8102	8109	8116	8122	1	1	2	3	3	4	5	5	6
65	8129	8136	8142	8149	8156	8162	8169	8176	8182	8189	1	1	2	3	3	4	5	5	6
66	8195	8202	8209	8215	8222	8228	8235	8241	8248	8254	1	1	2	3	3	4	5	5	6
67	8261	8267	8274	8280	8287	8293	8299	8306	8312	8319	1	1	2	3	3	4	5	5	6
68	8325	8331	8338	8344	8351	8357	8363	8370	8376	8382	1	1	2	3	3	4	4	5	6
69	8388	8395	8401	8407	8414	8420	8426	8432	8439	8445	1	1	2	3	3	4	4	5	6
70	8451	8457	8463	8470	8476	8482	8488	8494	8500	8506	1	1	2	3	3	4	4	5	6
71	8513	8519	8525	8531	8537	8543	8549	8555	8561	8567	1	1	2	3	2	4	4	5	6
72	8573	8579	8585	8591	8597	8603	8609	8615	8621	8627	1	1	2	3	3	4	4	5	6
73	8633	8639	8645	8651	8657	8663	8669	8675	8681	8686	1	1	2	2	3	4	4	5	5
74	8692	8698	8704	8710	8716	8722	8727	8733	8739	8745	1	1	2	2	3	4	4	5	5
75	8751	8756	8762	8768	8774	8779	8785	8791	8797	8802	1	1	2	2	3	3	4	5	5
76	8808	8814	8820	8825	8831	8837	8842	8848	8854	8859	1	1	2	2	3	3	4	4	5
77	8865	8871	8876	8882	8887	8893	8899	8904	8910	8915	1	1	2	2	3	3	4	4	5
78	8921	8927	8932	8938	8943	8949	8954	8960	8965	8971	1	1	2	2	3	3	4	4	5
79	8976	8982	8987	8993	8998	9004	9009	9015	9020	9025	1	1	2	2	3	3	4	4	5
80	9031	9036	9042	9047	9053	9058	9063	9069	9074	9079	1	1	2	2	3	3	4	4	5
81	9085	9090	9096	9101	9106	9112	9117	9122	9128	9133	1	1	2	2	3	3	4	4	5
82	9138	9143	9149	9154	9159	9165	9170	9175	9180	9186	1	1	2	2	3	3	4	4	5
83	9191	9196	9201	9206	9212	9217	9222	9227	9232	9238	1	1	2	2	3	3	4	4	5
84	9243	9248	9253	9258	9263	9269	9274	9279	9284	9289	1	1	2	2	3	3	4	4	5
85	9294	9299	9304	9309	9315	9320	9325	9330	9335	9340	1	1	2	2	3	3	4	4	5
86	9345	9350	9355	9360	9365	9370	9375	9380	9385	9390	1	1	2	2	3	3	4	4	5
87	9395	9400	9405	9410	9415	9420	9425	9430	9435	9440	1	1	2	2	3	3	4	4	5
88	9445	9450	9455	9460	9465	9469	9474	9479	9484	9489	0	1	1	2	2	3	3	4	4
89	9494	9499	9504	9509	9513	9518	9523	9528	9533	9538	0	1	1	2	2	3	3	4	4
90	9542	9547	9552	9557	9562	9566	9571	9576	9581	9586	0	1	1	2	2	3	3	4	4
91	9590	9595	9600	9605	9609	9614	9619	9624	9628	9633	0	1	1	2	2	3	3	4	4
92	9638	9643	9647	9652	9657	9661	9666	9671	9675	9680	0	1	1	2	2	3	3	4	4
93	9685	9689	9694	9699	9703	9708	9713	9717	9722	9727	0	1	1	2	2	3	3	4	4
94	9731	9736	9741	9745	9750	9754	9759	9763	9768	9773	0	1	1	2	2	3	3	4	4
95	9777	9782	9786	9791	9795	9800	9805	9809	9814	9818	0	1	1	2	2	3	3	4	4
96	9823	9827	9832	9836	9841	9845	9850	9854	9859	9863	0	1	1	2	2	3	3	4	4
97	9868	9872	9877	9881	9886	9890	9894	9899	9903	9908	0	1	1	2	2	3	3	4	4
98	9912	9917	9921	9926	9930	9934	9939	9943	9948	9952	0	1	1	2	2	3	3	3	4
99	9956	9961	9965	9969	9974	9978	9983	9987	9991	9996	0	1	1	2	2	3	3	3	4
N	0	1	2	3	4	5	6	7	8	9	1	2	3	4	5	6	7	8	9

ing the proportional part corresponding to the fourth figure to the tabular number corresponding to the first three figures. There may be an error of 1 in the last place.

307

Table 1025—Natural Logarithms

No.	0	1	2	3	4	5	6	7	8	9	Diff.
1.00	.0000	.0010	.0020	.0030	.0040	.0050	.0060	.0070	.0080	.0090	10
1.01	.0100	.0109	.0119	.0129	.0139	.0149	.0159	.0169	.0178	.0188	10–9
1.02	.0198	.0208	.0218	.0227	.0237	.0247	.0257	.0266	.0276	.0286	10–9
1.03	.0296	.0305	.0315	.0325	.0334	.0344	.0354	.0363	.0373	.0383	10–9
1.04	.0392	.0402	.0411	.0421	.0431	.0440	.0450	.0459	.0469	.0478	10–9
1.05	.0488	.0497	.0507	.0516	.0526	.0535	.0545	.0554	.0564	.0573	10–9
1.06	.0583	.0592	.0602	.0611	.0620	.0630	.0639	.0649	.0658	.0667	10–9
1.07	.0677	.0686	.0695	.0705	.0714	.0723	.0733	.0742	.0751	.0760	10–9
1.08	.0770	.0779	.0788	.0797	.0807	.0816	.0825	.0834	.0843	.0853	10–9
1.09	.0862	.0871	.0880	.0889	.0898	.0908	.0917	.0926	.0935	.0944	10–9
1.10	.0953	.0962	.0971	.0980	.0989	.0998	.1007	.1017	.1026	.1035	10–9
1.11	.1044	.1053	.1062	.1071	.1080	.1089	.1098	.1106	.1115	.1124	9–8
1.12	.1133	.1142	.1151	.1160	.1169	.1178	.1187	.1196	.1204	.1213	9–8
1.13	.1222	.1231	.1240	.1249	.1258	.1266	.1275	.1284	.1293	.1302	9–8
1.14	.1310	.1319	.1328	.1337	.1345	.1354	.1363	.1371	.1380	.1389	9–8
1.15	.1398	.1406	.1415	.1424	.1432	.1441	.1450	.1458	.1467	.1476	9–8
1.16	.1484	.1493	.1501	.1510	.1519	.1527	.1536	.1544	.1553	.1561	9–8
1.17	.1570	.1579	.1587	.1596	.1604	.1613	.1621	.1630	.1638	.1647	9–8
1.18	.1655	.1664	.1672	.1681	.1689	.1697	.1706	.1714	.1723	.1731	9–8
1.19	.1740	.1748	.1756	.1765	.1773	.1781	.1790	.1798	.1807	.1815	9–8
1.20	.1823	.1832	.1840	.1848	.1856	.1865	.1873	.1881	.1890	.1898	9–8
1.21	.1906	.1914	.1923	.1931	.1939	.1947	.1956	.1964	.1972	.1980	9–8
1.22	.1989	.1997	.2005	.2013	.2021	.2029	.2038	.2046	.2054	.2062	9–8
1.23	.2070	.2078	.2086	.2095	.2103	.2111	.2119	.2127	.2135	.2143	9–8
1.24	.2151	.2159	.2167	.2175	.2183	.2191	.2199	.2207	.2215	.2223	8
1.25	.2231	.2239	.2247	.2255	.2263	.2271	.2279	.2287	.2295	.2303	8
1.26	.2311	.2319	.2327	.2335	.2343	.2351	.2359	.2367	.2374	.2382	8–7
1.27	.2390	.2398	.2406	.2414	.2422	.2429	.2437	.2445	.2453	.2461	8–7
1.28	.2469	.2476	.2484	.2492	.2500	.2508	.2515	.2523	.2531	.2539	8–7
1.29	.2546	.2554	.2562	.2570	.2577	.2585	.2593	.2601	.2608	.2616	8–7
1.30	.2624	.2631	.2639	.2647	.2654	.2662	.2670	.2677	.2685	.2693	8–7
1.31	.2700	.2708	.2716	.2723	.2731	.2738	.2746	.2754	.2761	.2769	8–7
1.32	.2776	.2784	.2791	.2799	.2807	.2814	.2822	.2829	.2837	.2844	8–7
1.33	.2852	.2859	.2867	.2874	.2882	.2889	.2897	.2904	.2912	.2919	8–7
1.34	.2927	.2934	.2942	.2949	.2957	.2964	.2971	.2979	.2986	.2994	8–7
1.35	.3001	.3008	.3016	.3023	.3031	.3038	.3045	.3053	.3060	.3067	8–7
1.36	.3075	.3082	.3090	.3097	.3104	.3112	.3119	.3126	.3133	.3141	8–7
1.37	.3148	.3155	.3163	.3170	.3177	.3185	.3192	.3199	.3206	.3214	8–7
1.38	.3221	.3228	.3235	.3243	.3250	.3257	.3264	.3271	.3279	.3286	8–7
1.39	.3293	.3300	.3307	.3315	.3322	.3329	.3336	.3343	.3350	.3358	8–7
1.40	.3365	.3372	.3379	.3386	.3393	.3400	.3407	.3415	.3422	.3429	8–7
1.41	.3436	.3443	.3450	.3457	.3464	.3471	.3478	.3485	.3492	.3500	8–7
1.42	.3507	.3514	.3521	.3528	.3535	.3542	.3549	.3556	.3563	.3570	7
1.43	.3577	.3584	.3591	.3598	.3605	.3612	.3619	.3626	.3633	.3639	7–6
1.44	.3646	.3653	.3660	.3667	.3674	.3681	.3688	.3695	.3702	.3709	7
1.45	.3716	.3723	.3729	.3736	.3743	.3750	.3757	.3764	.3771	.3778	7–6
1.46	.3784	.3791	.3798	.3805	.3812	.3819	.3825	.3832	.3839	.3846	7–6
1.47	.3853	.3859	.3866	.3873	.3880	.3887	.3893	.3900	.3907	.3914	7–6
1.48	.3920	.3927	.3934	.3941	.3947	.3954	.3961	.3968	.3974	.3981	7–6
1.49	.3988	.3994	.4001	.4008	.4015	.4021	.4028	.4035	.4041	.4048	7–6

Table 1025 (*continued*)—Natural Logarithms

No.	0	1	2	3	4	5	6	7	8	9	Diff.
1.50	.4055	.4061	.4068	.4075	.4081	.4088	.4095	.4101	.4108	.4114	7–6
1.51	.4121	.4128	.4134	.4141	.4148	.4154	.4161	.4167	.4174	.4181	7–6
1.52	.4187	.4194	.4200	.4207	.4213	.4220	.4226	.4233	.4240	.4246	7–6
1.53	.4253	.4259	.4266	.4272	.4279	.4285	.4292	.4298	.4305	.4311	7–6
1.54	.4318	.4324	.4331	.4337	.4344	.4350	.4357	.4363	.4370	.4376	7–6
1.55	.4383	.4389	.4395	.4402	.4408	.4415	.4421	.4428	.4434	.4440	7–6
1.56	.4447	.4453	.4460	.4466	.4472	.4479	.4485	.4492	.4498	.4504	7–6
1.57	.4511	.4517	.4523	.4530	.4536	.4543	.4549	.4555	.4562	.4568	7–6
1.58	.4574	.4581	.4587	.4593	.4600	.4606	.4612	.4618	.4625	.4631	7–6
1.59	.4637	.4644	.4650	.4656	.4662	.4669	.4675	.4681	.4688	.4694	7–6
1.60	.4700	.4706	.4713	.4719	.4725	.4731	.4737	.4744	.4750	.4756	7–6
1.61	.4762	.4769	.4775	.4781	.4787	.4793	.4800	.4806	.4812	.4818	7–6
1.62	.4824	.4830	.4837	.4843	.4849	.4855	.4861	.4867	.4874	.4880	7–6
1.63	.4886	.4892	.4898	.4904	.4910	.4916	.4923	.4929	.4935	.4941	7–6
1.64	.4947	.4953	.4959	.4965	.4971	.4977	.4983	.4990	.4996	.5002	7–6
1.65	.5008	.5014	.5020	.5026	.5032	.5038	.5044	.5050	.5056	.5062	6
1.66	.5068	.5074	.5080	.5086	.5092	.5098	.5104	.5110	.5116	.5122	6
1.67	.5128	.5134	.5140	.5146	.5152	.5158	.5164	.5170	.5176	.5182	6
1.68	.5188	.5194	.5200	.5206	.5212	.5218	.5224	.5230	.5235	.5241	6–5
1.69	.5247	.5253	.5259	.5265	.5271	.5277	.5283	.5289	.5295	.5300	6–5
1.70	.5306	.5312	.5318	.5324	.5330	.5336	.5342	.5347	.5353	.5359	6–5
1.71	.5365	.5371	.5377	.5382	.5388	.5394	.5400	.5406	.5412	.5417	6–5
1.72	.5423	.5429	.5435	.5441	.5446	.5452	.5458	.5464	.5470	.5475	6–5
1.73	.5481	.5487	.5493	.5499	.5504	.5510	.5516	.5522	.5527	.5533	6–5
1.74	.5539	.5545	.5550	.5556	.5562	.5568	.5573	.5579	.5585	.5590	6–5
1.75	.5596	.5602	.5608	.5613	.5619	.5625	.5630	.5636	.5642	.5647	6–5
1.76	.5653	.5659	.5664	.5670	.5676	.5682	.5687	.5693	.5698	.5704	6–5
1.77	.5710	.5715	.5721	.5727	.5732	.5738	.5744	.5749	.5755	.5761	6–5
1.78	.5766	.5772	.5777	.5783	.5789	.5794	.5800	.5805	.5811	.5817	6–5
1.79	.5822	.5828	.5833	.5839	.5844	.5850	.5856	.5861	.5867	.5872	6–5
1.80	.5878	.5883	.5889	.5895	.5900	.5906	.5911	.5917	.5922	.5928	6–5
1.81	.5933	.5939	.5944	.5950	.5955	.5961	.5966	.5972	.5977	.5983	6–5
1.82	.5988	.5994	.5999	.6005	.6010	.6016	.6021	.6027	.6032	.6038	6–5
1.83	.6043	.6049	.6054	.6060	.6065	.6070	.6076	.6081	.6087	.6092	6–5
1.84	.6098	.6103	.6109	.6114	.6119	.6125	.6130	.6136	.6141	.6146	6–5
1.85	.6152	.6157	.6163	.6168	.6173	.6179	.6184	.6190	.6195	.6200	6–5
1.86	.6206	.6211	.6217	.6222	.6227	.6233	.6238	.6243	.6249	.6254	6–5
1.87	.6259	.6265	.6270	.6275	.6281	.6286	.6292	.6297	.6302	.6307	6–5
1.88	.6313	.6318	.6323	.6329	.6334	.6339	.6345	.6350	.6355	.6360	6–5
1.89	.6366	.6371	.6376	.6382	.6387	.6392	.6397	.6403	.6408	.6413	6–5
1.90	.6419	.6424	.6429	.6434	.6440	.6445	.6450	.6455	.6461	.6466	6–5
1.91	.6471	.6476	.6481	.6487	.6492	.6497	.6502	.6508	.6513	.6518	6–5
1.92	.6523	.6528	.6534	.6539	.6544	.6549	.6554	.6560	.6565	.6570	6–5
1.93	.6575	.6580	.6586	.6591	.6596	.6601	.6606	.6611	.6617	.6622	6–5
1.94	.6627	.6632	.6637	.6642	.6647	.6653	.6658	.6663	.6668	.6673	6–5
1.95	.6678	.6683	.6689	.6694	.6699	.6704	.6709	.6714	.6719	.6724	6–5
1.96	.6729	.6735	.6740	.6745	.6750	.6755	.6760	.6765	.6770	.6775	6–5
1.97	.6780	.6785	.6790	.6796	.6801	.6806	.6811	.6816	.6821	.6826	6–5
1.98	.6831	.6836	.6841	.6846	.6851	.6856	.6861	.6866	.6871	.6876	5
1.99	.6881	.6886	.6891	.6896	.6901	.6906	.6911	.6916	.6921	.6926	5

Table 1025 (*continued*)—Natural Logarithms

No.	0	1	2	3	4	5	6	7	8	9	Diff.
2.0	.6931	.6981	.7031	.7080	.7129	.7178	.7227	.7275	.7324	.7372	50–48
2.1	.7419	.7467	.7514	.7561	.7608	.7655	.7701	.7747	.7793	.7839	48–46
2.2	.7885	.7930	.7975	.8020	.8065	.8109	.8154	.8198	.8242	.8286	45–44
2.3	.8329	.8372	.8416	.8459	.8502	.8544	.8587	.8629	.8671	.8713	44–42
2.4	.8755	.8796	.8838	.8879	.8920	.8961	.9002	.9042	.9083	.9123	42–40
2.5	.9163	.9203	.9243	.9282	.9322	.9361	.9400	.9439	.9478	.9517	40–39
2.6	.9555	.9594	.9632	.9670	.9708	.9746	.9783	.9821	.9858	.9895	39–37
2.7	.9933	.9969	1.0006	1.0043	1.0080	1.0116	1.0152	1.0188	1.0225	1.0260	37–35
2.8	1.0296	1.0332	1.0367	1.0403	1.0438	1.0473	1.0508	1.0543	1.0578	1.0613	36–35
2.9	1.0647	1.0682	1.0716	1.0750	1.0784	1.0818	1.0852	1.0886	1.0919	1.0953	35–33
3.0	1.0986	1.1019	1.1053	1.1086	1.1119	1.1151	1.1184	1.1217	1.1249	1.1282	34–32
3.1	1.1314	1.1346	1.1378	1.1410	1.1442	1.1474	1.1506	1.1537	1.1569	1.1600	32–31
3.2	1.1632	1.1663	1.1694	1.1725	1.1756	1.1787	1.1817	1.1848	1.1878	1.1909	31–30
3.3	1.1939	1.1969	1.2000	1.2030	1.2060	1.2090	1.2119	1.2149	1.2179	1.2208	31–29
3.4	1.2238	1.2267	1.2296	1.2326	1.2355	1.2384	1.2413	1.2442	1.2470	1.2499	30–28
3.5	1.2528	1.2556	1.2585	1.2613	1.2641	1.2669	1.2698	1.2726	1.2754	1.2782	29–28
3.6	1.2809	1.2837	1.2865	1.2892	1.2920	1.2947	1.2975	1.3002	1.3029	1.3056	28–27
3.7	1.3083	1.3110	1.3137	1.3164	1.3191	1.3218	1.3244	1.3271	1.3297	1.3324	27–26
3.8	1.3350	1.3376	1.3403	1.3429	1.3455	1.3481	1.3507	1.3533	1.3558	1.3584	27–25
3.9	1.3610	1.3635	1.3661	1.3686	1.3712	1.3737	1.3762	1.3788	1.3813	1.3838	26–25
4.0	1.3863	1.3888	1.3913	1.3938	1.3962	1.3987	1.4012	1.4036	1.4061	1.4085	25–24
4.1	1.4110	1.4134	1.4159	1.4183	1.4207	1.4231	1.4255	1.4279	1.4303	1.4327	25–24
4.2	1.4351	1.4375	1.4398	1.4422	1.4446	1.4469	1.4493	1.4516	1.4540	1.4563	24–23
4.3	1.4586	1.4609	1.4633	1.4656	1.4679	1.4702	1.4725	1.4748	1.4770	1.4793	24–22
4.4	1.4816	1.4839	1.4861	1.4884	1.4907	1.4929	1.4951	1.4974	1.4996	1.5019	23–22
4.5	1.5041	1.5063	1.5085	1.5107	1.5129	1.5151	1.5173	1.5195	1.5217	1.5239	22
4.6	1.5261	1.5282	1.5304	1.5326	1.5347	1.5369	1.5390	1.5412	1.5433	1.5454	22–21
4.7	1.5476	1.5497	1.5518	1.5539	1.5560	1.5581	1.5602	1.5623	1.5644	1.5665	22–21
4.8	1.5686	1.5707	1.5728	1.5748	1.5769	1.5790	1.5810	1.5831	1.5851	1.5872	21–20
4.9	1.5892	1.5913	1.5933	1.5953	1.5974	1.5994	1.6014	1.6034	1.6054	1.6074	21–20
5.0	1.6094	1.6114	1.6134	1.6154	1.6174	1.6194	1.6214	1.6233	1.6253	1.6273	20–19
5.1	1.6292	1.6312	1.6332	1.6351	1.6371	1.6390	1.6409	1.6429	1.6448	1.6467	20–19
5.2	1.6487	1.6506	1.6525	1.6544	1.6563	1.6582	1.6601	1.6620	1.6639	1.6658	19
5.3	1.6677	1.6696	1.6715	1.6734	1.6752	1.6771	1.6790	1.6808	1.6827	1.6845	19–18
5.4	1.6864	1.6882	1.6901	1.6919	1.6938	1.6956	1.6974	1.6993	1.7011	1.7029	19–18
5.5	1.7047	1.7066	1.7084	1.7102	1.7120	1.7138	1.7156	1.7174	1.7192	1.7210	19–18
5.6	1.7228	1.7246	1.7263	1.7281	1.7299	1.7317	1.7334	1.7352	1.7370	1.7387	18–17
5.7	1.7405	1.7422	1.7440	1.7457	1.7475	1.7492	1.7509	1.7527	1.7544	1.7561	18–17
5.8	1.7579	1.7596	1.7613	1.7630	1.7647	1.7664	1.7681	1.7699	1.7716	1.7733	18–17
5.9	1.7750	1.7766	1.7783	1.7800	1.7817	1.7834	1.7851	1.7867	1.7884	1.7901	17–16
6.0	1.7918	1.7934	1.7951	1.7967	1.7984	1.8001	1.8017	1.8034	1.8050	1.8066	17–16
6.1	1.8083	1.8099	1.8116	1.8132	1.8148	1.8165	1.8181	1.8197	1.8213	1.8229	17–16
6.2	1.8245	1.8262	1.8278	1.8294	1.8310	1.8326	1.8342	1.8358	1.8374	1.8390	17–16
6.3	1.8405	1.8421	1.8437	1.8453	1.8469	1.8485	1.8500	1.8516	1.8532	1.8547	16–15
6.4	1.8563	1.8579	1.8594	1.8610	1.8625	1.8641	1.8656	1.8672	1.8687	1.8703	16–15
6.5	1.8718	1.8733	1.8749	1.8764	1.8779	1.8795	1.8810	1.8825	1.8840	1.8856	16–15
6.6	1.8871	1.8886	1.8901	1.8916	1.8931	1.8946	1.8961	1.8976	1.8991	1.9006	15
6.7	1.9021	1.9036	1.9051	1.9066	1.9081	1.9095	1.9110	1.9125	1.9140	1.9155	15–14
6.8	1.9169	1.9184	1.9199	1.9213	1.9228	1.9242	1.9257	1.9272	1.9286	1.9301	15–14
6.9	1.9315	1.9330	1.9344	1.9359	1.9373	1.9387	1.9402	1.9416	1.9430	1.9445	15–14

Table 1025 (*continued*)—**Natural Logarithms**

No.	0	1	2	3	4	5	6	7	8	9	Diff.
7.0	1.9459	1.9473	1.9488	1.9502	1.9516	1.9530	1.9544	1.9559	1.9573	1.9587	15–14
7.1	1.9601	1.9615	1.9629	1.9643	1.9657	1.9671	1.9685	1.9699	1.9713	1.9727	14
7.2	1.9741	1.9755	1.9769	1.9782	1.9796	1.9810	1.9824	1.9838	1.9851	1.9865	14–13
7.3	1.9879	1.9892	1.9906	1.9920	1.9933	1.9947	1.9961	1.9974	1.9988	2.0001	14–13
7.4	2.0015	2.0028	2.0042	2.0055	2.0069	2.0082	2.0096	2.0109	2.0122	2.0136	14–13
7.5	2.0149	2.0162	2.0176	2.0189	2.0202	2.0215	2.0229	2.0242	2.0255	2.0268	14–13
7.6	2.0281	2.0295	2.0308	2.0321	2.0334	2.0347	2.0360	2.0373	2.0386	2.0399	14–13
7.7	2.0412	2.0425	2.0438	2.0451	2.0464	2.0477	2.0490	2.0503	2.0516	2.0528	13–12
7.8	2.0541	2.0554	2.0567	2.0580	2.0592	2.0605	2.0618	2.0631	2.0643	2.0656	13–12
7.9	2.0669	2.0681	2.0694	2.0707	2.0719	2.0732	2.0744	2.0757	2.0769	2.0782	13–12
8.0	2.0794	2.0807	2.0819	2.0832	2.0844	2.0857	2.0869	2.0882	2.0894	2.0906	13–12
8.1	2.0919	2.0931	2.0943	2.0956	2.0968	2.0980	2.0992	2.1005	2.1017	2.1029	13–12
8.2	2.1041	2.1054	2.1066	2.1078	2.1090	2.1102	2.1114	2.1126	2.1138	2.1150	13–12
8.3	2.1163	2.1175	2.1187	2.1199	2.1211	2.1223	2.1235	2.1247	2.1258	2.1270	12–11
8.4	2.1282	2.1294	2.1306	2.1318	2.1330	2.1342	2.1353	2.1365	2.1377	2.1389	12–11
8.5	2.1401	2.1412	2.1424	2.1436	2.1448	2.1459	2.1471	2.1483	2.1494	2.1506	12–11
8.6	2.1518	2.1529	2.1541	2.1552	2.1564	2.1576	2.1587	2.1599	2.1610	2.1622	12–11
8.7	2.1633	2.1645	2.1656	2.1668	2.1679	2.1691	2.1702	2.1713	2.1725	2.1736	12–11
8.8	2.1748	2.1759	2.1770	2.1782	2.1793	2.1804	2.1815	2.1827	2.1838	2.1849	12–11
8.9	2.1861	2.1872	2.1883	2.1894	2.1905	2.1917	2.1928	2.1939	2.1950	2.1961	12–11
9.0	2.1972	2.1983	2.1994	2.2006	2.2017	2.2028	2.2039	2.2050	2.2061	2.2072	12–11
9.1	2.2083	2.2094	2.2105	2.2116	2.2127	2.2138	2.2148	2.2159	2.2170	2.2181	11–10
9.2	2.2192	2.2203	2.2214	2.2225	2.2235	2.2246	2.2257	2.2268	2.2279	2.2289	11–10
9.3	2.2300	2.2311	2.2322	2.2332	2.2343	2.2354	2.2364	2.2375	2.2386	2.2396	11–10
9.4	2.2407	2.2418	2.2428	2.2439	2.2450	2.2460	2.2471	2.2481	2.2492	2.2502	11–10
9.5	2.2513	2.2523	2.2534	2.2544	2.2555	2.2565	2.2576	2.2586	2.2597	2.2607	11–10
9.6	2.2618	2.2628	2.2638	2.2649	2.2659	2.2670	2.2680	2.2690	2.2701	2.2711	11–10
9.7	2.2721	2.2732	2.2742	2.2752	2.2762	2.2773	2.2783	2.2793	2.2803	2.2814	11–10
9.8	2.2824	2.2834	2.2844	2.2854	2.2865	2.2875	2.2885	2.2895	2.2905	2.2915	11–10
9.9	2.2925	2.2935	2.2946	2.2956	2.2966	2.2976	2.2986	2.2996	2.3006	2.3016	11–10
10.0	2.3026										

x	$\text{Log}_e\, x$	x	$\text{Log}_e\, x$
10	2.3026	.1	$\overline{3}.6974$
100	4.6052	.01	$\overline{5}.3948$
1 000	6.9078	.001	$\overline{7}.0922$
10 000	9.2103	.000 1	$\overline{10}.7897$
100 000	11.5129	.000 01	$\overline{12}.4871$
1 000 000	13.8155	.000 001	$\overline{14}.1845$
. . . .		. . . .	

For a large table of natural logarithms, see Ref. 55d.

Table 1030—Exponential and Hyperbolic Functions

x	e^x Value	e^x Log₁₀	e^{-x} Value	Sinh x Value	Sinh x Log₁₀	Cosh x Value	Cosh x Log₁₀	Tanh x Value
0.00	1.0000	.00000	1.0000	0.0000	$-\infty$	1.0000	.00000	.00000
0.01	1.0101	.00434	.99005	0.0100	.00001	1.0001	.00002	.01000
0.02	1.0202	.00869	.98020	0.0200	.30106	1.0002	.00009	.02000
0.03	1.0305	.01303	.97045	0.0300	.47719	1.0005	.00020	.02999
0.04	1.0408	.01737	.96079	0.0400	.60218	1.0008	.00035	.03998
0.05	1.0513	.02171	.95123	0.0500	.69915	1.0013	.00054	.04996
0.06	1.0618	.02606	.94176	0.0600	.77841	1.0018	.00078	.05993
0.07	1.0725	.03040	.93239	0.0701	.84545	1.0025	.00106	.06989
0.08	1.0833	.03474	.92312	0.0801	.90355	1.0032	.00139	.07983
0.09	1.0942	.03909	.91393	0.0901	.95483	1.0041	.00176	.08976
0.10	1.1052	.04343	.90484	0.1002	.00072	1.0050	.00217	.09967
0.11	1.1163	.04777	.89583	0.1102	.04227	1.0061	.00262	.10956
0.12	1.1275	.05212	.88692	0.1203	.08022	1.0072	.00312	.11943
0.13	1.1388	.05646	.87810	0.1304	.11517	1.0085	.00366	.12927
0.14	1.1503	.06080	.86936	0.1405	.14755	1.0098	.00424	.13909
9.15	1.1618	.06514	.86071	0.1506	.17772	1.0113	.00487	.14889
0.16	1.1735	.06949	.85214	0.1607	.20597	1.0128	.00554	.15865
0.17	1.1853	.07383	.84366	0.1708	.23254	1.0145	.00625	.16838
0.18	1.1972	.07817	.83527	0.1810	.25762	1.0162	.00700	.17808
0.19	1.2092	.08252	.82696	0.1911	.28136	1.0181	.00779	.18775
0.20	1.2214	.08686	.81873	0.2013	.30392	1.0201	.00863	.19738
0.21	1.2337	.09120	.81058	0.2115	.32541	1.0221	.00951	.20697
0.22	1.2461	.09554	.80252	0.2218	.34592	1.0243	.01043	.21652
0.23	1.2586	.09989	.79453	0.2320	.36555	1.0266	.01139	.22603
0.24	1.2712	.10423	.78663	0.2423	.38437	1.0289	.01239	.23550
0.25	1.2840	.10857	.77880	0.2526	.40245	1.0314	.01343	.24492
0.26	1.2969	.11292	.77105	0.2629	.41986	1.0340	.01452	.25430
0.27	1.3100	.11726	.76338	0.2733	.43663	1.0367	.01564	.26362
0.28	1.3231	.12160	.75578	0.2837	.45282	1.0395	.01681	.27291
0.29	1.3364	.12595	.74826	0.2941	.46847	1.0423	.01801	.28213
0.30	1.3499	.13029	.74082	0.3045	.48362	1.0453	.01926	.29131
0.31	1.3634	.13463	.73345	0.3150	.49830	1.0484	.02054	.30044
0.32	1.3771	.13897	.72615	0.3255	.51254	1.0516	.02187	.30951
0.33	1.3910	.14332	.71892	0.3360	.52637	1.0549	.02323	.31852
0.34	1.4049	.14766	.71177	0.3466	.53981	1.0584	.02463	.32748
0.35	1.4191	.15200	.70469	0.3572	.55290	1.0619	.02607	.33638
0.36	1.4333	.15635	.69768	0.3678	.56564	1.0655	.02755	.34521
0.37	1.4477	.16069	.69073	0.3785	.57807	1.0692	.02907	.35399
0.38	1.4623	.16503	.68386	0.3892	.59019	1.0731	.03063	.36271
0.39	1.4770	.16937	.67706	0.4000	.60202	1.0770	.03222	.37136
0.40	1.4918	.17372	.67032	0.4108	.61358	1.0811	.03385	.37995
0 41	1.5068	.17806	.66365	0.4216	.62488	1.0852	.03552	.38847
0.42	1.5220	.18240	.65705	0.4325	.63594	1.0895	.03723	.39693
0.43	1.5373	.18675	.65051	0.4434	.64677	1.0939	.03897	.40532
0.44	1.5527	.19109	.64404	0.4543	.65738	1.0984	.04075	.41364
0.45	1.5683	.19543	.63763	0.4653	.66777	1.1030	.04256	.42190
0.46	1.5841	.19978	.63128	0.4764	.67797	1.1077	.04441	.43008
0.47	1.6000	.20412	.62500	0.4875	.68797	1.1125	.04630	.43820
0.48	1.6161	.20846	.61878	0.4986	.69779	1.1174	.04822	.44624
0.49	1.6323	.21280	.61263	0.5098	.70744	1.1225	.05018	.45422
0.50	1.6487	.21715	.60653	0.5211	.71692	1.1276	.05217	.46212

312

Table 1030 (*continued*)—Exponential and Hyperbolic Functions

x	e^x Value	e^x Log$_{10}$	e^{-x} Value	Sinh x Value	Sinh x Log$_{10}$	Cosh x Value	Cosh x Log$_{10}$	Tanh x Value
0.50	1.6487	.21715	.60653	0.5211	.71692	1.1276	.05217	.46212
0.51	1.6653	.22149	.60050	0.5324	.72624	1.1329	.05419	.46995
0.52	1.6820	.22583	.59452	0.5438	.73540	1.1383	.05625	.47770
0.53	1.6989	.23018	.58860	0.5552	.74442	1.1438	.05834	.48538
0.54	1.7160	.23452	.58275	0.5666	.75330	1.1494	.06046	.49299
0.55	1.7333	.23886	.57695	0.5782	.76204	1.1551	.06262	.50052
0.56	1.7507	.24320	.57121	0.5897	.77065	1.1609	.06481	.50798
0.57	1.7683	.24755	.56553	0.6014	.77914	1.1669	.06703	.51536
0.58	1.7860	.25189	.55990	0.6131	.78751	1.1730	.06929	.52267
0.59	1.8040	.25623	.55433	0.6248	.79576	1.1792	.07157	.52990
0.60	1.8221	.26058	.54881	0.6367	.80390	1.1855	.07389	.53705
0.61	1.8404	.26492	.54335	0.6485	.81194	1.1919	.07624	.54413
0.62	1.8589	.26926	.53794	0.6605	.81987	1.1984	.07861	.55113
0.63	1.8776	.27361	.53259	0.6725	.82770	1.2051	.08102	.55805
0.64	1.8965	.27795	.52729	0.6846	.83543	1.2119	.08346	.56490
0.65	1.9155	.28229	.52205	0.6967	.84308	1.2188	.08593	.57167
0.66	1.9348	.28663	.51685	0.7090	.85063	1.2258	.08843	.57836
0.67	1.9542	.29098	.51171	0.7213	.85809	1.2330	.09095	.58498
0.68	1.9739	.29532	.50662	0.7336	.86548	1.2402	.09351	.59152
0.69	1.9937	.29966	.50158	0.7461	.87278	1.2476	.09609	.59798
0.70	2.0138	.30401	.49659	0.7586	.88000	1.2552	.09870	.60437
0.71	2.0340	.30835	.49164	0.7712	.88715	1.2628	.10134	.61068
0.72	2.0544	.31269	.48675	0.7838	.89423	1.2706	.10401	.61691
0.73	2.0751	.31703	.48191	0.7966	.90123	1.2785	.10670	.62307
0.74	2.0959	.32138	.47711	0.8094	.90817	1.2865	.10942	.62915
0.75	2.1170	.32572	.47237	0.8223	.91504	1.2947	.11216	.63515
0.76	2.1383	.33006	.46767	0.8353	.92185	1.3030	.11493	.64108
0.77	2.1598	.33441	.46301	0.8484	.92859	1.3114	.11773	.64693
0.78	2.1815	.33875	.45841	0.8615	.93527	1.3199	.12055	.65271
0.79	2.2034	.34309	.45384	0.8748	.94190	1.3286	.12340	.65841
0.80	2.2255	.34744	.44933	0.8881	.94846	1.3374	.12627	.66404
0.81	2.2479	.35178	.44486	0.9015	.95498	1.3464	.12917	.66959
0.82	2.2705	.35612	.44043	0.9150	.96144	1.3555	.13209	.67507
0.83	2.2933	.36046	.43605	0.9286	.96784	1.3647	.13503	.68048
0.84	2.3164	.36481	.43171	0.9423	.97420	1.3740	.13800	.68581
0.85	2.3396	.36915	.42741	0.9561	.98051	1.3835	.14099	.69107
0.86	2.3632	.37349	.42316	0.9700	.98677	1.3932	.14400	.69626
0.87	2.3869	.37784	.41895	0.9840	.99299	1.4029	.14704	.70137
0.88	2.4109	.38218	.41478	0.9981	.99916	1.4128	.15009	.70642
0.89	2.4351	.38652	.41066	1.0122	.00528	1.4229	.15317	.71139
0.90	2.4596	.39087	.40657	1.0265	.01137	1.4331	.15627	.71630
0.91	2.4843	.39521	.40252	1.0409	.01741	1.4434	.15939	.72113
0.92	2.5093	.39955	.39852	1.0554	.02341	1.4539	.16254	.72590
0.93	2.5345	.40389	.39455	1.0700	.02937	1.4645	.16570	.73059
0.94	2.5600	.40824	.39063	1.0847	.03530	1.4753	.16888	.73522
0.95	2.5857	.41258	.38674	1.0995	.04119	1.4862	.17208	.73978
0.96	2.6117	.41692	.38289	1.1144	.04704	1.4973	.17531	.74428
0.97	2.6379	.42127	.37908	1.1294	.05286	1.5085	.17855	.74870
0.98	2.6645	.42561	.37531	1.1446	.05864	1.5199	.18181	.75307
0.99	2.6912	.42995	.37158	1.1598	.06439	1.5314	.18509	.75736
1.00	2.7183	.43429	.36788	1.1752	.07011	1.5431	.18839	.76159

Table 1030 (*continued*)—Exponential and Hyperbolic Functions

x	e^x Value	e^x Log₁₀	e^{-x} Value	Sinh x Value	Sinh x Log₁₀	Cosh x Value	Cosh x Log₁₀	Tanh x Value
1.00	2.7183	.43429	.36788	1.1752	.07011	1.5431	.18839	.76159
1.01	2.7456	.43864	.36422	1.1907	.07580	1.5549	.19171	.76576
1.02	2.7732	.44298	.36060	1.2063	.08146	1.5669	.19504	.76987
1.03	2.8011	.44732	.35701	1.2220	.08708	1.5790	.19839	.77391
1.04	2.8292	.45167	.35345	1.2379	.09268	1.5913	.20176	.77789
1.05	2.8577	.45601	.34994	1.2539	.09825	1.6038	.20515	.78181
1.06	2.8864	.46035	.34646	1.2700	.10379	1.6164	.20855	.78566
1.07	2.9154	.46470	.34301	1.2862	.10930	1.6292	.21197	..78946
1.08	2.9447	.46904	.33960	1.3025	.11479	1.6421	.21541	.79320
1.09	2.9743	.47338	.33622	1.3190	.12025	1.6552	.21886	.79688
1.10	3.0042	.47772	.33287	1.3356	.12569	1.6685	.22233	.80050
1.11	3.0344	.48207	.32956	1.3524	.13111	1.6820	.22582	.80406
1.12	3.0649	.48641	.32628	1.3693	.13649	1.6956	.22931	.80757
1.13	3.0957	.49075	.32303	1.3863	.14186	1.7093	.23283	.81102
1.14	3.1268	.49510	.31982	1.4035	.14720	1.7233	.23636	.81441
1.15	3.1582	.49944	.31664	1.4208	.15253	1.7374	.23990	.81775
1.16	3.1899	.50378	.31349	1.4382	.15783	1.7517	.24346	.82104
1.17	3.2220	.50812	.31037	1.4558	.16311	1.7662	.24703	.82427
1.18	3.2544	.51247	.30728	1.4735	.16836	1.7808	.25062	.82745
1.19	3.2871	.51681	.30422	1.4914	.17360	1.7957	.25422	.83058
1.20	3.3201	.52115	.30119	1.5095	.17882	1.8107	.25784	.83365
1.21	3.3535	.52550	.29820	1.5276	.18402	1.8258	.26146	.83668
1.22	3.3872	.52984	.29523	1.5460	.18920	1.8412	.26510	.83965
1.23	3 4212	.53418	.29229	1.5645	.19437	1.8568	.26876	.84258
1.24	3.4556	.53853	.28938	1.5831	.19951	1.8725	.27242	.84546
1.25	3.4903	.54287	.28650	1.6019	.20464	1.8884	.27610	.84828
1.26	3.5254	.54721	.28365	1.6209	.20975	1.9045	.27979	.85106
1.27	3.5609	.55155	.28083	1.6400	.21485	1.9208	.28349	.85380
1.28	3.5966	.55590	.27804	1.6593	.21993	1.9373	.28721	.85648
1.29	3.6328	.56024	.27527	1.6788	.22499	1.9540	.29093	.85913
1.30	3.6693	.56458	.27253	1.6984	.23004	1.9709	.29467	.86172
1.31	3.7062	.56893	.26982	1.7182	.23507	1.9880	.29842	.86428
1.32	3.7434	.57327	.26714	1.7381	.24009	2.0053	.30217	.86678
1.33	3.7810	.57761	.26448	1.7583	.24509	2.0228	.30594	.86925
1.34	3.8190	.58195	.26185	1.7786	.25008	2.0404	.30972	.87167
1.35	3.8574	.58630	.25924	1.7991	.25505	2.0583	.31352	.87405
1.36	3.8962	.59064	.25666	1.8198	.26002	2.0764	.31732	.87639
1.37	3.9354	.59498	.25411	1.8406	.26496	2.0947	.32113	.87869
1.38	3.9749	.59933	.25158	1.8617	.26990	2.1132	.32495	.88095
1.39	4.0149	.60367	.24908	1.8829	.27482	2.1320	.32878	.88317
1.40	4.0552	.60801	.24660	1.9043	.27974	2.1509	.33262	.88535
1.41	4.0960	.61236	.24414	1.9259	.28464	2.1700	.33647	.88749
1.42	4.1371	.61670	.24171	1.9477	.28952	2.1894	.34033	.88960
1.43	4.1787	.62104	.23931	1.9697	.29440	2.2090	.34420	.89167
1.44	4.2207	.62538	.23693	1.9919	.29926	2.2288	.34807	.89370
1.45	4.2631	.62973	.23457	2.0143	.30412	2.2488	.35196	.89569
1.46	4.3060	.63407	.23224	2.0369	.30896	2.2691	.35585	.89765
1.47	4.3492	.63841	.22993	2.0597	.31379	2.2896	.35976	.89958
1.48	4.3929	.64276	.22764	2.0827	.31862	2.3103	.36367	.90147
1.49	4.4371	.64710	.22537	2.1059	.32343	2.3312	.36759	.90332
1.50	4.4817	.65144	.22313	2.1293	.32823	2.3524	.37151	.90515

Table 1030 (*continued*)—Exponential and Hyperbolic Functions

x	e^x Value	e^x Log$_{10}$	e^{-x} Value	Sinh x Value	Sinh x Log$_{10}$	Cosh x Value	Cosh x Log$_{10}$	Tanh x Value
1.50	4.4817	.65144	.22313	2.1293	.32823	2.3524	.37151	.90515
1.51	4.5267	.65578	.22091	2.1529	.33303	2.3738	.37545	.90694
1.52	4.5722	.66013	.21871	2.1768	.33781	2.3955	.37939	.90870
1.53	4.6182	.66447	.21654	2.2008	.34258	2.4174	.38334	.91042
1.54	4.6646	.66881	.21438	2.2251	.34735	2.4395	.38730	.91212
1.55	4.7115	.67316	.21225	2.2496	.35211	2.4619	.39126	.91379
1.56	4.7588	.67750	.21014	2.2743	.35686	2.4845	.39524	.91542
1.57	4.8066	.68184	.20805	2.2993	.36160	2.5073	.39921	.91703
1.58	4.8550	.68619	.20598	2.3245	.36633	2.5305	.40320	.91860
1.59	4.9037	.69053	.20393	2.3499	.37105	2.5538	.40719	.92015
1.60	4.9530	.69487	.20190	2.3756	.37577	2.5775	.41119	.92167
1.61	5.0028	.69921	.19989	2.4015	.38048	2.6013	.41520	.92316
1.62	5.0531	.70356	.19790	2.4276	.38518	2.6255	.41921	.92462
1.63	5.1039	.70790	.19593	2.4540	.38987	2.6499	.42323	.92606
1.64	5.1552	.71224	.19398	2.4806	.39456	2.6746	.42725	.92747
1.65	5.2070	.71659	.19205	2.5075	.39923	2.6995	.43129	.92886
1.66	5.2593	.72093	.19014	2.5346	.40391	2.7247	.43532	.93022
1.67	5.3122	.72527	.18825	2.5620	.40857	2.7502	.43937	.93155
1.68	5.3656	.72961	.18637	2.5896	.41323	2.7760	.44341	.93286
1.69	5.4195	.73396	.18452	2.6175	.41788	2.8020	.44747	.93415
1.70	5.4739	.73830	.18268	2.6456	.42253	2.8283	.45153	.93541
1.71	5.5290	.74264	.18087	2.6740	.42717	2.8549	.45559	.93665
1.72	5.5845	.74699	.17907	2.7027	.43180	2.8818	.45966	.93786
1.73	5.6407	.75133	.17728	2.7317	.43643	2.9090	.46374	.93906
1.74	5.6973	.75567	.17552	2.7609	.44105	2.9364	.46782	.94023
1.75	5.7546	.76002	.17377	2.7904	.44567	2.9642	.47191	.94138
1.76	5.8124	.76436	.17204	2.8202	.45028	2.9922	.47600	.94250
1.77	5.8709	.76870	.17033	2.8503	.45488	3.0206	.48009	.94361
1.78	5.9299	.77304	.16864	2.8806	.45948	3.0492	.48419	.94470
1.79	5.9895	.77739	.16696	2.9112	.46408	3.0782	.48830	.94576
1.80	6.0496	.78173	.16530	2.9422	.46867	3.1075	.49241	.94681
1.81	6.1104	.78607	.16365	2.9734	.47325	3.1371	.49652	.94783
1.82	6.1719	.79042	.16203	3.0049	.47783	3.1669	.50064	.94884
1.83	6.2339	.79476	.16041	3.0367	.48241	3.1972	.50476	.94983
1.84	6.2965	.79910	.15882	3.0689	.48698	3.2277	.50889	.95080
1.85	6.3598	.80344	.15724	3.1013	.49154	3.2585	.51302	.95175
1.86	6.4237	.80779	.15567	3.1340	.49610	3.2897	.51716	.95268
1.87	6.4883	.81213	.15412	3.1671	.50066	3.3212	.52130	.95359
1.88	6.5535	.81647	.15259	3.2005	.50521	3.3530	.52544	.95449
1.89	6.6194	.82082	.15107	3.2341	.50976	3.3852	.52959	.95537
1.90	6.6859	.82516	.14957	3.2682	.51430	3.4177	.53374	.95624
1.91	6.7531	.82950	.14808	3.3025	.51884	3.4506	.53789	.95709
1.92	6.8210	.83385	.14661	3.3372	.52338	3.4838	.54205	.95792
1.93	6.8895	.83819	.14515	3.3722	.52791	3.5173	.54621	.95873
1.94	6.9588	.84253	.14370	3.4075	.53244	3.5512	.55038	.95953
1.95	7.0287	.84687	.14227	3.4432	.53696	3.5855	.55455	.96032
1.96	7.0993	.85122	.14086	3.4792	.54148	3.6201	.55872	.96109
1.97	7.1707	.85556	.13946	3.5156	.54600	3.6551	.56290	.96185
1.98	7.2427	.85990	.13807	3.5523	.55051	3.6904	.56707	.96259
1.99	7.3155	.86425	.13670	3.5894	.55502	3.7261	.57126	.96331
2.00	7.3891	.86859	.13534	3.6269	.55953	3.7622	.57544	.96403

315

Table 1030 (continued)—Exponential and Hyperbolic Functions

x	e^x Value	e^x Log₁₀	e^{-x} Value	Sinh x Value	Sinh x Log₁₀	Cosh x Value	Cosh x Log₁₀	Tanh x Value
2.00	7.3891	.86859	.13534	3.6269	.55953	3.7622	.57544	.96403
2.01	7.4633	.87293	.13399	3.6647	.56403	3.7987	.57963	.96473
2.02	7.5383	.87727	.13266	3.7028	.56853	3.8355	.58382	.96541
2.03	7.6141	.88162	.13134	3.7414	.57303	3.8727	.58802	.96609
2.04	7.6906	.88596	.13003	3.7803	.57753	3.9103	.59221	.96675
2.05	7.7679	.89030	.12873	3.8196	.58202	3.9483	.59641	.96740
2.06	7.8460	.89465	.12745	3.8593	.58650	3.9867	.60061	.96803
2.07	7.9248	.89899	.12619	3.8993	.59099	4.0255	.60482	.96865
2.08	8.0045	.90333	.12493	3.9398	.59547	4.0647	.60903	.96926
2.09	8.0849	.90768	.12369	3.9806	.59995	4.1043	.61324	.96986
2.10	8.1662	.91202	.12246	4.0219	.60443	4.1443	.61745	.97045
2.11	8.2482	.91636	.12124	4.0635	.60890	4.1847	.62167	.97103
2.12	8.3311	.92070	.12003	4.1056	.61337	4.2256	.62589	.97159
2.13	8.4149	.92505	.11884	4.1480	.61784	4.2669	.63011	.97215
2.14	8.4994	.92939	.11765	4.1909	.62231	4.3085	.63433	.97269
2.15	8.5849	.93373	.11648	4.2342	.62677	4.3507	.63856	.97323
2.16	8.6711	.93808	.11533	4.2779	.63123	4.3932	.64278	.97375
2.17	8.7583	.94242	.11418	4.3221	.63569	4.4362	.64701	.97426
2.18	8.8463	.94676	.11304	4.3666	.64015	4.4797	.65125	.97477
2.19	8.9352	.95110	.11192	4.4116	.64460	4.5236	.65548	.97526
2.20	9.0250	.95545	.11080	4.4571	.64905	4.5679	.65972	.97574
2.21	9.1157	.95979	.10970	4.5030	.65350	4.6127	.66396	.97622
2.22	9.2073	.96413	.10861	4.5494	.65795	4.6580	.66820	.97668
2.23	9.2999	.96848	.10753	4.5962	.66240	4.7037	.67244	.97714
2.24	9.3933	.97282	.10646	4.6434	.66684	4.7499	.67668	.97759
2.25	9.4877	.97716	.10540	4.6912	.67128	4.7966	.68093	.97803
2.26	9.5831	.98151	.10435	4.7394	.67572	4.8437	.68518	.97846
2.27	9.6794	.98585	.10331	4.7880	.68016	4.8914	.68943	.97888
2.28	9.7767	.99019	.10228	4.8372	.68459	4.9395	.69368	.97929
2.29	9.8749	.99453	.10127	4.8868	.68903	4.9881	.69794	.97970
2.30	9.9742	.99888	.10026	4.9370	.69346	5.0372	.70219	.98010
2.31	10.074	.00322	.09926	4.9876	.69789	5.0868	.70645	.98049
2.32	10.176	.00756	.09827	5.0387	.70232	5.1370	.71071	.98087
2.33	10.278	.01191	.09730	5.0903	.70675	5.1876	.71497	.98124
2.34	10.381	.01625	.09633	5.1425	.71117	5.2388	.71923	.98161
2.35	10.486	.02059	.09537	5.1951	.71559	5.2905	.72349	.98197
2.36	10.591	.02493	.09442	5.2483	.72002	5.3427	.72776	.98233
2.37	10.697	.02928	.09348	5.3020	.72444	5.3954	.73203	.98267
2.38	10.805	.03362	.09255	5.3562	.72885	5.4487	.73630	.98301
2.39	10.913	.03796	.09163	5.4109	.73327	5.5026	.74056	.98335
2.40	11.023	04231	.09072	5.4662	.73769	5.5569	.74484	.98367
2.41	11.134	.04665	.08982	5.5221	.74210	5.6119	.74911	.98400
2.42	11.246	.05099	.08892	5.5785	.74652	5.6674	.75338	.98431
2.43	11.359	.05534	.08804	5.6354	.75093	5.7235	.75766	.98462
2.44	11.473	.05968	.08716	5 6929	.75534	5.7801	.76194	.98492
2.45	11.588	.06402	.08629	5.7510	.75975	5.8373	.76621	.98522
2.46	11.705	.06836	.08543	5.8097	.76415	5.8951	.77049	.98551
2.47	11.822	.07271	.08458	5.8689	.76856	5.9535	.77477	.98579
2.48	11.941	.07705	.08374	5.9288	.77296	6.0125	.77906	.98607
2.49	12.061	.08139	.08291	5.9892	.77737	6.0721	.78334	.98635
2.50	12.182	.08574	.08208	6.0502	.78177	6.1323	.78762	.98661

316

Table 1030 (*continued*)—Exponential and Hyperbolic Functions

x	e^x Value	e^x Log$_{10}$	e^{-x} Value	Sinh x Value	Sinh x Log$_{10}$	Cosh x Value	Cosh x Log$_{10}$	Tanh x Value
2.50	12.182	.08574	.08208	6.0502	.78177	6.1323	.78762	.98661
2.51	12.305	.09008	.08127	6.1118	.78617	6.1931	.79191	.98688
2.52	12.429	.09442	.08046	6.1741	.79057	6.2545	.79619	.98714
2.53	12.554	.09877	.07966	6.2369	.79497	6.3166	.80048	.98739
2.54	12.680	.10311	.07887	6.3004	.79937	6.3793	.80477	.98764
2.55	12.807	.10745	.07808	6.3645	.80377	6.4426	.80906	.98788
2.56	12.936	.11179	.07730	6.4293	.80816	6.5066	.81335	.98812
2.57	13.066	.11614	.07654	6.4946	.81256	6.5712	.81764	.98835
2.58	13.197	.12048	.07577	6.5607	.81695	6.6365	.82194	.98858
2.59	13.330	.12482	.07502	6.6274	.82134	6.7024	.82623	.98881
2.60	13.464	.12917	.07427	6.6947	.82573	6.7690	.83052	.98903
2.61	13.599	.13351	.07353	6.7628	.83012	6.8363	.83482	.98924
2.62	13.736	.13785	.07280	6.8315	.83451	6.9043	.83912	.98946
2.63	13.874	.14219	.07208	6.9008	.83890	6.9729	.84341	.98966
2.64	14.013	.14654	.07136	6.9709	.84329	7.0423	.84771	.98987
2.65	14.154	.15088	.07065	7.0417	.84768	7.1123	.85201	.99007
2.66	14.296	.15522	.06995	7.1132	.85206	7.1831	.85631	.99026
2.67	14.440	.15957	.06925	7.1854	.85645	7.2546	.86061	.99045
2.68	14.585	.16391	.06856	7.2583	.86083	7.3268	.86492	.99064
2.69	14.732	.16825	.06788	7.3319	.86522	7.3998	.86922	.99083
2.70	14.880	.17260	.06721	7.4063	.86960	7.4735	.87352	.99101
2.71	15.029	.17694	.06654	7.4814	.87398	7.5479	.87783	.99118
2.72	15.180	.18128	.06587	7.5572	.87836	7.6231	.88213	.99136
2.73	15.333	.18562	.06522	7.6338	.88274	7.6991	.88644	.99153
2.74	15.487	.18997	.06457	7.7112	.88712	7.7758	.89074	.99170
2.75	15.643	.19431	.06393	7.7894	.89150	7.8533	.89505	.99186
2.76	15.800	.19865	.06329	7.8683	.89588	7.9316	.89936	.99202
2.77	15.959	.20300	.06266	7.9480	.90026	8.0106	.90367	.99218
2.78	16.119	.20734	.06204	8.0285	.90463	8.0905	.90798	.99233
2.79	16.281	.21168	.06142	8.1098	.90901	8.1712	.91229	.99248
2.80	16.445	.21602	.06081	8.1919	.91339	8.2527	.91660	.99263
2.81	16.610	.22037	.06020	8.2749	.91776	8.3351	.92091	.99278
2.82	16.777	.22471	.05961	8.3586	.92213	8.4182	.92522	.99292
2.83	16.945	.22905	.05901	8.4432	.92651	8.5022	.92953	.99306
2.84	17.116	.23340	.05843	8.5287	.93088	8.5871	.93385	.99320
2.85	17.288	.23774	.05784	8.6150	.93525	8.6728	.93816	.99333
2.86	17.462	.24208	.05727	8.7021	.93963	8.7594	.94247	.99346
2.87	17.637	.24643	.05670	8.7902	.94400	8.8469	.94679	.99359
2.88	17.814	.25077	.05613	8.8791	.94837	8.9352	.95110	.99372
2.89	17.993	.25511	.05558	8.9689	.95274	9.0244	.95542	.99384
2.90	18.174	.25945	.05502	9.0596	.95711	9.1146	.95974	.99396
2.91	18.357	.26380	.05448	9.1512	.96148	9.2056	.96405	.99408
2.92	18.541	.26814	.05393	9.2437	.96584	9.2976	.96837	.99420
2.93	18.728	.27248	.05340	9.3371	.97021	9.3905	.97269	.99431
2.94	18.916	.27683	.05287	9.4315	.97458	9.4844	.97701	.99443
2.95	19.106	.28117	.05234	9.5268	.97895	9.5791	.98133	.99454
2.96	19.298	.28551	.05182	9.6231	.98331	9.6749	.98565	.99464
2.97	19.492	.28985	.05130	9.7203	.98768	9.7716	.98997	.99475
2.98	19.688	.29420	.05079	9.8185	.99205	9.8693	.99429	.99485
2.99	19.886	.29854	.05029	9.9177	.99641	9.9680	.99861	.99496
3.00	20.086	.30288	.04979	10.018	.00078	10.068	.00293	.99505

Table 1030 (*continued*)—Exponential and Hyperbolic Functions

x	e^x Value	e^x Log₁₀	e^{-x} Value	Sinh x Value	Sinh x Log₁₀	Cosh x Value	Cosh x Log₁₀	Tanh x Value
3.00	20.086	.30288	.04979	10.018	.00078	10.068	.00293	.99505
3.05	21.115	.32460	.04736	10.534	.02259	10.581	.02454	.99552
3.10	22.198	.34631	.04505	11.076	.04440	11.122	.04616	.99595
3.15	23.336	.36803	.04285	11.647	.06620	11.689	.06779	.99633
3.20	24.533	.38974	.04076	12.246	.08799	12.287	.08943	.99668
3 25	25.790	.41146	.03877	12.876	.10977	12.915	.11108	.99700
3.30	27.113	.43317	.03688	13.538	.13155	13.575	.13273	.99728
3.35	28.503	.45489	.03508	14.234	.15332	14.269	.15439	.99754
3.40	29.964	.47660	.03337	14.965	.17509	14.999	.17605	.99777
3.45	31.500	.49832	.03175	15.734	.19685	15.766	.19772	.99799
3.50	33.115	.52003	.03020	16.543	.21860	16.573	.21940	.99818
3.55	34.813	.54175	.02872	17.392	.24036	17.421	.24107	.99835
3.60	36.598	.56346	.02732	18.286	.26211	18.313	.26275	.99851
3.65	38.475	.58517	.02599	19.224	.28385	19.250	.28444	.99865
3.70	40.447	.60689	.02472	20.211	.30559	20.236	.30612	.99878
3.75	42.521	.62860	.02352	21.249	.32733	21.272	.32781	.99889
3.80	44.701	.65032	.02237	22.339	.34907	22.362	.34951	.99900
3.85	46.993	.67203	.02128	23.486	.37081	23.507	.37120	.99909
3.90	49.402	.69375	.02024	24.691	.39254	24.711	.39290	.99918
3.95	51.935	.71546	.01925	25.958	.41427	25.977	.41459	.99926
4.00	54.598	.73718	.01832	27.290	.43600	27.308	.43629	.99933
4.10	60.340	.78061	.01657	30.162	.47946	30.178	.47970	.99945
4.20	66.686	.82404	.01500	33.336	.52291	33.351	.52310	.99955
4.30	73.700	.86747	.01357	36.843	.56636	36.857	.56652	.99963
4.40	81.451	.91090	.01227	40.719	.60980	40.732	.60993	.99970
4.50	90.017	.95433	.01111	45.003	.65324	45.014	.65335	.99975
4.60	99.484	.99775	.01005	49.737	.69668	49.747	.69677	.99980
4.70	109.95	.04118	.00910	54.969	.74012	54.978	.74019	.99983
4.80	121.51	.08461	.00823	60.751	.78355	60.759	.78361	.99986
4.90	134.29	.12804	.00745	67.141	.82699	67.149	.82704	.99989
5.00	148.41	.17147	.00674	74.203	.87042	74.210	.87046	.99991
5.10	164.02	.21490	.00610	82 008	.91386	82.014	.91389	.99993
5.20	181.27	.25833	.00552	90.633	.95729	90.639	.95731	.99994
5.30	200.34	.30176	.00499	100.17	.00074	100.17	.00074	.99995
5.40	221.41	.34519	.00452	110.70	.04415	110.71	.04417	.99996
5.50	244.69	.38862	.00409	122.34	.08758	122.35	.08760	.99997
5.60	270.43	.43205	.00370	135.21	.13101	135.22	.13103	.99997
5.70	298.87	.47548	.00335	149.43	.17444	149.44	.17445	.99998
5.80	330.30	.51891	.00303	165.15	.21787	165.15	.21788	.99998
5.90	365.04	.56234	.00274	182.52	.26130	182.52	.26131	.99998
6.00	403.43	.60577	.00248	201.71	.30473	201.72	.30474	.99999
6.25	518.01	.71434	.00193	259.01	.41331	259.01	.41331	.99999
6.50	665.14	.82291	.00150	332.57	.52188	332.57	.52189	1.0000
6.75	854.06	.93149	.00117	427.03	.63046	427.03	.63046	1.0000
7.00	1096.6	.04006	.00091	548.32	.73903	548.32	.73903	1.0000
7.50	1808.0	.25721	.00055	904.02	.95618	904.02	.95618	1.0000
8.00	2981.0	.47436	.00034	1490.5	.17333	1490.5	.17333	1.0000
8.50	4914.8	.69150	.00020	2457.4	.39047	2457.4	.39047	1.0000
9.00	8103.1	.90865	.00012	4051.5	.60762	4051.5	.60762	1.0000
9.50	13360.	.12580	.00007	6679.9	.82477	6679.9	.82477	1.0000
10.00	22026.	.34294	.00005	11013.	.04191	11013.	.04191	1.0000

Table 1030 (*continued*)—**Exponential and Hyperbolic Functions**

x	e^{-x}	x	e^x	e^{-x}
1	0.367879	11	5.9874×10^4	1.6702×10^{-5}
2	0.135335	12	1.6275×10^5	6.1442×10^{-6}
3	0.049787	13	4.4241×10^5	2.2603×10^{-6}
4	0.018316	14	1.2026×10^6	8.3153×10^{-7}
5	6.7379×10^{-3}	15	3.2690×10^6	3.0590×10^{-7}
6	2.4788×10^{-3}	16	8.8861×10^6	1.1254×10^{-7}
7	9.1188×10^{-4}	17	2.4155×10^7	4.1399×10^{-8}
8	3.3546×10^{-4}	18	6.5660×10^7	1.5230×10^{-8}
9	1.2341×10^{-4}	19	1.7848×10^8	5.6028×10^{-9}
10	4.5400×10^{-5}	20	4.8517×10^8	2.0612×10^{-9}

x	e^x	e^{-x}
0.001	1.00100	0.99900
0.002	1.00200	0.99800
0.003	1.00300	0.99700
0.004	1.00401	0.99601
0.005	1.00501	0.99501
0.006	1.00602	0.99402
0.007	1.00702	0.99302
0.008	1.00803	0.99203
0.009	1.00904	0.99104

Interpolation for the last two columns can be done by inspection.

For tables of exponential and hyperbolic functions, see References 30, 55b and 55c.

Note. For large values of x use e^x = natural anti-logarithm of x, which may be obtained from a table of natural logarithms. When x is large, subtract multiples of 2.3026 from x. Note also that

$$e^{-x} = 1/e^x$$
$$\sinh x = \tfrac{1}{2}(e^x - e^{-x})$$
$$\cosh x = \tfrac{1}{2}(e^x + e^{-x})$$
$$\tanh x = \frac{e^{2x} - 1}{e^{2x} + 1}$$
$$= 1 - \frac{2}{e^{2x}} + \frac{2}{e^{4x}} - \frac{2}{e^{6x}} + \cdots.$$

The quantity e^x is equal to the common anti-logarithm of $0.4342945\,x$. For example, if $x = 7$, $0.4342945 \times 7 = 3.04006$. The common anti-logarithm of 0.04006 is 1.0966 and that of 3.04006 is $1.0966 \times 10^3 = 1096.6 = e^7$. Also, $-3.04006 = -4 + 0.95994 = \bar{4}.95994$. The common anti-logarithm of 0.95994 is 9.1188 and that of $\bar{4}.95994$ is $9.1188 \times 10^{-4} = e^{-7}$, as in the table. This is useful chiefly where a 7-place logarithm table is used, to obtain accuracy.

Note.—Tables 1020 and 1030 are from *The Macmillan Mathematical Tables*.

Table 1040—Complete Elliptic Integrals of the First Kind

$$K = \int_0^{\pi/2} \frac{d\phi}{\sqrt{(1 - \sin^2\theta \sin^2\phi)}} \quad \text{[See 773]}$$

θ Deg.	K	Diff.	θ Deg.	K	Diff.	θ Deg.	K	Diff.	θ Deg.	K	Diff.	θ Deg.	K	Diff.	θ Deg.	K	Diff.
0	1.571	0	25	1.649	7	50	1.936	9	62.5	2.228	16	72.0	2.600	10	77.0	2.903	14
1	1.571	0	26	1.656	7	50.5	1.945	9	63	2.244	15	72.2	2.610	10	77.2	2.917	15
2	1.571	1	27	1.663	7	51	1.954	9	63.5	2.259	16	72.4	2.620	11	77.4	2.932	15
3	1.572	1	28	1.670	8	51.5	1.963	10	64	2.275	17	72.6	2.631	10	77.6	2.947	16
4	1.573	1	29	1.678	8	52	1.973	10	64.5	2.292	17	72.8	2.641	11	77.8	2.963	16
5	1.574	1	30	1.686	8	52.5	1.983	10	65	2.309	17	73.0	2.652	11	78.0	2.979	16
6	1.575	2	31	1.694	9	53	1.993	10	65.5	2.326	18	73.2	2.663	11	78.2	2.995	16
7	1.577	1	32	1.703	9	53.5	2.003	10	66	2.344	18	73.4	2.674	11	78.4	3.011	17
8	1.578	3	33	1.712	9	54	2.013	11	66.5	2.362	19	73.6	2.685	12	78.6	3.028	16
9	1.581	2	34	1.721	10	54.5	2.024	11	67	2.381	19	73.8	2.697	11	78.8	3.044	18
10	1.583	2	35	1.731	10	55	2.035	11	67.5	2.400	20	74.0	2.708	12	79.0	3.062	17
11	1.585	3	36	1.741	11	55.5	2.046	11	68	2.420	20	74.2	2.720	12	79.2	3.079	18
12	1.588	3	37	1.752	11	56	2.057	12	68.5	2.440	21	74.4	2.732	12	79.4	3.097	19
13	1.591	4	38	1.763	12	56.5	2.069	11	69	2.461	21	74.6	2.744	12	79.6	3.116	18
14	1.595	3	39	1.775	12	57	2.080	12	69.5	2.482	23	74.8	2.756	12	79.8	3.134	19
15	1.598	4	40	1.787	12	57.5	2.092	13	70.0	2.505	9	75.0	2.768	13	80.0	3.153	20
16	1.602	4	41	1.799	13	58	2.105	12	70.2	2.514	9	75.2	2.781	12	80.2	3.173	20
17	1.606	4	42	1.812	14	58.5	2.117	13	70.4	2.523	9	75.4	2.793	13	80.4	3.193	20
18	1.610	5	43	1.826	14	59	2.130	13	70.6	2.532	9	75.6	2.806	13	80.6	3.213	21
19	1.615	5	44	1.840	14	59.5	2.143	14	70.8	2.541	10	75.8	2.819	14	80.8	3.234	21
20	1.620	5	45	1.854	15	60	2.157	13	71.0	2.551	9	76.0	2.833	13	81.0	3.255	22
21	1.625	6	46	1.869	16	60.5	2.170	14	71.2	2.560	10	76.2	2.846	14	81.2	3.277	22
22	1.631	5	47	1.885	16	61	2.184	15	71.4	2.570	10	76.4	2.860	14	81.4	3.299	23
23	1.636	7	48	1.901	17	61.5	2.199	14	71.6	2.580	10	76.6	2.874	14	81.6	3.322	24
24	1.643	6	49	1.918	18	62	2.213	15	71.8	2.590	10	76.8	2.888	15	81.8	3.346	24

Table 1040 (continued)—Complete Elliptic Integrals of the First Kind

θ Degrees	K	Diff.	θ Degrees	K	Diff.
82.0	3.370	12	84.5	3.738	18
82.1	3.382	13	84.6	3.756	18
82.2	3.395	12	84.7	3.774	19
82.3	3.407	13	84.8	3.793	19
82.4	3.420	13	84.9	3.812	20
82.5	3.433	13	85.0	3.832	20
82.6	3.446	13	85.1	3.852	20
82.7	3.459	14	85.2	3.872	21
82.8	3.473	14	85.3	3.893	21
82.9	3.487	13	85.4	3.914	22
83.0	3.500	15	85.5	3.936	22
83.1	3.515	14	85.6	3.958	23
83.2	3.529	14	85.7	3.981	23
83.3	3.543	15	85.8	4.004	24
83.4	3.558	15	85.9	4.028	25
83.5	3.573	15	86.0	4.053	25
83.6	3.588	16	86.1	4.078	26
83.7	3.604	16	86.2	4.104	26
83.8	3.620	16	86.3	4.130	27
83.9	3.636	16	86.4	4.157	28
84.0	3.652	16	86.5	4.185	29
84.1	3.668	17	86.6	4.214	30
84.2	3.685	17	86.7	4.244	30
84.3	3.702	18	86.8	4.274	32
84.4	3.720	18	86.9	4.306	33

θ Degrees	Min.	K	Diff.
87.0		4.339	33
87.1		4.372	35
87.2		4.407	37
87.3		4.444	37
87.4		4.481	39
87.5		4.520	42
87.6		4.561	41
87.7		4.603	45
87.8		4.648	46
87.9		4.694	49
88.0		4.743	51
88.1		4.794	54
88.2		4.848	57
88.3		4.905	60
88.4		4.965	65
88.5		5.030	69
88.6		5.099	74
88.7		5.173	80
88.8		5.253	87
88.9		5.340	95
89	0	5.435	34
	2	5.469	35
	4	5.504	36
	6	5.540	38
	8	5.578	39

θ Degrees	Min.	K	Diff.
89	10	5.617	41
	12	5.658	42
	14	5.700	45
	16	5.745	46
	18	5.791	49
	20	5.840	51
	22	5.891	55
	24	5.946	57
	26	6.003	60
	28	6.063	65
	30	6.128	69
	32	6.197	74
	34	6.271	80
	36	6.351	87
	38	6.438	95
	40	6.533	51
	41	6.584	55
	42	6.639	57
	43	6.696	60
	44	6.756	65
	45	6.821	69
	46	6.890	74
	47	6.964	80
	48	7.044	87
	49	7.131	95

θ Degrees	Min.	K	Diff.
89	50	7.226	106
	51	7.332	117
	52	7.449	134
	53	7.583	154
	54	7.737	182
	55	7.919	224
	56	8.143	287
	57	8.430	406
	58	8.836	693
	59	9.529	
90	0	∞	

For values of θ greater than about 89° 50' it is often better to use series 773.3 than to interpolate from tables.

321

Table 1041—Complete Elliptic Integrals of the Second Kind

$$E = \int_0^{\pi/2} \sqrt{(1 - \sin^2 \theta \sin^2 \phi)} \, d\phi \quad \text{[See 774]}$$

θ Degrees	E	Diff.	θ Degrees	E	Diff.	θ Degrees	E	Diff.	θ Degrees	E	Diff.	θ Degrees	E	Diff.	θ Degrees	E	Diff.
0	1.571	0	15	1.544	−3	30	1.467	−6	45	1.351	−9	60	1.211	−9	75	1.076	−7
1	1.571	−1	16	1.541	−4	31	1.461	−7	46	1.342	−9	61	1.202	−10	76	1.069	−8
2	1.570	0	17	1.537	−4	32	1.454	−7	47	1.333	−9	62	1.192	−9	77	1.061	−7
3	1.570	−1	18	1.533	−5	33	1.447	−7	48	1.324	−9	63	1.183	−10	78	1.054	−7
4	1.569	−1	19	1.528	−4	34	1.440	−8	49	1.315	−9	64	1.173	−9	79	1.047	−7
5	1.568	−2	20	1.524	−5	35	1.432	−7	50	1.306	−10	65	1.164	−9	80	1.040	−6
6	1.566	−1	21	1.519	−5	36	1.425	−8	51	1.296	−9	66	1.155	−10	81	1.034	−6
7	1.565	−2	22	1.514	−5	37	1.417	−8	52	1.287	−9	67	1.145	−9	82	1.028	−6
8	1.563	−2	23	1.509	−5	38	1.409	−8	53	1.278	−10	68	1.136	−9	83	1.022	−5
9	1.561	−2	24	1.504	−6	39	1.401	−8	54	1.268	−9	69	1.127	−9	84	1.017	−4
10	1.559	−3	25	1.498	−6	40	1.393	−8	55	1.259	−10	70	1.118	−8	85	1.013	−4
11	1.556	−2	26	1.492	−6	41	1.385	−8	56	1.249	−9	71	1.110	−9	86	1.009	−4
12	1.554	−3	27	1.486	−6	42	1.377	−9	57	1.240	−10	72	1.101	−8	87	1.005	−2
13	1.551	−3	28	1.480	−6	43	1.368	−9	58	1.230	−9	73	1.093	−9	88	1.003	−2
14	1.548	−4	29	1.474	−7	44	1.359	−8	59	1.221	−10	74	1.084	−8	89	1.001	−1
															90	1.000	

For tables of elliptic integrals, see Ref. 47, 26, 31, 35, 36, 45, and 48.

Table 1045—Normal Probability Integral

$$\frac{1}{\sqrt{(2\pi)}}\int_{-x}^{x} e^{-t^2/2}dt \quad [\text{See 585}]$$

x	0	1	2	3	4	5	6	7	8	9	Diff.
.0	.0000	.0080	.0160	.0239	.0319	.0399	.0478	.0558	.0638	.0717	79–80
.1	.0797	.0876	.0955	.1034	.1113	.1192	.1271	.1350	.1428	.1507	78–79
.2	.1585	.1663	.1741	.1819	.1897	.1974	.2051	.2128	.2205	.2282	76–78
.3	.2358	.2434	.2510	.2586	.2661	.2737	.2812	.2886	.2961	.3035	73–76
.4	.3108	.3182	.3255	.3328	.3401	.3473	.3545	.3616	.3688	.3759	70–74
.5	.3829	.3899	.3969	.4039	.4108	.4177	.4245	.4313	.4381	.4448	67–70
.6	.4515	.4581	.4647	.4713	.4778	.4843	.4907	.4971	.5035	.5098	63–66
.7	.5161	.5223	.5285	.5346	.5407	.5467	.5527	.5587	.5646	.5705	58–62
.8	.5763	.5821	.5878	.5935	.5991	.6047	.6102	.6157	.6211	.6265	54–58
.9	.6319	.6372	.6424	.6476	.6528	.6579	.6629	.6680	.6729	.6778	49–53
1.0	.6827	.6875	.6923	.6970	.7017	.7063	.7109	.7154	.7199	.7243	44–48
1.1	.7287	.7330	.7373	.7415	.7457	.7499	.7540	.7580	.7620	.7660	39–43
1.2	.7699	.7737	.7775	.7813	.7850	.7887	.7923	.7959	.7995	.8029	34–38
1.3	.8064	.8098	.8132	.8165	.8198	.8230	.8262	.8293	.8324	.8355	30–34
1.4	.8385	.8415	.8444	.8473	.8501	.8529	.8557	.8584	.8611	.8638	26–30
1.5	.8664	.8690	.8715	.8740	.8764	.8789	.8812	.8836	.8859	.8882	22–26
1.6	.8904	.8926	.8948	.8969	.8990	.9011	.9031	.9051	.9070	.9090	19–22
1.7	.9109	.9127	.9146	.9164	.9181	.9199	.9216	.9233	.9249	.9265	16–19
1.8	.9281	.9297	.9312	.9328	.9342	.9357	.9371	.9385	.9399	.9412	13–16
1.9	.9426	.9439	.9451	.9464	.9476	.9488	.9500	.9512	.9523	.9534	11–13
2.0	.9545	.9556	.9566	.9576	.9586	.9596	.9606	.9615	.9625	.9634	9–11
2.1	.9643	.9651	.9660	.9668	.9676	.9684	.9692	.9700	.9707	.9715	7–9
2.2	.9722	.9729	.9736	.9743	.9749	.9756	.9762	.9768	.9774	.9780	6–7
2.3	.9786	.9791	.9797	.9802	.9807	.9812	.9817	.9822	.9827	.9832	4–6
2.4	.9836	.9840	.9845	.9849	.9853	.9857	.9861	.9865	.9869	.9872	4–5
2.5	.9876	.9879	.9883	.9886	.9889	.9892	.9895	.9898	.9901	.9904	3–4
2.6	.9907	.9909	.9912	.9915	.9917	.9920	.9922	.9924	.9926	.9929	2–3
2.7	.9931	.9933	.9935	.9937	.9939	.9940	.9942	.9944	.9946	.9947	1–2
2.8	.9949	.9950	.9952	.9953	.9955	.9956	.9958	.9959	.9960	.9961	1–2
2.9	.9963	.9964	.9965	.9966	.9967	.9968	.9969	.9970	.9971	.9972	1
3.0	.9973	.9974	.9975	.9976	.9976	.9977	.9978	.9979	.9979	.9980	0–1
3.1	.9981	.9981	.9982	.9983	.9983	.9984	.9984	.9985	.9985	.9986	0–1
3.2	.9986	.9987	.9987	.9988	.9988	.9988	.9989	.9989	.9990	.9990	0–1
3.3	.9990	.9991	.9991	.9991	.9992	.9992	.9992	.9992	.9993	.9993	0–1
3.4	.9993	.9994	.9994	.9994	.9994	.9994	.9995	.9995	.9995	.9995	0–1
3.5	.9995	.9996	.9996	.9996	.9996	.9996	.9996	.9996	.9997	.9997	0–1
3.6	.9997	.9997	.9997	.9997	.9997	.9997	.9997	.9998	.9998	.9998	0–1
3.7	.9998	.9998	.9998	.9998	.9998	.9998	.9998	.9998	.9998	.9998	0–1
3.8	.9999	.9999	.9999	.9999	.9999	.9999	.9999	.9999	.9999	.9999	0
3.9	.9999	.9999	.9999	.9999	.9999	.9999	.9999	.9999	.9999	.9999	0
4.0	.9999	.9999	.9999	.9999	.9999	.9999	1.0000	1.0000	1.0000	1.0000	0–1

For a large table of 15 decimal places, see Ref. 55e, "Tables of Probability Functions," Vol. II, A. N. Lowan, Technical Director, Work Projects Administration for the City of New York, 1942, sponsored by the National Bureau of Standards.

Table 1050—Bessel Functions
For tables of Bessel Functions of real arguments see References 12 and 50

ber $x + i$ bei $x = J_0(xi\sqrt{i}) = I_0(x\sqrt{i})$

x	ber x	bei x	ber' x	bei' x
0	+1.0	0	0	0
0.1	+0.999 998 438	+0.002 500 000	−0.000 062 500	+0.049 999 974
0.2	+0.999 975 000	+0.009 999 972	−0.000 499 999	+0.099 999 167
0.3	+0.999 873 438	+0.022 499 684	−0.001 687 488	+0.149 993 672
0.4	+0.999 600 004	+0.039 998 222	−0.003 999 911	+0.199 973 334
0.5	+0.999 023 464	+0.062 493 218	−0.007 812 076	+0.249 918 621
0.6	+0.997 975 114	+0.089 979 750	−0.013 498 481	+0.299 797 507
0.7	+0.996 248 828	+0.122 448 939	−0.021 433 032	+0.349 562 345
0.8	+0.993 601 138	+0.159 886 230	−0.031 988 623	+0.399 146 758
0.9	+0.989 751 357	+0.202 269 363	−0.045 536 553	+0.448 462 528
1.0	+0.984 381 781	+0.249 566 040	−0.062 445 752	+0.497 396 511
1.1	+0.977 137 973	+0.301 731 269	−0.083 081 791	+0.545 807 563
1.2	+0.967 629 156	+0.358 704 420	−0.107 805 642	+0.593 523 499
1.3	+0.955 428 747	+0.420 405 966	−0.136 972 169	+0.640 338 102
1.4	+0.940 075 057	+0.486 733 934	−0.170 928 324	+0.686 008 176
1.5	+0.921 072 184	+0.557 560 062	−0.210 011 017	+0.730 250 674
1.6	+0.897 891 139	+0.632 725 677	−0.254 544 638	+0.772 739 922
1.7	+0.869 971 237	+0.712 037 292	−0.304 838 207	+0.813 104 947
1.8	+0.836 721 794	+0.795 261 955	−0.361 182 125	+0.850 926 951
1.9	+0.797 524 167	+0.882 122 341	−0.423 844 516	+0.885 736 950
2.0	+0.751 734 183	+0.972 291 627	−0.493 067 125	+0.917 013 613
2.1	+0.698 685 001	+1.065 388 161	−0.569 060 755	+0.944 181 339
2.2	+0.637 690 457	+1.160 969 944	−0.652 000 244	+0.966 608 614
2.3	+0.568 048 926	+1.258 528 975	−0.742 018 947	+0.983 606 691
2.4	+0.489 047 772	+1.357 485 476	−0.839 202 721	+0.994 428 643
2.5	+0.399 968 417	+1.457 182 044	−0.943 583 409	+0.998 268 847
2.6	+0.300 092 090	+1.556 877 774	−1.055 131 815	+0.994 262 944
2.7	+0.188 706 304	+1.655 742 407	−1.173 750 173	+0.981 488 365
2.8	+0.065 112 108	+1.752 850 564	−1.299 264 112	+0.958 965 456
2.9	−0.071 367 826	+1.847 176 116	−1.431 414 136	+0.925 659 305
3.0	−0.221 380 249	+1.937 586 785	−1.569 846 632	+0.880 482 324
3.1	−0.385 531 455	+2.022 839 042	−1.714 104 430	+0.822 297 688
3.2	−0.564 376 430	+2.101 573 388	−1.863 616 954	+0.749 923 691
3.3	−0.758 407 012	+2.172 310 131	−2.017 689 996	+0.662 139 131
3.4	−0.968 038 995	+2.233 445 750	−2.175 495 175	+0.557 689 801
3.5	−1.193 598 180	+2.283 249 967	−2.336 059 130	+0.435 296 178
3.6	−1.435 305 322	+2.319 863 655	−2.498 252 527	+0.293 662 421
3.7	−1.693 259 984	+2.341 297 714	−2.660 778 962	+0.131 486 760
3.8	−1.967 423 273	+2.345 433 061	−2.822 163 850	−0.052 526 621
3.9	−2.257 599 466	+2.330 021 882	−2.980 743 427	−0.259 654 097
4.0	−2.563 416 557	+2.292 690 323	−3.134 653 964	−0.491 137 441
4.1	−2.884 305 732	+2.230 942 780	−3.281 821 353	−0.748 166 860
4.2	−3.219 479 832	+2.142 167 987	−3.419 951 224	−1.031 862 169
4.3	−3.567 910 863	+2.023 647 069	−3.546 519 744	−1.343 251 997
4.4	−3.928 306 621	+1.872 563 796	−3.658 765 306	−1.683 250 947
4.5	−4.299 086 552	+1.686 017 204	−3.753 681 326	−2.052 634 662
4.6	−4.678 356 937	+1.461 036 836	−3.828 010 348	−2.452 013
4.7	−5.063 885 587	+1.194 600 797	−3.878 239 739	−2.881 799
4.8	−5.453 076 175	+0.883 656 854	−3.900 599 216	−3.342 181
4.9	−5.842 942 442	+0.525 146 811	−3.891 060 511	−3.833 085
5.0	−6.230 082 479	+0.116 034 382	−3.845 339 473	−4.354 141

Table 1050 (*continued*)—Bessel Functions

x	ber x	bei x	ber' x	bei' x
5.1	−6.610 653 357	−0.346 663 218	−3.758 900 943	−4.904 641
5.2	−6.980 346 403	−0.865 839 727	−3.626 966 748	−5.483 505
5.3	−7.334 363 435	−1.444 260 151	−3.444 527 187	−6.089 232
5.4	−7.667 394 351	−2.084 516 693	−3.206 356 389	−6.719 859
5.5	−7.973 596 451	−2.788 980 155	−2.907 031 958	−7.372 913
5.6	−8.246 575 962	−3.559 746 593	−2.540 959 318	−8.045 365
5.7	−8.479 372 252	−4.398 579 111	−2.102 401 197	−8.733 576
5.8	−8.664 445 263	−5.306 844 640	−1.585 512 696	−9.433 252
5.9	−8.793 666 753	−6.285 445 623	−0.984 382 394	−10.139 389
6.0	−8.858 315 966	−7.334 746 541	−0.293 079 967	−10.846 224
6.1	−8.849 080 413	−8.454 495 269	+0.494 289 242	−11.547 179
6.2	−8.756 062 474	−9.643 739 286	+1.383 522 213	−12.234 815
6.3	−8.568 792 593	−10.900 736 825	+2.380 248 360	−12.900 779
6.4	−8.276 249 873	−12.222 863 128	+3.489 851 325	−13.535 755
6.5	−7.866 890 928	−13.606 512 001	+4.717 382 012	−14.129 423
6.6	−7.328 687 885	−15.046 992 991	+6.067 462 487	−14.670 413
6.7	−6.649 176 464	−16.538 424 538	+7.544 180 362	−15.146 266
6.8	−5.815 515 115	−18.073 623 609	+9.150 973 359	−15.543 406
6.9	−4.814 556 200	−19.643 992 365	+10.890 503 759	−15.847 109
7.0	−3.632 930 243	−21.239 402 580	+12.764 522 560	−16.041 489
7.1	−2.257 144 280	−22.848 078 597	+14.773 723 174	−16.109 484
7.2	−0.673 695 379	−24.456 479 797	+16.917 584 633	−16.032 856
7.3	+1.130 799 653	−26.049 183 639	+19.194 204 342	−15.792 207
7.4	+3.169 457 312	−27.608 770 523	+21.600 120 535	−15.367 001
7.5	+5.454 962 184	−29.115 711 867	+24.130 124 710	−14.735 602
7.6	+7.999 382 494	−30.548 262 965	+26.777 064 473	−13.875 334
7.7	+10.813 965 476	−31.882 362 359	+29.531 637 360	−12.762 551
7.8	+13.908 911 711	−33.091 539 670	+32.382 176 399	−11.372 739
7.9	+17.293 127 645	−34.146 833 988	+35.314 428 336	−9.680 623
8.0	+20.973 955 611	−35.016 725 165	+38.311 325 701	−7.660 318
8.1	+24.956 880 800	−35.667 080 514	+41.352 754 078	−5.285 490
8.2	+29.245 214 796	−36.061 119 681	+44.415 316 208	−2.529 555
8.3	+33.839 755 432	−36.159 400 616	+47.472 094 831	+0.634 098
8.4	+38.738 422 961	−35.919 829 830	+50.492 416 438	+4.231 841
8.5	+43.935 872 751	−35.297 700 300	+53.441 618 430	+8.289 519
8.6	+49.423 084 977	−34.245 760 640	+56.280 822 496	+12.832 116
8.7	+55.186 932 099	−32.714 319 308	+58.966 717 374	+17.883 387
8.8	+61.209 725 224	−30.651 387 879	+61.451 354 516	+23.465 444
8.9	+67.468 740 848	−28.002 867 538	+63.681 960 575	+29.598 302
9.0	+73.935 729 857	−24.712 783 168	+65.600 770 999	+36.299 384
9.1	+80.576 411 145	−20.723 569 533	+67.144 889 467	+43.582 976
9.2	+87.349 952 674	−15.976 414 197	+68.246 178 293	+51.459 634
9.3	+94.208 443 358	−10.411 661 917	+68.831 185 381	+59.935 547
9.4	+101.096 359 718	−3.969 285 324	+68.821 113 743	+69.011 850
9.5	+107.950 031 881	+3.410 573 282	+68.131 840 035	+78.683 888
9.6	+114.697 114 173	+11.786 984 189	+66.673 989 017	+88.940 434
9.7	+121.256 066 255	+21.217 531 810	+64.353 071 286	+99.762 855
9.8	+127.535 651 521	+31.757 530 896	+61.069 692 033	+111.124 240
9.9	+133.434 460 262	+43.459 152 933	+56.719 839 030	+122.988 479
10.0	+138.840 465 942	+56.370 458 554	+51.195 258 394	+135.309 302

For x up to 20, see Ref. 45 and 51.

Table 1050 (*continued*)—Bessel Functions

$$\ker x + i \ker x = K_0(x\sqrt{i})$$

x	ker x	kei x	ker$'$ x	kei$'$ x
0	$+ \quad \infty$	−0.785 398 2	$- \quad \infty$	0
0.1	+2.420 474 0	−0.776 850 6	−9.960 959 3	+0.145 974 8
0.2	+1.733 142 7	−0.758 124 9	−4.922 948 5	+0.222 926 8
0.3	+1.337 218 6	−0.733 101 9	−3.219 865 2	+0.274 292 1
0.4	+1.062 623 9	−0.703 800 2	−2.352 069 9	+0.309 514 0
0.5	+0.855 905 9	−0.671 581 7	−1.819 799 8	+0.333 203 8
0.6	+0.693 120 7	−0.637 449 5	−1.456 538 6	+0.348 164 4
0.7	+0.561 378 3	−0.602 175 5	−1.190 943 3	+0.356 309 5
0.8	+0.452 882 1	−0.566 367 6	−0.987 335 1	+0.359 042 5
0.9	+0.362 514 8	−0.530 511 1	−0.825 868 7	+0.357 443 2
1.0	+0.286 706 2	−0.494 994 6	−0.694 603 9	+0.352 369 9
1.1	+0.222 844 5	−0.460 129 5	−0.585 905 3	+0.344 521 0
1.2	+0.168 945 6	−0.426 163 6	−0.494 643 2	+0.334 473 9
1.3	+0.123 455 4	−0.393 291 8	−0.417 227 4	+0.322 711 8
1.4	+0.085 126 0	−0.361 664 8	−0.351 055 1	+0.309 641 6
1.5	+0.052 934 9	−0.331 395 6	−0.294 181 6	+0.295 608 1
1.6	+0.026 029 9	−0.302 565 5	−0.245 114 7	+0.280 903 8
1.7	+0.003 691 1	−0.275 228 8	−0.202 681 8	+0.265 777 2
1.8	−0.014 696 1	−0.249 417 1	−0.165 942 4	+0.250 438 5
1.9	−0.029 661 4	−0.225 142 2	−0.134 128 2	+0.235 065 7
2.0	−0.041 664 5	−0.202 400 1	−0.106 601 0	+0.219 807 9
2.1	−0.051 106 5	−0.181 172 6	−0.082 823 4	+0.204 789 7
2.2	−0.058 338 8	−0.161 430 7	−0.062 337 3	+0.190 113 7
2.3	−0.063 670 5	−0.143 135 7	−0.044 747 9	+0.175 863 8
2.4	−0.067 373 5	−0.126 241 5	−0.029 712 3	+0.162 106 9
2.5	−0.069 688 0	−0.110 696 1	−0.016 929 8	+0.148 895 4
2.6	−0.070 825 7	−0.096 442 9	−0.006 135 8	+0.136 268 9
2.7	−0.070 973 6	−0.083 421 9	+0.002 904 3	+0.124 255 8
2.8	−0.070 296 3	−0.071 570 7	+0.010 399 0	+0.112 874 8
2.9	−0.068 939 0	−0.060 825 5	+0.016 534 2	+0.102 136 2
3.0	−0.067 029 2	−0.051 121 9	+0.021 476 2	+0.092 043 1
3.1	−0.064 678 6	−0.042 395 5	+0.025 373 8	+0.082 592 2
3.2	−0.061 984 8	−0.034 582 3	+0.028 360 3	+0.073 775 2
3.3	−0.059 032 9	−0.027 619 7	+0.030 555 4	+0.065 579 4
3.4	−0.055 896 6	−0.021 446 3	+0.032 066 2	+0.057 988 1
3.5	−0.052 639 3	−0.016 002 6	+0.032 988 6	+0.050 982 1
3.6	−0.049 315 6	−0.011 231 1	+0.033 408 7	+0.044 539 4
3.7	−0.045 971 7	−0.007 076 7	+0.033 403 0	+0.038 636 4
3.8	−0.042 646 9	−0.003 486 7	+0.033 040 0	+0.033 248 0
3.9	−0.039 373 61	−0.000 410 81	+0.032 380 46	+0.028 348 32
4.0	−0.036 178 85	+0.002 198 40	+0.031 478 49	+0.023 910 62
4.1	−0.033 084 40	+0.004 385 82	+0.030 381 79	+0.019 908 04
4.2	−0.030 107 58	+0.006 193 61	+0.029 132 42	+0.016 313 67
4.3	−0.027 261 77	+0.007 661 27	+0.027 767 30	+0.013 100 84
4.4	−0.024 556 89	+0.008 825 62	+0.026 318 68	+0.010 243 31
4.5	−0.021ʹ999 88	+0.009 720 92	+0.024 814 54	+0.007 715 43
4.6	−0.019 595 03	+0.010 378 86	+0.023 279 08	+0.005 492 26
4.7	−0.017 344 41	+0.010 828 72	+0.021 733 00	+0.003 549 76
4.8	−0.015 248 19	+0.011 097 40	+0.020 193 91	+0.001 864 78
4.9	−0.013 304 90	+0.011 209 53	+0.018 676 61	+0.000 415 22
5.0	−0.011 511 73	+0.011 187 59	+0.017 193 40	−0.000 819 98

Table 1050 (*continued*)—Bessel Functions

x	ker x	kei x	ker' x	kei' x
5.1	−0.009 864 74	+0.011 052 01	+0.015 754 36	−0.001 860 79
5.2	−0.008 359 11	+0.010 821 28	+0.014 367 57	−0.002 726 05
5.3	−0.006 989 28	+0.010 512 06	+0.013 039 35	−0.003 433 49
5.4	−0.005 749 13	+0.010 139 29	+0.011 774 46	−0.003 999 69
5.5	−0.004 632 16	+0.009 716 31	+0.010 576 33	−0.004 440 16
5.6	−0.003 631 56	+0.009 254 96	+0.009 447 17	−0.004 769 28
5.7	−0.002 740 38	+0.008 765 72	+0.008 388 18	−0.005 000 41
5.8	−0.001 951 58	+0.008 257 74	+0.007 399 67	−0.005 145 84
5.9	−0.001 258 12	+0.007 739 02	+0.006 481 21	−0.005 216 89
6.0	−0.000 653 04	+0.007 216 49	+0.005 631 71	−0.005 223 92
6.1	−0.000 129 53	+0.006 696 06	+0.004 849 57	−0.005 176 37
6.2	+0.000 319 05	+0.006 182 75	+0.004 132 75	−0.005 082 83
6.3	+0.000 699 12	+0.005 680 77	+0.003 478 86	−0.004 951 05
6.4	+0.001 016 83	+0.005 193 58	+0.002 885 23	−0.004 788 03
6.5	+0.001 278 080	+0.004 723 992	+0.002 348 995	−0.004 600 032
6.6	+0.001 488 446	+0.004 274 219	+0.001 867 130	−0.004 392 632
6.7	+0.001 653 215	+0.003 845 947	+0.001 436 521	−0.004 170 782
6.8	+0.001 777 354	+0.003 440 398	+0.001 053 999	−0.003 938 849
6.9	+0.001 865 512	+0.003 058 385	+0.000 716 382	−0.003 700 651
7.0	+0.001 922 022	+0.002 700 365	+0.000 420 510	−0.003 459 509
7.1	+0.001 950 901	+0.002 366 486	+0.000 163 267	−0.003 218 285
7.2	+0.001 955 861	+0.002 056 629	−0.000 058 386	−0.002 979 421
7.3	+0.001 940 312	+0.001 770 454	−0.000 247 403	−0.002 744 978
7.4	+0.001 907 373	+0.001 507 429	−0.000 406 628	−0.002 516 671
7.5	+0.001 859 888	+0.001 266 868	−0.000 538 787	−0.002 295 904
7.6	+0.001 800 431	+0.001 047 959	−0.000 646 478	−0.002 083 800
7.7	+0.001 731 326	+0.000 849 790	−0.000 732 165	−0.001 881 234
7.8	+0.001 654 654	+0.000 671 373	−0.000 798 170	−0.001 688 855
7.9	+0.001 572 275	+0.000 511 664	−0.000 846 677	−0.001 507 120
8.0	+0.001 485 834	+0.000 369 584	−0.000 879 724	−0.001 336 313
8.1	+0.001 396 782	+0.000 244 032	−0.000 899 210	−0.001 176 567
8.2	+0.001 306 386	+0.000 133 902	−0.000 906 891	−0.001 027 888
8.3	+0.001 215 743	+0.000 038 090	−0.000 904 388	−0.000 890 168
8.4	+0.001 125 797	−0.000 044 491	−0.000 893 190	−0.000 763 209
8.5	+0.001 037 349	−0.000 114 902	−0.000 874 656	−0.000 646 733
8.6	+0.000 951 070	−0.000 174 175	−0.000 850 022	−0.000 540 398
8.7	+0.000 867 511	−0.000 223 306	−0.000 820 407	−0.000 443 813
8.8	+0.000 787 120	−0.000 263 248	−0.000 786 819	−0.000 356 543
8.9	+0.000 710 249	−0.000 294 910	−0.000 750 159	−0.000 278 127
9.0	+0.000 637 164	−0.000 319 153	−0.000 711 231	−0.000 208 079
9.1	+0.000 568 055	−0.000 336 788	−0.000 670 745	−0.000 145 903
9.2	+0.000 503 046	−0.000 348 579	−0.000 629 326	−0.000 091 093
9.3	+0.000 442 203	−0.000 355 236	−0.000 587 517	−0.000 043 145
9.4	+0.000 385 540	−0.000 357 420	−0.000 545 789	−0.000 001 559
9.5	+0.000 333 029	−0.000 355 743	−0.000 504 544	+0.000 034 158
9.6	+0.000 284 604	−0.000 350 768	−0.000 464 122	+0.000 064 485
9.7	+0.000 240 168	−0.000 343 010	−0.000 424 806	+0.000 089 887
9.8	+0.000 199 598	−0.000 332 940	−0.000 386 830	+0.000 110 811
9.9	+0.000 162 751	−0.000 320 983	−0.000 350 379	+0.000 127 684
10.0	+0.000 129 466	−0.000 307 524	−0.000 315 597	+0.000 140 914

See Report of the British Assoc. for the Advancement of Science, 1912, p. 56; 1915, p. 36; and 1916, p. 122.

Table 1050 (*continued*)—Bessel Functions

$$\text{ber}_n x + i \,\text{bei}_n x = J_n(xi\sqrt{i}) = i^n I_n(x\sqrt{i})$$

$$\text{ber}_n' x = \frac{d}{dx}\,\text{ber}_n x$$

x	$\text{ber}_1 x$	$\text{bei}_1 x$	$\text{ber}_1' x$	$\text{bei}_1' x$
1	$-0.395\ 868$	$+0.307\ 557$	$-0.476\ 664$	$+0.212\ 036$
2	$-0.997\ 078$	$+0.299\ 775$	$-0.720\ 532$	$-0.305\ 845$
3	$-1.732\ 64$	$-0.487\ 45$	$-0.635\ 99$	$-1.364\ 13$
4	$-1.869\ 25$	$-2.563\ 82$	$+0.658\ 74$	$-2.792\ 83$
5	$+0.359\ 78$	$-5.797\ 91$	$+4.251\ 33$	$-3.327\ 80$
6	$+7.462\ 20$	$-7.876\ 68$	$+10.206\ 52$	$+0.235\ 45$
7	$+20.368\ 9$	$-2.317\ 2$	$+14.677\ 5$	$+12.780\ 7$
8	$+32.506\ 9$	$+21.673\ 5$	$+5.866\ 4$	$+36.882\ 2$
9	$+20.719\ 2$	$+72.054\ 3$	$-37.108\ 0$	$+61.749\ 0$
10	-59.478	$+131.879$	-132.087	$+45.127$

x	$\text{ber}_2 x$	$\text{bei}_2 x$	$\text{ber}_2' x$	$\text{bei}_2' x$
1	$+0.010\ 411$	$-0.124\ 675$	$+0.041\ 623$	$-0.248\ 047$
2	$+0.165\ 279$	$-0.479\ 225$	$+0.327\ 788$	$-0.437\ 789$
3	$+0.808\ 37$	$-0.891\ 02$	$+1.030\ 93$	$-0.286\ 47$
4	$+2.317\ 85$	$-0.725\ 36$	$+1.975\ 73$	$+0.853\ 82$
5	$+4.488\ 43$	$+1.422\ 10$	$+2.049\ 97$	$+3.785\ 30$
6	$+5.242\ 91$	$+7.432\ 44$	$-1.454\ 56$	$+8.368\ 74$
7	$-0.950\ 4$	$+17.592\ 4$	$-12.493\ 0$	$+11.015\ 1$
8	$-22.889\ 0$	$+25.438\ 9$	$-32.589\ 1$	$+1.300\ 6$
9	$-65.869\ 2$	$+10.134\ 8$	$-50.963\ 2$	$-38.551\ 6$
10	-111.779	-66.610	-28.840	-121.987

x	$\text{ber}_3 x$	$\text{bei}_3 x$	$\text{ber}_3' x$	$\text{bei}_3' x$
1	$+0.013\ 788$	$+0.015\ 629$	$+0.039\ 433$	$+0.048\ 634$
2	$+0.085\ 612$	$+0.144\ 210$	$+0.093\ 575$	$+0.239\ 418$
3	$+0.130\ 44$	$+0.565\ 38$	$+0.072\ 00$	$+0.636\ 27$
4	$-0.282\ 63$	$+1.437\ 76$	$-0.914\ 09$	$+1.073\ 55$
5	$-2.094\ 35$	$+2.454\ 41$	$-2.922\ 76$	$+0.695\ 57$
6	$-6.430\ 04$	$+1.901\ 46$	$-5.747\ 81$	$-2.498\ 96$
7	$-12.876\ 5$	$-4.407\ 2$	$-6.249\ 2$	$-11.222\ 9$
8	$-15.420\ 4$	$-22.575\ 0$	$+3.979\ 6$	$-25.707\ 4$
9	$\cdot +3.166\ 6$	$-54.538\ 7$	$+38.354\ 6$	$-35.563\ 4$
10	$+72.253$	-81.423	$+104.463$	-7.513

x	$\text{ber}_4 x$	$\text{bei}_4 x$	$\text{ber}_4' x$	$\text{bei}_4' x$
1	$-0.002\ 60$	$-0.000\ 13$	$-0.010\ 40$	$-0.000\ 78$
2	$-0.040\ 97$	$-0.008\ 30$	$-0.080\ 56$	$-0.024\ 83$
3	$-0.193\ 27$	$-0.093\ 02$	$-0.234\ 32$	$-0.183\ 52$
4	$-0.493\ 10$	$-0.499\ 85$	$-0.323\ 71$	$-0.716\ 65$
5	$-0.628\ 67$	$-1.727\ 62$	$+0.248\ 34$	$-1.834\ 36$
6	$+0.648\ 3$	$-4.230\ 2$	$+2.770\ 0$	$-3.071\ 1$
7	$+6.083\ 5$	$-7.116\ 9$	$+8.745\ 2$	$-1.921\ 9$
8	$+19.094\ 7$	$-5.288\ 8$	$+17.319\ 5$	$+7.703\ 5$
9	$+38.667$	$+14.082$	$+19.140$	$+34.545$
10	$+46.579$	$+70.500$	-12.148	$+80.465$

Table 1050 (*continued*)—Bessel Functions

x	$\mathrm{ber}_5\, x$	$\mathrm{bei}_5\, x$	$\mathrm{ber}_5'\, x$	$\mathrm{bei}_5'\, x$
1	+0.000 19	−0.000 18	+0.000 97	−0.000 87
2	+0.006 80	−0.004 84	+0.017 84	−0.011 00
3	+0.058 59	−0.025 54	+0.104 78	−0.028 32
4	+0.273 08	−0.033 53	+0.360 76	+0.046 69
5	+0.851 04	+0.211 43	+0.815 11	+0.565 64
6	+1.830 5	+1.475 6	+1.007 4	+2.220 0
7	+2.209 0	+5.242 3	−0.847 2	+5.589 6
8	−1.821 3	+12.812 8	−8.623 9	+9.233 7
9	−18.619	+21.384	−26.955	+5.504
10	−58.722	+15.193	−53.427	−24.511

Table 1050 (*continued*)—Bessel Functions

$$\mathrm{ker}_n\, x + i\,\mathrm{kei}_n\, x = i^{-n} K_n(x\sqrt{i})$$

x	$\mathrm{ker}_1\, x$	$\mathrm{kei}_1\, x$	$\mathrm{ker}_1'\, x$	$\mathrm{kei}_1'\, x$
1	−0.740 322	−0.241 996	+0.887 604	+0.794 742
2	−0.230 806	+0.080 049	+0.287 983	+0.073 632
3	−0.049 898	+0.080 270	+0.100 178	−0.038 005
4	+0.005 351 3	+0.039 166 0	+0.022 690 0	−0.036 928 3
5	+0.012 737 4	+0.011 577 8	−0.002 318 3	−0.018 366 4
6	+0.007 676 09	+0.000 288 35	−0.005 920 41	−0.005 612 66
7	+0.002 743 59	−0.002 148 90	−0.003 660 46	+0.000 156 61
8	+0.000 322 857	−0.001 566 975	−0.001 352 336	+0.000 985 180
9	−0.000 355 78	−0.000 650 05	−0.000 185 34	+0.000 748 45
10	−0.000 322 80	−0.000 123 52	+0.000 158 19	+0.000 321 35

	$\mathrm{ker}_2\, x$	$\mathrm{kei}_2\, x$	$\mathrm{ker}_2'\, x$	$\mathrm{kei}_2'\, x$
1	+0.418 03	+1.884 20	−0.141 46	−4.120 77
2	+0.261 472	+0.309 001	−0.154 871	−0.528 809
3	+0.128 391	+0.036 804	−0.107 070	−0.116 579
4	+0.048 134 2	−0.017 937 6	−0.055 545 6	−0.014 941 8
5	+0.011 183 7	−0.018 064 9	−0.021 666 9	+0.008 046 0
6	−0.001 088 3	−0.009 093 7	−0.005 268 9	+0.008 255 2
7	−0.002 910 45	−0.002 820 51	+0.000 411 05	+0.004 265 37
8	−0.001 819 91	−0.000 149 65	+0.001 334 70	+0.001 373 73
9	−0.000 683 40	+0.000 477 20	+0.000 863 10	+0.000 102 03
10	−0.000 101 28	+0.000 370 64	+0.000 335 85	−0.000 215 04

	$\mathrm{ker}_3\, x$	$\mathrm{kei}_3\, x$	$\mathrm{ker}_3'\, x$	$\mathrm{kei}_3'\, x$
1	+4.887 27	−6.269 71	−16.289 7	+17.772 4
2	+0.298 022	−0.886 821	−0.850 418	+1.296 62
3	−0.036 451	−0.236 018	−0.080 360	+0.300 78
4	−0.052 071 1	−0.060 518 2	+0.017 701 2	+0.092 108 5
5	−0.029 282 9	−0.007 685 2	+0.022 435 5	+0.025 293 0
6	−0.011 449 9	+0.004 511 5	+0.012 924 7	+0.003 405 0
7	−0.002 707 2	+0.004 464 6	+0.005 212 6	−0.001 977 0
8	+0.000 267 67	+0.002 263 32	+0.001 292 32	−0.002 029 80
9	+0.000 720 5	+0.000 714 8	−0.000 094 4	−0.001 059 0
10	+0.000 456 3	+0.000 047 3	−0.000 327 3	−0.000 347 9

329

Table 1050 (*continued*)—Bessel Functions

x	$\ker_4 x$	$\text{kei}_4 x$	$\ker_4' x$	$\text{kei}_4' x$
1	−47.753 1	+3.981 0	+191.990	−8.035
2	−2.774 90	+0.940 03	+5.966 15	−1.042 25
3	−0.410 62	+0.348 52	+0.740 16	−0.323 58
4	−0.057 09	+0.137 36	+0.136 71	−0.131 38
5	+0.007 143	+0.049 433	+0.020 426	−0.054 819
6	+0.012 375	+0.014 000	−0.003 344	−0.020 620
7	+0.007 257	+0.001 780	−0.005 361	−0.006 088
8	+0.002 878 3	−0.001 192 6	−0.003 228 8	−0.000 814 8
9	+0.000 680 7	−0.001 153 8	−0.001 317 5	+0.000 516 8
10	−0.000 072 2	−0.000 584 3	−0.000 327 2	+0.000 522 9
	$\ker_5 x$	$\text{kei}_5 x$	$\ker_5' x$	$\text{kei}_5' x$
1	+287.76	+253.88	−1407.9	−1306.0
2	+10.209 4	+6.076 6	−24.226 0	−17.818 4
3	+1.467 9	+0.353 1	−2.402 6	−1.125 3
4	+0.327 07	−0.052 99	−0.465 59	−0.071 26
5	+0.077 13	−0.056 32	−0.117 13	+0.026 42
6	+0.012 982	−0.029 378	−0.029 468	+0.023 332
7	−0.001 719	−0.011 767	−0.005 162	+0.011 279
8	−0.003 146 2	−0.003 455 3	+0.000 774 4	+0.005 038 1
9	−0.001 873 6	−0.000 417 5	+0.001 375 4	+0.001 529 2
10	−0.000 746 0	+0.000 324 1	+0.000 837 2	+0.000 200 1

[Ref. 14]

Table 1060—Some Numerical Constants

$$\sqrt{2} = 1.41421\ 35624$$
$$\sqrt{3} = 1.73205\ 08076$$
$$\sqrt{5} = 2.23606\ 79775(^-)$$
$$\sqrt{6} = 2.44948\ 97428$$
$$\sqrt{7} = 2.64575\ 13111$$
$$\sqrt{8} = 2.82842\ 71247$$
$$\sqrt{10} = 3.16227\ 76602$$
$$\pi = 3.14159\ 26536$$
$$\log_{10}\pi = 0.49714\ 98727$$
$$\pi^2 = 9.86960\ 44011$$
$$\frac{1}{\pi} = 0.31830\ 98862$$
$$\sqrt{\pi} = 1.77245\ 38509$$
$$\epsilon = 2.71828\ 18285(^-)$$
$$M = \log_{10}\epsilon = 0.43429\ 44819$$
$$1/M = \log_{\epsilon}10 = 2.30258\ 50930$$
$$\log_{\epsilon}2 = 0.69314\ 71806$$

Table 1070—Greek Alphabet

α	A	Alpha	ν	N	Nu
β	B	Beta	ξ	Ξ	Xi
γ	Γ	Gamma	o	O	Omicron
δ	Δ	Delta	π	Π	Pi
ϵ	E	Epsilon	ρ	P	Rho
ζ	Z	Zeta	$\sigma\ s$	Σ	Sigma
η	H	Eta	τ	T	Tau
$\theta\ \vartheta$	Θ	Theta	υ	Υ	Upsilon
ι	I	Iota	$\varphi\ \phi$	Φ	Phi
κ	K	Kappa	χ	X	Chi
λ	Λ	Lambda	ψ	Ψ	Psi
μ	M	Mu	ω	Ω	Omega

B. REFERENCES

1. *Integral Tables*, by Meyer Hirsch; Wm. Baynes & Son, London, 1823.
2. *Integraltafeln*, by F. Minding; C. Reimarus, Berlin, 1849.
3. *Synopsis of Elementary Results in Pure Mathematics*, by G. S. Carr; F. Hodgson, London, 1886.
4. *Sammlung von Formeln der Mathematik*, by W. Láska; F. Vieweg und Sohn, Braunschweig, 1894.
5. *A Short Table of Integrals*, by B. O. Peirce and Ronald M. Foster; 4th ed., Ginn & Co., Boston, 1957.
6. *Elementary Treatise on the Integral Calculus*, by Benj. Williamson; Longmans, Green & Co., London, 1896.
7. *A Treatise on the Integral Calculus*, by I. Todhunter; Macmillan & Co., London, 1921.
8. *Elements of the Infinitesimal Calculus*, by G. H. Chandler; J. Wiley & Sons, New York, 1907.
9. *Advanced Calculus*, by E. B. Wilson; Ginn & Co., Boston, 1912.
10. *Functions of a Complex Variable*, by J. Pierpont; Ginn & Co., Boston, 1914.
11. *Advanced Calculus*, by F. S. Woods; Ginn & Co., Boston, 1926.
12. *Treatise on Bessel Functions*, by Gray, Mathews, and MacRobert; Macmillan & Co., London, 1931.
13. *Theory of Bessel Functions*, by G. N. Watson; Cambridge University Press, 1944.
14. *Bessel Functions for Alternating-Current Problems*, by H. B. Dwight; Transactions of American Institute of Electrical Engineers, July, 1929.
15. *Tables d'Intégrales Définies*, by B. de Haan; ed. of 1858–1864.
16. *Nouvelles Tables d'Intégrales Définies*, by B. de Haan; Hafner Publ. Co., New York, 1957; originally published in Leyden in 1867.
17. *Tables of Functions with Formulas and Curves*, by E. Jahnke and F. Emde; 4th ed., Dover Publications, New York, 1945 (1st ed. published by B. G. Teubner, Leipzig, 1909).
18. *Elementary Integrals—A Short Table*, by T. J. I'a. Bromwich; Macmillan & Co., London, 1911.
19. *Logarithmic and Trigonometric Tables*, by E. R. Hedrick (*The Macmillan Mathematical Tables*); The Macmillan Co., New York, 1935.
20. *The Calculus*, by E. W. Davis and W. C. Brenke (*The Macmillan Mathematical Tables*); The Macmillan Co., New York, 1930.
21. *Synopsis of Applicable Mathematics*, by L. Silberstein; D. Van Nostrand Co., New York, 1923.
22. *Fourier's Series and Spherical, Cylindrical, and Ellipsoidal Harmonics*, by W. E. Byerly; Ginn & Co., Boston, 1893.
23. *Sag Calculations*, by J. S. Martin, ed. of 1931; Copperweld Steel Co., Glassport, Pa.
24. *Principles of Electric Power Transmission*, by L. F. Woodruff; J. Wiley & Sons, New York, 1938.
25. "The Magnetic Field of a Circular Cylindrical Coil", by H. B. Dwight; *Philosophical Magazine*, Vol. XI, April, 1931.
26. *Scientific Paper 169 of the Bureau of Standards*, Washington, D.C., by E. B. Rosa and F. W. Grover; ed. of 1916, also published as *Bulletin of the Bureau of Standards*, Vol. 8, No. 1, 1912.

27. *Traité Élémentaire des Nombres de Bernoulli*, by N. Nielsen; Gauthier-Villars & Cie., Paris, 1923.
28. *American Standard Mathematical Symbols*, Report Z10 of American Engineering Standards Committee, 1928.
29. *Higher Trigonometry*, by J. B. Lock; Macmillan & Co., London, 1899.
30. *Smithsonian Mathematical Tables—Hyperbolic Functions*, by G. F. Becker and C. E. Van Orstrand; Smithsonian Institution, Washington, D.C., 1909.
31. *Smithsonian Mathematical Formulae and Tables of Elliptic Functions*, by E. P. Adams and R. L. Hippisley; Smithsonian Institution, Washington, D.C., 1922.
32. *A Course in Mathematics*, 2 vols., by F. S. Woods and F. H. Bailey; Ginn & Co., Boston, 1907–09.
33. *Elliptic Functions*, by A. Cayley; G. Bell & Sons, London, 1895.
34. *Introduction to the Theory of Infinite Series*, by T. J. I'a. Bromwich; Macmillan & Co., London, 1908.
35. *Theory of Elliptic Functions*, by H. Hancock; J. Wiley & Sons, New York, 1910.
36. *Elliptic Integrals*, by H. Hancock; J. Wiley & Sons, New York, 1917.
37. *Application of Hyperbolic Functions to Electrical Engineering Problems*, by A. E. Kennelly; University of London Press, 1912.
38. *Differential Equations for Electrical Engineers*, by P. Franklin; J. Wiley & Sons, New York, 1933.
39. *Methods of Advanced Calculus*, by P. Franklin; McGraw-Hill Book Co., New York, 1944.
40. *Integral Calculus*, by W. E. Byerly; Ginn & Co., Boston, 1898.
41. *College Algebra*, by H. B. Fine; Ginn & Co., Boston, 1905.
42. *Advanced Calculus*, by W. B. Fite; The Macmillan Co., New York, 1938.
43. *Application of the Method of Symmetrical Components*, by W. V. Lyon; McGraw-Hill Book Co., New York, 1937.
44. *Tables of the Higher Mathematical Functions*, by H. T. Davis; Principia Press, Bloomington, Indiana, Vol. 1, 1933, Vol. 2, 1935.
45. *Mathematical Tables of Elementary and Some Higher Mathematical Functions*, by H. B. Dwight; 2nd ed., Dover Publications, New York, 1958.
46. *Electrical Coils and Conductors*, by H. B. Dwight; McGraw-Hill Book Co., New York, 1945.
47. *Traité des Fonctions Elliptiques*, Vol. 2, by A. M. Legendre; Huzard-Courcier, Paris, 1825.
48. *Ten-Figure Table of the Complete Elliptic Integrals*, by L. M. Milne-Thomson; Proc. Lond. Math. Soc., Ser. 2, Vol. 33, 1931.
49. *Bessel Functions for Engineers*, by N. W. McLachlan; Clarendon Press, Oxford, 1934.
50. *Bessel Functions*, Vol. VI, part I, *Mathematical Tables of the British Association for the Advancement of Science;* Cambridge University Press, 1937.
51. *Values of the Bessel Functions ber x and bei x and Their Derivatives*, by H. B. Dwight; Transactions of American Institute of Electrical Engineers, 1939, p. 787.
52. *Tables of the Spherical Function $P_n(x)$ and Its Derived Functions*, by H. Tallquist; Acta Soc. Sci. Fennicae, Finland, Vol. 32, 1906, p. 5, and Vol. 33, No. 9, 1908.
53. *Six-Place Tables of the 16 First Surface Zonal Harmonics $P_n(x)$*, by H. Tallquist; Acta Soc. Sci. Fennicae, Finland, 1937.
54. *Six-Place Tables of the 32 First Surface Zonal Harmonics $P_n(\cos \theta)$*, by H. Tallquist; Acta Soc. Sci. Fennicae, Finland, 1938.
55a. *Tables of Sines and Cosines for Radian Arguments*, 1940.
   b. *Tables of Circular and Hyperbolic Sines and Cosines for Radian Arguments*, 1939.
   c. *Tables of the Exponential Function*, 1939.
   d. *Table of Natural Logarithms*, Vols. I–IV, 1941.

e. *Tables of Probability Functions*, Vols. I and II, 1941.

f. *Tables of Sine, Cosine, and Exponential Integrals*, Vols. I and II, 1940.

g. *Table of the Bessel Functions $J_0(z)$ and $J_1(z)$ for Complex Arguments* (giving values of $I_0(x)$ and $I_1(x)$ on pp. 362–381).
A. N. Lowan, Technical Director; Work Projects Administration for the City of New York, sponsored by the National Bureau of Standards, Washington, D.C.

56. *Tafeln der Besselschen, Theta-, Kugel-, und anderer Funktionen*, by K. Hayashi; Julius Springer, Berlin, 1930.

57. *Tafeln für die Differenzenrechnung*, by K. Hayashi; Julius Springer, Berlin, 1933.

58. *Seven-Place Values of Trigonometric Functions for Every Thousandth of a Degree*, by J. Peters; D. Van Nostrand Co., New York, 1942.

59. *Zehnstellige Logarithmen der Zahlen von 1 bis 100 000* (vol. 1), by J. Peters and J. Stein; Preussische Landesaufnahme, 1922.

60. *Zehnstellige Logarithmen der Trigonometrischen Funktionen von 0° bis 90° für jedes Tausendstel des Grades* Vol. 2, by J. Peters; Preussische Landesaufnahme, 1919.

61. *Transmission Circuits for Telephone Communication*, by K. S. Johnson; D. Van Nostrand Co., New York, 1939.

62. *Electromagnetic Theory*, by J. A. Stratton; McGraw-Hill Book Co., New York, 1941.

63. *Tables of the Bessel Functions of the First Kind*, by the Staff of the Harvard Computation Laboratory; Harvard University Press, Cambridge, Mass., 1947–48.

64. *An Index of Mathematical Tables*, 2nd ed., by A. Fletcher, J. C. P. Miller, L. Rosenhead, and L. J. Comrie; Addison-Wesley Pub. Co., Inc., Reading, Mass., 1962.

65. *Barlow's Tables of Squares, Cubes, Square Roots, Cube Roots, and Reciprocals*, ed. L. J. Comrie; Chemical Pub. Co., New York, 1952.

66. *The Application of Elliptic Functions*, by A. G. Greenhill; Macmillan & Co., London, 1892.

67. *Tables of Integral Transforms*, Vol. I, by A. Erdélyi, W. Magnus, F. Oberhettinger, and F. G. Tricomi; McGraw-Hill Book Co., New York, 1954.

68. *Integraltafeln, Sammlung Unbestimmter Integrale*, by W. Meyer zur Capellen; Springer Verlag, Berlin, 1950.

69. *Integraltafel, Vol. I, Unbestimmte Integrale, Vol. II, Bestimmte Integrale*, by W. Gröbner and N. Hofreiter; Springer Verlag, Vienna, 1950.

70. *Handbook of Elliptic Integrals*, by P. W. Byrd and M. D. Friedman; Springer Verlag, Berlin, 1954.

71. *Treatise on the Integral Calculus*, Vol. II, by J. Edwards; Macmillan & Co., London, 1922.

72. *Vorlesungen über die Theorie der Bestimmten Integrale zwischen reellen Greuzen*, by G. F. Meyer; B. G. Teubner, Leipzig, 1871.

73. *Traité de Calcul Différentiel et de Calcul Intégral*, Vol. II, by J. Bertrand; Gauthier-Villars, Paris, 1870.

74. *Functions of Complex Variables*, by P. Franklin; Prentice-Hall, Inc., Englewood Cliffs, N.J., 1958.

75. *A Course of Modern Analysis*, by E. T. Whittaker and G. N. Watson; 4th ed., Cambridge University Press, 1958.

76. *Examen des Nouvelles Tables D'Intégrales Définies de M. Bierens de Haan*, by C. F. Lindman; G. E. Stechert & Co., New York, 1944.
A survey, originally published in Stockholm (1891), giving comments and corrections for Reference 16.

INDEX